Impacts of Oil Spill Disasters on Marine Habitats and Fisheries in North America

CRC

MARINE BIOLOGY

SERIES

The late Peter L. Lutz, Founding Editor

David H. Evans, Series Editor

PUBLISHED TITLES

Biology of Marine Birds
E.A. Schreiber and Joanna Burger

Biology of the Spotted Seatrout
Stephen A. Bortone

Early Stages of Atlantic Fishes: An Identification Guide for the Western Central North Atlantic
William J. Richards

Biology of the Southern Ocean, Second Edition
George A. Knox

Biology of the Three-Spined Stickleback
Sara Östlund-Nilsson, Ian Mayer, and Felicity Anne Huntingford

Biology and Management of the World Tarpon and Bonefish Fisheries
Jerald S. Ault

Methods in Reproductive Aquaculture: Marine and Freshwater Species
Elsa Cabrita, Vanesa Robles, and Paz Herráez

Sharks and Their Relatives II: Biodiversity, Adaptive Physiology, and Conservation
Jeffrey C. Carrier, John A. Musick, and Michael R. Heithaus

Artificial Reefs in Fisheries Management
Stephen A. Bortone, Frederico Pereira Brandini, Gianna Fabi, and Shinya Otake

Biology of Sharks and Their Relatives, Second Edition
Jeffrey C. Carrier, John A. Musick, and Michael R. Heithaus

The Biology of Sea Turtles, Volume III
Jeanette Wyneken, Kenneth J. Lohmann, and John A. Musick

The Physiology of Fishes, Fourth Edition
David H. Evans, James B. Claiborne, and Suzanne Currie

Interrelationships Between Coral Reefs and Fisheries
Stephen A. Bortone

Impacts of Oil Spill Disasters on Marine Habitats and Fisheries in North America
J. Brian Alford, PhD, Mark S. Peterson, and Christopher C. Green

Impacts of Oil Spill Disasters on Marine Habitats and Fisheries in North America

EDITED BY

J. Brian Alford • Mark S. Peterson

Christopher C. Green

CRC Press
Taylor & Francis Group
Boca Raton London New York

CRC Press is an imprint of the
Taylor & Francis Group, an **informa** business

Front cover: Oiled Spartina following oil spill in Dixon Bay, Louisiana, 1995. Courtesy of the National Oceanic and Atmospheric Administration/Department of Commerce. Photographer: Dr. Terry McTigue, NOAA, NOS, and ORR.

CRC Press
Taylor & Francis Group
6000 Broken Sound Parkway NW, Suite 300
Boca Raton, FL 33487-2742

Printed on acid-free paper
Version Date: 20140825

International Standard Book Number-13: 978-1-4665-5720-8 (Hardback)

Visit the Taylor & Francis Web site at
http://www.taylorandfrancis.com

and the CRC Press Web site at
http://www.crcpress.com

Contents

Preface

Background

The 2010 *Deepwater Horizon* (DWH) oil spill was one of the largest maritime disasters in the world (McNutt et al. 2011; NAS 2013). Even the most conservative estimate of leaked oil from DWH makes it the most extensive oil disaster ever to occur in a North American marine environment (see Figure I.1). An estimated 794 million liters to 1.11 billion liters of crude oil leaked into the northern Gulf of Mexico for 84 days. The estimated peak flow was 155,200,000 L/day (Ryerson et al. 2012). At its maximum, the surface expression of the discharge covered 62,159 km^2 (Norse and Amos 2010). This response included the release of 2.9 million liters of the dispersant Corexit® (Place et al. 2010), which was applied at the surface and injected at the leaking wellhead 1500 m below the surface. Doubtless, there was great concern that ecosystem services to the Gulf of Mexico, and to human interests far beyond the region, could be dramatically curtailed (NAS 2013).

At the 2010 American Fisheries Society (AFS) meeting in Pittsburgh, Pennsylvania, a symposium was convened to address the response, recovery, and research efforts initiated by state and federal agencies following this historic spill. Subsequently, the editors of this book organized another symposium at the 2011 AFS meeting in Seattle, Washington, this time bringing in experts researching the *Exxon Valdez, Ixtoc I,* and DWH spills to discuss impacts to coastal fisheries resources following such large-scale oil disasters. This book contains recent research findings and reviews of spill-related science as it pertains to North American marine fisheries and their habitats. There are three major sections: (I) ecotoxicology of oil and oil-related compounds to fishes, (II) impacts of oil and oil response measures to coastal habitats, primarily that of salt marsh vegetation, and (III) short- and long-term impacts of oil to fish and shellfish populations and communities.

Section I: Ecotoxicology of Fishes Impacted by Oil-Derived Compounds

Section I of this book details the physiological effects of oil-derived compounds on fishes with the specific intent of illustrating both field and laboratory investigations. Chapter 1 is a comprehensive review of the effects of oil and specific compounds on fishes, and it begins by identifying chemicals of potential concern (COPC). The COPC and their effects include crude oil, aliphatic hydrocarbons, monoaromatic hydrocarbons, and polycyclic aromatic hydrocarbons. The chapter is divided into sections detailing the aforementioned compounds, and the author reviews the current literature on a variety of topics including, but not limited to, acute toxicity, biochemical markers, genotoxicity, histopathology, and developmental, immunological, and reproductive impacts to fishes.

The acute toxicity of dioctyl sodium sulfosuccinate (DOSS), a surfactant included in Corexit, is investigated in Chapter 2 along with an examination of the physiological effect of DOSS on osmoregulation within an estuarine species (Gulf killifish, *Fundulus grandis*).

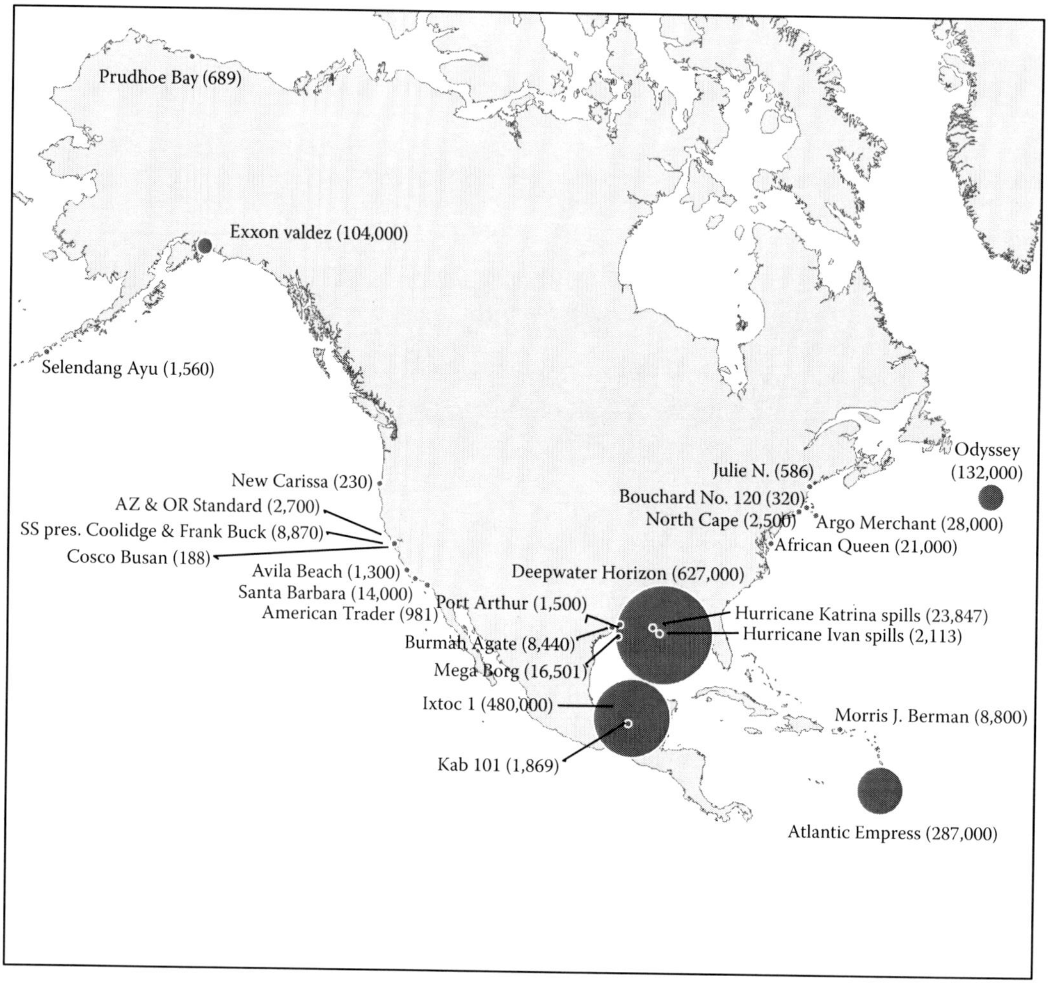

FIGURE I.1
Locations of oil spills and their relative volumes that have impacted North American marine environments. Bubble sizes represent the relative estimated maximum tons of oil spilled (data obtained from en.wikipedia.org/wiki/List_of_oil_spills). Hurricane Katrina spills and Hurricane Ivan spills are the sum of all reported spills from platforms, pipelines, and port facilities that occurred as a result of storm damage.

The authors demonstrate the influence of salinity on the acute toxicity of DOSS. Using immunohistochemistry at the site of the gills in these larval fish, the authors demonstrated that this surfactant appeared to alter ion transport protein activities. This study integrates acute toxicity assays across a salinity gradient, while highlighting the physiological role of ion regulation and possible inhibition by the surfactant DOSS.

Chapter 3 details several investigations centered on acute and chronic exposures of dissolved hydrocarbon fractions (DHF) that demonstrate impairment to the physiological stress response, immune function, and swimming performance in Pacific herring (*Clupea harengus pallasi*). Differential responses to the generalized stress response were monitored by a variety of physiologically oriented experiments. Further in vitro examinations indicated that interrenal cells chronically exposed to DHF treatments did not produce cortisol when exposed to adrenocorticotropic hormone. Disease challenges with *Listonella anguillarum*, a gram-negative marine bacteria, demonstrated that while acute DHF exposure produced a stress response that assisted in pathogen resistance, chronically exposed individuals with an inhibited stress response were more susceptible to the pathogen. This chapter demonstrates a variety of physiological effects of endocrine disruption originating from different types of exposures.

Considerations in the collection and handling of biological samples from fishes are addressed in Chapter 4. This volume also focuses on the preservation and curation of blood and spermatozoa, which are easily collected by sublethal sampling techniques. Biological samples procured by sublethal techniques such as these can be very important when working with endangered fishes and monitoring of potential impacted populations. The authors perform a number of experiments that demonstrate the influence of handling variables that can alter sample quality and viability such as temperature, enzymatic degradation, and bacteria.

Section II: Oil Impacts to Physical Habitat in Coastal Ecosystems

Section II addresses the science of assessing oil spill impacts to coastal habitats, with an emphasis on salt marsh ecosystems in the Gulf of Mexico. Chapter 5 encompasses a review of oil impacts to salt marsh vegetation, a habitat that provides a disproportionate amount of living space and food to early life stages of most recreational and commercial fishery species. In addition, this chapter discusses advantages of using a relatively new aerial imaging technique called Synthetic Aperture Radar applied by the U. S. Geological Survey after the DWH spill to detect oil that was not visible by observations from boats or by aerial photography. Chapter 6 offers a review of current research findings on impacts from oil compounds and chemically dispersed oil (e.g., with Corexit™ 9500) on wetland plants and microbial degradation in wetland soils. The authors provide suggestions for future research on this topic, which include (1) determining the sensitivity of wetland plants to specific oil hydrocarbons, (2) quantifying the effects of wetland plant community changes on hydrocarbon degradation by microbes (although, see Bik et al. 2012), and (3) developing improved methods for measuring hydrocarbons that will in turn provide more accurate indicators of toxicity to the flora and fauna of coastal wetlands. In Chapter 7, research findings using controlled field experiments show that DWH oil had short-term negative effects on Smooth cordgrass (*Spartina alterniflora*) photosynthetic rates, but there was no evidence of long-term damage. However, the authors suggest that recovery of marsh vegetation function is confounded by extraneous environmental factors such as exposure of marsh to wave energy that acts to disturb and expose oiled sediments to degradation processes. Finally, in Chapter 8, a case study is presented on oil prevention measures taken to protect marsh habitat for fish and shellfish in Ocean Springs, Mississippi, following the DWH oil spill. A systematic process of collaboration among local, state, and federal agencies, in concert with private entities, helped to find appropriate methods of preventing oil from entering valuable estuarine habitat.

Overall, from this section we learn that no two oil spills are alike regarding the degree of impact and recovery of coastal habitats to crude oil and the responses taken to prevent oil inundation of sensitive fisheries habitats. Likewise, at smaller scales, oiled sites within the same geographic location (e.g., a tidal basin) may show different rates of recovery given inherent environmental conditions (e.g., wave energy, salinity, human response, and prevention measures), the type of oil (e.g., heavy or light crude, degree of weathering), and the types of microbes available to degrade oil-derived compounds. Finally, it is critical that more research is conducted to adequately monitor and restore coastal fisheries habitats that have been exposed to oil (see NAS 2013 for further discussion of this topic).

Section III: Population and Community Dynamics Following Oil Spill Disasters

Section III focuses on quantified and potential impacts of oil spills on population and community dynamics of commercial and recreational fishery species. The topics covered in this section are wide ranging, yet they illustrate the difficulty of assessing impacts from oil spill disasters to fisheries resources at historic and contemporary time scales. This section contains reviews of long-term impacts to nearshore vertebrate populations more than 20 years after the *Exxon Valdez* spill (Chapter 9), in addition to a reexamination of the collapse of the Pacific herring (*Clupea pallasi*) stock in Prince William Sound, Alaska (Chapter 10). In addition, Chapter 11 presents newly released fishery and benthic invertebrate data in the aftermath of the *Ixtoc I* spill off of the coast of Mexico in 1979. The authors discuss short-term changes in community structure and commercial catches to shellfish and finfish fisheries in the southern Gulf of Mexico.

Chapters 12 and 13 provide new data on population dynamics of two recreationally and commercially important species in the northern Gulf of Mexico: Spotted seatrout (*Cynoscion nebulosus*) and Blue crab (*Callinectes sapidus*), respectively. Using before–after indices of larval recruitment (Blue crab) and reproductive output (Spotted seatrout), these chapters discuss potential long-term, negative population effects following DWH (e.g., high larval mortality rates and reduced settlement pulses in Blue crab). The chapters in this section also illustrate some immediate, short-term impacts to communities and populations followed by either long-term potential declines (or recovery) and illustrate the need for long-term monitoring to facilitate a more complete understanding of realized impacts of oil spills on fisheries resources. For example, after *Ixtoc I*, fish community composition, species richness, and commercial landings changed immediately after the spill, but recovered to prespill levels after just 2–3 years.

Impacts on population abundances may take short periods to quantify (e.g., with before-impact experimental designs over 4–5 years). In contrast, documenting evidence of long-term population declines (e.g., through lower recruitment) of some species may be caused by oil impacts that require longer monitoring time scales (e.g., decades) to be able to detect true population or community changes as a result of oil exposure. Short-term changes in population metrics, such as reduced survival, growth, and reproduction of commercially, recreationally, and ecologically important species, may be indicative of future population declines that may not be evident under current time ranges for completing most research projects. For example, developmental delays and lower percentages of spawning-capable Spotted seatrout (Chapter 12) collected 1 year post-DWH in Louisiana and Mississippi suggest future negative impacts on reproduction of this prize recreational species compared to historical data. These data sets illustrate that, at least in estuarine resident fishes, impacts to abundance may be short-lived, but if sensitive metrics are periodically examined (e.g., reproduction), these short-term impacts can translate into long-term declines in populations. Moreover, in Chapter 10, the authors suggest that the collapse of the Pacific herring stock following *Exxon Valdez* could be detected using 20 years of acoustic surveys of herring recruitment (i.e., mile-days of spawn) in contrast to more traditional age-based stock assessment models. Chapter 14 presents evidence from controlled field experiments of a lack of short-term effects of DWH oil toxicity on the Eastern oyster (*Crassostrea virginica*) in Louisiana. However, these authors show that oysters in nonimpacted sites with low salinity (7 ppt) coupled with warm summer water temperature had higher mortality than oysters inhabiting oiled sites with higher salinity and warm temperature. Using results from biomarker assays, they suggest potential confounding factors of salinity and water temperature, in addition to *Perkinsus marinus* infections on oysters. In this case, oil prevention measures, such as diverting freshwater from the Mississippi River to keep oil out of estuaries, appeared to have a greater impact on oyster populations than the oil alone. Finally, the Lemon shark (*Negaprion brevirostris*), which has only recently been discovered to use shallow habitats associated with the Chandeleur Islands, Louisiana (Chapter 15), was found to have significantly reduced relative abundance after the DWH oil spill compared to prespill data (back to 2009) indicating a potential short-term decline in this population. However, these authors also suggest that oil prevention measures, in this case sand berm construction to keep oil off of the islands, may have had a more deleterious effect on juvenile lemon shark habitat than the oil itself.

References

Bik, H. M., Halanych, K. M., Sharma, J., and W. K. Thomas. 2012. Dramatic shifts in benthic microbial eukaryote communities following the Deepwater Horizon oil spill. *PLoS ONE* 7(6): e38550. doi:10.1371/journal.pone.0038550.

McNutt, M. K., E. Camilli, T. J. Crone, G. D. Guthrie, P. A. Hsieh, T. B. Ryerson et al. 2011. Review of flow rate estimates of the Deepwater Horizon oil spill. *Proceedings of the National Academy of Sciences of the United States of America* 10: 1073/pnas.1112139108.

NAS (National Academy of Sciences). 2013. *An ecosystem services approach to assessing the impacts of the Deepwater Horizon oil spill in the Gulf of Mexico.* Committee on the Effects of the Deepwater Horizon Mississippi Canyon-252 Oil Spill on Ecosystem Services in the Gulf of Mexico, Ocean Studies Board; Division on Earth and Life Studies. Washington DC: National Research Council. ISBN 978-0-309-28845-3.

Norse, E. A. and J. Amos. 2010. Impacts, Perception, and Policy Implications of the Deepwater Horizon Oil and Gas Disaster. Environmental Law Reporter, Washington, DC: Environmental Law Institute.

Place, B., B. Anderson, A. Mekebri, E. T. Furlong, J. L. Gray, R. Tjeerdema et al. 2010. A role for analytical chemistry in advancing our understanding of the occurrence, fate, and effects of Corexit oil dispersants. *Environmental Science and Technology* 44: 6016–6018.

Ryerson, T. B., R. Camilli, J. D. Kessler, E. B. Kujawinski, C. M. Reddy, D. L. Valentine et al. 2012. Chemical data quantify Deepwater Horizon hydrocarbon flow rate and environmental distribution. *Proceedings of the National Academy of Sciences of the United States of America* 10: 1073/pnas.1110564109.

J. Brian Alford
University of Tennessee

Christopher C. Green
Louisiana State University

Mark S. Peterson
University of Southern Mississippi

Acknowledgments

The editors thank the Oil Spill Recovery Institute in Cordova, Alaska, the U.S. Geological Survey National Wetlands Research Center, Lafayette, Louisiana, and the Louisiana Chapter of the American Fisheries Society for sponsoring the symposium that provided the impetus for this book. We are also very grateful for the many professional reviewers who took time to offer their critique of the chapters herein.

Contributors

J. Brian Alford
Department of Forestry, Wildlife, and Fisheries, Knoxville, Tennessee
The University of Tennessee
Knoxville, Tennessee

Felipe Amezcua
Instituto de Ciencias del Mar y Limnología, Unidad Académica Mazatlan
Universidad Nacional Autónoma de México
Mazatlán, Sinaloa, México

Felipe Amezcua-Linares
Instituto de Ciencias del Mar y Limnología
Universidad Nacional Autónoma de México. Ciudad Universitaria
México D. F., México

Matthew E. Andersen
U. S. Geological Survey
National Wetlands Research Center
Lafayette, Louisiana

Brenda E. Ballachey
U. S. Geological Survey
Alaska Science Center
Anchorage, Alaska

Patrick D. Biber
Department of Coastal Sciences
The University of Southern Mississippi
Ocean Springs, Mississippi

Charlotte Bodinier
Department of Biological Sciences
Louisiana State University
Baton Rouge, Louisiana

James L. Bodkin
U. S. Geological Survey
Alaska Science Center
Anchorage, Alaska

Rachel A. Brewton
Harte Research Institute for Gulf of Mexico Studies
Texas A&M University-Corpus Christi
Corpus Christi, Texas

Nancy J. Brown-Peterson
Department of Coastal Sciences
The University of Southern Mississippi
Ocean Springs, Mississippi

Sandra Casas
Department of Veterinary Science
Louisiana State University Agricultural Center
Baton Rouge, Louisiana

Dan E. Esler
U. S. Geological Survey
Alaska Science Center
Anchorage, Alaska

Richard S. Fulford
Ecosystem Assessment Branch
U. S. Environmental Protection Agency
Gulf Breeze, Florida

Fernando Galvez
Department of Biological Sciences
Louisiana State University
Baton Rouge, Louisiana

Katherine Gautreaux
Aquaculture Research Station
Louisiana State University Agricultural Center
Baton Rouge, Louisiana

Brigitte Gil-Manrique
Posgrado en Ciencias del Mar y Limnología
Universidad Nacional Autónoma de México
Ciudad Universitaria, México D. F., México

Christopher G. Green
Aquaculture Research Station
Louisiana State University Agricultural Center
Baton Rouge, Louisiana

Robert J. Griffitt
Department of Coastal Sciences
University of Southern Mississippi
Ocean Springs, Mississippi

Jill A. Jenkins
U. S. Geological Survey
National Wetlands Research Center
Lafayette, Louisiana

David Keith
Anchor QEA LLC
Ocean Springs, Mississippi

Christopher J. Kennedy
Department of Biological Sciences
Simon Fraser University
Burnaby, British Columbia, Canada

Zhanfei Liu
University of Texas Marine Science Institute
Port Aransas, Texas

Jonathan F. McKenzie
Pontchartrain Institute for Environmental Sciences
University of New Orleans
New Orleans, Louisiana

Scott Miles
Department of Environmental Sciences
Louisiana State University
Baton Rouge, Louisiana

J. Andrew Nyman
School of Renewable Natural Resources
Louisiana State University Agricultural Center
Baton Rouge, Louisiana

Martin T. O'Connell
Pontchartrain Institute for Environmental Sciences
University of New Orleans
New Orleans, Louisiana

Heather M. Olivier
U. S. Geological Survey
National Wetlands Research Center
Lafayette, Louisiana

Harriet Perry
Department of Coastal Sciences
The University of Southern Mississippi
Ocean Springs, Mississippi

Mark S. Peterson
Department of Coastal Sciences
The University of Southern Mississippi
Ocean Springs, Mississippi

Jerome La Peyre
Department of Veterinary Science
Louisiana State University Agricultural Center
Baton Rouge, Louisiana

Linh Thuy Pham
Department of Coastal Sciences
The University of Southern Mississippi
Ocean Springs, Mississippi

Joseph R. Pursley
Anchor QEA LLC
Seattle, Washington

Stanley D. Rice
Auke Bay Laboratory
National Marine Fisheries Service
Juneau, Alaska

Arianna Rivera
Department of Biological Sciences
Louisiana State University
Baton Rouge, Louisiana

Guillermo Sanchez-Rubio
Gulf Coast Research Laboratory
Center for Fisheries Research and Development
Ocean Springs, Mississippi

Christopher Schieble
Earth and Environmental Sciences Department
University of New Orleans
New Orleans, Louisiana

Patrick W. Smith
Pontchartrain Institute for Environmental Sciences
University of New Orleans
New Orleans, Louisiana

Gary L. Thomas
Rosenstiel School of Marine and Atmospheric Sciences
Miami, Florida

Richard E. Thorne
Prince William Sound Science Center
Cordova, Alaska

Wei Wu
Department of Coastal Sciences
The University of Southern Mississippi
Ocean Springs, Mississippi

Reviewers

The following is a list of professional scientists who provided critical reviews of the information presented in this book:

Amy B. Alford
Department of Wildlife, Fisheries, and Aquaculture
Mississippi State University
Starkville, Mississippi

J. Brian Alford
Department of Forestry, Wildlife, and Fisheries
University of Tennessee
Knoxville, Tennessee

Patrick D. Biber
Department of Coastal Sciences
The University of Southern Mississippi
Ocean Springs, Mississippi

Harry Blanchet
Marine Fisheries Section
Louisiana Department of Wildlife and Fisheries
Baton Rouge, Louisiana

Kevin Boswell
Department of Biological Sciences
Florida International University
Miami, Florida

Ernesto A. Chavez
Instituto Politécnico Nacional
Centro Interdisciplinario de Ciencias Marinas
La Paz, BCS, Mexico

Robert T. Cooney (deceased)
School of Fisheries and Ocean Sciences
University of Alaska-Fairbanks
Fairbanks, Alaska

Tanya L. Darden
Marine Resources Research Institute
Hollings Marine Lab
South Carolina Department of Natural Resources
Charleston, South Carolina

Christopher C. Green
LSU Agricultural Center
Louisiana State University
Baton Rouge, Louisiana

Richard S. Grippo
Department of Biological Sciences
Arkansas State University
Jonesboro, Arkansas

Alf Haukenes
Aquaculture/Fisheries Center
University of Arkansas at Pine Bluff
Pine Bluff, Arkansas

Jake Heare
School of Aquatic and Fishery Sciences
University of Washington
Seattle, Washington

Fiona Hollinshead
Matamata Veterinary Services
Matamata, New Zealand

Adam Kuhl
Huntsman Corporation
The Woodlands, Texas

R. Paul Lang
Clinical Laboratory Sciences
Our Lady of the Lake College
Baton Rouge, Louisiana

Charles Martin
Department of Oceanography & Coastal Sciences
Louisiana State University
Baton Rouge, Louisiana

Josh Patterson
School of Renewable Natural Resources
Louisiana State University
Baton Rouge, Louisiana

Mark S. Peterson
Department of Coastal Sciences
The University of Southern Mississippi
Ocean Springs, Mississippi

Sean P. Powers
Department of Marine Sciences
Dauphin Island Sea Lab
University of South Alabama
Dauphin Island, Alabama

R. Glenn Thomas
Fisheries Management Section
Louisiana Department of Wildlife and Fisheries
Baton Rouge, Louisiana

Section I

Ecotoxicology of Fishes Impacted by Oil-Derived Compounds

1

Multiple Effects of Oil and Its Components in Fish

Christopher J. Kennedy

CONTENTS

Introduction

The effects of oil pollution on fisheries have been of concern for decades, mainly due to the visible impacts of large-scale oil spills into freshwater and marine environments. This concern is not misplaced, but equal or greater impacts on fish may occur from the slow persistent contamination of water bodies through the use of petroleum products. Each of these exposure scenarios can result in very different effects, of both a short- and long-term nature. The evaluation of ecological risks from exposure to chemical mixtures such as petroleum presents one of the most difficult challenges of our time for toxicological research and risk assessment for two reasons. First exposure to petroleum mixtures occur across multimedia environments (sediment, water, and air) and capturing relevant and realistic exposure concentrations is difficult. Second, petroleum is an inconsistent and complex chemical mixture that changes with time in the environment. Because of this, two evaluation and research approaches for such mixtures are usually followed: (1) evaluation of complex mixtures as though they are single entities and (2) a component-based approach in which certain individual chemicals in a mixture are considered to estimate the toxicity of the mixture. This chapter takes both approaches and provides a comprehensive overview to understanding the toxicity of oil and its individual components to fish.

Contaminants of Potential Concern in Crude Oil

Petroleum or crude oil is a complex mixture of hydrocarbons of various molecular weights and other liquid organic compounds. Not all of these chemicals have been evaluated for their toxicity to aquatic organisms, but of those which have, only some have high inherent toxicity or are found in high enough concentrations to be toxic to aquatic organisms. The following classes of compounds have been identified as contaminants of potential concern (COPCs): aliphatic hydrocarbons (which includes all non-aromatic hydrocarbons [non-AHs]), naphthenic acids, monoaromatic hydrocarbons (MAHs—specifically benzene, toluene, ethylbenzene, and xylene [BTEX]), and polycyclic aromatic hydrocarbons (PAHs).

The *alkanes*, also known as *paraffins or saturated hydrocarbons*, are saturated hydrocarbons with straight or branched chains that contain only carbon and hydrogen bonded exclusively by single bonds. *Naphthenes* are also known as *cycloalkanes*, and are a type of alkane having one or more rings of carbon atoms in their chemical structure. Like alkanes, they have only single chemical bonds in their chemical structure. Cycloalkanes consist of only carbon and hydrogen atoms and are saturated.

The AHs, or arenes (aryl hydrocarbons), are unsaturated hydrocarbons that have one or more planar six-carbon benzene rings, to which hydrogen atoms are attached. An AH has alternating double and single bonds between carbon atoms. The configuration of six carbon atoms in aromatic compounds is the simplest possible such hydrocarbon, benzene. AHs can be monocyclic (MAH) or polycyclic (PAH). Benzene, toluene, ethylbenzene, and xylene (*o*-, *m*-, and *p*-), also known as BTEX, are naturally occurring MAH compounds in crude oil. BTEX is among the volatile organic compounds (VOCs) found in petroleum derivatives such as gasoline.

Polyaromatic hydrocarbons, also known as polycyclic aromatic hydrocarbons or polynuclear aromatic hydrocarbons, are components of the aromatic fraction of crude oil that consist of more fused aromatic rings and do not contain heteroatoms or carry substituents. The resulting structure is a molecule where all carbon and hydrogen atoms lie in one plane. Naphthalene, the simplest example of a PAH, is formed from two benzene rings fused together, and has the lowest molecular weight of all PAHs. The environmentally significant PAHs are those molecules that contain two (e.g., naphthalene) to seven benzene rings (e.g., coronene).

The main focus of this chapter is on cold-water fish species, although data for warmer water fish species is included when cold-water fish species data were limited or unavailable. Generally, only in vivo toxicity studies are included, particularly for general crude oil and PAH toxicity where a large number of studies are available.

Crude Oil Toxicity

Crude oil is a complex mixture of thousands of hydrocarbons and non-hydrocarbon compounds with individual effects, and in addition, there can be potential additive, synergistic, or antagonistic effects between the various components (see Table 1.1). Moreover, the chemical composition of each crude oil can vary significantly and can have diverse effects on different organisms within the same ecosystem (Overton et al. 1994). These differences in toxic effects are due to qualitative compositional and concentration differences of the chemical constituents (Albert 1995, Sverdrup et al. 2003). The toxicity of crude oil is largely attributed to the aryl hydrocarbon component, although other components of oil may also be toxic (BTEX is volatile and exposure may be limited). Often, the toxicity of oil is reported in terms of the total PAH content or as weight of total dissolved hydrocarbons per liter of water (mg/L), particularly when referring to the water-soluble fraction (WSF) of crude oil.

Acute Toxicity

The acute toxicity of crude oil to fish varies with water temperature (Korn et al. 1979) and salinity (Ramachandran et al. 2006). Crude oil toxicity tends to be the highest at lower water temperatures presumably because oil persistence is higher, and biotransformation lower. The tolerance to oil is similar among salmonid species and depends on the life stage (Moles et al. 1979). For example, LC_{50} values in pink salmon are approximately 1.2–1.7 mg/L total aromatics (Moles et al. 1979).

Carls and Korn (1985) tested the sensitivity of Arctic cod and sculpin to WSF of Cook Inlet crude oil (97% MAHs, 3% 2-ring PAHs) or naphthalene using a flow-through exposure system for up to 40 days. The LC_{50} was reached in 8 days; for Arctic cod it was found to be 1.6 mg/L total AHs and for sculpin it was >1.7 mg/L total AHs. For naphthalene, the 96-hour LC_{50} for both Arctic cod and sculpin was found to be 1.4 mg/L. This study also investigated the effects of different temperatures on acute lethality, but no consistent patterns could be identified and effects seemed to be species specific. Results for the cold-water species were also compared to acute effects for warm-water species, and acute lethality for both types of fish was similar.

TABLE 1.1

Summary of Toxicological Effects of Crude Oil on Various Fish Species

Species	Lab or Field	Toxicological Effect									Reference
		Acute/ Lethality	Biochemical	Growth/ General	Immunological	Genotoxicity	Histopathology	Reproductive	Developmental	Behavioral	
Salmonids											
Chinook salmon	Lab		X								Lin et al. (2009)
Chinook salmon	Lab		X								Van Scoy et al. (2010)
Coho salmon	Lab									X	Folmar et al. (1981)
Pacific salmon (6)	Lab	X									Moles et al. (1979)
Pink salmon	Lab			X							Thomas and Rice (1975)
Pink salmon	Lab						X				Hawkes et al. (1980)
Pink salmon	Lab	X							X		Moles et al. (1987)
Pink salmon	Lab			X							Wang et al. (1993)
Pink salmon	Field	X	X				X				Wiedmer et al. (1996)
Pink salmon	Lab	X	X	X			X		X		Marty et al. (1997)

Pink salmon	Lab	X									X		Heintz et al. (1999)
Pink salmon	Field				X						X		Heintz et al. (2000)
Rainbow trout	Lab	X	X										Lockhart et al. (1996)
Other Fish Species													
Atlantic cod, turbot	Lab						X						Barsiene et al. (2006)
Atlantic herring	Lab				X						X		Ingvardsottir et al. (2011)
Arctic cod, sculpin	Lab	X											Carls and Korn (1985)
English sole		X			X				X				McCain et al. (1978)
Flounder	Lab		X										Alkindi et al. (1996)
Pacific herring	Lab	X			X						X	X	Carls (1987)
Pacific herring	Both						X		X		X		Hose et al. (1996)
Pacific herring	Field							X		X			Kocan et al. (1996)
Pacific herring	Field			X		X				X			Norcross et al. (1996)

(Continued)

TABLE 1.1

Summary of Toxicological Effects of Crude Oil on Various Fish Species (*Continued*)

Species	Lab or Field	Toxicological Effect									Reference
		Acute/ Lethality	Biochemical	Growth/ General	Immunological	Genotoxicity	Histopathology	Reproductive	Developmental	Behavioral	
Pacific herring	Both		X		X		X				Carls et al. (1998)
Pacific herring	Lab	X				X		X			Carls et al. (1999)
Pacific herring	Field		X				X				Marty et al. (1999)
Pacific herring	Lab		X					X	X		Carls et al. (2000)
Pacific herring	Lab		X								Kennedy and Farrell (2005)
Pacific herring	Lab		X							X	Kennedy and Farrell (2006)
Pacific herring	Lab		X		X						Kennedy and Farrell (2008)
Polar cod	Lab		X		X						Narhgang et al. (2010)
Pollock	Lab	X		X					X		
Sole, Pacific halibut	Lab									X	Moles et al. (1994)
Sole, Pacific halibut	Lab	X		X	X		X				Moles and Norcross (1998)
Zebrafish	Lab		X						X	X	Hicken et al. (2011)

Biochemical Indictors

Wiedmer et al. (1996) conducted a field study in the 8–26 months following the Exxon Valdez oil spill (EVOS) to determine if pink salmon were being impacted by oil remaining in the spawning grounds. They collected samples of pink salmon from four sites that had been oiled and five reference sites that had not been oiled. Most pink salmon alevins, but not eggs, from oiled sites exhibited CYP1A induction. Histopathological lesions were also noted in alevins from oiled sites, but the differences were not significant. The use of CYP1A induction as a biomarker indicated that PAHs from crude oil were still bioavailable to the pink salmon, but the biological implications were less certain.

Several studies were completed with Chinook salmon parr (Van Scoy et al. 2010) and smolts (Lin et al. 2009) following exposure to crude oil, using metabolomics to identify metabolic processes that were impacted by oil exposure. After 96 hours of exposure to the WSF of Prudhoe Bay crude oil (3.5–8.7 mg/L total petroleum hydrocarbons), liver and muscle tissues were examined for their metabolic profiles. In smolts, increases in amino acid and decreases in organic osmolytes were observed, suggesting the fish were shifting their energy sources in response to stress. Increased amino acids (to synthesize proteins) may also be required to help repair cellular damage, and an imbalance in amino acids can lead to reduced development, reproduction, or ability to adapt to additional stressors (Lin et al. 2009). Alterations in osmolyte profile may also make it more difficult for fish to adjust to osmotic stress during seaward migration (Lin et al. 2009).

In a parallel study with Chinook salmon parr, Van Scoy et al. (2010) found that the WSF of Prudhoe Bay crude oil (4.2–11.2 mg/L total petroleum hydrocarbons) also changed the metabolic profile in muscle of salmon. Decreases in lactate and ATP content were noted, whereas some amino acids and organic osmolytes increased. Some of these changes persisted for up to 3 months after exposure, but did not result in changes to growth. The alterations may be bioindicators of cellular repair processes, changes in cellular structure, or responses to overall stress (Van Scoy et al. 2010).

In a study with Pacific herring, Kennedy and Farrell (2005) demonstrated the effects of the WSF of North Slope crude oil on a range of responses following both acute (96 h) and chronic (9 week) exposures. Initial concentrations of total PAHs were 9.7, 37.9, and 99.3 μg/L in the acute exposures, with concentrations declining over time and shifting toward larger and more substituted PAHs. Ethoxyresorufin-O-deethylation (EROD) activity was induced indicating PAHs were bioavailable, and plasma cortisol, lactate, and glucose were elevated indicating a stress response to the exposure. Plasma ions (chloride, sodium, and potassium) were all elevated in the highest exposure group within 96 hours and remained elevated through the chronic exposures. Overall, this study provided evidence that exposure to the WSF of crude oil can alter ion homeostasis chronically and induce a short-term stress response in Pacific herring. Similarly, Alkindi et al. (1996) monitored a range of responses of flounder to the WSF of Omani crude oil (6 mg/L total AHs) for up to 48 hours. Plasma cortisol and noradrenalin were elevated within 3 hours of exposure as was plasma glucose, indicating a generalized stress response, and thyroid hormone (T4) was decreased by WSF of crude oil treatment. Plasma potassium content increased within 24 hours of treatment while blood hematocrit and hemoglobin decreased. Blood oxygen was profoundly impacted and showed a significant decline in treated fish. Overall, this study demonstrated that flounder exposed to WSF of crude oil show multiple physiological disturbances, which may have implications for overall organism health. Similar findings regarding elevated cortisol and depressed thyroid hormone (T4) responses were observed in larval and juvenile turbot exposed to WSF of crude oil (Stephens et al. 1997).

Growth Impairment or Somatic Indicators of Toxicity

Wang et al. (1993) conducted a study where juvenile pink salmon were fed with crude oil—contaminated food. Fish that received 34.83 mg crude oil/g of food experienced much lower growth after 6 weeks compared to unexposed fish. Similarly, Lockhart et al. (1996) reported that juvenile rainbow trout exposed to Norman Wells crude oil (0.15–1.5 mg/L total dissolved hydrocarbons) experienced a decrease in growth as measured by length of fish after 55 days. These fish also experienced fin erosion and imbalances in water content, which increased over time. In a different type of study, Thomas and Rice (1975) examined the opercular rate (respiration rate) of pink salmon exposed to the WSF (1.05–3.46 mg/L dissolved total hydrocarbon) of Prudhoe Bay crude oil. They found that at concentrations of 2.83 mg/L or more, opercular rate was elevated within 3 hours of exposure and remained elevated through at least 9–12 hours of exposure, before returning to normal at 23 hours of exposure. This response may be adaptive in the short term, but in the long term may place additional energy demands on the fish.

Moles and Norcross (1998) conducted a study to assess the effects of Alaska North Slope crude oil–contaminated sediments on juvenile yellowfin sole, rock sole, and Pacific halibut. Oil was artificially weathered to remove MAHs and then two treatments were tested: sediment oil contamination giving 1600–1800 μg/g total petroleum hydrocarbons (low group, similar to levels detected in sediments following EVOS) and sediment oil contamination giving 4300–4700 μg/g (high group). The growth rate of all three fish decreased after 90 days in both oiled sediment treatment groups. Condition factors (a length/weight relationship, which is a general indicator of health) were also decreased in both rock sole and halibut by oil-sediment treatments, and in yellowfin sole by the high treatment. Other general health measures were also affected following 90 days of exposure including increases in fin erosion, decreases in macrophage aggregates, and increases in parasite loads.

Immunological Toxicity

The effect of the WSF of North Slope crude oil on Pacific herring immune system and disease susceptibility was assessed by Kennedy and Farrell (2008). In herring exposed to WSF (total PAH concentration of 111–133 μg/L), respiratory burst activity was elevated in acute exposures (1–4 days) but decreased in chronic exposures (29 or 57 days), whereas lysozyme activity was decreased in acute exposures only. Mortality following exposure to a pathogen (*Listonella anguillarum*) was decreased in acute exposures (1 day), but elevated in chronic exposures, indicating that WSF of crude oil has the ability to alter immune responses, including the response to pathogens.

Genotoxicity

Aas et al. (2000) demonstrated that exposure to crude oil (0.04, 0.14, or 0.95 mg/L) WSF containing total PAH concentrations of 0.33, 1.14, and 7.81 μg/L was genotoxic to Atlantic cod. DNA adducts are one of the potential consequences following PAH exposure due to the reactivity of some oxidative metabolites (Varanasi et al. 1993). In fish from the highest WSF-treatment group, DNA adducts were noted within 3 days of exposure, and fish exposed to the middle WSF-treatment group exhibited elevated DNA adduct formation by day 30 of exposure. An additional 7-day depuration period was not enough for the genomic damage in the highest treatment group to be reversed. Formation of DNA adducts at such low exposure concentrations would be of concern in field settings, where these concentrations may be achieved. Similar findings were made by Nahrgang et al. (2010) in polar cod exposed to WSF of North Sea crude oil where genomic damage was detected in all treatment groups (≥15 μg/L total PAH) after 4 weeks, which was correlated with the concentrations of benzo[a]pyrene and pyrene metabolites in the bile. Barsiene et al. (2006) demonstrated the genotoxicity of crude oil to turbot and Atlantic cod following exposure to 0.5 mg/L North Sea crude oil. Various types of genomic abnormalities were detected in turbot either in peripheral blood or in kidney. No abnormalities were noted in cod, showing that there are interspecies differences in sensitivity to genomic damage from crude oil.

Histopathology

Pink salmon fry that were exposed to the WSF (predominantly MAHs and naphthalene) of Alaska North Slope crude oil for a period of 10 days were examined for histological abnormalities. Exposure concentrations were either 25–54 μg/L or 178–348 μg/L total dissolved hydrocarbons. WSF-exposed salmon exhibited greater histological abnormalities in the liver (steatosis, nuclear pleomorphism, megalocytosis, and necrosis), head kidney (increased interrenal cell diameter), and gill tissue (epithelial lifting, fusion, mucus cell hyperplasia, and vascular constriction).

Hawkes et al. (1980) conducted a study in which Chinook salmon were fed a model mixture of petroleum hydrocarbons including equal amounts of various substituted thiophenes and naphthalenes, fluorine, phenanthrene, and several aliphatic hydrocarbons, with eight chemicals in total. They found that, while the gut mucosal cells remained intact at the macroscopic level in the hydrocarbon-fed fish, the cells themselves underwent changes at the microscopic level. These changes were described as alterations in the columnar cells of the mucosa and development of inclusions in the cells, which were not observed in untreated fish. The feeding experiment lasted 28 days and no effects were observed on growth of the fish; however, the duration of the experiment may not have been long enough for growth impairments to become apparent.

Three weeks after the EVOS, Pacific herring were collected from oiled sites in Prince William Sound (PWS), Alaska, as well as several reference sites (Marty et al. 1999). Fish livers were examined histologically and showed evidence of multifocal hepatic necrosis in fish from oiled sites but not reference sites. Fish from oiled sites also had elevated levels of tissue PAHs, predominantly naphthalenes. The authors suggest that the hepatic lesions may, in part, be due to the reactivation of a virus (viral hemorrhagic septicemia virus [VHSV]). This is supported by a study done by Carls et al. (1998), which demonstrated that histopathological lesions (hepatic necrosis) due to VHSV were correlated with total PAH concentration in Pacific herring exposed to weathered crude oil. Tissue damage was also detected in English sole exposed to Alaska North Slope crude oil–contaminated sediments (McCain et al. 1978). Oil was added to the sediments at the beginning of the experiment (700 μg/g), which decreased to 400 μg/g over the course of the experiment. PAH concentrations were monitored in fish, and English sole experienced both lower growth and significant lesions (hepatocellular lipid vacuolization) associated with periods of high tissue PAH concentrations.

Reproductive Toxicity

Adult pink salmon (*Oncorhynchus gorbuscha*) returned to PWS during late summer and early fall 1989 to spawn, and spawning often occurred near heavily oiled habitats from the EVOS. Terrestrial anadromous spawning habitat is limited in PWS because this region is geologically immature, with numerous but short streams suitable for salmon spawning, so pink salmon have adapted to spawn in the intertidal segments of streams there. These stream segments were mostly protected from direct oiling from the EVOS by the freshwater streamflow that diverted oil away from the incised stream channels on these beaches. However, at some streams, the adjacent beaches were heavily oiled at elevations just above the stream grade, and oil-contaminated water could flow into these streams and affect salmon eggs incubating there (Carls et al. 2003). Studies that compared the survival of salmon embryos in streams near heavily oiled beaches and in streams on unoiled beaches found patterns of mortality that persisted through 4 successive years of pink salmon spawning events (Bue et al. 1996, 1998). Adult Pacific herring were collected 3 years after the EVOS from a site that had been previously oiled and were assessed for reproductive performance and histopathology (Kocan et al. 1996). Embryos derived from females from oiled sites had lower percent hatch and had greater morphological abnormalities than embryos from females from non-oiled reference sites. Histopathology (granulomatous inflammation and splenic congestion) was correlated to lower embryo success in the females from previously oiled sites. There are multiple hypotheses to explain the findings, one of which is due to residual effects of EVOS exposure.

Developmental Toxicity

Heintz et al. (1999) looked at exposure of pink salmon embryos to three types of oil contamination: direct contact with oil-coated gravel, effluent (containing dissolved PAHs) from oil-coated gravel, and direct contact with gravel coated with very weathered oil. They found that mortality of pink salmon embryos increased, as did PAH accumulation, under all three scenarios, indicating that it is the PAHs dissolved in water that are being taken up. A total PAH concentration of 1.0 μg/L derived from the fresher oil resulted in mortality, but the same amount of total PAH did not affect mortality when it came from the very weathered oil as it was associated with the higher molecular weight PAHs.

Marty et al. (1997) found that development of pink salmon was impaired when concentrations of Prudhoe Bay crude oil were as low as 55.2 μg/g gravel. Toxicity was observed at concentrations of total PAHs in the water of 4.4 μg/L. Examples of the effects included induction of CYP1A, development of ascites, and increased mortality. There was evidence of premature emergence in oil-exposed pink salmon including greater amounts of yolk and liver glycogen stores compared to unexposed control fish. Moles et al. (1987) examined the sensitivity of pink salmon alevins to the WSF of Cook Inlet crude oil using a simulated tidal cycle (switching from fresh to salt water). Alevins exposed to the simulated tidal cycle were more sensitive to oil, had lower yolk sac reserves, and accumulated more hydrocarbons than fish in freshwater. Older alevins (60 days) were more sensitive to toxic effects than younger alevins (5 days posthatch).

Heintz et al. (2000) incubated pink salmon eggs in water that had percolated through gravel contaminated with crude oil. As the water passed through the gravel, it became contaminated with PAHs, which were predominantly substituted naphthalenes and larger PAHs. Some fish were tagged and then released to the marine environment to complete their lifecycle. When those salmon returned at maturity 2 years later, those that had been exposed to as little as 5.4 μg/L total PAHs had a 15% decrease in marine survival relative to the control group. Following exposure, some salmon were retained to assess the effects of early life stage exposure on subsequent developmental stages. Fish exposed to more than 18 μg/L total PAH experienced decreased growth, which became apparent in the juvenile stage.

In contrast to the study done by Heintz et al. (2000), Birtwell et al. (1999) conducted exposures of pink salmon to the WSF of North Slope crude oil, using sublethal concentrations of 25–54 μg/L or 178–349 μg/L for 10 days. The WSF consisted mainly of MAHs (BTEX). After the exposures, tagged pink salmon were released to the marine environment to complete their lifecycle. There was no apparent treatment effect of the oil on pink salmon growth before release or on the proportion of adults returning to their natal stream to spawn. By comparing these findings to those of Heintz et al. (2000), it appears that exposure to PAHs, particularly those of higher molecular weight, is required before long-term consequences of early life stage exposure become apparent.

Taken together, the various studies that examined the effects of crude oil exposure to pink salmon, both in the short term and in the long term, suggest that toxicity can occur at low concentrations of PAHs, which would be expected to occur in the environment. The types of toxicity observed (mortality, growth, histopathology, poor marine survival, and lower adult returns) suggest that these early life stage exposures to crude oil could result in declines at the population level. This is supported by a study using population modeling for pink salmon that found that simulated exposure to 18 nL/L aqueous PAH could result in significant declines in population productivity and an 11% probability of population extinctions (Heintz 2007).

It should be noted that there is some disagreement about the impact of crude oil and PAHs on the development of pink salmon. Research done by U.S. government scientists (NOAA), which included most of the studies cited in this section, shows that pink salmon are impacted by low-level exposure crude oil. Other studies done by predominantly academic or industry-funded scientists have opposite findings (for example, see Brannon et al. [2001] for a review or Brannon et al. [2006]). In these studies, either the effects of crude oil are not observed at all or they occur at much higher concentrations of toxicant. Disparity in findings may be due to differences in sources of fish or oil, experimental methodologies, assay sensitivity, statistical methods, or data interpretation.

A study was done with Pacific herring to investigate the importance of exposure route in the toxicity of the WSF of Cook Inlet crude oil (Carls 1987). Herring were either exposed to WSF directly in the water or indirectly by feeding of oil-contaminated prey for up to 28 days. Direct exposure to 0.9 mg/L WSF caused high mortality and reduced both swimming ability and feeding rates. Larval weight and length were also reduced at concentrations of 0.3 mg/L or more. In contrast, the eating of highly contaminated food items resulted in mortality, but did not affect swimming, feeding, or growth. The author concluded that the eating of oil-contaminated prey is not likely to be an important exposure route for oil-induced toxicity and that direct exposure to the WSF is the more likely source of low-molecular weight PAHs in the environment.

The EVOS occurred just before Pacific herring (*Clupea pallasi*) deposited eggs on inter- and subtidal shorelines within PWS. Approximately half of the eggs were spawned within the oil slick trajectory and subsequently, a higher level of egg mortality was documented in the 1989 year class. In the summer following the EVOS (1989), larval Pacific herring were collected from potentially impacted areas and assessed for a range of parameters (Norcross et al. 1996). Larvae collected from both oiled and non-oiled areas exhibited symptoms consistent with those observed in previous laboratory studies of oil toxicity. Effects included morphological malformations, genetic damage, and small size. Fish collected in May 1995 did not show any of these symptoms. Other field observations of larvae indicated significantly greater abnormality rates among larvae hatched in oiled areas as compared to unoiled areas (Brown et al. 1996). These effects would be consistent with our understanding of hydrocarbon toxicity to fish; however, links to population-level effects have never been clearly established. Subsequently (4 years after EVOS), Pacific herring populations in PWS collapsed.

Much of the developmental toxicity associated with crude oil has been attributed to the PAH component, although the PAHs alone may not be the only constituent responsible (Gonzalez-Doncel et al. 2008). Fish embryos exposed to crude oil or PAHs develop a series of symptoms and abnormalities typically observed in blue-sac disease, including induction of CYP1A1, cardiac dysfunction, edema, spinal curvature, and craniofacial deformities (Incardona et al. 2004). This can ultimately lead to membrane damage, circulatory system failure, and impaired development (Bauder et al. 2005). Other effects from crude oil exposure (which may or may not be due to PAHs) include genetic damage, histopathological abnormalities, fin erosion, and light pigmentation. Various studies of crude oil toxicity in early developmental stages have made findings consistent with these effects in Pacific herring (Hose et al. 1996, Carls et al. 1999, 2000), Atlantic herring (Ingvarsdottir et al. 2011), and pollock (Carls and Rice 1990).

Delayed effects of early life stage crude oil exposure can also occur. In a study using zebrafish, exposure to partially weathered Alaskan North Slope crude oil (containing 24–36 μg/L total PAH) during embryonic development led to sublethal, delayed toxicity that became apparent in adult fish (Hicken et al. 2011). Adult zebrafish had changes in heart shape (morphology) and had reduced swimming performance, which is indicative of reduced cardiac output. This type of delayed toxicity can be very difficult to detect in the natural environment, but can have significant negative implications at the population level because it may impact fish survivability.

Behavioral Toxicity

Folmar et al. (1981) reported on the effects of oil exposure on predatory behavior of Coho salmon. Coho were exposed to the WSF of Cook Inlet crude oil (230–530 μg/L), and their ability to capture prey items (small rainbow trout) was evaluated. The authors noted that behavioral changes (lethargy, little interest in prey items) could be observed by 10 days of exposure to the WSF, which was associated with reduced predation by the WSF-exposed Coho.

A study using various Alaskan fish (rock sole, yellowfin sole, and Pacific Halibut) examined the avoidance behavior of the fish to oiled sediments and the dependence of the response on grain size of sediments (Moles et al. 1994). It was found that the fish could detect and avoid heavily oiled sediments (2%), but did not avoid sediments with lower oil contamination (0.05%). Oiled sediments, in general, were preferred over oiled sediments that also consisted of undesirable grain sizes. The authors suggest that the lack of avoidance is of concern in contaminated environments, where fish may be exposed in the long term to low levels of oil-derived hydrocarbons.

Kennedy and Farrell (2006) investigated the impacts of the WSF of North Slope crude oil on swimming performance in Pacific herring. Fish were exposed to three different concentrations of WSF of crude oil, namely 0.2 μg/L (low), 9.6 μg/L (medium), and 120 μg/L (high) total PAH concentrations, and concentrations declined over time. Sustained swimming performance of herring was impaired by exposure to the high WSF-treated fish within 24 hours and in both medium and high WSF treatments after 96 hours, 4 weeks, and 8 weeks of exposure. Mortality was elevated following swim performance testing in treated fish, indicating an inability to recover from sustained swimming.

Aliphatic Hydrocarbons

Aliphatic constituents of oil are not generally considered to be of significant toxicological concern to fish species (Payne et al. 1995, see Table 1.2). They are more readily biodegraded in the environment and are less water soluble than other oil constituents such as MAHs or PAHs so would be less bioavailable to fish. In addition, the lower molecular weight compounds are quite volatile and would not persist in water for very long before partitioning to the air, thus chronic toxicity is unlikely to be a problem. At high enough concentrations, aliphatic hydrocarbons would be expected to cause toxicity through narcosis; however, these levels are rarely attained for long enough in the environment to cause toxicity (Evans and Rice 1974).

Naphthenic acids consist of a class of carboxylic acid derivatives of naphthenes (cyclic aliphatic hydrocarbons) that may be found in crude oil (up to 3%). They are commonly associated with bitumen-derived oil because some of the processes used to extract the oil may generate these acidic derivatives (Headley and McMartin, 2004, Clemente and Fedorak, 2005). Naphthenic acids are thought to be one of the main toxicological concerns associated with oil sands processing water, although there are relatively few studies of their effects on fish. Naphthenic acids (both pure chemicals and those from oil sands processing waters) have been associated with mortality and developmental and histopathological abnormalities in fish such as yellow perch (Nero et al. 2006, Peters et al. 2007). However, their contribution to the overall toxicity of crude oil has not been specifically investigated.

There are very few studies available on the toxicological effects of oil-derived aliphatic hydrocarbons on salmonids. Morrow et al. (1975) found that juvenile Coho salmon exposed to various aliphatic hydrocarbons exhibited no lethal toxicity and either no or mild, reversible effects on behavior (e.g., coughing or erratic swimming) at the highest concentration tested (100 mg/L). Payne et al. (1995) exposed winter flounder to sediments contaminated with predominantly aliphatic hydrocarbons. They found that there were no concentration–response relationships and that generally there were no effects on the various endpoints even at the highest concentrations. They concluded that the aliphatic hydrocarbon component of complex hydrocarbon mixtures is relatively nontoxic.

Aromatic Hydrocarbons

It has been clearly demonstrated that AHs do not have one single toxic mechanism of action (see Table 1.3). Different toxic mechanisms play a role depending on the compound, the exposure (acute or chronic), the organism, and the environment. A number of toxic

TABLE 1.2

Summary of Toxicological Effects of Aliphatic Hydrocarbons and Monocyclic Aromatic Hydrocarbons on Various Fish Species

Species	Lab or Field	Toxicological Effect: Acute/ Lethality	Biochemical	Growth/ Somatic	Immunological	Genotoxicity	Histopathology	Reproductive	Developmental	Behavioral	Reference
Aliphatic Hydrocarbons											
Coho salmon	Lab	X	X							X	Morrow et al. (1974)
Winter flounder	Lab		X				X				Payne et al. (1995)
Monocyclic Aromatic Hydrocarbons											
Chinook salmon	Lab			X							Brocksen and Bailey (1973)
Coho salmon	Lab	X			X						Moles (1980)
Coho salmon	Lab									X	Maynard and Weber (1981)
Fathead minnow	Lab	X		X					X		Devlin et al. (1982)
Medaka	Lab	X							X		Teushcler et al. (2005)
Pacific herring	Lab			X					X		Eldridge et al. (1977)
Pacific salmon	Lab	X									Moles et al. (1979)
Pacific salmon	Field									X	Weber et al. (1981)
Striped bass	Lab	X									Benville and Korn (1977)

mechanisms have been linked to AHs, including nonpolar narcosis, phototoxicity, biochemical activation that in turn may result in mutagenicity, carcinogenicity and teratogenesis, immunotoxicity, reproductive and developmental toxicity, and behavioral effects among others. In general, the literature suggests that acute toxicity of crude oil components arises from exposure to monocyclic aromatic hydrocarbons (e.g. BTEX—benzene, toluene, ethylbenzene, and xylenes) or lower molecular weight PAH (LPAH) (e.g., naphthalenes and a number of its methylated derivatives). Higher molecular weight PAH (HPAH >3 rings) are less acutely toxic within limits of their solubility (Anderson et al. 1974). Longer term chronic toxicity is believed to be caused by HPAH (>3-ringed compounds) by a variety of mechanisms affecting multiple physiological systems.

One mechanism has been proposed that purports to accurately address the toxicity of hydrocarbon mixtures such as those found in crude oil: the nonpolar narcosis model or Target Lipid Model (Di Toro et al. 2007). Nonpolar narcosis is a nonspecific mechanism of toxicity of some nonpolar organic compounds and is thought to result from the physical accumulation of these compounds in biological membranes. It has been used in ecological risk assessment using species-sensitivity-distribution approaches, and has been applied in generating water and sediment benchmarks for PAH, crude oil, and gasoline (Di Toro et al. 2007). Under short-term conditions of exposure, nonpolar narcosis may be a mechanism of toxicity that can be used in preliminary ecological risk assessment; however, it has severe limitations. AHs may be more reactive and have a more chemical-specific mode of action than nonpolar narcosis (e.g., alkylated PAHs) particularly under chronic exposure regimes (e.g., benzo[a]pyrene).

Monoaromatic Hydrocarbons (BTEX)

The four MAHs that are considered to be the most important to oil toxicity are benzene, toluene, ethylbenzene, and xylene (BTEX). These chemicals are volatile, undergo photodegradation, and do not generally bioaccumulate, and thus are not generally associated with chronic effects. Most aquatic studies to date have therefore addressed acute toxicity issues.

Acute Toxicity

Moles et al. (1979) tested the sensitivity of six salmonid species, three spine stickleback, and slimy sculpin at different life stages to benzene. They found that the six salmonid species had similar response and had the greatest sensitivity, based on median tolerance limits, of the species tested. Emergent fry and out-migrant smolts were the most sensitive life stage to benzene, whereas egg stages were relatively resistant. Out-migrant smolts tested in saltwater were twice as sensitive as those tested in freshwater. Rainbow trout and Coho salmon represent two of the more sensitive fish species among tested fish, with 96-hour LC_{50} values ranging from 5 to 10 mg/L (Toxic Substances Data Base, 2006). Moles (1980) also found that Coho salmon previously infected with a fin parasite (glonchidia) were more susceptible to acute lethality with subsequent exposure to toluene. Sensitivity to toluene increased in a linear manner with number of parasites.

Benville and Korn (1977) conducted acute lethality tests for benzene, toluene, ethylbenzene, and all three isomers of xylene in striped bass, finding that the 24-hour and 96-hour LC_{50} values (range 4.3–11 mg/L) were similar for all chemicals. Similarly, Devlin et al. (1982) reported on acute toxicity of toluene to different developmental stages of fathead minnow, with LC_{50}s ranging from 18 to 36 mg/L for larvae (30-day-old fish and protolarvae).

TABLE 1.3
Summary of Toxicological Effects of Polycyclic Aromatic Hydrocarbons on Various Fish Species

Species	Lab or Field	Toxicological Effect									Reference
		Acute/ Lethality	Biochemical	Growth/ General	Immunological	Genotoxicity	Histopathology	Reproductive	Developmental	Behavioral	
Salmonids											
Chinook salmon	Field		X				X				Stehr et al. (1997)
Chinook salmon	Field				X						Arkoosh et al. (1998)
Chinook salmon	Lab				X						Palm et al. (2003)
Chinook salmon	Lab		X	X							Meador et al. (2006)
Chinook salmon	Field		X								Blanc et al. (2010)
Coho salmon	Lab			X							Moles et al. (1981)
Coho salmon	Lab									X	Purdy (1989)
Coho salmon	Lab					X					Barbee et al. (2008)
Pink salmon	Lab	X	X	X					X		Carls et al. (2005)
Rainbow trout	Lab		X						X		Billiard et al. (1999)
Rainbow trout	Lab		X								Fragoso et al. (2006)

Rainbow trout	Lab	X		X	X				Bravo et al. (2011)
Rainbow trout	Lab	X							Gesto et al. (2006)
Rainbow trout	Lab	X							Gesto et al. (2008)
Rainbow trout	Lab	X							Gesto et al. (2009)
Other fish species									
Fathead minnow	Lab	X				X	X		Hall and Oris (1991)
Fathead minnow	Lab							X	Farr et al. (1995)
Gilthead seabream	Lab							X	Goncalves et al. (2008)
Greenling, gunnel	Field	X							Jewett et al. (2002)
Medaka	Lab						X		Fallahtafti et al. (2012)
Sole	Lab	X			X				Wessel et al. (2010)
Winter flounder	Lab	X	X						Payne et al. (1988)
Zebrafish	Lab						X		Incardona et al. (2004)
Zebrafish	Lab						X		Carls et al. (2008)

Embryos were less sensitive than the larval stages, with LC_{50} values from 55 to 72 mg/L. In addition, growth of fathead minnow larvae was decreased with increasing toluene concentrations, with effects noticeable at concentrations >6 mg/L.

Estimates of acute toxicity of benzene for marine fish ranged from 0.7 mg/L, which caused delayed mortality in herring *(Clupea harengus)* larvae at 17 days postfertilization, to 40–50 mg/L, which caused 50% mortality in herring eggs (Struhsaker et al. 1974, Struhsaker 1977). The acute toxicity of benzene to juvenile striped bass (*Morone saxalilis*) as the 96-hour LC_{50} was 10.9 μl/L (Meyerhoff 1975).

Energetics

Brocksen and Bailey (1973) investigated the effects of benzene on respiration in juvenile Chinook salmon, with exposure periods ranging from 1 to 96 hours. Respiration rates were found to increase relative to preexposure levels, peaking at 48 hours of exposure. Fish respiration rates recovered to preexposure levels after 6 days.

Behavioral Effects

Maynard and Weber (1981) tested the avoidance response of Coho salmon exposed to either a mixture of MAHs (toluene, *o*-xylene, benzene, 1,2,4-trimethylbenzene, *p*-xylene, *m*-xylene, ethylbenzene) or several MAHs (benzene, toluene, and *o*-xylene) individually. For the MAH mixture, they found that Coho olfactory bulb electrophysiological responses corresponded to the observed avoidance behavior at concentrations of less than 2 mg/L for smolt Coho and 2–4 mg/L in pre-smolt (parr) Coho. The avoidance responses to individual MAHs occurred at lower concentrations, with *o*-xylene being the most potent for evoking responses at 0.2 mg/L and toluene and benzene exerting effects at 1.9 mg/L. Similarly, a study by Weber et al. (1981) demonstrated that mature Pacific salmon that are migrating upstream to spawn will avoid a mixture of MAHs (BTEX) at concentrations of 3.2 mg/L or higher. Detection and avoidance of contamination, while beneficial to the salmon in that it decreases exposure, may be detrimental overall as it may prevent the fish from completing the migration and the fish may never arrive at the spawning grounds.

Early Life Stage Toxicity

Eldridge et al. (1977) investigated the effects of benzene exposure on the energetic processes of early life stages in Pacific herring. Sublethal benzene exposures (0.04–2.1 μL saturated benzene solution/L of water) resulted in less embryonic tissue growth, differences in oxygen consumption in embryos, and greater assimilation in feeding larvae.

Teuschler et al. (2005) used medaka as a model fish species to investigate the effects of binary mixtures of benzene and toluene of heart development using heart rate, heart rate progression, and mortality as endpoints. Heart rate was decreased by toluene exposure alone, unaffected by benzene, and synergistically decreased by combination treatments after 72 hours. Heart rate was decreased by both toulene and benzene and additively decreased by the combination treatment at 96 hours. Heart rate progression was decreased by toluene and to a lesser extent for benzene, and increased by combination treatments (antagonism). Only one chronic study was found for marine fish. Korn et al. (1976) noted decreased dry weight in striped bass exposed to 5.3 mg/L benzene. Studies by Struhsaker (1977) suggest that benzene may exert some effect on egg fertilization.

General PAH Toxicity

The toxicity of PAHs to aquatic organisms is determined by several factors including (1) the PAH type (e.g., molecular weight and alkyl substitution), (2) the species of the organism exposed, and (3) the duration and the type of exposure to a given PAH. Most of the literature on acute and lethal toxic effects in estuarine and marine environments is related to LPAHs containing three or less benzene rings in their structure. These compounds are relatively more soluble in water than an HPAH; at saturation, their concentrations in water can exceed LC_{50} values, unlike the HPAH compounds (e.g., benz[a]anthracene and benzo[a]pyrene), which have limited water solubility. In addition, alkyl homologues of PAHs are generally more toxic to aquatic life than the parent compound. The minimum LC_{50} for relatively more soluble and lower molecular weight PAHs, containing three or less aromatic (benzene) rings in their structure, was found for rainbow trout (*Oncorhynchus mykiss*) exposed to phenanthrene (LC_{50} = 30 μg/L) (NRCC 1983).

Role of Metabolism in PAH Toxicity

Once PAHs have been taken up from the environment, they can be metabolized via various pathways in the liver. Unfortunately, some of the metabolites generated during this process are reactive and can cause damage to cellular macromolecules including DNA, lipids, and proteins. Interaction with lipids can result in lipid peroxidation and subsequent membrane damage and cell death. The formation of DNA adducts, where PAH metabolites bind to the DNA, can be of particular concern as adducts can be mutagenic and lead to formation of cancer and tumors (Akcha et al. 2003). This is an important concern, particularly for HPAHs such as benz[a]anthracene, benzo[b]fluoranthene, cholanthene, indeno(1,2,3-cd)pyrene, and the model carcinogenic PAH, benzo[a]pyrene (Menzie and Potoki 1992).

Photooxidation and Photosensitization

Sunlight can greatly enhance the toxicity of PAH. Photosensitization reactions (e.g., generation of singlet-state oxygen) and photomodification reactions (e.g., photooxidation of PAHs to more toxic species) are both pathways of photo-induced toxicity of PAHs. PAH can become more bioactive after photooxidation or photosensitization, which leads to greater toxicological effects in marine organisms. Both mechanisms are important pathways for cellular damage (Lin et al. 2008). Some PAHs, such as anthracene, have been shown to be more toxic when exposure to solar ultraviolet radiation (SUVR) occurs concurrent with or subsequent to PAH exposure. The mechanism for this enhancement of PAH toxicity in the presence of UV may be due to oxidative stress (Arey and Atkinson 2003).

The phototoxicity of a PAH is a function of several factors: (1) PAH concentration in tissue, (2) length of exposure to and absorption of SUVR by the organism, (3) the efficiency of conversion of ground-state molecules to the excited triplet state, and (4) the probability of the excited intermediate reacting with a target molecule (Newsted and Giesy 1987). For instance, Bowling et al. (1983) found that 12.7 μg/L anthracene was fatal to bluegill sunfish (*Lepomis macrochirus*) in 48 hours in an outdoor channel in bright sunlight. No mortality was noted in fish exposed to PAH in the shaded area of the channel. But, when shading was removed after day 4 (when anthracene concentration in water had dropped to zero and fish were allowed to depurate for 24 hours), all fish previously in the shaded area died within 24 hours. It was concluded that direct sunlight exposure of anthracene-contaminated fish, and not the toxic anthracene photoproducts in the water, was responsible for the mortality of the bluegill.

While modeling photo-induced acute (96-hour LC_{50}) and chronic toxic effects in bluegill sunfish (*Lepomis macrochirus*) exposed to anthracene, Oris and Giesy (1987) found that both acute and chronic toxicities were dependent on the length of exposure to SUVR. As the daily exposure to SUVR was increased, the threshold concentration for predicted effects decreased. Furthermore, the absolute difference between acute and chronic threshold concentrations decreased as the daily exposure to SUVR increased (e.g., at daily exposure of 5 hours, the predicted acute concentration = 55 μg/L and chronic concentration = 17 μg/L, and at daily exposure of 20 hours, the acute concentration = 7.8 μg/L and chronic concentration = 2.2 μg/L).

The evidence for the photo-induced toxicity of PAHs is recent and its significance in natural aquatic systems is yet to be understood. There is some debate as to whether PAH phototoxicity is relevant in the environment, but it should be considered in a site-specific manner as part of ecological risk assessments (McDonaldm and Chapman 2002). Aquatic organisms in deep and turbid waters and shaded areas may not be affected by this phenomenon. However, juveniles of most fish are found in the shallow areas of the littoral zone and are subject to photo-induced toxicity of PAHs.

Acute Toxicity

Most HPAHs are not acutely toxic at concentrations that reflect their water solubilities. However, as mentioned, two- and three-ringed PAHs do have acute toxicity at concentrations that would be encountered in water. Given are several examples to illustrate concentration ranges that are acutely lethal in both salmonids and non-salmonids: the lowest concentration of naphthalene that was toxic to rainbow trout (*O. mykiss*) was 11 μg/L (geometric mean of the two lowest observed effect levels (LOELs) [Black et al. 1983]). A value for the 648-hour LC_{50} of phenanthrene was 30 μg/L for rainbow trout (*O. mykiss*) (Millemann et al. 1984). Black et al. (1983) found an LOEL of 4 μg/L for rainbow trout (at the embryo-larval stages) exposed to phenanthrene. Acenaphthene toxicity as represented by the 96-hour LC_{50} value was of 580 μg/L for fathead minnow (*P. promelas*) (Holcombe et al. 1983). Finger et al. (1985) reported 12% mortality in bluefish exposed to 500 μg/L fluorene for 30 days. The 96-hour LC_{50} value for anthracene in the fathead minnow was 11.9 μg/L (Oris et al. 1987).

Biochemical or Somatic Indictors of Toxicity

PAHs are known to interact with the aryl hydrocarbon receptor (AhR), which is a nuclear transcription factor present in most species examined, including fish. Interaction of PAH (or other ligands such as dioxin) with the AhR leads to transcription of various genes that contain a dioxin- or xenobiotic-response element (DRE or XRE). Some of these genes are important for developmental processes or adaptive responses to xenobiotic exposure such as biotransformation (Beischlag et al. 2008, Zhou et al. 2010).

Cytochrome P450s (CYP) are enzymes responsible for Phase I metabolism reactions in fish (and other species), which add functional groups to both endogenous and exogenous (xenobiotic) compounds to increase their polarity and allow these chemicals to be excreted more readily. CYP1A1 is one of these enzymes that are induced following AhR interaction with DRE in the gene. Because CYP1A1 activity, which can be measured using the EROD assay, is induced following exposure to AhR ligands (such as PAHs), it is frequently used as a biomarker of exposure to these chemicals in the environment.

Indeed, there have been a number of studies that have looked at CYP1A1 activity in either gill or liver tissue following collection of wild Pacific salmon from PAH-contaminated aquatic environments (field studies), following caging of salmonids in potentially PAH-contaminated areas (in situ studies) in the Pacific Northwest, or in laboratory studies using contaminated sediments. Gill or liver EROD activity was induced in these fish, even with low levels of dissolved (aquatic) or sediment-associated PAHs (Stehr 2000, Fragoso 2006, Blanc et al. 2010, Bravo et al. 2011). Carls et al. (2005) demonstrated in their study and meta-analysis with pink salmon that CYP1A1 induction can be used as a good biomarker for predicting other sublethal effects of PAHs, including poor marine survival, reduced growth, and developmental abnormalities.

Growth Impairments and Somatic Indicators of Toxicity

Moles et al. (1981) conducted a study to assess the impact of toluene (MAH) and naphthalene (PAH) exposure in freshwater on Coho salmon growth. They report that dry weights, wet weights, and lengths of fry exposed to concentrations of 3.2 μL/L or more for toluene, or 0.7 mg/L or more for naphthalene, are decreased, as is daily growth rate.

Meador et al. (2006) conducted a 56-day study in which juvenile Chinook salmon were fed a diet contaminated with a mixture of both LPAHs and HPAHs that was intended to mimic the types and concentrations of these PAHs that the fish would encounter in the natural aquatic environment. The feeding of contaminated food was intended to mimic contaminated prey items that the fish would consume and was based on a previous study examining PAH levels in stomach contents of field-collected fish (Varanasi et al. 1993). They found that fish had accumulated PAHs in their tissues to a limited degree (concentrations in tissue were lower than that in food) and that fish were able to metabolize the PAHs for excretion through the bile. Exposure to dietary PAHs also led to decreased growth of fish (both wet and dry weight measurements), decreased the overall whole-body lipid content, influenced the distribution of different lipid classes (specifically decreasing the triacylglycerol content) and altered various plasma chemistry parameters (e.g., albumin, amylase, cholesterol, creatinine, glucose, lipase). Overall, this study demonstrated the detrimental effects of PAH on growth and metabolism, which the authors termed "toxicant-induced starvation," because the effects were similar to what would be observed in starved fish.

A 4-month study was done of winter flounder using sediments contaminated with petroleum-derived PAHs, containing a mix of LPAHs and HPAHs (Payne et al. 1988). These fish live in close contact with sediments in the aquatic environment and so could be at risk from PAHs that might partition to the sediments. The authors found that general health indices, such as liver somatic index and spleen somatic index, were altered (increased and decreased, respectively). Muscle protein content decreased, while glycogen stores in the liver increased. EROD activity, a measure of CYP1A1 induction, and liver lipids were elevated in livers of exposed fish. Some effects on fish were notable at total PAH concentrations as low as 1 μg/g in the sediment, which would be expected to occur in a variety of natural aquatic environments that are impacted by anthropogenic activities.

Histopathological Effects

The most sensitive chronic effects of naphthalene were observed by DiMichele and Taylor (1978) while studying histopathological and physiological responses in mummichog (*Fundulus heteroclitus*). These investigators found gill hyperplasia in 80% of the fish after a 15-day exposure to 2 μg/L naphthalene; only 30% of the control fish showed the effect. All of the fish exposed to 20 μg/L demonstrated necrosis of taste buds, a change not observed in the control fish.

Immunotoxicity

Field-collected salmonids (Chinook) that spend time in contaminated estuaries have been shown to be more susceptible to infectious diseases caused by bacteria such as *Listonella anguillarum* (Arkoosh et al. 1998). However, one of the difficulties in interpretation of field studies is that contaminants and other factors do not occur individually or in isolation, so it can be difficult to assign causation to particular chemicals or other factors.

Bravo et al. (2011) had similar findings of increased mortality following disease challenge (with *Aeromonas salmonicida*) in the lab following feeding of juvenile rainbow trout with food contaminated with predominantly HPAHs. In contrast, Palm et al. (2003) found that Chinook salmon fed a diet containing both LPAH and HPAH had no changes in disease susceptibility (*Listonella anguillarum*) relative to control groups. The study done by Bravo et al. (2011) was double the duration (50 days) compared to the Palm et al. (2003) study (28 days), and effects on disease susceptibility were not apparent until day 50 of exposure, suggesting that both duration of exposure and type of PAHs may affect the overall immunotoxicity.

A review of the immunotoxicity of PAH in fish is available that summarizes the diverse impacts on both the innate and adaptive immune system (Reynaud and Deschaux 2006). Effects on enzyme and cellular functions may vary depending on species, chemical(s) used in exposures, and method of exposure, and some studies have found that susceptibility to pathogens is increased following PAH exposure.

Genotoxicity

An additional biomarker of PAH exposure that may be useful in field settings is based on the measurement of genomic damage. Once metabolized in the liver, some PAHs, particularly those with higher molecular weight (four or more rings), can form reactive intermediates or metabolites. These metabolites can react with DNA to form DNA adducts that can lead to DNA strand breaks, mutations, and ultimately cancer or tumors, if not repaired. Measurement of the various types of DNA damage (strand breaks or fragmentation, DNA content changes, and micronuclei) can be a biomarker for PAH exposure, uptake, and metabolism. For example, Barbee et al. (2008) demonstrated that juvenile Coho salmon, caged in situ in an area with sediment PAH contamination, had higher chromosomal damage in both peripheral blood and liver, measured using several different assay methods. The level of damage correlated with the sediment PAH concentration, but not the aquatic PAH concentration. This is consistent with the preferential partitioning of HPAHs, which are more typically associated with genomic damage, to the sediment.

A study by Wessel et al. (2010) looked at the relationship in juvenile sole between various biomarkers commonly used to measure PAH exposure in the environment: bioavailability (CYP1A1 activity/EROD), biotransformation (bile PAH metabolites), and genotoxicity (DNA strand breaks). Fish were given food containing a mixture of three HPAHs (>4 rings; benzo[a] pyrene, fluoranthene, and pyrene), followed by 1 week of depuration. The production of metabolites confirmed the biotransformation of PAHs. EROD activity was slightly elevated, whereas formation of DNA strand breaks was significantly elevated in PAH-exposed fish. There was good correlation between the concentration of metabolites in bile and the formation of DNA strand breaks (genotoxicity), but not for EROD. This study highlights the importance of using multiple biomarkers when assessing PAH exposure, and the correlation between bile metabolites and DNA damage further strengthens the hypothesis that metabolism of PAHs is required to generate reactive intermediates that can form DNA adducts.

Reproductive Toxicity

Hall and Oris (1991) conducted a study using adult fathead minnows to assess the effects of anthracene (three-ring PAH) exposure on reproduction and larval/fry survivability. Anthracene is one of the PAHs that are associated with potential for phototoxicity or photosensitization. Adult fish were exposed to anthracene for up to 6 weeks. Eggs that were laid were collected and transferred to fresh, clean (no anthracene) water. Some of the eggs were exposed to SUVR, whereas others were not. Eggs were monitored for hatchability, survivorship, and abnormalities/deformities. The authors report that anthracene bioconcentrated in eggs, gonads, and carcasses of exposed fish. Fewer eggs were laid by anthracene-exposed fish. Egg survivorship was impaired in all groups and teratogenicity (larval/fry deformities) was increased by maternal anthracene exposure when eggs were subsequently exposed to UV radiation. This study demonstrates the potential for maternal transfer of contaminant effects, indicating that PAH exposure does not need to occur to the eggs themselves or during developmental stages of the egg, larvae, or fry before detrimental effects are observed. HPAHs can also be chronically toxic to sand sole (*Psettichthys melanostichus*) (Hose et al. 1982). These investigators observed that the average hatching success in sand sole exposed to 0.10 μg/L B[a]P was reduced by about 29% compared to the control.

Developmental Toxicity

While much of the developmental toxicity resulting from crude oil exposure has been attributed to PAHs, causation cannot be established from these studies alone because crude oil consists of a variety of components that change with time and weathering. However, studies that examine the toxicity of various fractions of crude oil can be invaluable in assessing which component(s) of crude oil is causing the observed effects. An experiment done by Sundberg et al. (2006) used special separation columns to investigate the effects of three different sediment-derived crude oil fractions (aliphatic hydrocarbons/MAHs, dicyclic AHs, and PAHs) on fertilized egg development and larval deformities in rainbow trout. Exposure (through microinjections, to mimic maternal transfer and environmental uptake in the egg) to the PAH fraction led to increased mortality in the developing eggs and elevation of deformities such as asymmetrical yolk sacs in embryos and hemorrhage in larvae. No effects were observed from the other two fractions of crude oil, suggesting that PAHs, and particularly HPAHs, contribute to the embryo toxicity of crude oil observed in both lab and field studies.

Billiard et al. (1999) demonstrated that exposure of rainbow trout to retene (32–320 μg/L) during egg and posthatch stages resulted in increased incidence of blue-sac disease. Some symptoms were observed at the lowest concentration tested and effects included induction of CYP1A1, edema, hemorrhaging, craniofacial malformation, reduced growth, mortality, fin erosion, and opercular sloughing. In a follow-up study to the Billiard et al. (1999) investigation, a study with rainbow trout was done to evaluate retene (320 μg/L) for its ability to produce blue-sac disease through an oxidative stress mechanism (Bauder et al. 2005). Retene-exposed fish had increased prevalence of blue-sac disease and decreased vitamin E and glutathione concentrations in the tissues. Co-exposures with retene and vitamin E resulted in reduced incidence of blue-sac disease and increased tissue concentrations of vitamin E, but did not affect glutathione concentrations. The authors concluded that a portion of the effects of retene are related to oxidative stress, but there may be additional mechanisms of toxicity such as formation of retene adducts in DNA, lipids, or protein.

The minimum concentrations of LPAHs (naphthalene, acridine, and phenanthrene) causing gross developmental anomalies in rainbow trout were found to be much higher than B[a]P at 230, 410, and 85 μg/L, respectively (Black et al. 1983). Teratogenic effects during organogenesis (7–24 days postfertilization) were studied by Hannah et al. (1982) and Hose et al. (1984) in rainbow trout (*O. mykiss*) exposed to B[a]P-contaminated sand (1–500 μg/g). Gross anomalies (e.g., microphthalmia) were noted in a significant proportion of fish (6.8%) exposed to the contaminated sand; the average aqueous concentration was 0.2 μg/L (Hose et al. 1984).

The zebrafish is commonly used as a model organism for assessing the developmental consequences of many chemicals because zebrafish are small, readily grow and reproduce in the lab setting, and have a short lifecycle allowing full lifecycle assessments. Using zebrafish, Incardona et al. (2004) demonstrated that different PAHs cause different effects in developing embryos. In his study, embryos were exposed to seven different individual PAHs (naphthalene, fluorene, dibenzothiophene, phenanthrene, anthracene, pyrene, or chrysene), or two mixtures of those PAHs. All of these PAHs, except pyrene and anthracene, are common components of Alaskan North Slope crude oil. Exposures to dibenzothiophene or phenanthrene alone were enough to cause the PAH-associated blue-sac disease previously described. The relative toxicity of the mixtures was proportional to either the amount of phenanthrene or dibenzothiophene plus phenanthrene present in the mixture. Pyrene, an HPAH (four rings), was found to cause a different suite of symptoms that included anemia, peripheral vascular defects, and neuronal cell death. The change in effects between LPAHs and HPAHs has implications for oil-contaminated aquatic environments, because crude oil will typically undergo a weathering process in which the LPAHs are lost, leaving the higher molecular weight compounds behind. This shift in PAH composition may have consequences for fish embryos in terms of the types of effects that might be observed.

Also using zebrafish embryos, Carls et al. (2008) investigated the effects of either dissolved PAHs only, or total PAHs (dissolved plus oil droplets) derived from Alaskan North Slope crude oil. Embryos were assessed for physiological effects including pericardial edema, abnormal heart development, and intracranial hemorrhaging following 2 days of exposure and all parameters were altered by exposures. The authors found that the effects of total PAHs (including oil droplets) were not different than the effects of dissolved PAHs (no oil droplets) on any of the physiological alterations assessed, indicating that the embryos need not come into direct contact with oil droplets for toxicological consequences to occur.

Japanese medaka is a warm-water fish species that is also a commonly used fish model in toxicological studies. Fallahtafti et al. (2012) used this species to examine the effects of a series of alkyl-phenanthrenes and their hydroxylated metabolites on the development and severity of blue-sac disease (edema, heart defects, deformities, hemorrhaging) in early life stages following a 17-day exposure period. Generally, the metabolites of the alkyl-phenanthrenes were found to cause more severe blue-sac disease symptoms, confirming the importance of metabolism in the generation of more toxic PAH intermediates. Among PAHs studied, B[a]P is typically found to be the most toxic. For example, 5% of sand sole (*P. melanostichus*) eggs exposed to B[a]P at 0.1 μg/L in water showed gross anomalies such as overgrowth of tissue originating from the somatic musculature, and arrested development (as compared to 0% in control fish) (Hose et al. 1982). Also, the hatching success of eggs exposed to 0.1 μg B[a]P/L was significantly lower than that of controls.

Neurotoxicity

In a series of studies, Gesto and colleagues demonstrated that PAH can be neurotoxic to rainbow trout. Gesto et al. (2009) showed that both naphthalene and benzo[a]pyrene, following intraperitoneal injections, can disrupt the functioning of the pineal gland, altering the release of melatonin and other hormones responsible for regulation of biological rhythms. In addition, exposure to naphthalene for up to 5 days (via injection or implants) altered the levels of the monoaminergic neurotransmitters (dopamine, serotonin, noradrenalin) and metabolites in the brain of immature rainbow trout (Gesto et al. 2006). An earlier study had demonstrated that naphthalene also decreases plasma cortisol and other plasma analytes in rainbow trout (Gesto et al. 2008). Taken together, these studies suggest that naphthalene (and potentially other PAHs) may modify neuroendocrine interactions, which may have widespread physiological implications for fish.

Behavioral Toxicity

Purdy (1989) conducted a study in which Coho salmon were exposed to a mixture of LPAHs and HPAHs. Effects on feeding and avoidance behavior were evaluated at the end of the 24-hour exposure period and periodically following recovery from that exposure. It was found that exposure to the mixture of PAHs resulted in impaired feeding behavior, as well as loss of a learned avoidance response. In addition, the time taken for fish to respond in the avoidance assay increased, indicating reaction times were slowed. These effects persisted for 1–10 days after the exposure was withdrawn.

Goncalves et al. (2008) carried out a study using gilthead sea bream (warm-water fish species) to examine the effects of three PAHs (fluorene, pyrene, and phenanthracene) individually and in mixtures. Each of the PAHs impaired swim performance and increased lethargy. Phenanthracene exposure also impaired social performance of the fish. Pyrene was the most potent, and measurement of lethargy was the most sensitive endpoint. The mixture of the three PAHs had similar effects and mixture of chemicals appeared to have an additive effect. Farr et al. (1995) used fathead minnow to demonstrate that fish are able to detect, respond to, and avoid plumes of fluoranthene at high concentrations, but not at lower concentrations. Thus, fish might be unable to detect fluoranthene at environmentally relevant concentrations, resulting in further exposure of the fish to an environment where toxic effects may occur.

Summary and Conclusions

Crude oil contains COPCs that have been well studied in many fish species. These components can be acutely lethal, or can result in chronic sublethal effects that may persist for the life of the animal or perhaps through generations. Toxic effects from COPCs in oil can include death, morphological and histopathological effects, genotoxicity, immunotoxicity, and developmental and reproductive effects, among others. There are limitations to decisions based on scientific databases for chemical mixtures like oil, particularly because most studies have focused on high doses of a few constituents, using experimentally expedient compositions. Most real-world exposures are to low doses and to a complex range of chemicals. More novel approaches must be used to determine the toxicity and risks of mixtures; these must use low doses, computer modeling, and mechanistic modeling, among other tools (e.g., Hazard Index, Target-Organ Toxicity Dose, Weight of Evidence, and Toxic Equivalences). These approaches have their limitations, primarily with respect to the prediction of potentially unforeseen interactions between the mixture constituents that may affect the resulting toxic outcome.

REFERENCES

Aas, E., T. Baussant, L. Balk, B. Liewenborg, and O. D. Andersen. 2000. PAH metabolites in bile, cytochrome P4501A and DNA adducts as environmental risk parameters for chronic oil exposure: A laboratory experiment with Atlantic cod. *Aquatic Toxicology* 51:241–258.

Akcha, F., T. Burgeot, J. F. Narbonne, and P. Garrigues. 2003. Metabolic activation of PAHs: Role of DNA adduct formation in induced carcinogenesis. In *PAHs: An Ecotoxicological Perspective*, edited by P. E. T. Douben, 64–79. United Kingdom: Wiley and Sons.

Albert, P. H. 1995. Petroleum and individual polycyclic aromatic hydrocarbons. In *Handbook of Ecotoxicology*, edited by D. J. Hoffman, B. A. Rattner, G. A. Burton Jr., and J. Cairons Jr., 330–355. Boca Raton, FL: CRC Press.

Alkindi, A. Y. A., J. A. Brown, C. P. Waring, and J. E. Collins. 1996. Endocrine, osmoregulatory, respiratory and haematological parameters in flounder exposed to the water soluble fraction of crude oil. *Journal of Fish Biology* 49:1291–1305.

Anderson, J., J. Neff, B. Cox, H. Tatem, and G. Hightower. 1974. Characteristics of dispersions and water-soluble extracts of oxide and refined oils and their toxicity to estuarine crustaceans and fish. *Marine Biology* 27:75–88.

Arey, J., and R. Atkinson. 2003. Photochemical reactions of PAHs in the atmosphere. In *PAHs: An Ecotoxicological Perspective*, edited by P. E. T. Douben, 46–73. United Kingdom: Wiley and Sons.

Arkoosh, M. R., E. Casillas, P. Huffman, E. Clemons, U. Varanasi, J. Evered et al. 1998. Increased susceptibility of juvenile Chinook salmon from a contaminated estuary to *Vibrio anguillarum*. *Transactions of the American Fisheries Society* 127:360–374.

Barbee, G. C., J. Barich, B. Duncan et al. 2008. In situ biomonitoring of PAH-contaminated sediments using juvenile coho salmon (*Oncorhynchus kisutch*). *Ecotoxicology and Environmental Safety* 71:454–464.

Barsiene, J., V. Dedonyte, A. Rybakova, L. Andreikenaite, and O. K. Andersen. 2006. Investigation of micronuclei and other nuclear abnormalities in peripheral blood and kidney of marine fish treated with crude oil. *Aquatic Toxicology* 78:S99–S104.

Bauder, M. B., V. P. Palace, and P. V. Hodson. 2005. Is oxidative stress the mechanism of blue sac disease in retene-exposed trout larvae? *Environmental Toxicology and Chemistry* 24:694–702.

Beischlag, T. V., J. L. Morales, B. D. Hollingshead, and G. H. Perdew 2008. The aryl hydrocarbon receptor complex and the control of gene expression. *Critical Reviews in Eukaryotic Gene Expression* 18:207–250.

Benville, P. E. and S. Korn. 1977. The acute toxicity of six monocyclic aromatic crude oil components to striped bass (*Morone saxatilus*) and bay shrimp (*Crago franciscorum*). *California Fish and Game* 63:204–209.

Billiard, S. M., K. Querbach, and P. V. Hodson. 1999. Toxicity of retene to early life stages of two freshwater fish species. *Environmental Toxicology and Chemistry* 18:2070–2077.

Birtwell, I. K., R. Fink, D. Brand, R. Alexander, and C. D. McAllister. 1999. Survival of pink salmon (*Oncorhynchus gorbuscha*) fry to adulthood following a 10-day exposure to the aromatic hydrocarbon water-soluble fraction of crude oil and release to the Pacific Ocean. *Canadian Journal of Fisheries and Aquatic Sciences* 56:2087–2098.

Black, J. A., W. J. Birge, and A. G. Westerman. 1983. Comparative aquatic toxicology of aromatic hydrocarbons. *Fundamental and Applied Toxicology* 3:353–358.

Blanc, A. M., L. G. Holland, S. D. Rice, and C. J. Kennedy. 2010. Anthropogenically sourced low concentration PAHs: In situ bioavailability to juvenile Pacific salmon. *Ecotoxicology and Environmental Safety* 73:849–857.

Bowling, J. W., G. J. Leversee, P. F. Landrum, and J. P. Giesy. 1983. Acute mortality of anthracene-contaminated fish exposed to sunlight. *Aquatic Toxicology* 3:79–90.

Brannon, E. L., K. C. M. Collins, L. L. Moulton, and K. R. Parker. 2001. Resolving allegations of oil damage to incubating pink salmon eggs in Prince William Sound. *Canadian Journal of Fisheries and Aquatic Sciences* 58:1070–1076.

Brannon, E. L., K. M. Collins, J. S. Brown, J. M. Neff, K. R. Parker, and W. A. Stubblefield. 2006. Toxicity of weathered *Exxon Valdez* crude oil to pink salmon embryos. *Environmental Toxicology and Chemistry* 25:962–972.

Bravo, C. F., L. R. Curtis, M. S. Myers, J. P. Meador, L. L. Johnson, J. Buzitis et al. 2011. Biomarker responses and disease susceptibility in juvenile rainbow trout (*Oncorhynchus mykiss*) fed a high molecular weight PAH mixture. *Environmental Toxicology and Chemistry* 30:704–714.

Brocksen, R. W. and H. T. Bailey. 1973. Respiratory responses of juvenile Chinook salmon and striped bass exposed to benzene, a water-soluble component of crude oil. In *Prevention and Control of Oil Spills*, Proceedings of joint conference, 783–791, Washington, D.C.

Bue, B. G., S. Sharr, S. D. Moffitt, and D. Craig. 1996. Effects of *Exxon Valdez* oil spill on pink salmon embryos and pre-emergent fry. In *Proceedings of the Exxon Valdez Oil Spill Symposium*, edited by S. D. Rice, R. B. Spies, D. A. Wolfe, and B. A. Wright, 619–627. Bethesda, MD: American Fisheries Society.

Bue, B. G., S. Sharr, and J. E. Seeb. 1998. Evidence of damage to pink salmon populations inhabiting Prince William Sound, Alaska, two generations after the *Exxon Valdez* oil spill. *Transactions of the American Fisheries Society* 127:35–43.

Carls, M. G. 1987. Effects of dietary and water-borne oil exposure on larval Pacific herring (*Clupea harengus pallasi*). *Marine Environmental Research* 22:253–270.

Carls, M. G., R. A. Heintz, G. D. Marty and S. D. Rice. 2005. Cytochrome P4501A induction in oil-exposed pink salmon (*Oncorhynchus gorbuscha*) embryos predicts reduced survival potential. *Marine Ecological Progress Series* 301:253–265.

Carls, M. G., L. Holland, M. Larsen, T. K. Collier, N. L. Scholz, and J. P. Incardona. 2008. Fish embryos are damaged by dissolved PAHs, not oil particles. *Aquatic Toxicology* 88:121–127.

Carls, M. G., J. E. Hose, R. E. Thomas, and S. D. Rice. 2000. Exposure of Pacific herring to weathered crude oil: assessing effects on ova. *Environmental Toxicology and Chemistry* 19:649–1659.

Carls, M. G. and S. Korn. 1985. Sensitivity of arctic marine amphipods and fish to petroleum hydrocarbons. In *Proceedings of the Tenth Annual Aquatic Toxicology Workshop*, edited by P. G. Wells and R. F. Addison. Halifax: Canadian Technical Reports of Fisheries and Aquatic Sciences.

Carls, M. G., G. D. Marty, T. R. Meyers, R. E. Thomas, and S. D. Rice. 1998. Expression of viral hemorrhagic septicemia virus in pre-spawning Pacific herring (*Clupea pallasi*) exposed to weathered crude. *Canadian Journal of Fisheries and Aquatic Sciences* 55:2300–2309.

Carls, M. G. and S. D. Rice. 1990. Abnormal development and growth reductions of Pollock (*Theragra chalcogramma*) embryos exposed to water-soluble fractions of crude oil. *Fishery Bulletin* 88:29–37.

Carls, M. G., S. D. Rice, and J. E. Hose. 1999. Sensitivity of fish embryos to weathered crude oil: Part I. Low level exposure during incubation causes malformations, genetic damage and mortality in larval Pacific herring (*Clupea pallasi*). *Environmental Toxicology and Chemistry* 18:481–493.

Carls, M. G., R. E. Thomas, and S. D. Rice. 2003. Mechanism for transport of oil-contaminated water into pink salmon redds. *Marine Ecological Progress Series* 248:245–255.

Clemente, J. S. and P. M. Fedorak. 2005. A review of the occurrence, analyses, toxicity and biodegradation of naphthenic acids. *Chemosphere* 60:585–600.

Devlin, E. W., J. D. Brammer, and R. L. Puyear. 1982. Acute toxicity of toluene to three age groups of fathead minnow (*Pimephales promelas*). *Bulletin of Environmental Contamination and Toxicology* 29:12–17.

Di Toro, D. M., J. A. McGrath, and W. A. Stubblefield. 2007. Predicting the toxicity of neat and weathered crude oil: toxic potential and the toxicity of saturated mixtures. *Environmental Toxicology and Chemistry* 26:24–36.

DiMichele, L. and M. H. Taylor. 1978. Histopathological and physiological responses of *Fundulus heteroclitus* to naphthalene exposure. *Journal of the Fisheries Research Board of Canada* 35:1060–1066.

Eldridge, M. B., T. Echeverria, and J. A. Whipple. 1977. Energetics of Pacific herring (*Clupea harengus pallasi*) embryos and larvae exposed to low concentrations of benzene, a monoaromatic component of crude oil. *Transactions of the American Fisheries Society* 106:452–461.

Evans, D. R. and S. D. Rice. 1974. Effects of oil on marine ecosystems: A review for administrators and policy makers. *Fishery Bulletin* 72:625–638.

Fallahtafti, S., T. Rantanen, R. S. Brown, V. Snieckus, and P. V. Hodson. 2012. Toxicity of hydroxylated alkyl-phenanthrenes to early life stages of Japanese medaka (*Oryzias latipes*). *Aquatic Toxicology* 106–107:56–64.

Farr, A. J., C. C. Chabot, and D. H. Taylor. 1995. Behavioral avoidance of fluoranthene by fathead minnow (*Pimephales promelas*). *Neurotoxicology and Teratology* 17:265–271.

Finger, S. E., E. F. Little, M. G. Henry, J. F. Fairchild, and T. P. Boyle. 1985. Comparison of laboratory and field assessment of fluorene. Part I. Effects of fluorene on the survival, growth, reproduction, and behavior of aquatic organisms in laboratory tests. In *Validation and Predictability of Laboratory Methods for Assessing the Fate and Effects of Contaminants in Aquatic Ecosystems*, ASTM STP 865, edited by T. P. Boyle. Philadelphia, PA.

Folmar, L. C., D. R. Craddock, J. W. Blackwell, G. Joyce, and H. O. Hodgins. 1981. Effects of petroleum exposure on predatory behaviour of coho salmon (*Oncorhynchus kisutch*). *Environmental Toxicology and Chemistry* 27:458–462.

Fragoso, N. M., P. V. Hodson, and S. Zambon. 2006. Evaluation of an exposure assay to measure uptake of sediment PAH by fish. *Environmental Monitoring and Assessment* 116:481–511.

Gesto, M., J. L. Soengas, and J. M. Miguez. 2008. Acute and prolonged stress response of brain monoaminergic activity and plasma cortisol levels in rainbow trout are modified by PAHs (naphthalene, β-naphthoflavone and benzo(a)pyrene) treatment. *Aquatic Toxicology* 86:341–351.

Gesto, M., A. Tintos, J. L. Soengas, and J. M. Miguez. 2006. Effects of acute and prolonged naphthalene exposure on brain monoaminergic neurotransmitters in rainbow trout (*Oncorhynchus mykiss*). *Comparative Biochemistry and Physiology: Part C* 144:173–183.

Gesto, M., A. Tintos, A. Rodriguez-Illamola, J. L. Soengas, and J. M. Miguez. 2009. Effects of naphthalene, β-naphthoflavone and benzo(a)pyrene on the diurnal and nocturnal indoleamine metabolism and melatonin content in the pineal organ of rainbow trout, *Oncorhynchus mykiss*. *Aquatic Toxicology* 92:1–8.

Goncalves, R., M. Scholze, A. M. Ferreira, M. Martins, and A. D. Correia. 2008. The joint effect of polycyclic aromatic hydrocarbons on fish behaviour. *Environmental Research* 108:205–213.

Gonzalez-Doncel, M., L. Gonzalez, C. Fernandez-Torija, J. M. Navas, and J. V. Tarazona. 2008. Toxic effects of an oil spill on fish early life stages may not be exclusively associated with PAHs: studies with *Prestige* oil and medaka (*Oryzias latipes*). *Aquatic Toxicology* 87:280–288.

Hall, A. T. and J. T. Oris 1991. Anthracene reduces reproductive potential and is maternally transferred during long-term exposure in fathead minnows. *Aquatic Toxicology* 19:249–264.

Hannah, J. B., J. E. Hose, M. L. Landolt, B. S. Miller, S. P. Felton, and W. T. Iwaoka. 1982. Benzo(a)-pyrene induced morphologic and developmental abnormalities in rainbow trout. *Archives of Environmental Contamination and Toxicology* 11:727–734.

Hawkes, J. W., E. H., Gruger Jr., and O. P. Olson. 1980. Effects of petroleum hydrocarbons and chlorinated biphenyls on the morphology of the intestine of Chinook salmon (*Oncorhynchus tshawytscha*). *Environmental Research* 23:149–161.

Headley, J. V. and D. W. McMartin. 2004. A review of the occurrence and fate of naphthenic acids in aquatic environments. *Journal of Environmental Science and Health* A39:1989–2004.

Heintz, R. A. 2007. Chronic exposure to polynuclear aromatic hydrocarbons in natal habitats leads to decreased equilibrium size, growth, and stability of pink salmon populations. *Integrated Environmental Assessment and Management* 3:351–363.

Heintz, R. A., S. D. Rice, A. C. Wertheimer, R. F. Bradshaw, F. P. Thrower, J. E. Joyce et al. 2000. Delayed effects on growth and marine survival of pink salmon (*Oncorhynchus gorbuscha*) after exposure to crude oil during embryonic development. *Marine Ecological Progress Series* 208:205–216.

Heintz, R. A., J. W. Short, and S. D. Rice. 1999. Sensitivity of fish embryos to weathered crude oil: Part II. Increased mortality of pink salmon (*Oncorhynchus gorbuscha*) embryos incubating downstream from weathered *Exxon Valdez* crude oil. *Environmental Toxicology and Chemistry* 18:494–503.

Hicken, C. E., T. L. Linbo, D. H. Baldwin, M. L. Willis, M. S. Myers, L. Holland et al. 2011. Sublethal exposure to crude oil during embryonic development alters cardiac morphology and reduces aerobic capacity in adult fish. *Proceedings of the National Academy of Sciences* 108:7086–7090.

Holcombe, G. W., G. L. Phipps, and J. T. Fiandt. 1983. Toxicity of selected priority pollutants to various aquatic organisms. *Ecotoxicology and Environmental Safety* 7:400–409.

Hose, J. E., J. B. Hannah, D. DiJulio, M. L. Landolt, B. S. Miller, W. T. Iwaoka et al. 1982. Effects of benzo(a)pyrene on early development of flatfish. *Archives of Environmental Contamination and Toxicology* 11:167–171.

Hose, J. E., J. B. Hannah, W. H. Puffer, and M. L. Landoit.1984. Histologic and skeletal abnormalities in benzo(a)pyrene treated rainbow trout alevins. *Archives of Environmental Contamination and Toxicology* 13: 675–684.

Hose, J. E., M. D. McGurk, G. D. Marty, D. E. Hinton, E. D. Brown, and T. T. Baker. 1996. Sublethal effects of the *Exxon Valdez* oil spill on herring embryos and larvae: Morphological, cytogenetic, and histopathological assessments 1989–1991. *Canadian Journal of Fisheries and Aquatic Sciences* 53:2355–2365.

Incardona, J. P., T. K. Collier, and N. L. Scholz. 2004. Defects in cardiac function precede morphological abnormalities in fish embryos exposed to polycyclic aromatic hydrocarbons. *Toxicology and Applied Pharmacology* 196:191–205.

Ingvarsdottir, A., C. Bjorkblom, E. Ravagnan, B. F. Godal, M. Arnberg, D. L. Joachim et al. 2011. Effects of different concentrations of crude oil on first feeding larvae of Atlantic herring (*Clupea harengus*). *Journal of Marine Systems* 93:69–76.

Kennedy,C. J. and A. P. Farrell. 2005. Ion homeostasis and interrenal stress responses in juvenile Pacific herring, *Clupea pallasi*, exposed to the water-soluble fraction of oil. *Journal of Experimental Marine Biology and Ecology* 323:43–56.

Kennedy, C. J. and A. P. Farrell. 2006. Effects of exposure to the water-soluble fraction of crude oil on the swimming performance and the metabolic and ionic recovery postexercise in Pacific herring (*Clupea pallasi*). *Environmental Contamination and Toxicology* 25:2715–2724.

Kennedy, C. J. and A. P. Farrell. 2008. Immunological alterations in juvenile Pacific herring, *Clupea pallasi*, exposed to aqueous hydrocarbons derived from oil. *Environmental Pollution* 153:638–648.

Kocan, R. M., G. D. Marty, M. S. Okihiro, E. D. Brown, and T. T. Baker. 1996. Reproductive success and histopathology of individual Prince William Sound Pacific herring 3 years after the *Exxon Valdez* oil spill. *Canadian Journal of Fisheries and Aquatic Sciences* 53:2388–2393.

Korn, S., D. A. Moles, and S. D. Rice. 1979. Effects of temperature on the median tolerance limit of pink salmon and shrimp exposed to toluene, naphthalene, and Cook Inlet crude oil. *Bulletin of Environmental Contamination and Toxicology* 21:521–525.

Korn, S., J. W. Struhsaker, and P. Benville Jr. 1976. Effects of benzene on growth, fat content, and caloric content of striped bass, *Morone saxatilis*. *Fishery Bulletin* 74: 694–698.

Lin, C. Y., B. S. Anderson, B. M. Phillips, A. C. Peng, S. Clark, J. Voorhees et al. 2009. Characterization of the metabolic actions of crude versus dispersed oil from salmon smolts via NMR-based metabolomics. *Aquatic Toxicology* 95:230–238.

Lockhart, W., D. A. Duncan, B. N. Billeck, R. A. Danell, and M. J. Ryan. 1996. Chronic toxicity of the 'water soluble fraction' of Norman Wells crude oil to juvenile fish. *Spill Science Technical Bulletin* 3:259–262.

Marty, G. D., M. S. Okihiro, E. D. Brown, D. Hanes, and D. E. Hinton. 1999. Histopathology of adult Pacific herring in Prince William Sound, Alaska, after the *Exxon Valdez* oil spill. *Canadian Journal of Fisheries and Aquatic Sciences* 56:419–426.

Marty, G. D., J. W. Short, D. M. Dambach, N. H. Willits, R. A. Heintz, S. D. Rice et al. 1997. Ascites, premature emergence, increased gonadal cell apoptosis, and cytochrome P4501A induction in pink salmon larvae continuously exposed to oil-contaminated gravel during development. *Canadian Journal of Zoology* 75:989–1007.

Maynard, D. J. and D. D. Weber 1981. Avoidance reactions of juvenile coho salmon (*Oncorhynchus kisutch*) to monocyclic aromatics. *Canadian Journal of Fisheries and Aquatic Sciences* 38:772–778.

McCain, B. B., H. O. Hodgins, W. D. Gronlund, J. W. Hawkes, D. W. Brown, M. S. Myers et al. 1978. Bioavailability of crude oil from experimentally oiled sediments to English sole (*Paraphrys vetulus*), and pathological consequences. *Journal of the Fisheries Research Board of Canada* 35:657–664.

McDonaldm, B. G. and P. M. Chapman. 2002. PAH phototoxicity: An ecological irrelevant phenomenon? *Marine Pollution Bulletin* 44:1321–1326.

Meador, J. P., F. C. Sommers, G. M. Ylitalo, and C. A. Sloan. 2006. Altered growth and related physiological responses in juvenile Chinook salmon (*Oncorhynchus tshawytscha*) from dietary exposure to polycyclic aromatic hydrocarbons (PAHs). *Canadian Journal of Fisheries and Aquatic Sciences* 63:2364–2376.

Menzie, C. A. and B. B. Potocki. 1992. Exposure to carcinogenic PAHs in the environment. *Environmental Science and Technology* 26:1278–1284.

Meyerhoff, R. D. 1975. Acute toxicity of benzene, a component of crude oil, to juvenile striped bass (*Morone suxufilis*). *Journal of the Fisheries Research Board of Canada* 32:1864–1866.

Millemann, R. E., W. J. Birge, J. A. Black, R. M. Cushman, K. L. Daniels, P. J. Franco et al. 1984. Comparative acute toxicity to aquatic organisms of components of coal-derived synthetic fuels. *Transactions of the American Fisheries Society* 113:74–85.

Moles, A. 1980. Sensitivity of parasitized coho salmon fry to crude oil, toluene, and naphthalene. *Transactions of the American Fisheries Society* 109:293–297.

Moles, A., M. M. Babcock, and S. D. Rice. 1987. Effects of oil exposure on pink salmon, *Oncorhynchus gorbuscha*, alevins in a simulated intertidal environment. *Marine Environmental Research* 21:49–58.

Moles, A., S. Bates, S. D. Rice, and S. Korn. 1981. Reduced growth of coho salmon fry exposed to two petroleum components, toluene and naphthalene, in freshwater. *Transactions of the American Fisheries Society* 110:430–436.

Moles, A, and B. L. Norcross. 1998. Effects of oil-laden sediments on growth and health of juvenile flatfishes. *Canadian Journal of Fisheries and Aquatic Sciences* 55:605–610.

Moles, A., S. D. Rice, and S. Korn. 1979. Sensitivity of Alaskan freshwater and anadromous fishes to Prudhoe Bay crude oil and benzene. *Transactions of the American Fisheries Society* 108:408–414.

Moles, A., S. Rice, and B. L. Norcross. 1994. Non-avoidance of hydrocarbon laden sediments by juvenile flatfishes. *Netherlands Journal of Sea Research* 34:361-367.

Morrow, J. E., R. L. Gritz, and M. P. Kirton 1975. Effects of some components of crude oil on young coho salmon. *Copeia* 2:326–331.

Nahrgang, J., L. Camus, M. G. Carls, P. Gonzalez, M. Jonsson, I. C. Taban et al. 2010. Biomarker responses in polar cod (*Boreogadus saida*) exposed to the water soluble fraction of crude oil. *Aquatic Toxicology* 97:234–242.

Nero, V., A. Farwell, L. E. J. Lee, T. Van Meer, M. D. MacKinnon, and D. G. Dixon. 2006. The effects of salinity on naphthenic acid toxicity to yellow perch: gill and liver histopathology. *Ecotoxicology and Environmental Safety* 65:252–264.

Norcross, B. L., J. E. Hose, M. Frandsen, and E. D. Brown. 1996. Distribution, abundance, morphological condition, and cytogenetic abnormalities of larval herring in Prince William Sound, Alaska, following the *Exxon Valdez* oil spill. *Canadian Journal of Fisheries and Aquatic Sciences* 53:2376–2387.

Oris, J. T., and J. P. Giesy Jr. 1987. The photo-induced toxicity of polycyclic aromatic hydrocarbons to larvae of the fathead minnow (*Pimephales promelas*). *Chemosphere* 16:1395–1404.

Overton, E. B., W. D. Sharp, and P. Roberts 1994. Toxicity of petroleum. In *Basic Environmental Toxicology*, 133–156, edited by L. G. Cockerham and B. S. Shane. Boca Raton, FL: CRC Press.

Palm Jr., R, D. B. Powell, A. Skillman, and K. Godtfredsen. 2003. Immunocompetence of juvenile chinook salmon against *Listonella anguillarum* following dietary exposure to polycyclic aromatic hydrocarbons. *Environmental Toxicology and Chemistry* 22(12):2986–2994.

Payne, J. F., L. L. Fancey, and J. Hellou. 1995. Aliphatic hydrocarbons in sediments: a chronic toxicity study with winter flounder (*Pleuronectes americanus*) exposed to oil well drill cuttings. *Canadian Journal of Fisheries and Aquatic Sciences* 52:2724–2735.

Payne, J. F., J. Kiceniuk, L. L. Fancey, and U. William.1988. What is a safe level of polycyclic aromatic hydrocarbons for fish: Subchronic toxicity study on winter flounder (*Pseudopleuronectes amercanus*). *Canadian Journal of Fisheries and Aquatic Sciences* 45:1983–1993.

Peters, L. E., M. MacKinnon, T. Van Meer, M. R. van den Heuvel, and D. G. Dixon. 2007. Effects of oil sands process-affected waters and naphthenic acids on yellow perch (*Perca flavescens*) and Japanese medaka (*Orizias latipes*) embryonic development. *Chemosphere* 67:2177–2183.

Purdy, J. E. 1989. The effects of brief exposure to aromatic hydrocarbons on feeding and avoidance behaviour in coho salmon, *Oncorhynchus kisutch*. *Journal of Fish Biology* 34:621–629.

Ramachandran, S. D., M. J. Sweezey, P. Y. Hodson, M. Boudreau, S. M. Courtenay, K. Lee et al. 2006. Influence of salinity and fish species on PAH uptake from dispersed crude oil. *Marine Pollution Bulletin* 52:1182–1189.

Reynaud, S. and P. Deschaux. 2006. The effects of polycyclic aromatic hydrocarbons on the immune system of fish: a review. *Aquatic Toxicology* 77:229–238.

Stephens, S. M., A. Y. A. Alkindi, C. P. Waring, and J. A. Brown. 1997. Corticosteroid and thyroid responses of larval and juvenile turbot exposed to the water-soluble fraction of crude oil. *Journal of Fish Biology* 50:953–964.

Struhsaker, J. W., M. B. Eldrige, and T. Echeverria. 1974. Effects of benzene (a water soluble component of crude oil) on eggs and larvae of Pacific herring and northern anchovy. In *Pollution and physiology of marine organisms,* edited by F. J. Vernberg and W. B. Vernberg. New York: Academic Press.

Struhsaker, J. W. 1977. Effects of benzene (a toxic component of petroleum) on spawning Pacific herring, *Clupea harengus pallasi*. *Fishery Bulletin* 75(1):43–49.

Sverdrup, L. E., P. H. Krogh, T. Nielsen, C. Kjær, and J. Stenersen. 2003. Toxicity of eight polycyclic aromatic compounds to red clover (*Trifolium pretense*), ryegrass (*Lolium perenne*), and mustard (*Sinapsis alba*). *Chemosphere* 53:993–1003.

Teuschler, L., C. Gennings, W. Hartley, H. Carter, A. Thiyagarajah, R. Schoeny et al. 2005. The interaction effects of binary mixtures of benzene and toluene on the developing heart of medaka (*Oryzias latipes*). *Chemosphere* 58: 1283–1291.

Thomas, R. E. and S. D. Rice. 1975. Increased opercular rates of pink salmon (*Oncorhynchus gorbuscha*) fry after exposure to the water-soluble fraction of Prudhoe Bay crude oil. *Journal of the Fisheries Research Board of Canada* 32:2221–2224.

Toxic Substances Data Base. 2006. http://toxnet.nlm.nih.gov.

Van Scoy, A. R., C. Y. Lin, B. S. Anderson, B. M. Philips, M. J. Martin, J. McCall et al. 2010. Metabolic responses produced by crude versus dispersed oil in Chinook salmon pre-smolts via NMR-based metabolomics. *Ecotoxicology and. Environmental Safety* 73:710–717.

Varanasi, U., E. Casillas, M. R. Arkoosh, T. Hom, D. A. Misitano, D. W. Brown et al. 1993. Contaminant exposure and associated biological effects in juvenile Chinook salmon (*Oncorhynchus tshawytshca*) from urban and nonurban estuaries of Puget Sound. NOAA Technical Memorandum NMFS NWFSC-8.

Wang, S. Y., J. L. Lum, M. G. Carls, and S. D. Rice. 1993. Relationship between growth and total nucleic acids in juvenile pink salmon, *Oncorhynchus gorbuscha*, fed crude oil contaminated food. *Canadian Journal of Fisheries and Aquatic Sciences* 50:996–1001.

Weber, D. D., D. J. Maynard, W. D. Gronlund, and V. Konchin. 1981. Avoidance reactions of migrating adult salmon to petroleum hydrocarbons. *Canadian Journal of Fisheries and Aquatic Sciences* 38:779–781.

Wessel, N., R. Santos, D. Menard, K. Le Menach, V. B uchet, N. Lebayon et al. 2010. Relationship between PAH biotransformation as measured by biliary metabolites and EROD activity, and genotoxicity in juveniles of sole (*Solea solea*). *Marine Environmental Research* 69:S71–73.

Wiedmer, M., M. J. Fink, J. J. Stegeman, R. Smolowitz, G. D. Marty, and D. E. Hinton. 1996. Cytochrome P-450 induction and histopathology of pre-emergent pink salmon from oiled spawning sites in Prince William Sound. In *Proceedings of the Exxon Valdez Oil Spill Symposium*, edited by S. D. Rice, R. B. Spies, D. A. Wolfe, and B. A. Wright, 509–517. Bethesda, MD: American Fisheries Society.

Zhou, H., H. Wu, C. Kia, X. Diao, L. Chen, and Q. Xue. 2010. Toxicology mechanism of the persistent organic pollutants (POPs) in fish through the AhR pathway. *Toxicology Mechanisms and Methods* 20:279–286.

2

Toxicological and Physiological Effects of the Surfactant Dioctyl Sodium Sulfosuccinate at Varying Salinities during Larval Development of the Gulf Killifish (Fundulus grandis)

**Charlotte Bodinier, Fernando Galvez, Katherine Gautreaux,
Arianna Rivera, and Christopher C. Green**

CONTENTS

Introduction

Formulations of Corexit®, a family of oil dispersants used in large-scale oil spills, were the primary emulsification agents utilized to enhance the biodegradation of Macondo-252 oil following the *Deepwater Horizon* spill (DWH) in the Gulf of Mexico (Deepwater Horizon Unified Command [DHUC] 2011). An estimated 6.9 million liters of the dispersant Corexit 9500A was applied in the northern Gulf of Mexico in response to the DWH, with smaller amounts of Corexit 9527 also applied (www.restorethegulf.gov). Although precise formulations are proprietary, Corexit 9527 consists of approximately 48% nonionic surfactants (ethoxylated sorbitan monooleate, ethoxylated sorbitan trioleate, and sorbitan monooleate), 35% of the anionic surfactant, dioctyl sodium sulfosuccinate (DOSS), and 17% hydrocarbon solvents (Wheelock et al. 2002). Although exact percentages are unknown, the primary components of Corexit 9500A as released by the manufacturer (Nalco) include sorbitan, butanedioic acid, propanol, hydro-treated petroleum distillates, and DOSS. The presence of anionic surfactants, such as DOSS, in these mixtures serves to decrease the surface tension between the water–oil interface, facilitating the formation of small (~100 μm) oil-surfactant micelles (National Research Council [NRC] 2005). The resulting droplets are maintained in the water column due to their greater surface-area-to-volume ratio, which increases physical and microbial degradation. Unfortunately, surfactants are also able to interact with biological membranes, where they can exert toxic effects in aquatic organisms (Abel and Skidmore 1975, Attwood and Florence 1983). Acute exposures of the surfactant, sodium lauryl sulfate (SLS), in rainbow trout (*Oncorhynchus mykiss*) resulted in damage to the gill epithelial tissue and was characterized by swelling of filament epithelium, increase and subsequent cell death of mitochondrial rich cells (MRC), and sloughing of dead epithelial cell (Abel and Skidmore 1975).

Several studies have characterized the sensitivity of aquatic invertebrates and vertebrates to DOSS using acute lethality tests to assess relative species sensitivities to the surfactant, however few investigations have been conducted to compare DOSS toxicity across a range of salinities. Goodrich et al. (1991) reported an acute 96-hour median lethal concentration (LC_{50}) for DOSS in fingerling rainbow trout, *Oncorhynchus mykiss,* of 28 $mg{\cdot}L^{-1}$ in fresh water. In comparison, the European eel, *Anguilla anguilla,* and sand goby, *Pomatoschistus minutus,* had 96-hour LC_{50} values of 9.1 and 8.1 $mg{\cdot}L^{-1}$, respectively, in seawater (Maggi and Cossa 1973). A comparative study of DOSS toxicity between the California mysid, *Acanthomysis sculpta,* and the Gulf coast mysid, *Mysidopsis bahia,* indicated that *A. sculpta* was more sensitive to DOSS than was *M. bahia* with 72-hour LC_{50} values of 0.94 and 4.5 $mg{\cdot}L^{-1}$, respectively (Tatem and Portzer 1985). The species described as more sensitive, *A. sculpta,* were maintained at a salinity of 30‰ while *M. bahia* were maintained at a salinity of 20‰ (Tatem and Portzer 1985). Based on this limited number of studies it appears that hypo- and hyperosmotic conditions could, in part, influence observed acute toxicity values of DOSS.

DOSS is an industrial chemical agent used for dispersing, wetting, or emulsifying and is a common consumer product within stool softeners, acrylic adhesives, cleaning products, water-thinned interior paints, and cosmetic products (Anderson et al. 2011). Within pharmaceutical applications, DOSS enhances absorption of varying compounds by altering the permeability of biological membranes (Engle and Riggi 1969, Khalafallah et al. 1975, El-Laithy 2003). Adult Atlantic cod, *Gadus morhua,* exposed to 25 $mg{\cdot}L^{-1}$ Corexit 9527 for 96 hours resulted in gill epithelial ruptures, hyperplasia, and telangiectasis, common signs of direct effects on biological membranes (Khan and Payne 2005). Adult turbot, *Scophthalmus maximus,* exposed to DOSS presented lesions on gill lamellae and on filaments, altering respiratory and osmoregulatory functions (Rosety-Rodriguez et al. 2002). Generally, exposure to oil and/or dispersant–oil mixtures has been shown to alter the main function of the gills such as respiration and osmoregulation (Engelhardt et al. 1981, Alkindi et al. 1996, Duarte et al. 2010).

Few studies have examined the direct effect of DOSS on osmoregulation, a complex process of homeostatic mechanisms coordinating various organs, cells, and molecular machinery. Specialized MRC are involved in the transport of specific ions at the site the branchial cavity, gill filaments, and opercular membrane (Katoh et al. 2000, Evans et al. 2005, Evans 2009). There are a number of specific transmembrane proteins responsible for osmotic regulation through the plasma membrane of these MRCs (Marshall 2002, Evans et al. 2005, Hiroi and McCormick 2007). Among them, the Na^+/K^+-ATPase (NKA) is consistently located in the basolateral membrane (Evans 1993, Hirose et al. 2003), and generates an electrochemical gradient. Following this gradient, ions are exchanged, according to the expression, localization, and abundance of different transmembrane proteins (Hirose et al. 2003) such as the $Na^+/K^+/2Cl^-$ cotransporter (NKCC) (Lorin-Nebel et al. 2006), and the cystic fibrosis transmembrane regulator channel (CFTR) (Marshall et al. 1999, 2002, Bodinier et al. 2009). Despite the limited number of studies using these proteins as indicators of osmotic stress, they seem to be well correlated with this phenomenon. Cotou et al. (2001) suggested that NKA activity could be used an indicator of toxic stress from surfactant-based oil dispersants on *Artemia* sp.

The Gulf killifish, *Fundulus grandis*, is a euryhaline cyprinodontid native to the coastal salt marshes of the Gulf of Mexico and represents an ideal model species for assessing toxicity across a range of osmotic conditions. Throughout the year, populations of *F. grandis* reside and spawn in marshes where salinities range from 4‰ to 39‰ (Nordlie 2000, 2006). Under laboratory conditions, *F. grandis* embryos have been successfully incubated and hatched at salinities between 0‰ and 80‰ (Perschbacher et al. 1990). The ability for these larvae and juveniles to occupy a wide range of salinity allows for the examination of DOSS to determine lethal and sublethal effects across a salinity gradient. A large body of work exists for the congener species, *F. heteroclitus*, which has demonstrated relative sensitivity to a variety of environmental pollutants (Van Veld and Nacci 2008, Bugel et al. 2010). Captive reproduction and embryo incubation rearing parameters have been described for *F. grandis* allowing for this species to be available as a cultured bioassay organism (Green et al. 2010, Brown et al. 2011). *Fundulus* sp. play critical ecological roles within their respective habitats and have an increasing amount of available molecular tools (Burnett et al. 2007). Custom oligonucleotide microarrays and immunohistochemical analysis have been used to determine the effect of site-specific impacts from the DWH on marshes using *F. grandis* (Whitehead et al. 2012).

The acute toxicity of Corexit oil dispersants used during the past 25 years is well established (Singer et al. 2000; Khan and Payne 2005, Hemmer et al. 2011); however, the specific actions and impacts of a major anionic surfactant are not well known. To date, no acute toxicity information exists for DOSS across a salinity gradient. Because of the ability of DOSS to alter absorption rates, we postulate that DOSS toxicity will be altered as salinity ranges from freshwater to full-strength seawater. The goal of this study is to evaluate the effects of DOSS on acute toxicity, embryo development, and gill transport protein expression at varying salinities for larval Gulf killifish.

Materials and Methods

Test Organisms

Broodstock Gulf killifish, which were originally derived from a commercial baitfish operation (Gulf Coast Minnows, Houma, Louisiana), were maintained in a 7000 L indoor recirculation system at the Louisiana State University AgCenter's Aquaculture Research Station (Baton Rouge, Louisiana). Recirculating water was kept at 24.0 ± 0.5°C at a salinity of approximately 12‰, and was monitored routinely for optimal oxygen concentrations using a YSI 85 multiparameter meter (YSI Inc., Yellow Springs, Ohio). Total ammonia nitrogen (TAN; salicylate method) and nitrite concentrations were determined using a Hach DR 4000 spectrophotometer (Hach Co., Loveland, Colorado). These fish were used to obtain embryos, which were collected from Spawntex® spawning mats (Blocksom and

Co., Michigan City, Indiana) every 2 days using propagation methods described in Brown et al. (2011). Upon verification of viability, embryos were transferred to a separate recirculation system at the same temperature and salinity where they were maintained until hatch, which typically requires 10 days at these environmental conditions (Brown et al. 2011). Newly hatched larvae were maintained at a salinity of 10‰ and fed *Artemia* sp. nauplii twice daily within a 1500 L recirculation system divided into 8–150 L circular fiberglass tanks. Five days after hatching, larvae were gradually acclimated (~5‰–6‰ decrease or increase every 2 days) until the final salinities of 0.6‰, 5‰, 15‰, 20‰, 25‰, and 30‰ were achieved. A third group of larvae were maintained at 10‰ throughout, although water changes similar to those conducted in the other treatment were performed. Salinity treatments were mixed using Coralife® Salt Mix (Energy Savers Unlimited Inc., Carson, California) and verified using the YSI 85 multiparameter meter.

Acute Toxicity

DOSS (CAS 577-11-7) was purchased from Sigma Aldrich™ (St. Louis, Missouri). Stock solutions were mixed in 2 L volumetric flasks and diluted volumetrically to nominal concentrations of 1.0, 2.5, 10, 25, 50 and 100 $mg \cdot L^{-1}$ DOSS. An aliquot of each concentration at each salinity (*see below*) was frozen immediately and sent to Columbia Analytical Services Inc. (Kelso, Washington) for analysis of concentration using liquid chromatography with tandem mass spectrometry. The remaining waters were used in acute toxicity tests.

Static renewal acute toxicity tests for DOSS were conducted using 8- to 10-day posthatch larvae at the 0.6‰, 5‰, 12‰, 18‰, 24‰, and 30‰ acclimation salinities using a standard 96-hour LC_{50} test. Additional 96-hour LC_{50} tests for DOSS were performed on stage 19 embryos, characterized by formation of eye buds on a well-defined embryonic keel as per Armstrong and Child (1965), at salinities of 5‰ and 12‰. Each definitive 96-hour LC_{50} test was conducted with five test concentrations selected from a prior range finding examination and a control for each salinity with 20 individuals per concentration. Individuals were placed in 24-well tissue culture plates containing 2 mL of treatment water and replaced every 24 hours. Water quality variables of pH, salinity, TAN, nitrite, alkalinity, and hardness were recorded at the initiation of each bioassay and pH, TAN, and nitrite were monitored at each 24-hour water change by pooling the water from each well within its respective treatment group. Temperature was recorded at 10-minute intervals with a LogTag temperature recorder (LogTag Recorders, Auckland, New Zealand) immersed in a water bath. Mortality was recorded every 24 hours and determined by failure to respond to gentle probing, lack of opercular movement, and the absence of heartbeat confirmed by examination with a dissecting stereomicroscope. Each surviving individual was fed approximately 5–10 *Artemia* sp. nauplii at 48 hours, with remaining nauplii removed after 1 hour.

Larval Exposure for Histology and Histological Methods

A second series of exposures were conducted to obtain killifish larvae for microscopic assessment of gill protein levels by immunohistochemistry. Briefly, larvae were exposed to DOSS at concentrations of 0, 25, and 50 $mg \cdot L^{-1}$ for 96 hours under the same conditions as the bioassays but only for individuals reared at salinities of 12‰ and 18‰. After 96 hours, individuals from each DOSS concentration and salinity treatments were fixed in Bouin's solution for 48 hours, then washed and dehydrated in ascending grades of ethanol. Following immersion in butanol and Histochoice Clearing Agent (Amersco® Solon, Ohio), tissues were impregnated, embedded in Paraplast Sigma Aldrich (St. Louis, Missouri), and finally cut into 4 μm-thick sections using a Leitz microtome; the sections were transferred to microscope slides.

For triple immunostaining of each section, three different primary antibodies were used: 8 μg·mL^{-1} of rabbit polyclonal antihuman Na^+/K^+-ATPase (Santa Cruz Biotechnology®, Santa Cruz, California); 10 μg·mL^{-1} of mouse monoclonal antihuman CFTR (R&D Systems, Minneapolis, Minnesota); and 12 μg·mL^{-1} of goat antiserum antihuman NKCC1 (Santa Cruz Biotechnology, Santa Cruz, California). These heterologous antibodies cross-react with corresponding proteins in Gulf killifish. Control slides were placed in similar conditions, but without primary antibody. Slides were then exposed to a solution containing 12 μg·mL^{-1} of the AlexaFluor® 546-conjugated antirabbit (Invitrogen™, Grand Island, New York), 12 μg·mL^{-1} of the AlexaFluor 488-conjugated antimouse, and 15 μg·mL^{-1} of the AlexaFluor 633-conjugated antigoat. Slides were then mounted in an antibleaching medium Fluoro-Gel II, with DAPI (Electron Microscopy Sciences, Hatfield, Pennsylvania).

Slides were observed with a Leica DMIRE2 fluorescence microscope (Leica Microsystems, Bannockburn, Illinois) with special filters for fluorescence (Texas Red, CY5, GFP, and DAPI filters, Leica Microsystems). The fluorescence microscope was interfaced with a SensiCam QE digital camera (Cooke Corp., Romulus, Michigan) for digital image capture.

Image Analysis

The MRC morphometric parameters were taken using the software Image Tool® (version 3.0, University of Texas Health Science Center, San Antonio, Texas). A micrometer slide was used to calibrate the images with the length and the surface area parameters of each MRC taken using the images from the NKA immunofluorescence from 10 different cells per slide.

Embryo Development

Groups of 20 embryos at stage 19 were individually placed in DOSS solutions of 0, 1, 25, and 40 mg·L^{-1} at a salinity of 12‰ for 96 hours. Solution water was renewed every 24 hours and water quality monitored as previously described for 96-hour LC_{50} examinations. Embryos were incubated in water without DOSS after 96 hours. Mortality was assessed every 24 hours during and after the exposure period. The number of embryos and duration till hatch was recorded every 24 hours during exposure to each DOSS concentration. Upon hatch, embryos were placed in 10% neutral buffered formalin for 48 hours then digitally photographed with a Photometrics Coolsnap *cf*® camera (Photometrics®, Tuscon, Arizona) interfaced with a computer for analysis of morphometric parameters. Individuals were placed on their sagittal axis in a solution of glycerol for image capture. Morphometric data were collected to determine total length, depth at vent, and body cavity area.

Statistical Analysis

The 96-hour LC_{50} values for embryos were determined by estimating the concentration of DOSS in which half of the individuals died over the exposure period. Larval LC_{50} values were calculated for both 48- and 96-hour durations. The Spearman-Karber method was used to calculate LC_{50} values (Hamilton et al. 1977). Measured DOSS concentrations were used to calculate all LC_{50} values. The logistic regression procedure was used to determine the independence of hatch rate on DOSS treatment concentrations. A general linearized model analysis of variance (ANOVA) in randomized block design was performed to test for significant differences in embryo and MRC morphometrics, and time to hatch from fertilization. Assumptions of ANOVA were evaluated using diagnostic plots of residual versus predicted values and normal quintile plots. Dunnett's Test was utilized as a posthoc test to determine if differences existed between control and treatment groups. Hypotheses were tested at the $\alpha = 0.05$ significance level.

Results

Toxicity

The water quality parameters measured during the acute lethality examinations are presented in Table 2.1. Mean temperature (± SD) was 23.8 ± 0.55°C with minimum and maximum temperatures recorded at 22.1°C and 25°C, respectively. The mean daily total ammonia nitrate and unionized ammonia concentrations were at or below 2.2 $mg \cdot L^{-1}$ and ≤0.13 $mg \cdot L^{-1}$ at all salinities respectively, and nitrite was also low throughout toxicity testing. Only water hardness, as $CaCO_3$, increased with increasing salinity (Table 2.1).

The acute toxicity of DOSS to embryonic and 8- to 10-day posthatch Gulf killifish larvae was highly dependent on the concentration of DOSS and the water salinity of the test media, compared to the high survival (≥90%) at both life stages with no DOSS exposure (Figure 2.1). Complete mortality in the highest bioassay concentrations of DOSS often occurred within 24 hours for both embryos and larvae (data not shown). Concentration related mortality responses across time that appeared to diverge based on salinity conditions with increased

TABLE 2.1

Water Quality Parameters Measured during the Acute Lethality Examinations

Parameter ($mg \cdot L^{-1}$)[a]	Salinity (‰)					
	0.6	5	12	18	24	30
pH	8.1 ± 0.01 (8.0–8.2)	8.0 ± 0.04 (7.8–8.4)	8.2 ± 0.01 (8.1–8.2)	7.9 ± 0.03 (7.6–8.3)	8.1 ± 0.02 (8.0–8.2)	8.0 ± 0.04 (7.8–8.0)
TAN	2.2 ± 0.1 (1.4–2.8)	2.1 ± 0.1 (1.7–2.6)	1.5 ± 0.1 (0.1–2.0)	1.1 ± 0.1 (0.5–1.6)	1.2 ± 0.1 (0.5–1.7)	0.9 ± 0.1 (0.7–0.9)
NH_3[b]	0.13 ± 0.01 (0.06–0.21)	0.10 ± 0.01 (0.003–0.2)	0.10 ± 0.01 (0.007–0.18)	0.03 ± 0.01 (0.006–0.01)	0.05 ± 0.01 (0.03–0.09)	0.03 ± 0.01 (0.02–0.07)
NO_2	0.008 ± 0.01 (0.002–0.011)	0.04 ± 0.01 (0.003–0.08)	0.01 ± 0.01 (0.007–0.026)	0.003 ± 0.01 (0–0.008)	0.001 ± 0.01 (0.001–0.017)	0.006 ± 0.01 (0.001–0.03)
Alkalinity[c]	124	80	315	246	202	282
Hardness[c]	60	125	980	2060	2780	4860

Note: Mean water quality parameters (± SE) and ranges given in parentheses for daily pH, total ammonia nitrogen (TAN), NH_3, and nitrite (NO_2) during the dioctyl sodium sulfosuccinate LC_{50} examinations on Gulf killifish, *Fundulus grandis*, at six different salinities. Alkalinity and hardness were assessed at the initiation of each examination and are expressed as $mg \cdot L^{-1}$ $CaCO_3$.

[a] Exception: pH is in pH units.

[b] Unionized ammonia.

[c] As $CaCO_3$.

salinity producing greater toxicity in 8- to 10-day posthatch larvae (Figure 2.1). Few mortalities occurred in 8- to 10-day posthatch larvae between 48 and 96 hours, resulting in similar LC_{50} estimates between those observation periods (Table 2.2). Embryo 96-hour LC_{50} values (and 95% CI) for DOSS at incubation salinities of 5‰ and 12‰ were estimated at 50 mg·L^{-1} (27–125) and 43 mg·L^{-1} (21–134), respectively (Figure 2.2).

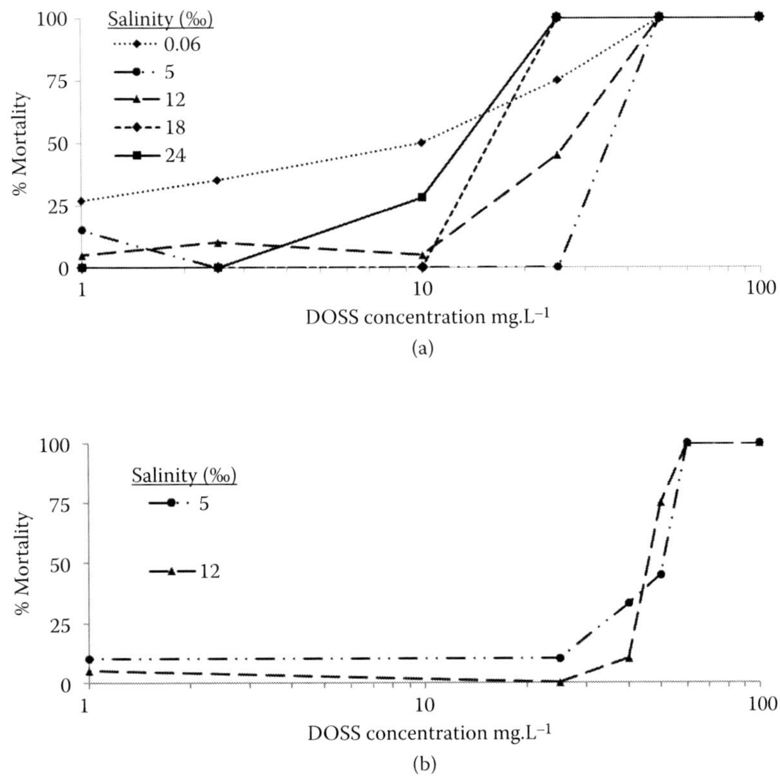

FIGURE 2.1
Concentration-response curves for (a) 8- to 10-day posthatch, and (b) embryos of Gulf killifish, *Fundulus grandis*, exposed to dioctyl sodium sulfosuccinate (DOSS) for 96 hours at salinities of 5‰, 12‰, 18‰, and 24‰.

TABLE 2.2
48- and 96-Hour Lethal Concentrations and Their 95% Confidence Intervals (CI) from Acute Toxicity Tests

Salinity (‰)	48-hour LC_{50} [95% CI]	96-hour LC_{50} [95% CI]
0.6	12.2 [6.3–26.1]	9.35 [5.3–19.3]
5	28.2 [15.9–62.4]	26.8 [13.4–79.6]
12	21.6 [14.5–34.6]	21.0 [13.3–37.4]
18	11.2 [7.0–19.3]	11.2 [7.0–19.3]
24	11.4 [7.2–19.6]	10.8 [6.8–18.2]
30	12.6 [8.5–20.6]	12.5 [8.5–20.2]

Note: Data are the 48- and 96-hour lethal concentrations presented in mg · L^{-1} and their 95% CI from acute toxicity tests with dioctyl sodium sulfosuccinate on 8- to 10-day posthatch Gulf killifish, *Fundulus grandis*, at six different salinities.

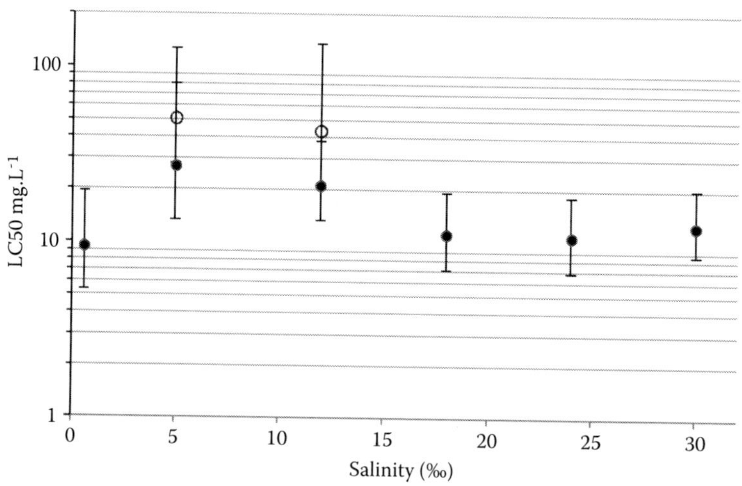

FIGURE 2.2
96-hour median lethal concentrations (LC_{50} values) and 95% confidence intervals for dioctyl sodium sulfosuccinate (DOSS) on 8- to 10-day posthatch larval Gulf killifish, *Fundulus grandis*, (closed circles, •) and stage 19 embryos (open circles, o) at salinities ranging between 0.6‰ and 30‰.

Physiological Effects

All negative slides, without the primary antibody, showed no immunostaining (not illustrated). The gills were completely formed with four branchial arches bearing filaments and lamellae. The localization and abundance of transport proteins on the killifish gills were monitored by immunohistochemistry to assess the putative actions of DOSS on membrane function. Staining for NKA, NKCC, and CFTR proteins was localized in the MRC on gill filaments (Figure 2.3a through e) and the operculum (Figure 2.3a) regardless of DOSS concentration, with additional CFTR staining in the lamellae. Within the gills, NKA and NKCC distributed to the basolateral membrane, whereas CFTR distributed predominantly to the apical membrane. The intensity of staining appeared to decrease from 0 to 25 mg·L^{-1} (Figure 2.3a through a3; 3b through b3) and 50 mg·L^{-1} DOSS (Figure 2.3c through c3; 2.3d through d3; 2.3e through e3). Mean cell diameter and the mean cross-sectional area of MRCs of fish gill exposed to 25 mg·L^{-1} DOSS at a salinity of 12‰ were reduced from those of MRCs analyzed from 0 mg·L^{-1}-exposed fish, and significantly lower at 18‰ for 25 and 50 mg·L^{-1} + DOSS-exposed fish compared to control fish (Figure 2.4). At a salinity of 18‰, the length and the surface area of the MRC were not significantly different from control at 25 and 50 mg·L^{-1} DOSS exposures.

Embryo Development

Percent hatch of embryos exposed to DOSS at a salinity of 12‰ for 96 hours was significantly dependent upon DOSS concentration. ANOVA analysis indicated time to hatch did not significantly differ between control (0 mg·L^{-1} DOSS) and DOSS treatments (Table 2.3). Regression analysis of time to hatch demonstrated no significant relationship with DOSS concentration ($r^2 = 0.02$, $p = .22$, $n = 67$). Morphometric data at hatch for total length, depth at vent, and body cavity area are presented in Table 2.3. Significant differences detected for total length of individuals exposed to 40 and 50 mg·L^{-1} DOSS were attributed to malformation of the tail. The instances of malformed newly hatched larvae increased from ≤33% at 0, 1, and 25 mg·L^{-1} to 72% at 40 mg·L^{-1} DOSS (Figure 2.5). With 40 mg·L^{-1} DOSS exposure,

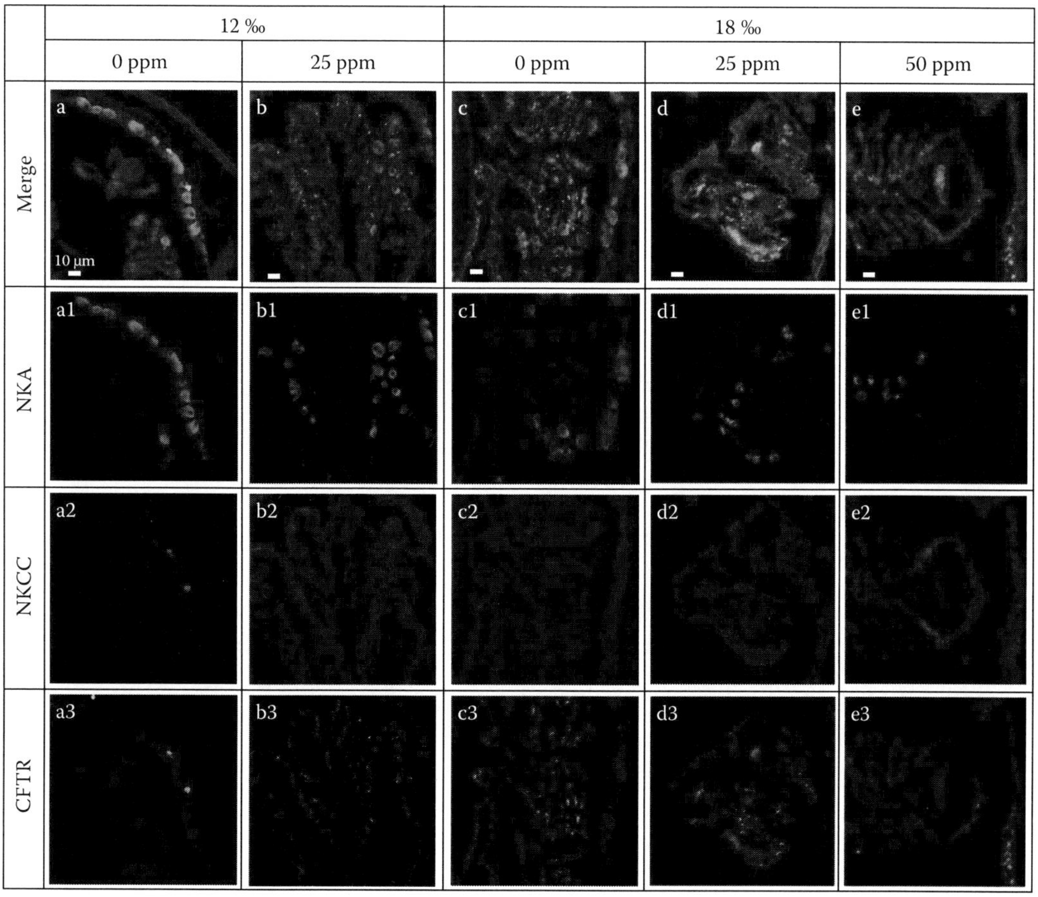

FIGURE 2.3
(See color insert.) Triple immunolocalization of Na^+/K^+-ATPase (NKA) (red), $Na^+/K^+/2Cl^-$ cotransporter (NKCC) (blue), and cystic fibrosis transmembrane regulator channel (CFTR) (green) in *Fundulus grandis* larvae after 96-hour exposure to 0 (a through a3; c through c3), 25 (b through b3; d through d3), and 50 ppm (e through e3) of dioctyl sodium sulfosuccinate (DOSS) at salinities of 12‰ (a through a3; b through b3) and 18‰ (c through c3; d through d3; e through e3).

cranial deformation and failure to develop a fully formed tail were the primary malformations in larvae.

Discussion

The present study describes a significant influence of variable environmental salinity on the acute toxicity of the surfactant DOSS, a primary component of crude oil dispersants, on a larval marsh fish. Salinity influences acute toxicity of a variety of compounds in aquatic organisms and as a result has been the focus of a number of studies with consideration to bioassay testing. Marine teleosts such as the silverside minnow (*Menidia beryllina*) and sheepshead minnow (*Cyprinodon variegatus*) demonstrate differential responses to changes in salinity, which could indicate altered susceptibility to toxicants under a range of test

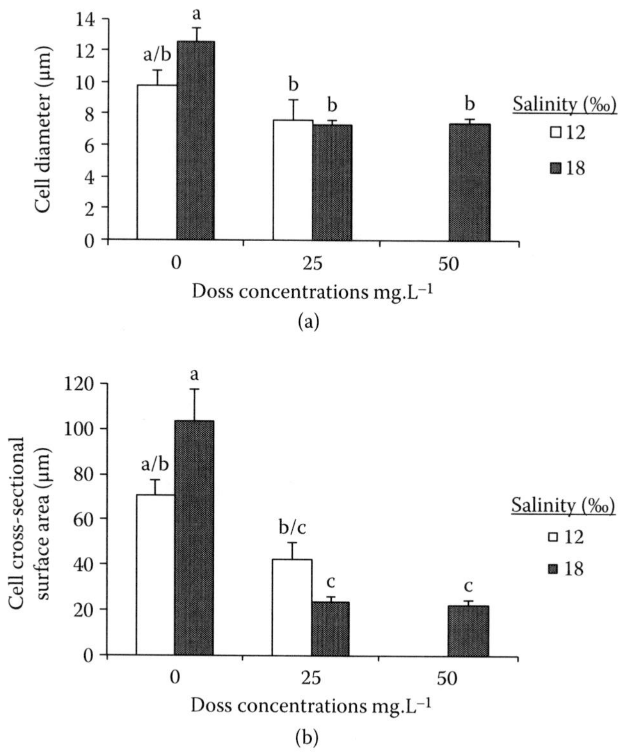

FIGURE 2.4
Mean cell diameter (a) and cross-sectional surface area (b) of mitochondria-rich cells for 8- to 10-day posthatch Gulf killifish, *Fundulus grandis*, after 96-hour exposure to 0, 25, and 50 mg·L^{-1} dioctyl sodium sulfosuccinate (DOSS) at salinities of 12‰ and 18‰. The absence of data at a salinity of 12‰ was due to mortality.

TABLE 2.3
Morphometric Data at Hatch for Total Length, Depth at Vent, and Body Cavity Area

DOSS Concentration (mg·L^{-1})	% Hatch	Time to Hatch (days)	Total Length (mm)	Depth at Vent (mm)	Body Cavity Area (mm^2)
0	94	14.3 ± 1.1	6.0 ± 0.06	0.53 ± 0.02	0.9 ± 0.06
1	88	14.3 ± 1.4	6.3 ± 0.08	0.53 ± 0.01	0.8 ± 0.03
25	77	15.5 ± 1.5	6.2 ± 0.10	0.52 ± 0.02	0.78 ± 0.04
40	83	15.1 ± 1.0	4.3 ± 0.50*	0.45 ± 0.04	0.9 ± 0.04
50	27	18.4 ± 4.4	3.7 ± 0.73*	0.47 ± 0.02	0.7 ± 0.15

Note: Mean (± SE) parameters for percent hatch, time to hatch in days, and morphometric parameters for Gulf killifish, *Fundulus grandis*, exposed to dioctyl sodium sulfosuccinate (DOSS) concentrations for 96 hours beginning at stage 19 at a salinity of 12‰. Values within columns with an asterisk are significantly different from controls (Dunnett's test; $p < .05$).
*Values within a column differ significantly from control.

salinities (Pillard et al. 1999). Specifically, Pillard et al. (1999) recommended that screening of specific ions could be valuable in conducting ion-related toxicity models for determining whole effluent toxicity tests for species such as the silverside minnow, which demonstrate a greater sensitivity to changes in ion composition when compared to other euryhaline

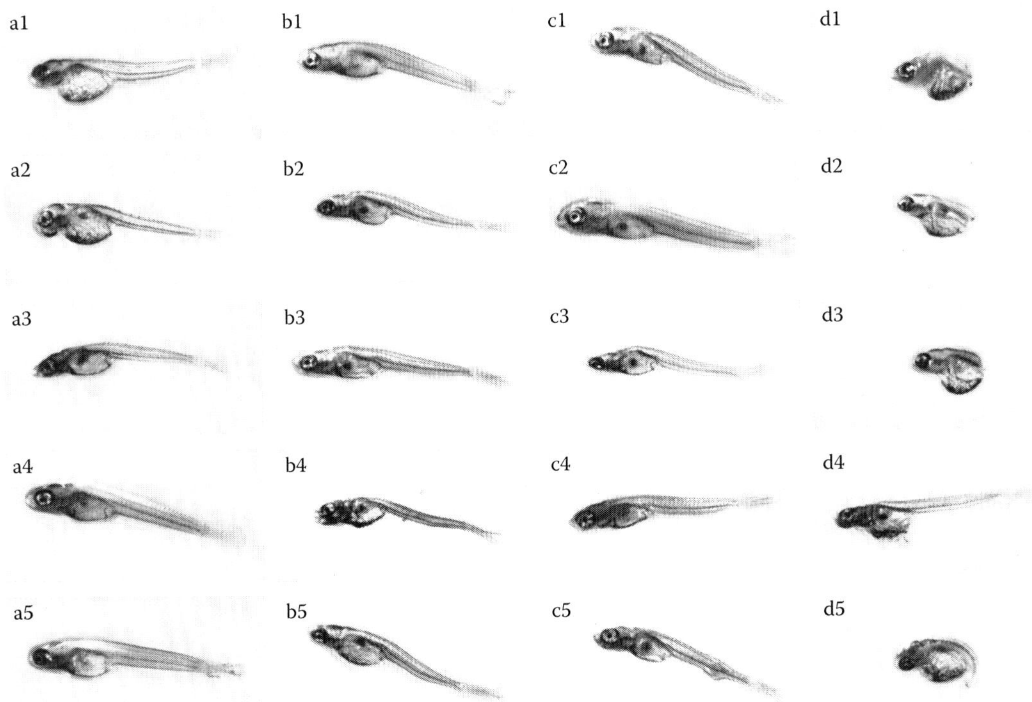

FIGURE 2.5
(See color insert.) Representative newly hatched Gulf killifish, *Fundulus grandis*, exposed to 0 (a1-5), 1 (b1-5), 25 (c1-5), and 40 (d1-5) mg·L^{-1} dioctyl sodium sulfosuccinate (DOSS) for 96 hours beginning at stage 19 at a salinity of 12‰.

teleost species. In this study, 96- and 48-hour median lethal concentrations for DOSS in 8- to 10-day posthatch larvae were lowest at 0.6‰ and salinities greater than 18‰, indicating increased toxicity at salinities furthest from isoosmotic conditions. The lowest salinity tested (0.6‰) resulted in the lowest 96-hour LC_{50} at 9.35 mg·L^{-1} and is not believed to be the result of salinity tolerance since survival of control (0 mg·L^{-1} DOSS) individuals for this exposure salinity was 90%. The ability to expose larval Gulf killifish to a wide range of salinity conditions for these bioassays with DOSS allowed us to investigate the role of salinity on acute mortality and localization of osmoregulatory proteins.

Brood fish and resulting embryos utilized in this study were maintained at a salinity of 12‰ prior to the acute assays and were acclimated to each salinity treatment (0.6‰–30‰) at a rate of salinity increase or decrease of 2.5‰ per day, resulting in control survival of ≥90% for all salinities tested. Although it is possible that differences in salinity acclimations or previous osmotic conditions could confound LC_{50} results, Gulf killifish embryos and larvae are shown to tolerate salinities up to 80‰, which are conditions much greater than employed in this study (Perschbacher et al. 1990). Gulf killifish embryos incubated in water at salinities ranging from 0‰ to 35‰ have demonstrated relatively high hatch rates (77%–91%) (Perschbacher et al. 1990), indicating that the salinities used for acute mortality exposures in this study did not confound our findings. Euryhaline teleost larvae, such as sea bream (*Sparus aurata*) and sea bass (*Dicentrarchus labrax*), survive exposures to a wide range of salinities, however, with a reduced capacity relative to juveniles and adults, which could, in turn, increase sensitivity to toxicant exposures (Varsamos et al.

2001, Bodinier et al 2010). Acclimation to varying salinities for euryhaline teleosts has been found to alter physiological functions that affect downstream processes related to acute toxicity responses (Lavado et al. 2011). Medaka adults and juveniles displayed no changes in acute toxicity values at varying salinities for aldicarb, a carbamate insecticide, however, 2-day posthatch larvae demonstrated increased sensitivity to acute exposures at salinities of 20‰ in comparison to 1.5‰ (El-Alfy et al. 2001). In this study, salinity affected DOSS toxicity with increased sensitivity at both hyper- and hypoosmotic conditions, which we presume resulted from an altered osmotic function as observed through a more detailed examination of ion-transport proteins.

ATPase activity has been previously evaluated as a biomarker for exposure to the herbicide thiobencarb, with significant decreases of the NKA expression observed in the gill tissue (Sancho et al. 2003). In this study, ion-transport proteins such as NKA, NKCC, and CFTR were used as osmotic markers to determine whether or not DOSS has an effect on osmoregulation of Gulf killifish. NKA and CFTR expression appeared to decrease while the DOSS concentration increased, and the NKCC expression appeared to be less specific to the MRC, and more diffuse in the gill epithelium. MRC morphometrics such as mean cell cross-sectional area and cell diameter were significantly reduced by DOSS concentrations of 25 and 50 $mg{\cdot}L^{-1}$. Previous research with DOSS has determined its actions on intestinal membranes, where DOSS increases tissue conductance and sodium secretion, and decreases chloride absorption in the rat cecum (Donowitz and Binder 1975). DOSS completely inhibits soluble phosphodiesterase activity in human colonic mucosa at concentrations at and above 0.1 $mmol{\cdot}L^{-1}$, indicating a potent inhibitory action for this anionic surfactant (Simon and Kather 1980). Similarly, significant reductions in NKA activity have been observed within the gills of brown trout (*Salmo trutta*) through the inhibition of phosphodiesterase by specific inhibitors (Tipsmark and Madsen 2001). Investigations with phosphodiesterase inhibitors have been responsible for determining the role of cyclic-AMP-mediated phosphorylation by protein kinases in the modulation of NKA activity as a critical osmoregulatory protein within many teleost fishes (Tipsmark and Madsen 2001, Tresguerres et al. 2003).

Anionic surfactants are used as strong denaturants for protein structures and are used for their properties as solubilizing agents for membrane bound proteins and lipids (Ananthapadmanabhan 1993; Loo et al. 1994; Lu et al. 2006). In the Gulf killifish, the action of DOSS on gill epithelia may alter the protein structure for NKA, NKCC, and CFTR, which may lead to protein dysfunction and so alter active ion flux through the epithelia. Experimental exposure to anionic surfactants such as DOSS and SLS by fish has resulted in damage to gill filaments. Juvenile turbot, *Scophthalmus maximus*, exposed to SLS concentrations between 3 and 7 $mg{\cdot}L^{-1}$ displayed histological lesions such as fusion of the secondary lamellae and severe degree of hyperplasia of the lamellar epithelium (Rosety-Rodriguezetal et al. 2002). Hemorrhaging of the branchial epithelium was observed at SLS concentrations of 10 $mg{\cdot}L^{-1}$ and produced 50% mortality of exposed juvenile turbot in 4 hours (Rosety-Rodriguezetal et al. 2002). In this study, the shrinking and the disappearance of the MRC could be due to an alteration of their membrane property, which may subsequently affect the osmotic function of these cells by decreasing the active ion flux.

Duarte et al. (2010) demonstrated the effects of the chemical dispersant Corexit 9500 (containing DOSS), crude oil, and chemically dispersed crude oil on the osmotic function of the gills of the Tambaqui, *Colossoma macropomum*, by measuring the unidirectional flux of Na^+, Cl^+, and K^+ ions when individuals were held in freshwater conditions (≤1‰). The largest degree of impairment was observed in the presence of the dispersant and chemically dispersed oil with a net efflux of Na^+ and Cl^- and a decrease of Na^+ and Cl^- concentration in the plasma within 6 and 12 hours (Duarte et al. 2010). The osmotic ability of Gulf killifish and probably respiratory exchanges in gills may be affected by the crude oil itself or by the DOSS itself as suggested by Engelhardt et al. (1981) and Alkindi et al. (1996), but chemically dispersed oil may have an even greater impact on the respiratory and the osmotic abilities.

Estuaries are extremely dynamic habitats in both space and time, where physical parameters such as temperature, salinity, and hypoxia can fluctuate widely both periodically and episodically. The effectiveness of dispersants in weathering oil has been observed to increase with increasing salinity, resulting in decreased weathering at estuarine salinities of 10‰ and 20‰ when compared to 34‰ (Chandrasekar et al. 2006). In this study, the lowest calculated 48-hour LC_{50} concentrations for larvae and percent viability of embryos at hatch was obtained at salinities between 18‰ and 30‰, considered to be the range of salinities most favorable for dispersant activities (Chandrasekar et al. 2006). In addition to affecting the physiology of aquatic animals inhabiting these coastal marshes, variations in salinity and oxygen may alter hydrocarbon toxicity by influencing physical and chemical dispersion and/or biodegradation. The most significant abiotic factors attributed to polycyclic aromatic hydrocarbon (PAH) degradation in aquatic sediments and water are temperature and salinity (Chen et al. 2008). The uptake of PAHs has been suspected to be linked to changes in osmotic conditions, as individuals utilize the gills as a major site of ion and osmoregulatory balance. Lower salinity water (0‰) has been observed to significantly increase PAH uptake in the euryhaline tilapia (*Oreochromis mossambicus*) within the tissues of the gills, gonad, and liver relative to higher salinities of 15‰ and 30‰ (Shukla et al. 2007). At a salinity of 34‰, larval topsmelt (*Atherinops affins*) had significantly increased rates of naphthalene uptake from water-accommodated fractions of Prudhoe Bay crude oil in the presence of dispersant (Wolfe et al. 2001). Clearly, there are two opposing forces occurring between the effects of dispersants on PAHs and the role of environmental salinity. As salinity decreases to estuarine conditions, PAH uptake potentially increases and the effectiveness of dispersants in weathering oil decreases.

Environmentally relevant concentrations of DOSS observed to produce acute lethal effects would likely only occur at the direct site of introduction of surfactant in the water column or water surface. Subsequent dilution of dispersant, and to a greater extent DOSS, would presumably reduce concentrations to the point of nondetection and noneffect. The initial site of dispersant surface application could be a point of concern as recent modeling indicates that approximately 12% of larval Atlantic bluefin tuna (*Thunnus thynnus*) were predicted to be located with contaminated waters of the northern Gulf of Mexico as the DWH was occurring (Muhling et al. 2012). Because of limitations in relative body length and swimming speed, larval fish would be unable to display sensory chemical avoidance behavior as observed in many adult fish when presented a potential toxicant in an open aquatic environment. Boundaries of anticyclonic regions in the northern Gulf of Mexico have demonstrated greater numbers of larvae such as Atlantic bluefin tuna and frigate mackerel (*Auxis* spp.) where these and other pelagic larvae are more dependent upon oceanic and wind currents for dispersal and locomotion due to their small size (Li et al. 2011, Lindo-Atichati et al. 2012).

To our knowledge no recorded DOSS concentrations associated with sampling in the northern Gulf of Mexico, nearshore, or inland locations have identified concentrations similar to values determined in this study to produce acute mortality or larval deformities. Calculations of DOSS concentrations near the wellhead indicate that deepwater/pelagic biota would have encountered DOSS concentrations between 1 and 10 $\mu g{\cdot}L^{-1}$ between 1 and 10 km from the site of subsurface application near the actively flowing wellhead (Kujawinski et al. 2011). Concentrations of DOSS detected in nearshore and inland locations across Alabama after the DWH have been linked to nonpoint source pollution from stormwater discharge, as DOSS is a common industrial surfactant with many commercial uses (Hayworth and Clement 2012). Based on the findings of this study, it appears that the toxicity of the main anionic surfactant in Corexit 9500 and 9527 would be the greatest at full saline conditions generally observed in the open waters in which it was applied during remediation events surrounding the DWH.

References

Abel, P. D. and J. F. Skidmore. 1976. Toxic action of an anionic detergent on the gills of rainbow trout. *Water Research* 9:759–765.

Alkindi, A. Y. A., J. A. Brown, C. P. Waring, and J. E. Collins. 1996. Endocrine, osmoregulatory, respiratory and haematological parameters in flounder exposed to the water soluble fraction of crude oil. *Journal of Fish Biology* 49:1291–1305.

Ananthapadmanabhan, K. P. 1993. *Interactions of Surfactants with Polymers and Proteins*, eds. E. D. Goddard and K. P. Ananthapadmanabhan. London: CRC Press.

Anderson, S. E., J. Franko, E. Lukomska, and B. J. Meade. 2011. Potential immunotoxicological health effects following exposure to COREXIT 9500A during cleanup of the *Deepwater Horizon* oil Spill. *Journal of Toxicology and Environmental Health*, Part A 74:1419–1430.

Armstrong, P. B. and J. S. Child. 1965. Stages in the normal development of *Fundulus heteroclitus*. *Biological Bulletin* 128:143–168.

Attwood, D. and A. T. Florence. 1983. *Surfactant systems: Their Chemistry, Pharmacy, and Biology*. London: Chapman and Hall.

Bodinier, C., C. Lorin-Nebel, G. Charmantier, and V. Boulo. 2009. Influence of salinity on the localization and expression of the CFTR chloride channel in the ionocytes of juvenile *Dicentrarchus labrax* exposed to seawater and freshwater. *Comparative Biochemistry and Physiology*, Part A 153:345–351.

Bodinier, C., E. Sucré, L. Lecurieux-Belfond, E. Blondeau-Bidet, and G. Charmantier. 2010. Ontogeny of osmoregulation and salinity tolerance in the gilthead sea bream *Sparus aurata*. *Comparative Biochemistry and Physiology*, Part A 157: 220–228.

Brown, C. A., C. T. Gothreaux, and C. C. Green. 2011. Effects of temperature and salinity during incubation on hatching and yolk utilization of Gulf killifish *Fundulus grandis* embryos. *Aquaculture* 315:335–339.

Bugel S. M., L. A. White, and K. R. Cooper. 2010. Impaired reproductive health of killifish (*Fundulus heteroclitus*) inhabiting Newark Bay, NJ, a chronically contaminated estuary. *Aquatic Toxicology* 96:182–193.

Burnett, K. G., L. J. Bain, W. S. Baldwin, G. V. Callard, S. Cohen, and R. T. Di Giulio, et al. 2007. *Fundulus* as the premier teleost model in environmental biology: Opportunities for new insights using genomics. *Comparative Biochemistry and Physiology* Part D Genomics & Proteomics 2:257–286.

Chandrasekar, S., G. A. Sorial, and J. W. Weaver. 2006. Dispersants effectiveness on oil spills – impact of salinity. *Journal of Marine Science* 63:1418–1430.

Chen, J., M. H. Wong, Y. S. Wong, and N. F. Tam. 2008. Multi-factors on biodegradation kinetics of polycyclic aromatic hydrocarbons (PAHs) by *Sphingomonas* sp. a bacterial strain isolated from mangrove sediment. *Marine Pollution Bulletin* 57:695–702.

Cotou, E., I. Castritsi-Catharios, and M. Moraitou-Apostolopoulou. 2001. Surfactant-based oil dispersant toxicity to developing nauplii of *Artemia*: Effects on ATPase enzymatic system. *Chemosphere* 42:959–964.

DHUC. 2011. Ongoing response timeline. Deepwater Horizon Unified Command. http://www.restorethegulf.gov/release/2010/08/02/ongoing-administration-wide-response-deepwater-bp-oil-spill (last accessed 21 September 2012).

Donowitz, M. and H. J. Binder. 1975. Effect of dioctyl sodium sulfosuccinate on colonic fluid and electrolyte movement. *Gastroenterology* 69:941–950.

Duarte, R. M., R. T. Honda, and A. L. Val. 2010. Acute effects of chemically dispersed crude oil on gill ion regulation, plasma ion levels and haematological parameters in tambaqui (*Colossoma macropomum*). *Aquatic Toxicology* 97:134–141.

El-Alfy, A. T., S. Grisle, and D. Schlenk. 2001. Characterization of salinity enhanced toxicity of aldicarb to Japanese medaka: Sexual and developmental differences. *Environmental Toxicology and Chemistry* 20:2093–2098.

El-Laithy, H. M. 2003. Preparation and physiochemical characterization of dioctyl sodium sulfosuccinate (aerosol OT) microemulsion for oral drug delivery. *AAPS Pharmacology Science and Technology* 4:1–10.

Engelhardt, F. R., M. P. Wong, and M. E. Duey. 1981. Hydromineral balance and gill morphology in rainbow trout *Salmo gairdneri*, acclimated to fresh and sea water as affected by petroleum exposure. *Aquatic Toxicology* 1:175–186.

Engle, R. H. and S. J. Riggi. 1969. Intestinal absorption of heparin facilitated by sulfated or sulfonated surfactants. *Journal of Pharmaceutical Sciences* 58:706–711.

Evans, D. H. 1993. Osmotic and ionic regulation. In *The Physiology of Fishes*, ed. D. H. Evans, 315–341. Boca Raton, FL: CRC Press.

Evans, D. H. 2009. *Osmotic and Ionic Regulation: Cells and Animals*. Boca Raton, FL: Taylor & Francis Group, LLC edition, CRC Press.

Evans, D. H., P. M. Piermarini, and K. P. Choe. 2005. The multifunctional fish gill: Dominant site of gas exchange, osmoregulation, acid-base regulation, and excretion of nitrogenous waste. *Physiological Reviews* 85:97–177.

Goodrich, M. S., M. J. Melancon, R. A. Davis, and J. J. Lech. 1991. The toxicity, bioaccumulation, metabolism and elimination of dioctyl sodium sulfosuccinate DSS in rainbow trout (*Oncorhynchus mykiss*). *Water Research* 25:119–124.

Green, C. C., C. T. Gothreaux, and G. C. Lutz. 2010. Reproductive output of gulf killifish *Fundulus grandis* at different stocking densities in static outdoor tanks. *North American Journal of Aquaculture* 72:321–331.

Hamilton, M. A., R. C. Russo, and R. V. Thurston. 1977. Trimmed Spearman-Karber method for estimating median lethal concentrations in toxicity bioassays. *Environmental Science and Technology* 11:714–719.

Hayworth, J. S. and T. P. Clement. 2012. Provenance of Corexit-related chemical constituents found in nearshore and inland Gulf coast waters. *Marine Pollution Bulletin*. http://dx.doi.org/10.1016/j.marpolbul.2012.06.031.

Hemmer, M. J., M. G. Barron, and R. M. Greene. 2011. Comparative toxicity of eight oil dispersants, Louisiana Sweet Crude Oil (LSC), and chemically dispersed LSC to two aquatic test species. *Environmental Toxicology and Chemistry* 30:2244–2252.

Hiroi, J. and S. D. McCormick. 2007. Variation in salinity tolerance, gill Na^+/K^+-ATPase, $Na^+/K^+/2Cl^-$ cotransporter and mitochondria-rich cell distribution in three salmonids *Salvelinus namaycush*, *Salvelinus fontinalis* and *Salmo salar*. *Journal of Experimental Biology* 210:1015–1024.

Hirose, S., T. Kaneko, N. Naito, and Y. Takei. 2003. Molecular biology of major components of chloride cells. *Comparative Biochemistry and Physiology*, Part B 136:593–620.

Katoh, F., A. Shimizu, K. Uchida, and T. Kaneko. 2000. Shift of chloride cell distribution during early life stages in seawater-adapted killifish, *Fundulus heteroclitus*. *Zoological Science* 17:11–18.

Khalafallah, N., M. W. Gouda, and S. A. Khalil, 1975. Effect of surfactants on absorption through membranes IV: Effects of dioctyl sodium sulfosuccinate on absorption of a poorly absorbable drug, phenolsulfonphthalein, in humans. *Journal of Pharmacuetical Sciences* 64:991–994.

Khan, R. A. and J. F. Payne. 2005. Influence of a crude oil dispersant, Corexit 9527, and dispersed oil on a capelin, Atlantic cod, longhorn sculpin, and cunner. *Bulletin of Environmental Contamination and Toxicology* 75:50–56.

Kujawinski, E. B., M. C. Kido Soule, D. L. Valentine, A. K. Boysen, K. Longnecker, and M. C. Redmond. 2011. Fate of dispersants associated with the *Deepwater Horizon* oil spill. *Environmental Science and Technology* 45:1298–1306.

Lavado, R., L. A. Maryoung, and D. Schlenk. 2011. Hypersalinity acclimation increases the toxicity of the insecticide phorate in Coho salmon (*Oncorhynchus kisutch*). *Environmental Science and Technology* 45:4623–4629.

Li, C., H. Roberts, G. W. Stone, E. Weeks, and Y. Luo. 2011. Wind surge and saltwater intrusion in Atchafalaya Bay during onshore winds prior to cold front passage. *Hydrobiologia* 658:27–39.

Lindo-Atichati, D., F. Bringas, G. Goni, B. Muhling, F. E. Muller-Karger, and S. Habtes. 2012. Varying mesoscale structures influence larval fish distribution in the northern Gulf of Mexico. *Marine Ecology Progress Series* 463:245–257.

Loo, R. R. O., N. Dales, and P. C. Andrews. 1994. Surfactant effects on protein structure examined by electrospray ionization mass spectrometry. *Protein Science* 3:1975–1983.

Lorin-Nebel, C., V. Boulo, C. Bodinier, and G. Charmantier. 2006. The $Na^+/K^+/2Cl$ ($^-$) cotransporter in the sea bass *Dicentrarchus labrax* during ontogeny: Involvement in osmoregulation. *Journal of Experimental Biology* 209:4908–4922.

Lu, R. C., A. N. Cao, L. H. Lai, and J. X. Xiao. 2006. Effect of anionic surfactant molecular structure on bovine serum albumin (BSA) fluorescence. *Colloids and Surfaces A: Physicochemical and Engineering Aspects* 278:67–73.

Maggi, P. and D. Cossa. 1973. Relative harmfulness of five anionic detergents in the sea. I. Acute toxicity with regard to fifteen organisms. *Revue des Travaux* 37: 411–417.

Marshall, W. S. 2002. Na^+, Cl^-, Ca^{2+} and Zn^{2+} transport by fish gills: Retrospective review and prospective synthesis. *Journal of Experimental Zoology* 293:264–283.

Marshall, W. S., T. R. Emberley, T. D. Singer, S. E. Bryon, and S. D. McCormick. 1999. Time course of salinity adaptation in a strongly euryhaline estuarine teleost, *Fundulus heteroclitus*: A multivariable approach. *Journal of Experimental Biology* 202:1535–1544.

Marshall, W. S., E. M. Lynch, and R. R. F. Cozzi. 2002. Redistribution of immunofluorescence of CFTR anion channel and NKCC cotransporter in chloride cells during adaptation of the killifish *Fundulus heteroclitus* to sea water. *Journal of Experimental Biology* 205:1265–1273.

Muhling, B. A., M. A. Roffer, J. T. Lamkin, G. W. Ingram, Jr, M. A. Upton, G. Gawlikowski, et al. 2012. Overlap between Atlantic bluefin tuna spawning grounds and observed *Deepwater Horizon* surface oil in the northern Gulf of Mexico. *Marine Pollution Bulletin* 64:679–687.

Nordlie, F. G. 2000. Patterns of reproduction and development of selected resident teleosts of Florida salt marshes. *Hydrobiologia* 434:165–182.

Nordlie, F. G. 2006. Physicochemical environments and tolerances of cyprinodontoid fishes found in estuaries and salt marshes of eastern North America. *Reviews in Fish Biology and Fisheries* 16:51–106.

NRC. 2005. *Oil Spill Dispersants: Efficacy and Effects*. Washington, D. C.: National Academies Press.

Perschbacher, P. W., D. V. Aldrich, and K. Strawn. 1990. Survival and growth of the early stages of Gulf killifish in various salinities. *Progressive Fish-Culturist* 52:109–111.

Pillard, D. A., D. L. DuFresne, J. E. Tietge, and J. M. Evans. 1999. Response of mysid shrimp (*Mysidopsis bahia*), sheepshead minnow (*Cyprinodon variegatus*), and inland silverside minnow (*Minidia beryllina*) to changes in artificial seawater salinity. *Environmental Toxicology and Chemistry* 18:430–435.

Rosety-Rodriguez, M., F. J. Ordonez, M. Rosety, J. M. Rosety, I. Rosety, A. Ribelles, et al. 2002. Morpho-histochemical changes in the gills of turbot, *Scophthalmus maximus* L., induced by sodium dodecyl sulfate. *Ecotoxicology and Environmental Safety* 51:223–228.

Sancho, E., C. Fernandez-Vega, M. D. Ferrando, and E. Andreu-Moliner 2003. Eel ATPase activity as biomarker of thiobencarb exposure. *Ecotoxicology and Environmental Safety* 56:434–441.

Shukla, P., M. Gopalani, D. S. Ramteke, and S. R. Wate. 2007. Influence of salinity on PAH uptake from water soluble fraction of crude oil in *Tilapia mossambica*. *Bulletin of Environmental Contamination and Toxicology* 79:601–605.

Simon, B. and H. Kather. 1980. Interaction of laxatives with enzymes of cyclic AMP metabolism from human colonic mucosa. *European Journal of Clinical Investigation* 10:231–234.

Singer, M. M., D. Aurand, G. E. Bragins, J. R. Clarks, G. M. Coelho, M. L. Sowby, et al. 2000. Standardization of the preparation and quantification of water-accommodated fractions of petroleum for toxicity testing. *Marine Pollution Bulletin*. 40:1007–1016.

Tatem, H. E. and A. S. Portzer. 1985. Culture and toxicity tests using Los Angeles district bioassay animals, *Acanthomysis* and *Neanthes*. Final Report. Miscellaneous Paper EL-85-6, US Army Engineer Waterways Experiment Station, Vicksburg, Mississippi.

Tipsmark, C. K. and S. S. Madsen. 2001. Rapid modulation of the Na^+/K^+-ATPase activity in osmoregulatory tissues of a salmonid fish. *Journal of Experimental Biology* 204:701–709.

Tresguerres, M., H. Onken, A. F. Perez, and C. M. Luquet. 2003. Electrophysiology of posterior, NaCl-absorbing gills of *Chasmagnathus granulates*: Rapid responses to osmotic variations. *Journal of Experimental Biology* 206:619–626.

Van Veld, P. A. and D. Nacci. 2008. Chemical tolerance: Acclimation and adaptations to chemical stress. In *The Toxicology of Fishes*, eds. R. T. Di Giulio and D. E. Hinton. Washington, D. C.: Taylor and Francis.

Varsamos, S., R. Connes, J. P. Diaz, G. Barnabe, and G. Charmantier. 2001. Ontogeny of osmoregulation in the European sea bass *Dicentrarchus labrax*. *Marine Biology* 138:909–915.

Wheelock, C. E., T. A. Baumgartner, J. W. Newman, M. F. Wolfe, and R. S. Tjeerdema. 2002. Effect of nutritional state on Hsp60 levels in the rotifer *Brachionus plicatilis* following toxicant exposure. *Aquatic Toxicology* 61: 98–93.

Whitehead, A., B. Dubansky, C. Bodinier, T. I. Garcia, S. Miles, C. Pilley, et al. 2012. Genomic and physiological footprint of the *Deepwater Horizon* oil spill on resident marsh fishes. *Proceedings of the National Academy of Sciences* 109:20298–20302.

Wolfe, M. F., G. J. B. Schwartz, S. Singaram, E. E. Miebrecht, R. S. Tjeerdema, M. L. Sowby. 2001. Influence of dispersants on the bioavailability and trophic transfer of petroleum hydrocarbons to larval topsmelt (*Atherinops affins*). *Aquatic Toxicology* 52, 49–60.

3

Endocrine Disruption as a Mechanism of Action Underlying Sublethal Effects in Pacific Herring (Clupea harengus pallasi) *Exposed to the Dissolved Hydrocarbon Fraction of Crude Oil*

Christopher J. Kennedy

CONTENTS

Introduction

Point and nonpoint sources of petroleum hydrocarbons contribute to their ubiquitous and pervasive nature in aquatic environments. Approximately 5 million tons of crude oil enters the marine environment each year from natural and anthropogenic sources (Neff 1990) and is the main impetus for research that is directed toward estimating the risk posed by these chemicals to marine organisms. Concentrations of total hydrocarbons in contaminated marine coastal waters typically reach 80 $\mu g \cdot L^{-1}$, with atypical reports reaching 500 $\mu g \cdot L^{-1}$ (Badawy and Al-Harthy 1991, Madany et al. 1994, Alkindi et al. 1996), values which are far above the aqueous solubility of most of the compounds found in this complex mixture.

Crude petroleum spills exhibit well-documented and highly visible surface effects on intertidal and benthic invertebrates and fish that spend at least some of their time in contact with the sediment/air/water interface. Less attention has been directed toward those organisms that inhabit the water column, mainly due to the assumption that exposure concentrations in water are low and well below toxicity thresholds. Recently however, the potential effects of purely dissolved hydrocarbons, which are the most available to marine biota (Neff and Anderson 1981), have been recognized as being of sufficient magnitude to cause measureable effects in pelagic species.

Oil is a mixture of various chemicals, but is comprised mainly of hydrocarbons with varied structures, chemical properties, and toxicity (see Chapter 1, *this volume*). In fish, the main uptake route of dissolved hydrocarbons from water is via the gills. With time however, many of these chemicals become associated with food and/or sediments and the predominant uptake route shifts toward the gastrointestinal tract. Regardless of the site of uptake, a chemical must first be delivered to the epithelium, where it moves from the environmental matrix (water or digestate) through several diffusion barriers including the cell membrane. Because of this, uptake can be generally determined by a chemical's log octanol–water partition coefficient (log K_{OW}) value. The uptake of chemicals with log $K_{OW} < 1$ is very low (McKim et al. 1985) even though these water-soluble chemicals are readily available. Molecules of low lipid solubility cannot easily pass through cell membranes and have difficulty entering the organism (Hansch and Clayton 1973). For chemicals with log K_{OW} values between 1 and 3, uptake will increase with increasing log K_{OW} as these compounds are both highly available (due to sufficient water solubility) and able to diffuse through cell membranes (due to sufficient lipid solubility). Uptake efficiencies remain relatively constant for chemicals with log K_{OW} values between 4 and 6. For hydrophobic chemicals with log $K_{OW} > 6$, delivery to uptake surfaces is limited by their very low water solubility. Compounds found in oil mixtures span the entire range of K_{OW} values discussed above, and therefore, their availability in the water column, and ability to move into organisms, is strictly chemical specific.

Earlier in Chapter 1, it was shown that oil exposure can be acutely lethal, or can result in short-term or chronic sublethal effects. Of particular interest in this chemical mixture are the polycyclic aromatic hydrocarbons (PAHs), which are known to produce a myriad of effects in a wide range of biota. The two hallmarks of PAH sublethal toxicity are carcinogenicity and immunotoxicity. In addition, other well-documented sublethal effects include morphological, histopathological, and genetic effects (Carls 1987, Brown et al. 1996, Hose et al. 1996, Kocan et al. 1996, McGurk and Brown 1996, Norcross et al. 1996, Carls et al. 1999, Heintz et al. 1999), physiological and stress effects (Thomas and Rice 1987, Kennedy and Farrell 2005, Kennedy and Farrell 2006), impaired swimming and reduced marine survival

(Marty et al. 1997, Carls et al. 1999, Heintz et al. 2000, Barron et al. 2003), and ecological effects (Heintz et al. 2000, Reddy et al. 2002). Predicting the toxicity of PAHs is a difficult challenge due to dynamic environmental profiles, as well as nonspecific, common, and chemical-specific mechanisms of action, and variable intra- and interspecies sensitivities. For example, individual PAHs have distinct and specific developmental consequences when fish are exposed at early life history stages (Incardona et al. 2004). Defects in zebrafish embryos induced by dibenzothiophene or phenanthrene have direct effects on cardiac conduction, leading to secondary consequences in cardiac morphogenesis and kidney and neural tube development. Yet, the etiology for pyrene exposure is completely different and consists of anemia, peripheral vascular defects, and neuronal cell death; these effects are similar to those described for potent aryl hydrocarbon receptor (AhR) ligands (Incardona et al. 2004).

There is clear evidence that some xenobiotics have the ability to interfere with the synthesis, secretion, transport, binding, and action or elimination of natural hormones in vertebrates, disrupting normal biological functions (Krimsky 2000) (Figure 3.1). The number of suspected or confirmed endocrine disrupting compounds (EDCs) is ever increasing and includes industrial intermediates, plasticizers, detergents/surfactants, pharmaceuticals, and personal care products as well as classic contaminants such as polychlorinated biphenyls, dioxins, certain pesticides, polybrominated diphenyl ethers, and even a number of trace elements (e.g., Cd or other metals) (Knudsen and Pottinger 1998, Denslow and Sepulveda 2007). EDCs are diverse in sources, structures (they tend to be synthetic and mostly have one or more phenyl ring), persistence, and effects, with perhaps little in common with one another, other than their endocrine system disruption. Evidence in mammals and in fish suggests that PAHs affect endocrine systems. Endocrine systems operate in a highly integrated manner with the nervous system to regulate and coordinate physiological processes as diverse as homeostasis, energy availability, growth, development, and reproduction. Therefore, effects on a single endocrine signaling pathway could potentially affect several physiological systems simultaneously and such wide-ranging effects have important implications regarding the presence of EDCs in the environment and subsequent effects on organisms.

The multifaceted acute and chronic toxicities demonstrated by organisms exposed to petroleum hydrocarbons present a unique challenge to assess sublethal toxicity. One potential solution to this problem is to use "performance indicators" that rely on the optimum functioning and integration of several key physiological systems, any or all of which may be targets for toxicity. If just one of these components is targeted, effects are seen with any measured performance. In a series of experiments, the effects of low dissolved hydrocarbons from crude oil on Pacific herring (*Clupea harengus pallasi*) were examined on several performance endpoints: the physiological stress response, immune function, and swimming performance. Several additional experiments were performed to assess the contribution of possible endocrine disruption to any observed effects on these particular endpoints.

An examination of the effects of the dissolved hydrocarbon fraction (DHF) for crude oil on the physiological stress response (by measuring plasma cortisol concentrations among several other parameters) in Pacific herring was undertaken for several reasons. First, saltwater fish are exposed to these dissolved hydrocarbons via the extensive respiratory surface of the gills and by drinking. The high bioavailability of some of these hydrocarbons in water, in combination with a variety of highly sensitive perceptive mechanisms in the integument, could potentially initiate a generalized stress response in fish in addition to more specific toxic effects. Second, the paradigm of the neuroendocrine stress response is

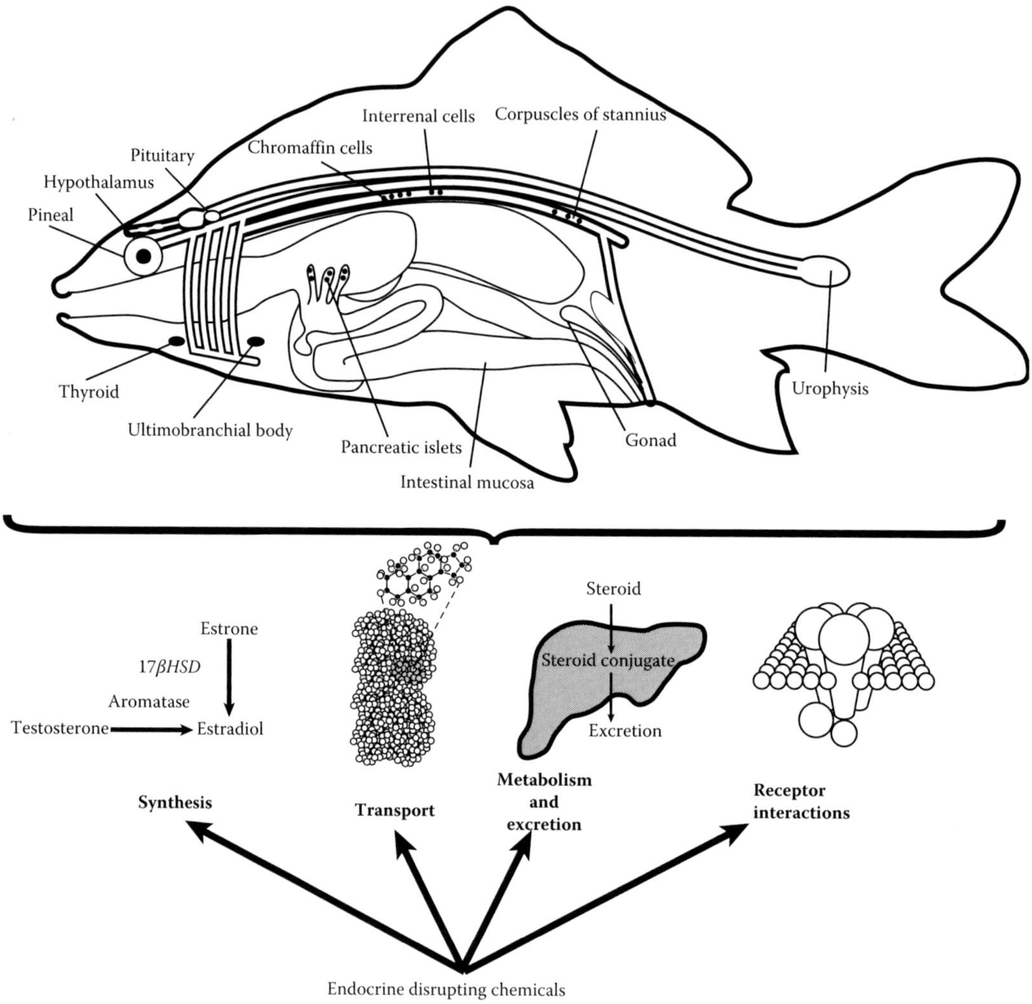

FIGURE 3.1
The main endocrine glands and tissues of teleost fishes. Of particular note are the interrenal cells located in the head kidney that secrete cortisol. In the lower panel, the four main categories of endocrine disruption (interference with hormone synthesis, transport, metabolism and excretion, and receptors) are highlighted.

well known and documented in scores of different teleost species. The response yields a consistent pattern for xenobiotic stressors, namely the activation of two hormonal axes, the hypothalamic–pituitary–interrenal (HPI) (Figure 3.2) and the sympathetico-chromaffin (SC) axes. Third, during stressful events, stimulation of these axes culminates in increased concentrations of circulating catecholamines and cortisol, from the HPI and SC axes, respectively. These increases result in the mobilization of free energy and homeostasis reparation (Wendelaar 1997, Mommsen et al. 1999). The final reason to examine the stress response is that the ability of fish to mount an appropriate stress response, and the negative consequences associated with chronic stress, has evolutionary and ecological significance.

Swimming performance influences the ability of fish to avoid predators, obtain food, find mates, and avoid unfavorable conditions (Plaut 2001) and therefore is suggested as a major trait affecting Darwinian fitness (Reidy et al. 2000). Swimming performance is usually

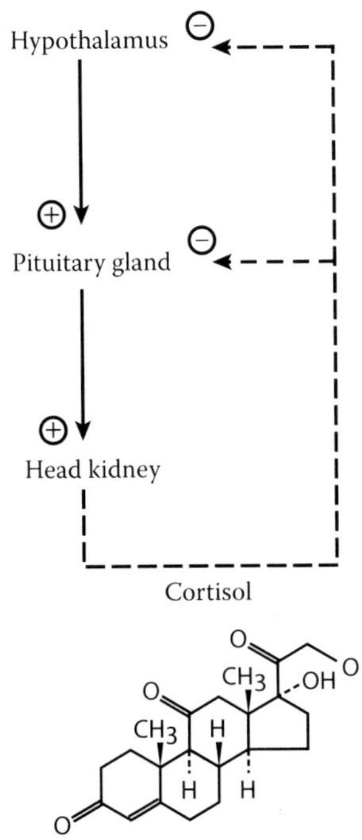

FIGURE 3.2
Schematic detailing the hypothalamic–pituitary–interrenal axis in fishes. (–) indicates negative feedback. The structure of the steroid cortisol is also shown.

classified into three categories: sustained, prolonged, and burst swimming (Beamish 1978). Sustained swimming is that which can be maintained indefinitely and is fueled aerobically. Prolonged swimming speed is also fueled aerobically, but is of shorter duration and ends in the fatigue of the fish. Burst swimming speed is the highest speed of which fish are capable, can be maintained only for short periods, and is likely fueled anaerobically (Beamish 1978). Fish usually use burst swimming when catching prey, avoiding a predator's attack, or in any other situation of sudden disturbance, and when maneuvering through strong currents (Reidy et al. 2000).

Locomotor performance is proving to be an important endpoint in the determination of the sublethal effects of toxicants on fish. In fact, toxicant-induced stress can result in the decreased performance of integrated systems such as locomotion. Swimming performance has been used extensively to define the thresholds of toxic effects of a wide range of contaminants (organics and metals) and environmental parameters (e.g., unfavorable water temperatures, hypoxia, and disease). In the present studies, both prolonged and burst swimming were used as endpoints, and the potential link between effects on swimming and exercise recovery and endocrine disruption were examined.

Disease agents in fish populations are believed to exert important effects on host population dynamics through enzootic or epizootic events. Enzootic diseases cause long-term impacts on physiological processes that can affect growth, reproduction, and survival. Epizootic diseases can affect population dynamics by reducing populations in short-term events, which if sufficient, may result in stochastic processes causing extinction (Gulland 1995).

Immune suppression is the mechanism by which toxicants are believed to increase disease incidence (Zelikoff 1993). Environmental xenobiotics, including metals, insecticides, fungicides, halogenated aromatic hydrocarbons, and PAHs (reviewed in Zelikoff 1994), can suppress teleost immune responses. Although PAH exposure could lead to increased host susceptibility to infectious diseases, the relationship between toxicant exposure and disease in aquatic organisms has not been clearly defined, and only a large body of circumstantial evidence exists linking various diseases in feral fish with chemical discharge (Mearns and Sherwood 1976, Waterman and Kranz 1992).

The *Exxon Valdez* oil spill (EVOS) occurred in March 1989, 3 weeks prior to the peak of spawning of Pacific herring, *Clupea pallasi*, in Prince William Sound (PWS) and was subsequently followed by drastic declines in spawning biomass in following years. Several hypotheses have been put forward to explain these declines and include the effects of oil on eggs, larval and juvenile fish, and adults. The impetus for examining the stress response, swimming performance, and immunotoxicity of oil exposure on Pacific herring in the present studies are as follows: (1) their economic importance as a commercial fishery, (2) their importance as a keystone fish species for the northwest Pacific, (3) they are the dominant ichthyofaunal species in PWS, Alaska, and (4) changes in their populations following EVOS.

Materials and Methods

Fish

Juvenile Pacific herring (10–20 g) were obtained locally in West Vancouver, British Columbia. Fish were transported to facilities at Fisheries and Oceans Canada, West Vancouver Laboratory, British Columbia, with a minimal use of nets to reduce trauma to the young fish. Fish were held in 500-L fiberglass tanks supplied with flowing filtered seawater, salinity 31 ppt, water temperature 10 ± 2°C, and dissolved O_2 content >95% saturation. Following transfer, mortality in the first week was approximately 6%, which declined to less than 0.3%. When the mortality rate had stabilized at this level, fish were acclimated for a further 4 weeks in experimental tanks before an experiment was performed. Fish were fed twice daily ad libitum with frozen krill until 1 day before an experiment.

Chemicals and Exposure

Herring were exposed to several concentrations (experiment-dependent) of the DHF of North Slope crude oil in 500-L fiberglass tanks for either 1–4 days (acute exposures) or 56–58 days (chronic exposures). DHFs were generated by seawater continuously passing through a modified apparatus (DHF cylinders) described in Kennedy and Farrell (2005). This method has previously generated water containing initial total polycyclic aromatic hydrocarbon (TPAH) concentrations of up to 120 $\mu g{\cdot}L^{-1}$, which declined to approximately 9 $\mu g{\cdot}L^{-1}$ 28 days after the initiation of water flow (Kennedy and Farrell 2005). In chronic experiments, due to reductions in TPAH concentrations with time, DHF cylinders were replaced at time 0, 28, and 56 days to maintain relatively constant hydrocarbon concentrations. All exposures were performed in duplicate or triplicate for statistical purposes.

Liver Ethoxyresorufin O-Deethylase Activity as a Measure of Hydrocarbon Bioavailability

Liver microsomes were prepared according to the method of Kennedy (1995) to assess hydrocarbon bioavailability through surrogate measurements of ethoxyresorufin O-deethylase (EROD: cytochrome P450-mediated oxidations) activity. Briefly, livers were dissected from fish at the end of experiments following euthanasia in 0.5 mg·L^{-1} MS222. The livers were homogenized and microsomal pellets prepared and stored (Kennedy 1995). Hepatic microsomal protein content was determined using the method of Bradford (1976). The activity of cytochrome P4501A1 (CYP1A) proteins was assessed by the EROD assay according to Burke and Mayer (1974). All assays were performed in duplicate.

Stress Physiology

During 4- and 58-day exposures to 0, 10, 40, and 100 μg·L^{-1} TPAH in the DHF of oil, fish were rapidly netted from tanks (n = 14 per exposure group) and immediately sacrificed by anesthetization with MS222 (Kennedy and Farrell 2005). Extreme care was taken to minimize any disturbance to fish in other treatment tanks. After weighing, blood samples were obtained from the caudal vasculature using heparinized microcapillary tubes and centrifuged at 13,000 × g. Plasma was stored at –86°C for no longer than 3 weeks before analysis. Plasma cortisol concentrations were determined using radioimmunoassays (IncStar Corp., Stillwater, Minnesota). Assays were performed in duplicate.

In a second set of stress physiology experiments, fish were exposed to either 0 or 100 μg·L^{-1} TPAH in the DHF for 58 days and then subjected to a handling stress according to Wilson et al. (1998) to determine if fish had lost the ability to respond to stress and increase plasma cortisol concentrations. Seventy herring from each treatment group as described above were subjected to handling disturbance involving netting and removal from tanks for 1 minute with occasional submersion for respiration. Terminal blood samples (n = 10 fish) were taken at 0 (prestress), 0.5, 1, 2, 4, 8, and 24 hours after handling and centrifuged, and the plasma fractions were frozen for the determination of cortisol concentrations.

In order to determine the potential cause of endocrine impairment and the inability of fish to secrete cortisol following DHF exposure, fish were exposed to either 0 or 100 μg·L^{-1} TPAH in the DHF for 58 days as described above. Isolated interrenal cells were exposed to several concentrations of adrenocorticotropic hormone (ACTH) as outlined in Wilson et al. (1998). Groups of six DHF-exposed or control fish were euthanized by an overdose of buffered MS222 and the fish bled by caudal puncture. Briefly, head kidney (containing the interrenal tissue) from each fish was excised immediately, minced and then divided between wells (50 mg tissue per wells in culture plates), with periodic shaking. Tissue pieces were preincubated (stabilization period), and the supernatant then discarded and replaced with fresh buffer. The buffer was removed and frozen for later determination of cortisol concentration. A sample well from each fish was incubated with ACTH at 0, 0.01, 0.05, 0.10, 0.50, or 1.00 mU·mL^{-1} buffer (porcine ACTH, Sigma) for 1 hour after which the supernatant was sampled and frozen for cortisol analysis. To overcome the differences in basal cortisol-release rates between tissue preparations (due to the heterogeneous distribution of interrenal tissue in head kidney), ACTH-stimulated cortisol release was expressed as percentage change over basal release, and corrected for the declining cortisol release with time as described by Arnold-Reed and Balment (1994).

Burst Swimming and Exercise Recovery

Following exposure to the DHF of oil for 1 and 56 days, fish (n = 6 for each exposure concentration) were burst exercised by chasing them around a large oval tank (Pagnotta et al. 1994, Milligan 1996, Kennedy and Farrell 2006) during which they swam in bursts for 5 minutes. This method of exercise has been well established to exhaust the fish, as indicated by the near total depletion of white muscle glycogen, ATP, and phosphocreatine stores (Schulte et al. 1992). At various time points after burst swimming, fish were euthanized in MS222 and a blood sample obtained from the caudal vasculature using heparinized microcapillary tubes and centrifuged at 13,000 × g. The plasma was separated and frozen in liquid nitrogen. A subsample of muscle from the left lateral side of the fish was removed and frozen until analysis. All samples were eventually stored at –86°C for no longer than 3 weeks before analysis. Plasma cortisol concentrations were determined using radioimmunoassays (IncStar Corp., Stillwater, Minnesota). Muscle glycogen was measured in selected liver samples according to standard procedures (Milligan et al. 2000).

Disease Challenges

Disease challenges were performed in herring following 1 and 57 days exposures to the DHF of oil as described in Kennedy and Farrell (2008). For disease challenges, three replicates of 50 fish each were randomly selected from each exposure group and exposed to the bacterium *Listonella anguillarum*, a gram-negative bacterium and an opportunistic pathogen commonly encountered in the marine environment. It is the main causative agent of Vibriosis, a syndrome characterized by hemorrhagic septicemia (Egidius 1987). The challenge dose of *L. anguillarum* was obtained from Fisheries and Oceans Canada, West Vancouver Laboratory. A subsample of the bacterial culture was used to make serial dilution plates to confirm the concentration of the bacterial culture solution used for the bath challenge. Disease challenges were performed as in Kennedy and Farrell (2008). Fish were observed twice daily for mortality and signs of disease for a 6-week period following the challenge. Each dead fish was evaluated for mass and length, and the head kidney was removed to test for the presence or absence of *L. anguillarum* using standard confirmatory microbiological techniques (Noga 1996).

In a separate set of experiments, fish were exposed for either 1 or 56 days to the DHF of oil at 0 and 100 $\mu g{\cdot}L^{-1}$ TPAH as described above. Control fish or fish that were exposed to DHF for 1 day were first implanted with silastic tubing containing either corn oil (controls) or dexamethasone (treated fish: 50 $mg{\cdot}kg^{-1}$) dissolved in corn oil. Control fish or fish that were exposed to DHF for 56 days were first implanted with silastic tubing containing either corn oil (controls) or cortisol (treated fish: 50 $mg{\cdot}kg^{-1}$) dissolved in corn oil (Bernier et al. 1999). Following DHF exposure, both groups of fish were then subjected to a disease challenge as described above.

Calculations and Statistics

Two-way analysis of variance (ANOVA) and a Holm-Sidak post hoc test were used to determine whether any measured hematological or biochemical parameters differed between replicate trials for any treatment group. As no differences were detected between replicates in any experiment, data were pooled and differences between treatment groups and exposure times were analyzed using two-way ANOVA and a Holm-Sidak post hoc test (Kennedy and Farrell 2005). All percent data were arcsine transformed prior to analysis. Disease challenge mortality curves were compared to controls using the Mantel–Cox and Breslow test statistics (Kennedy and Farrell 2008). All data analyses were conducted with a fiducial limit significance at $p < .05$, using SigmaStat version 3.0 (SPSS Scientific, Chicago, Illinois).

Results

Chemical Concentrations

Aqueous TPAH concentrations in exposure tanks at the beginning of any experiment were as follows: control (0.10–0.18 $\mu g{\cdot}L^{-1}$), low (7.4–13.4 $\mu g{\cdot}L^{-1}$), medium (30.7–42.5 $\mu g{\cdot}L^{-1}$) and high (111.3–133.5 $\mu g{\cdot}L^{-1}$). TPAH concentrations declined with time (Figure 3.3, Kennedy and Farrell 2006), and since declines were similar across treatments, each exposure treatment remained distinct (Carls et al. 1995). Using this design, alkane concentrations can

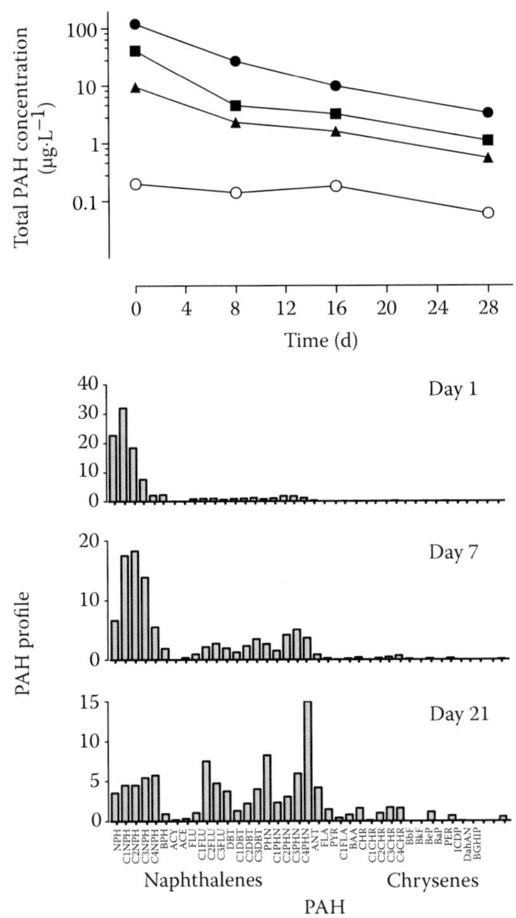

FIGURE 3.3
Total polycyclic aromatic hydrocarbon (TPAH) concentrations generated from cylinders for 28 days following initiation of water flow. Data points are single composite values for replicate tanks at each time period (white circle) control, (black triangle) 10 $\mu g{\cdot}L^{-1}$, (black square) 40 $\mu g{\cdot}L^{-1}$, (black circle) 100 $\mu g{\cdot}L^{-1}$. In the lower panels, individual PAHs as percent TPAH as a function of time following initiation of water flow through a cylinder are shown. On the left are lower molecular weight PAHs, which move to higher molecular weight PAHs on the right.

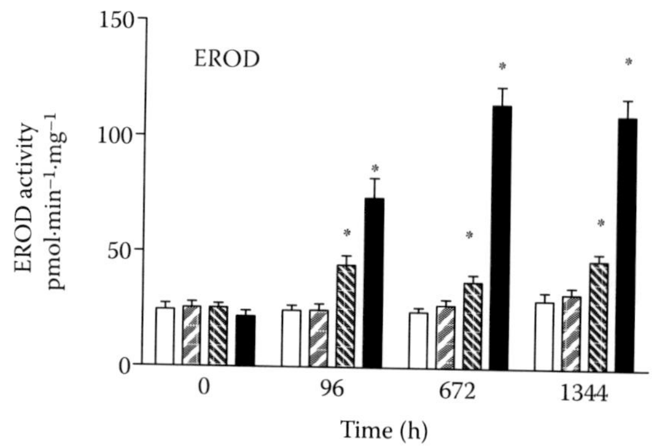

FIGURE 3.4
Ethoxyresorufin O-deethylase (EROD) activity (pmol·min^{-1}·mg^{-1} microsomal protein) with time in (white bar) control, (light diagonal hatch) 10 μg·L^{-1}, (dark diagonal hatch) 40 μg·L^{-1}, and (black bar) 100 μg·L^{-1} total polycyclic aromatic hydrocarbon.

range from 1.28 μg·L^{-1} (controls) to 119 μg·L^{-1} (high exposure) with concentration declines similar to those of TPAH (Carls et al. 1995). In this system, the predominant PAHs at the start of cylinder operation are the smaller and less substituted (e.g., naphthalenes) compounds, which progressively shift with time to proportionately larger and more substituted PAHs (e.g., phenanthrenes) with time (Kennedy and Farrell 2005).

Induction of Biotransformation Enzymes and Bioavailability

EROD activities were increased by exposure to aqueous hydrocarbons (Figure 3.4, Kennedy and Farrell 2005) indicating that these compounds were bioavailable. At the start of cylinder operation, no significant differences between exposure groups in any measured parameter were found. Significant induction of EROD activity was found in the medium and high concentration groups by day 4 and continued through the exposure period to day 56. Maximum induction occurred in the highest concentration group after 56 days (5.4-fold increase over controls).

Stress Physiology

Acute 4-day DHF exposure resulted in a significant stress response in herring while a chronic 58-day exposure resulted in an apparent abolition of the stress response (Figure 3.5). Control plasma cortisol concentrations did not change significantly over time in either the acute or chronic trials. In the acute exposure trials, plasma cortisol concentrations were significantly increased in DHF-exposed fish over controls and reached maximum concentrations between 4 and 12 hours in the medium and high exposure groups. By 4 days, cortisol concentrations in all groups were at baseline values.

In the chronic 58-day exposures, pulses of hydrocarbons and lower molecular weight PAHs occurred at 0, 28, and 56 days. Significant increases occurred in plasma cortisol concentrations in fish exposed to the highest concentration of DHF during the first introduction of aqueous hydrocarbons to the tanks (Figure 3.5). Concentrations of cortisol were similar to those seen in the 4-day acute exposure. Cortisol concentrations returned to baseline values by day 4. When a new column was brought online at 28 days, a second significant cortisol increase occurred, although 52% lower than the initial cortisol peak. No cortisol response occurred with a further cylinder replacement at 56 days (Figure 3.5).

When fish were exposed to DHF of oil for 58 days, and then subjected to a handling stress, control fish exhibited a typical increase in plasma cortisol concentrations, which peaked at approximately 2 hours following handling (175 ng·mL^{-1}), and decreased to control values by 24 hours (Figure 3.6). In fish exposed to 100 μg·L^{-1} TPAH of DHF for 58 days, no increase in plasma cortisol was seen following handling stress.

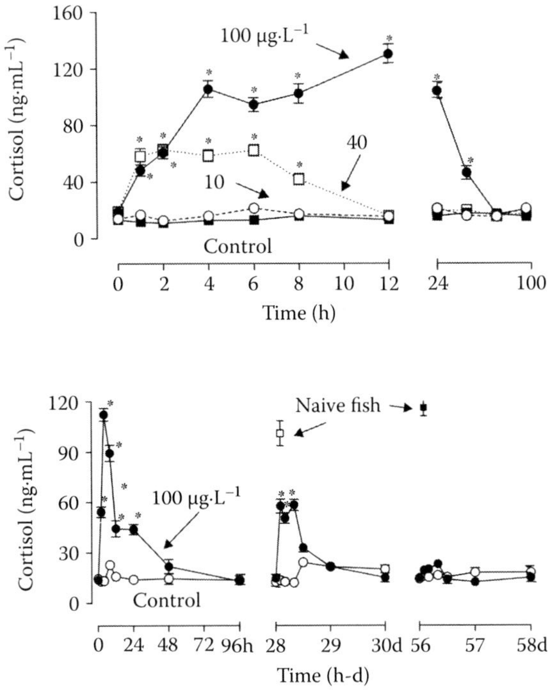

FIGURE 3.5
Plasma cortisol concentrations in control fish (white circles) and those exposed to 10 (black squares), 40 (white squares), and 100 μg·L^{-1} (black circles) concentrations. Values are means ± SE of n = 14 fish. *Denotes a significant difference (p < .05) from controls. In the lower figure, plasma cortisol concentrations in control fish (white circles) and those exposed to 100 μg·L^{-1} (black circles) TPAH concentration over 58 days. New pulses of DHF were brought online at 0, 28, and 56 days. Single symbols denote control values for fish that had been exposed to DHF at 28 days (white) and 56 days (black) only (naive fish). Time of exposure is given as hours (h) and days (d). Values are means ± SE of n = 14 fish. *Denotes a significant difference (p < .05) from controls.

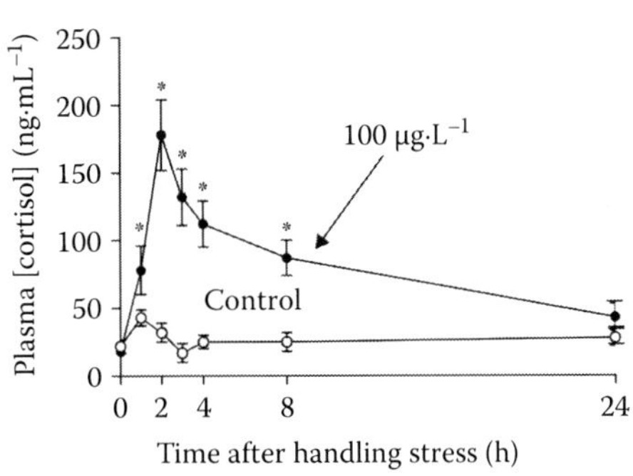

FIGURE 3.6
Plasma cortisol concentrations with time in control (white circles) and fish exposed to 100 μg·L^{-1} total polycyclic aromatic hydrocarbon following a handling stress. Values are means ± SE of n = 6 fish. *Denotes a significant difference (p < .05) from controls.

In vitro stimulation of a mixed interrenal cell preparation by ACTH was used to determine potential sites of HPI axis inactivation with chronic DHF exposure. The interrenal tissues of control fish responded positively to ACTH by producing cortisol in a concentration-dependent manner. Fish exposed to 100 μg·L^{-1} TPAH of the DHF were not able to produce cortisol with ACTH stimulation (Figure 3.7).

Burst Swimming and Recovery

The physiological effects of hydrocarbon exposure (1 day and 56 days) on the ability of herring to recover from exercise were examined in fish that were forced to burst swim for 5 minutes after DHF exposure. A description of the response of control fish to burst exercise is described here first because DHF exposure had significant effects on baseline parameters before fish swam, and these responses were further altered by the DHF exposure after swimming, forming a complex series of results. In control fish, plasma cortisol and muscle glycogen concentrations were measured for 24 hours after burst exercise (Figure 3.8). Plasma cortisol concentrations prior to exercise ranged from 14 to 25 ng·mL^{-1} and increased significantly postexercise, and then recovered to preexercise concentrations by 8 hours. The postexercise decrease in muscle glycogen concentrations was also complete by 12 hours (Figure 3.9). Therefore, all of the measured parameters had recovered to preexercise levels by 24 hours postexercise. The low DHF-exposure group did not differ significantly in its response to burst exercise compared with control fish.

The following results apply only to comparisons between the control and medium/high DHF-exposure groups. In fish exposed to DHF for 24 hours, high initial cortisol concentrations preceded burst swimming (85 ng·mL^{-1}, Figure 3.8). When these fish swam, cortisol values reached the highest concentrations due to the cumulative effects of DHF exposure and swimming (175 ng·mL^{-1}, Figure 3.8). In the high DHF-exposed group, cortisol concentrations had not returned to baseline values by 24 hours. In the 24-hour exposed fish, no difference was seen compared to controls in muscle glycogen use or repletion (Figure 3.9). In the 56-day DHF medium- and high-exposure groups, the typical peak responses of plasma cortisol following exercise were completely suppressed postexercise (Figure 3.8). Corresponding to this suppression, the postexercise depression of muscle glycogen had recovered significantly faster in DHF-exposed fish compared to that in control fish (Figure 3.9).

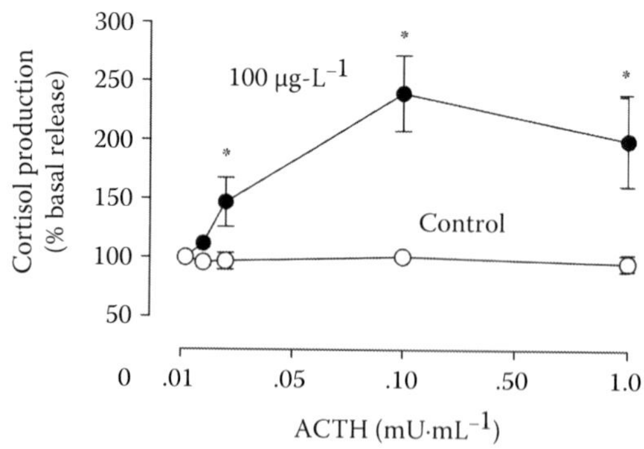

FIGURE 3.7
In vitro cortisol production in isolated interrenal cells exposed to varying concentrations of adrenocorticotropic hormone (ACTH). Interrenal cells were isolated from either control (white circles) or fish exposed to 100 μg·L^{-1} (black circles) total polycyclic aromatic hydrocarbon for 56 days. Values are means ± SE of $n = 6$ fish. *Denotes a significant difference ($p < .05$) from controls.

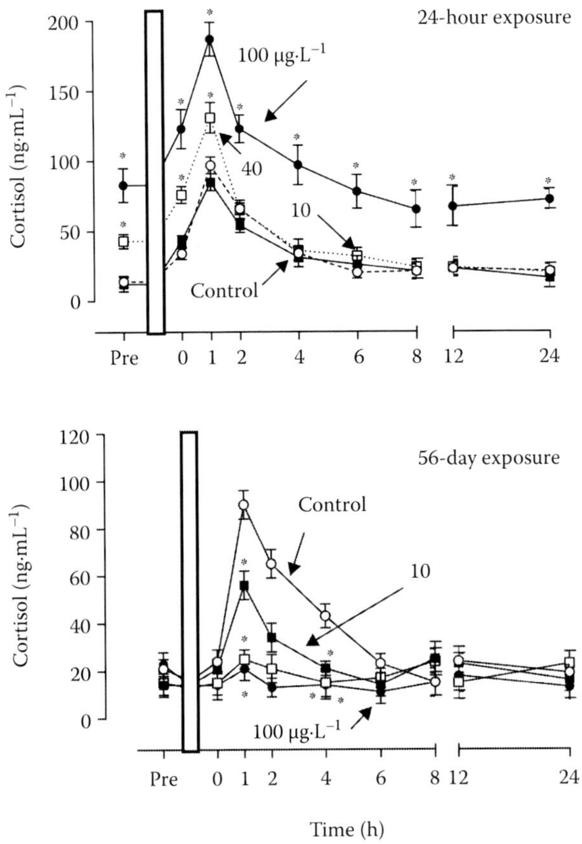

FIGURE 3.8
Plasma cortisol concentrations in control fish (white circles) and those exposed to 10 (black squares), 40 (white squares), and 100 (black circles) μg·L^{-1} concentrations for 24 hours before and following burst exercise (white bar). Similar graph in lower panel with fish exposed to the same concentrations for 57 days. Values are means ± SE of $n = 6$ fish. *Denotes a significant difference ($p < .05$) from controls.

Disease Challenges

Exposure of fish to aqueous hydrocarbons affected their susceptibility to the marine pathogen *L. anguillarum* (Figure 3.10). There were no significant differences in the rate of mortality between control fish and those in the low and medium treatment groups in any experiment described in this section; however, fish exposed to the highest concentration of hydrocarbons had a lower mortality rate and had less total fish die from the pathogen after a 1-day DHF exposure. After 56-day exposure, there was a significant increase in the susceptibility of fish in the high treatment group compared to all other groups (Figure 3.10). When fish exposed to DHF at 100 μg·L^{-1} for 24 hours, and implanted with a silastic implant containing the cortisol-antagonist dexamethasone, the increased tolerance to *L. anguillarum* challenge seen in DHF-exposed-only fish was abolished (Figure 3.11). Fish implanted with cortisol-containing silastic tubing and exposed to 100 μg·L^{-1} DHF for 57 days and then challenged showed a similar susceptibility to *L. anguillarum* (Figure 3.12) as seen in DHF-exposed-only fish (Figure 3.11).

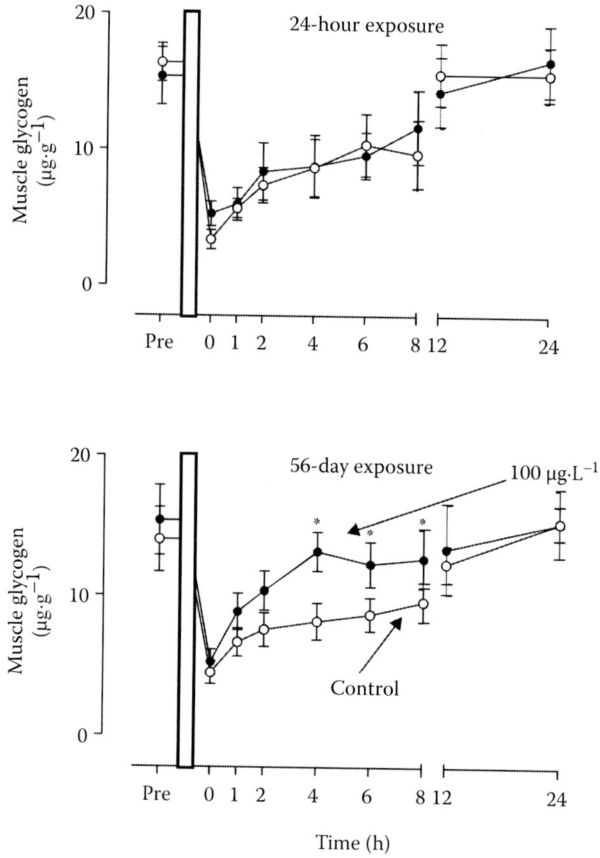

FIGURE 3.9
Muscle glycogen concentrations in control fish (white circles) and those exposed to 100 μg·L^{-1} (black circles) concentrations for 24 hours before and following burst exercise (white bar). Similar graph in lower panel with fish exposed to the same concentrations for 57 days. Values are means ± SE of $n = 6$ fish. *Denotes a significant difference ($p < .05$) from controls.

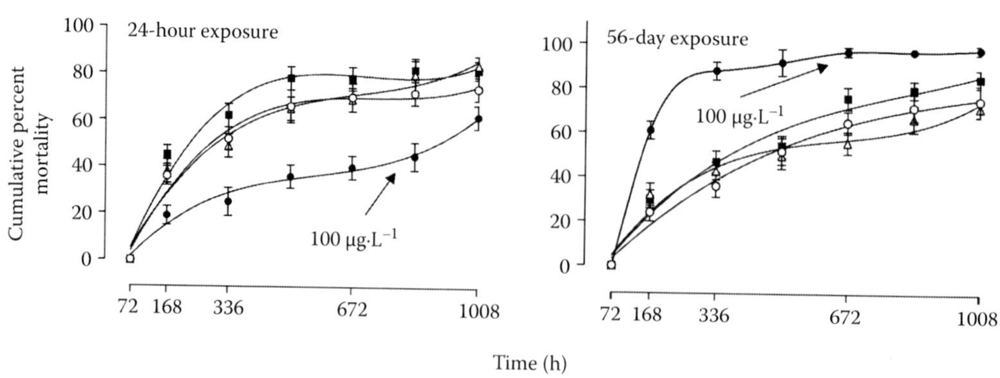

FIGURE 3.10
Cumulative percent mortality in herring exposed to DHF for 24 hours and 57 days following a challenge with *Listonella anguillarum* in control fish (white circles) and those exposed to 10 (black squares), 40 (white squares), and 100 (black circles) μg·L^{-1} total polycyclic aromatic hydrocarbon concentrations. Values are averages of duplicate tanks.

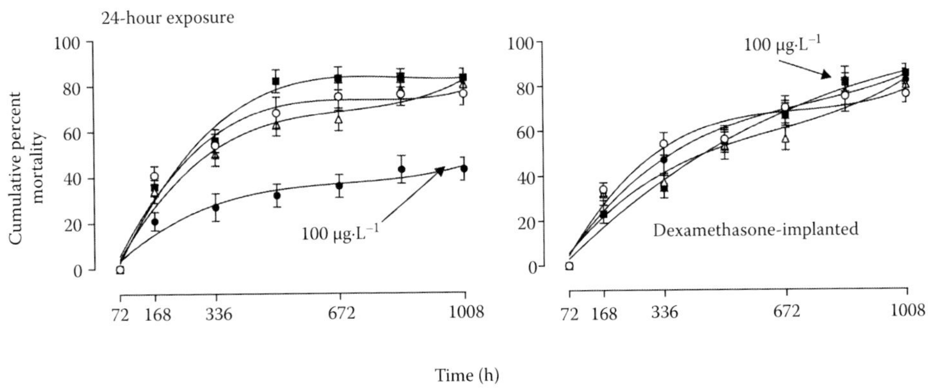

FIGURE 3.11
Cumulative percent mortality in herring exposed to DHF for 24 hours following a challenge with *Listonella anguillarum* in control fish (white circles) and those exposed to 10 (black squares), 40 (white squares), and 100 (black circles) µg·L^{-1} total polycyclic aromatic hydrocarbon concentrations. In the lower panel, fish were implanted with silastic implants containing dexamethasone. Values are averages of duplicate tanks.

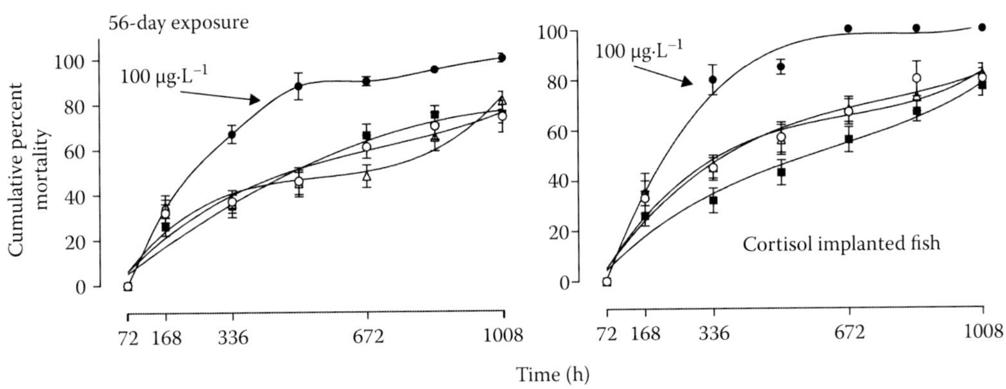

FIGURE 3.12
Cumulative percent mortality in herring exposed to DHF for 57 days following a challenge with *Listonella anguillarum* in control fish (white circles) and those exposed to 10 (black squares), 40 (white squares), and 100 (black circles) µg·L^{-1} total polycyclic aromatic hydrocarbon concentrations. In the lower panel, fish were implanted with silastic implants containing cortisol. Values are averages of duplicate tanks.

Discussion

Exposure System

One major challenge facing investigators examining the environmental effects of oil inputs into the aquatic environment is in using "realistic" or environmentally "relevant" exposures in laboratory and field studies. Oil is a complex mixture (see Chapter 1), and various oil types contain differing constituents and proportions of constituents. Depending on the release conditions, receiving environment, biological and abiotic factors, and so forth, the fate of each constituent will vary between releases and with time. What is known, however, is that the fraction of actual dissolved hydrocarbons will be very low due to the limited water solubility of most oil compounds in water, and that pelagic species of fish will be exposed to hydrocarbons in the low $\mu g{\cdot}L^{-1}$ range.

Many studies have attempted to obtain a truly dissolved fraction in exposure water in the laboratory (Carls et al. 1998, Kennedy and Farrell 2005). The present study has apparently achieved this goal, with the modified cylinders containing oil-soaked ceramic beads releasing only truly dissolved hydrocarbons into seawater, generating a reliable concentration range from background to 140 $\mu g{\cdot}L^{-1}$ TPAH, as well as the ability to expose fish in a concentration-dependent manner. The oil cylinders were particularly successful in the acute exposures (Kennedy and Farrell 2005, 2006, 2008), however, hydrocarbon concentration profiles declined and changed as time progressed. A changing profile may be a more realistic representation of chemical profiles in the environment following a spill which progressively change over time from predominantly low to high molecular weight hydrocarbons. In the chronic studies presented here, the employment of several DHF-generating cylinders was necessary to obtain a relatively consistent hydrocarbon exposure that remained distinct between treatment groups. Fluctuating hydrocarbon concentrations and profiles is not an optimal exposure scenario, but the exposure system has been the most successful application of a DHF-generating system in this regard to the author's knowledge.

Bioavailability

Compounds in a crude oil mixture span a wide range of log K_{OW} values, from those that are water soluble to somewhat hydrophobic, to hydrophobic, or super-hydrophobic. Varying log K_{OW}s result in disparate fates in the environment, uptake into aquatic organisms, and levels of accumulation in organisms. Many hydrocarbons including PAHs are known inducers of CYP1A in fish livers once absorbed. Measuring tissue concentrations from low concentration hydrocarbon exposures may not be good indicators of the bioavailability of these chemicals in fish due to low accumulation rates (and tissue concentrations below detection limits), and the ability of fish to easily biotransform these compounds. The results in the present study indicate that low aqueous hydrocarbon concentrations are bioavailable to fish, as indicated by significant inductions of hepatic CYP1A1 as measured by increased EROD activity (Gagnon and Holdway 2000).

Effects of Oil on Stress Physiology

PAHs have been previously implicated as endocrine disruptors in fish, specifically as modulators of steroidogenesis (Evansen and Van der Kraak 2001, Kennedy and Farrell 2005). The results from the present studies provide further evidence of hydrocarbon effects on the steroid cortisol, which has a number of cellular and physiological roles in vertebrates, the most significant being its involvement in stress (Mommsen et al. 1999). The exposure of fish to sublethal concentrations of a variety of contaminants can disturb homeostasis and impose stress on organisms. Well-documented stress responses in teleosts include biochemical (e.g., heat shock protein production), neuroendocrine (e.g., catecholamines and corticosteroid release), physiological (e.g., hyperlacticemia and hyperglycemia), and organismal responses (e.g., reduced growth, predisposition to disease, impaired reproduction [Adams 1990]), which depend on the individual stressor type and the exposure duration. Control or baseline cortisol values in herring (12–21 $ng{\cdot}mL^{-1}$) in these experiments were well within the baseline ranges reported for other unstressed teleosts (Morrow et al. 2004). An acute exposure of teleosts to dissolved hydrocarbons in the low ppb concentration range (40–140 μg·L^{-1}) initiated a stress response with activation of the HPI axis. The stimulated HPI axis resulted in a classical stress-related increase in plasma cortisol concentrations within the first hour of exposure, followed by a slow rise in plasma lactate and glucose concentrations (data not shown, Kennedy and Farrell 2005). Cortisol dynamics were concentration-dependent, with results that are similar to those reported for other teleosts (Thomas et al. 1980, Thomas and Rice 1987, Alkindi et al. 1996).

The volatile aromatic hydrocarbon fraction consisting of naphthalenes, benzene, toluene, ethylbenzene, and trimethybenzene (Thomas and Rice 1987) are acutely toxic to fish and have been suggested as the likely causative agents of the stress response to DHF exposure. For example, increases in plasma cortisol concentrations in other species have been documented following naphthalene exposure (DiMichelle and Taylor 1978, Thomas et al. 1980). One potential mechanism underlying this response by several classes of pollutants is through their irritant properties on surface epithelia (Schreck and Lorz 1978). Gill hyperplasia is a condition that is caused by naphthalene exposure in *Fundulus heteroclitus* and has been demonstrated for most other irritants as well (Gardner 1975). The stress response to new cylinder application and low molecular weight hydrocarbons was transient and plasma cortisol concentrations returned to baseline by 96 hours. The concentrations and proportions of these particular hydrocarbons in the dissolved fraction were lower at day 4 compared with the initiation of exposure. At day 4, the concentrations of naphthalene and other irritant hydrocarbons may have then been below the thresholds necessary for HPI activation. It is also possible that fish simply became "habituated" to the presence of these chemicals.

In the chronic DHF-exposure experiments, a different temporal pattern for plasma cortisol concentrations was seen, which suggests a possible endocrine disrupting mechanism underlying the results. In the chronic study, three cylinders were used to maintain relatively consistent hydrocarbon concentrations. As each new cylinder was brought online, total PAH concentrations and the proportions of lower molecular weight compounds (such as naphthalene) were highest and should have resulted in HPI axis activation and the release of cortisol by interrenal cells into the plasma (as was evident in any acute or naïve exposure to a new cylinder). Interestingly, the increase in plasma cortisol at 28 days with the first cylinder replacement was approximately 52% of the initial cortisol peak, and with the third replacement at 56 days, there was no plasma cortisol concentration peak at all, indicating an abnormal response to the stress of chemicals such as naphthalenes.

Several mechanisms present themselves to explain the reduction in the stress response to repeated pulse exposures to the more acutely toxic hydrocarbons at their highest concentrations. First, habituation of the corticosteroid response to these hydrocarbons may partially take place following 28 days of exposure, and more fully at 56 days. *F. heteroclitus*, however, exposed continuously to naphthalene for 15 days did not habituate to this chemical, and plasma cortisol concentrations were constantly elevated through the entire exposure period (DiMichelle and Taylor 1978). A second possibility is related to the induction of cytochrome P450 and related biotransformation enzymes by hydrocarbons. These remained induced despite fluctuating hydrocarbon concentrations. The induction of P450-associated enzymes by PAHs may have affected cortisol kinetics. Cytochrome P450 is associated with both steroid biosynthesis and biotransformation and may have led to increased cortisol metabolism and its clearance from plasma (Hansson and Lidman 1978). Third, some hydrocarbons, particularly high molecular weight PAHs (HPAHs), may act directly as endocrine disruptors, targeting the pituitary or adrenocortical tissues (Dorval et al. 2003). Multiple sites in the HPI axis may be affected including the perception of stimuli, the synthesis and secretion of the main secretagogue for cortisol (ACTH), and the synthesis (e.g., disruption of the steroidogenic acute regulatory protein expression [Stocco 2000, Walsh et al. 2000]) and secretion of cortisol itself. For example, in trout exposed to β-naphthoflavone, the sensitivity of interrenal cells to ACTH was abolished (Wilson et al. 1998). AhR agonists have also been shown to disrupt interrenal corticosteroidogenesis and target tissue responsiveness to glucocorticoid stimulation (Neelakanteswar and Vijayan 2004). Oil components in this experiment contain AhR agonists, so it is possible that the disruption of the cortisol response involves several mechanisms. Lastly, exposure of fish to PAHs such as naphthalene has been shown to result in histopathological effects in a number of tissues. For example, major necrosis occurred in interrenal tissues in mummichogs exposed to naphthalene resulting in a reduced ability to respond to secretagogues or to produce the hormone (DiMichelle and Taylor 1978).

Several further studies were performed to determine which of the above mechanisms were likely to be occurring. The specific objectives of these studies included investigations of the impact of hydrocarbon exposure on the response to a handling stress in herring and to examine the interrenal responses to ACTH stimulation of cortisol production in vitro following chronic hydrocarbon exposure. In in vitro experiments using interrenal cells isolated from control fish and those treated with hydrocarbons for 56 days (at 0 and 100 $\mu g \cdot L^{-1}$), interrenal sensitivity to ACTH stimulation of cortisol production was abolished with chemical exposure. The absence of a response to ACTH may be due to the effect of hydrocarbons on either ACTH receptor dynamics and/or the steroidogenic pathway in trout interrenal tissue.

An attenuation of the handling stress–induced increase in plasma cortisol in hydrocarbon-treated fish was also seen in herring. The lack of an in vivo plasma cortisol response to handling stress correlates with the in vitro lack of cortisol production with ACTH stimulation seen in hydrocarbon-treated fish. ACTH is the primary secretagogue for cortisol release with stressors such as handling in fish. These results, therefore, suggest that effects on the pituitary–interrenal axis are at the level of ACTH receptors. Several other studies have seen similar impairments. For example, capture did not increase in plasma cortisol concentrations in fish collected from heavily polluted areas (Hontela et al. 1992). It was suggested that prolonged exposure to some chemicals may lead to hyperactivity of the HPI axis, resulting in cellular "exhaustion" as evidenced by the presence of atrophied pituitary corticotropes. The results of the present study, however, support down-regulation and/or desensitization of the ACTH receptors as an additional mechanism for the suppressed cortisol response to cope with stressors in fish, which has been seen with β-naphthoflavone injection in rainbow trout (Wilson et al. 1998).

Interestingly, the loss of the physiological stress response on exposure to organics or metals (Lizardo-Daudt et al. 2007) have been well documented, but the effects of a diminished stress response and homeostatic ability on overall fish fitness when presented with biotic or abiotic challenges are unknown. Cortisol's putative role as a hypo-osmoregulatory hormone, and its role in reproduction, growth, and the immune response in fish (Maule et al. 1987, Donaldson 1990, Hontela et al. 1995, Shelley et al. 2012) suggest that chronic exposure to low sublethal levels of pollutants such as PAHs will affect a variety of physiological systems. Designing an appropriate test for the assessment of a loss of stress response in a complicated neuroendocrine system continues to be a challenge.

Effects of Dissolved Hydrocarbon Fraction on Swimming Performance

Prolonged swimming is not typically performed by some species (such as the Pacific herring) and can result in mortality (Kennedy and Farrell 2006). The burst swim exercise method used in this study exhausts fish completely, and a depletion of muscle glycogen stores is a reproducible effect seen in all fish (Milligan and Wood 1986, Pagnotta et al. 1994). This technique results in the same type of metabolic and endocrine disturbances (elevation in circulating levels of cortisol and catecholamines) as seen during studies in which fish were exercised to exhaustion in a swim tunnel (Schulte et al. 1992). Cortisol peaks approximately 1–2 hours postexercise (Gamperl et al. 1994) and is presumed to be caused by postexercise inactivity and not exercise per se (Milligan et al. 2000). Muscle glycogen is used for energy and is resynthesized approximately 4–12 hours after exhaustion (Milligan and Wood 1986).

In control fish, exercise increases plasma cortisol and decreases muscle glycogen concentrations. Cortisol and muscle glycogen concentrations returned to baseline values within typical times reported for other teleosts. When exposed acutely to DHF, cortisol values following swimming were the highest of any group but muscle glycogen concentrations showed a similar response to control fish. Thomas and Rice (1987) performed a similar study using prolonged swimming by salmon. They found that there was a cumulative stress response brought on by exposure followed by exercise. The stress response occurred in the acute exposures, but there was no such response in the chronic exposures. After 57 days of exposure to hydrocarbons, a postexercise cortisol release did not occur indicating a severely impaired cortisol response (Kennedy and Farrell 2005). In fish that produced limited or complete lack of cortisol in response to exercise, muscle glycogen levels recovered faster than in fish that show a typical cortisol increase in response to exercise. When cortisol concentrations are lowered after exercise through postexercise activity (Milligan et al. 2000) or pharmacological inhibition (Pagnotta et al. 1994), an enhancement of metabolic recovery occurs (i.e., a repletion of muscle glycogen). This relationship was also the case in the present study, as cortisol release was not seen in hydrocarbon-exposed fish, which resulted in a faster recovery of muscle glycogen stores. This study clearly shows that the endocrine disrupting effects of hydrocarbon-exposure affects the recovery of fish following burst swimming. In long-term exposures, an absence of cortisol secretion altered their response to exercise. Overall, these results indicate that exposure to oil can have effects on the locomotor performance of herring and is of considerable interest from management, physiological, environmental, ecological, and evolutionary perspectives.

Effects of Dissolved Hydrocarbon Fraction Exposure on Immune Performance

There exist numerous examples of both organic xenobiotics and metals affecting the immunological systems of fish through either general (i.e., via the physiological stress response, with resultant downstream effects brought on by neuroendocrine alterations) or chemical-specific (i.e., chemicals target different components of the immune system) mechanisms of action (Anderson 1990). In the current paradigm, the higher susceptibility of stressed individuals to diseases is due to stressor-induced modifications (Ruis and Bayne 1997), causing immune depression or immune suppression in fish. This (stress–immune relationship) phenomenon is not completely understood, but appears to be mediated through endocrine pathways that regulate and modulate immune system responses. Specifically, the activation of the HPI axis has been shown to affect both the innate and adaptive components of the immune system (Ruis and Bayne 1997). More current information now exists documenting how xenobiotic effects on the immune system of fish are mediated through endocrine disruption (Kennedy et al. 2013).

The population decline of Pacific herring in PWS following the EVOS has generated a number of hypotheses regarding causative agents. It has been suggested that oil exposure may have affected the immunocompetence (immunosuppression) of fish and the subsequent expression of viral hemorrhagic septicemia virus (VHSV) in the population. This hypothesis has some evidence to support it: many fish that returned to spawn exhibited external hemorrhages consistent with the signs of VHSV (Marty et al. 1998), and the North American strain of VHSV has been isolated from samples of Pacific herring (Meyers et al. 1994). In laboratory experiments, 97% of adult herring that died after exposure to PAH concentrations (28 $\mu g \cdot L^{-1}$) had lesions consistent with VHSV (Carls et al. 1995). Herring could be asymptomatic carriers of VHSV, and hydrocarbon exposure (eliciting a generalized stress or specific toxic effects) may reduce the ability of the immune system to contain it, resulting in outbreaks.

Some studies clearly show that PAHs are immunomodulatory in teleosts. For example, field studies have correlated immunosuppression with tissue PAH concentration, however the exposure history of fish in these studies is often unknown (Arkoosh et al. 1991, 1994). It has been suggested that exposure to PAHs in 1989 following the EVOS may have caused damage in developing cells destined to become functional immune cells and that abnormal function could be evident at a later stage in life if challenged with specific pathogens (Kocan et al. 1997). Several studies, however, have suggested through a weight-of-evidence approach that the available data does not support a relationship between VHSV outbreak in herring from PWS and permanent immune suppression caused by hydrocarbon exposure when fish were larvae or yearlings (Elston et al. 1997, Marty et al. 1998). These contradicting views were the reasons that these studies examining hydrocarbon immunotoxicity in both acute and chronic exposures in Pacific herring were undertaken.

The general principle that the physiological effects (in this case immunotoxicity) of environmental stressors depend on their duration and intensity (Ortuno et al. 2001) is highlighted by the current research. The susceptibility of Pacific herring to *L. anguillarum* in disease-challenge experiments was examined because these tests are considered the most important and comprehensive in immunotoxicological screening (Wester et al. 1994). These experiments are broad and direct tests of an organism's immune response that have true biological significance at the individual and population levels. In the present study, herring were more tolerant to *L. anguillarum* following exposure to the highest concentration of DHF for 1 day, however this enhanced tolerance was not maintained. By 57 days of hydrocarbon exposure, herring were significantly more susceptible to *L. anguillarum*.

The physiological outcomes following SC and HPI axes activation can be suppressors or stimulators of the immune system (Tort et al. 1996). The enhancement of fish immune responses, rather than immunosuppression, has been observed to follow acute stress (e.g., increase in nonspecific plasma proteins such as lysozyme and complement and enhanced yeast phagocytosis), leading to better protection against possible pathogens (Demers and Bayne 1997, Ruis and Bayne 1997, Ortuno et al. 2001). A practical example of immune enhancement was shown in juvenile spring Chinook salmon exposed to *L. anguillarum* following handling stress. Acute handling stress decreased the susceptibility of fish to this pathogen, in spite of a significantly depressed ability of lymphocytes to generate antibodies (specific immune response) (Maule et al. 1989). Many other studies, however, have shown contradictory results in that crowding, handling stress, and cortisol administration can result in immunosuppression and reduced disease resistance (Maule et al. 1987, Tripp et al. 1987). Cortisol is believed to play a major role in stress-induced immunostimulation and immunosuppression, but the underlying mechanisms are unclear.

Dexamethasone is a potent synthetic member of the glucocorticoid class of steroids and is known to suppress cortisol secretion and the pituitary–interrenal axis in response to stress in fish (Fagerlund and McBride 1969). It was the intent of the present study with cortisol and dexamethasone silastic implants to discern the role of cortisol in the acute stress response to hydrocarbon exposure, and the absence of a stress response with long-term hydrocarbon exposure, on the changing disease susceptibility in herring. It was believed that if cortisol was responsible for the enhancement of the immune system with short-term hydrocarbon exposure, then blocking the stress response with dexamethasone would cause these fish to become less resistant to the disease challenge with *L. anguillarum*. Interestingly, this reduced resistance is precisely what occurred in these studies. Dexamethasone-treated fish were as sensitive to the pathogen as control fish, that is, cortisol and the stress response is likely to enhance fish resistance to disease in the short term. Conversely, however, cortisol implantation (to reverse the EDC effects of chronic hydrocarbon exposure) did not reduce the susceptibility of fish to the pathogen as expected. Increased susceptibility due to chronic hydrocarbon exposure is therefore likely due to specific chemicals in the mixture that damaged the immune system.

PAHs are established mammalian immune toxicants with a large literature on their effects on specific immune components (Holladay et al. 1998). Benzo[a]pyrene (B[a]P) and 7,12-dimethyl-benz(a)anthracene (DMBA) are two commonly studied PAHs and known to be immunotoxic (Anderson et al. 1995). The first documentation of immunosuppression due to PAH exposure was shown in mice in 1952 (Malmgren et al. 1952). It has also been shown that chronic sublethal PAH exposure can cause immunosuppression and predispose fish to disease (Dunier and Siwicki 1993). For example, fish exposed to PAH in rivers show suppressed phagocytic activity (Weeks et al. 1986). Both DMBA and B[a]P decrease resistance to pathogens in fish (White et al. 1994) and exposure to DMBA by intraperitoneal injection is immunosuppressive (suppressed macrophage phagocytotic activity and complete inhibition of nonspecific cytotoxic cell activity) in the oyster toadfish (*Opsanus tau*, Seeley and Weeks-Perkins 1997). Juvenile European sea bass (*Dicentrarchus labrax*) showed decreased respiratory burst activity and phagocytosis following injection with B[a]P (Lemaire-Gony et al. 1995). Several possible mechanism have been brought forward to explain immune system suppression by PAHs and include the following: (1) binding to the Ah receptor and activation of the Ah gene complex signaling pathway, (2) perturbation of membranes, (3) altered interleukin production, (4) disruption of intracellular calcium mobilization, and (5) metabolic activation to reactive metabolites (White et al. 1994).

Conclusions

Knowledge of the subtle physiological changes associated with hormonal alteration, the subsequent effects on individual biology, and potentially population-level events are currently limited. There has never been a more important time to understand fish endocrinology than the present, with rising concern that certain environmental contaminants and naturally occurring compounds have the potential to impact the endocrine systems of animals.

EDCs were once thought to interact only with nuclear receptors that are present in most tissues throughout the organism. It is now known that they can exert their toxic effects through a multitude of mechanisms, and a single chemical may act through multiple pathways depending on life stage, gender, availability of receptors or targets, and concentration of the chemical. Effects can also occur as a result of interactions at higher levels, such as alteration in the HPI axis that regulates hormonal activity. In addition, EDCs have been found to alter transcription or activity of hormone receptors and enzymes involved in biotransformation or steroid biosynthesis pathways. EDCs have a myriad of adverse effects on eggs, larvae, and juveniles and adult laboratory and wild fish, which involve wide-ranging physiological systems such as reproduction, immunity, osmoregulation, metabolism, endocrine, development, and growth (Kime 1998).

The results of the present studies provide clear evidence of endocrine effects in juvenile herring at low $\mu g{\cdot}L^{-1}$ concentrations, levels that have previously been shown only to affect fish at the embryo/larval stages. The organism-level effects seen in herring on the stress response, swimming performance, exercise recovery, and functioning of the immune system are all related to endocrine effects that involve the steroid cortisol. Clearly, DHF caused stress in the short term, but long-term exposures resulted in an inhibition of these fish to mount a cortisol response to stressful situations and to normal physiological activities that release cortisol (e.g., exercise). The roles of cortisol in fish are numerous and have been extensively studied (Mommsen et al. 1999). Surprisingly, however, the complexities and multiplicity of its actions on many physiological systems make it difficult to understand the population-level relevance of oil exposure and its endocrine disrupting effects. What is clear is that extremely low concentrations of dissolved hydrocarbons from oil can affect the endocrine systems of fish and more research should be invested into understanding the potential population impacts of its release into the environment.

Acknowledgements

The research described here was supported by the *Exxon Valdez* Oil Spill Trustee Council through contracts with the Alaska Department of Fish to CJK. The findings and conclusions presented by the author are his own and do not necessarily reflect the view or position of the agency. The analytical chemistry support provided by the National Oceanic and Atmospheric Administration, National Marine Fisheries Service, Auke Bay Laboratory was greatly appreciated. Dr. Keith Tierney, Dr. Susan Sanders, and Joanne Precious are thanked for research assistance. North Slope crude oil was a gift from Dr. R. Kocan.

References

Adams, S. M. 1990. Status and use of biological indicators for evaluating the effects of stress in fish. In *Biological Indicators of Ecosystem Stress*, ed. S. M. Adams, 1–8. Bethesda: American Fisheries Society.

Alkindi, A. Y. A., J. A. Brown, C. P. Waring, and J. E. Collins. 1996. Endocrine, osmoregulatory, respiratory and haematological parameters in flounder exposed to the water-soluble fraction of crude oil. *Journal of Fish Biology* 49:1291–1305.

Anderson, C., A. Hehr, R. Robbins, R. Hasan, M. Athar, H. Mukhtar, et al. 1995. Metabolic requirements for induction of contact hypersensitivity to immunotoxic polycyclic aromatic hydrocarbons. *Journal of Immunology* 155:3530–3537.

Anderson, D. P. 1990. Immunological indicators: Effects of environmental stress on immune protection and disease outbreaks. In *Biological Indicators of Ecosystem Stress*, ed. S. M. Adams, 38–50. Bethesda: American Fisheries Society.

Arkoosh, M. R., E. Casillas, E. Clemens, B. B. McCain, and U. Varanasi. 1991. Suppression of immunological memory in juvenile Chinook salmon (*Oncoryhynchus tshawytscha*) from an urban estuary. *Fish and Shellfish Immunology* 1:261–277.

Arkoosh, M. R., E. Clemens, M. Myers, and E. Casillas. 1994. Suppression of ß-cell mediated immunity in juvenile Chinook salmon (*Onoryhynchus tshawytscha*) after exposure to either a polycyclic aromatic hydrocarbon or to polychlorinated biphenyls. *Immunopharmacology and Immunotoxicology* 16:293–314.

Arnold-Reed, D. E. and R. J. Balment. 1994. Peptide hormones influence in vitro interrenal secretion of cortisol in the trout, *Oncorhynchus mykiss*. *General and Comparative Endocrinology* 96:85–91.

Badawy, M. I. and F. Al-Harthy. 1991. Hydrocarbons in seawater, sediment and oyster from the Omani coastal water. *Bulletin of Environmental Contamination and Toxicology* 47:386–391.

Barron, M. G., M. G. Carls, J. W. Short, S. D. Rice. 2003. Photo-enhanced toxicity of aqueous phase and chemically dispersed weathered Alaska North Slope crude oil to Pacific herring eggs and larvae. *Environmental Toxicology and Chemistry* 22:650–660.

Beamish, F. W. H. 1978. Swimming Capacity. In *Fish Physiology*, Vol. 7, eds. W. S. Hoar and D. J. Randall, 101–187. New York: Academic Press.

Bernier, N. J., X. Lin, and R. E. Peter. 1999. Differential expression of corticotropin-releasing factor (CRF) and urotensin I precursor genes, and evidence of CRF gene expression regulated by cortisol in goldfish brain. *General and Comparative Endocrinology* 116:461–477.

Bradford, M. M. 1976. A rapid and sensitive method for the quantitation of microgram quantities of protein utilizing the principle of protein dye binding. *Analytical Biochemistry* 72:248–254.

Brown, E. D., T. T. Baker, J. E. Hose, R. M. Kocan, G. D. Marty, M. D. McGurk, et al. 1996. Injury to the early life history stages of Pacific herring in Prince William Sound after the *Exxon Valdez* oil spill. In *Proceedings of the Exxon Valdez Oil Spill Symposium*, eds. S. D. Rice, R. B. Spies, D. A. Wolfe, and B. A. Wright, 448–462. Bethesda: American Fisheries Society.

Burke, M. D. and R. T. Mayer. 1974. Ethoxyresorufin: Direct fluorometric assay of microsomal O-dealkylation which is preferentially induced by 3-methylcolanthrene. *Drug Metabolism and Disposition* 2:583–588.

Carls, M. G. 1987. Effects of dietary and water-borne oil exposure on larval Pacific herring (*Clupea harengus pallasi*). *Marine Environmental Research* 22:253–270.

Carls, M. G., G. D. Marty, T. R. Meyers, R. E. Thomas, and S. D. Rice. 1998. Expression of viral hemorrhagic septicemia virus in pre-spawning Pacific herring (*Clupea pallasi*) exposed to weathered crude. *Canadian Journal of Fisheries and Aquatic Sciences* 55:2300–2309.

Carls, M. G., S. D. Rice, and J. E. Hose. 1999. Sensitivity of fish embryos to weathered crude oil: Part I. Low level exposure during incubation causes malformations, genetic damage and mortality in larval Pacific herring (*Clupea pallasi*). *Environmental Toxicology and Chemistry* 18:481–493.

Carls, M. G., S. D. Rice, and R. E. Thomas. 1995. The impact of adult pre-spawn herring (*Clupea harengus pallasi*) on subsequent progeny. Restoration project 94166 annual report. National Oceanic and Atmospheric Administration. Auke Bay: National Marine Fisheries Service.

Demers, N. E. and C. J. Bayne. 1997. The immediate effects of stress on hormones and plasma lysozyme in rainbow trout. *Developmental and Comparative Immunology* 21:363–373.

Denslow, N. and Sepulveda M. 2007. Ecotoxicological effects of endocrine disrupting compounds on fish reproduction. In *The Fish Oocyte*, eds. P. J. Babin, J. Cerda, and E. Lubzens, 255–322. The Netherlands: Springer.

DiMichele, L. and M. H. Taylor. 1978. Histopathological and physiological responses on *Fundulus heteroclitus* to naphthalene exposure. *Journal of the Fisheries Research Board of Canada* 35:1060–1066.

Donaldson, E. M. 1990. Reproductive indices as measures of the effects of environmental stressors in fish. In *Biological Indicators of Ecosystem Stress*, ed. S. M. Adams, 109–122. Bethesda: American Fisheries Society.

Dorval J., L. S. Leblond, and A. Hontela. 2003. Oxidative stress and loss of cortisol secretion in adrenocortical cells of rainbow trout (*Oncorhynchus mykiss*) exposed in vitro to endosulfan, an organochlorine pesticide. *Aquatic Toxicology* 63:229–241.

Dunier, M. and A. K. Siwicki. 1993. Effects of pesticides and other organic pollutants in the aquatic environment on immunity of fish: A review. *Fish and Shellfish Immunology* 3:423–438.

Egidius, E. 1987. Vibriosis: Pathogenicity and Pathology. A review. *Aquaculture* 67:15–28.

Elston, R. A., A. S. Drum, W. H. Pearson, and K. Parker. 1997. Health and condition of Pacific herring, *Clupea pallasi*, from Prince William Sound, Alaska. *Diseases of Aquatic Organisms* 33:109–126.

Evanson, M. and G. J. Van Der Kraak. 2001. Stimulatory effects of selected PAHs on testosterone production in goldfish and rainbow trout and possible mechanisms of action. *Comparative Biochemistry and Physiology* 130:249–258.

Fagerlund, U. H. M. and J. R. McBride. 1969. Suppression by dexamethasone of interrenal activity in adult sockeye salmon (*Oncorhynchus nerka*). *General and Comparative Endocrinology* 12:651–657.

Gagnon, M. M. and D. A. Holdway. 2000. EROD induction and biliary metabolite excretion following exposure to the water accommodated fraction of crude oil and to chemically dispersed crude oil. *Archives of Environmental Contamination and Toxicology* 38:70–77.

Gamperl, A. K., M. M. Vijayan, and R. G. Boutilier. 1994. Experimental control of stress hormone levels in fishes: Techniques and applications. *Reviews in Fish Biology and Fisheries*. 4:215–255.

Gardner, G. R. 1975. Chemically induced lesions in estuarine or marine teleosts. In *The Pathology of Fishes*, eds. W. Ribelin and G. Migaki, 657–691. Madison: University of Wisconsin Press.

Gulland, F. M. D. 1995. The impact of infectious diseases on wild animal populations—a review. In *Ecology of Infectious Diseases in Natural Populations*, eds. B. T. Grenfella and A. P. Dobson, 20–51. Cambridge, United Kingdom: Cambridge University Press.

Hansch, C. and J. M. Clayton. 1973. Lipophilic character and biological activity of drugs II: The parabolic case. *Journal of Pharmaceutical Sciences* 62:1–21.

Hansson, T. and U. Lidman. 1978. Effects of cortisol administration on components of the hepatic microsomal mixed function oxidase system (MFO) of immature rainbow trout (*Salmo gairdneri* Rich.). *Acta Pharmacologica et Toxicologica* 43:6–12.

Heintz, R. A., S. D. Rice, A. C. Wertheimer, R. F. Bradshaw, F. P. Thrower, J. E. Joyce, et al. 2000. Delayed effects on growth and marine survival of pink salmon (*Oncorhychus gorbuscha*) after exposure to crude oil during embryonic development. *Marine Ecological Progress Series* 208:205–216.

Heintz, R. A., J. W. Short, and S. D. Rice. 1999. Sensitivity of fish embryos to weathered crude oil: Part II. Increased mortality of pink salmon (*Oncorhynchus gorbuscha*) embryos incubating downstream from weathered *Exxon Valdez* crude oil. *Environmental Toxicology and Chemistry* 18:494–503.

Holladay, S. D., S. A. Smith, E. G. Basteman, A. S. M. I. Deyab, R. M. Gogal, T. Hrubec, et al. 1998. Benzo[a]pyrene-induced hypocellularity of the pronephros in tilapia (*Oreochromis niloticus*) is accompanied by alterations in stromal and parenchymal cells and by enhanced immune cell apoptosis. *Veterinary Immunology and Immunopathology* 64:69–82.

Hontela, A., J. B. Rasmussen, C. Audet, G. Chevalier. 1992. Impaired cortisol stress response in fish from environments polluted by PAHs, PCBs, and mercury. *Archives of Environmental Contamination and Toxicology* 22:278–283.

Hontela, A. P., P. Dumont, D. Duclos, and R. Fortin. 1995. Endocrine and metabolic dysfunction in yellow perch, *Perca flavescens*, exposed to organic contaminants and heavy metals in the St. Lawrence River. *Environmental Toxicology and Chemistry* 14:725–731.

Hose, J. E., M. D. McGurk, G. D. Marty, D. E. Hinton, E. D. Brown, and T. T. Baker. 1996. Sublethal effects of the *Exxon Valdez* oil spill on herring embryos and larvae: Morphological, cytogenetic, and histopathological assessments 1989–1991. *Canadian Journal of Fisheries and Aquatic Sciences* 53:2355–2365.

Incardona, J. P., T. K. Collier, and N. L. Scholz. 2004. Defects in cardiac function precede morphological abnormalities in fish embryos exposed to polycyclic aromatic hydrocarbons. *Toxicology and Applied Pharmacology* 196:191–205.

Kennedy, C. J. 1995. Xenobiotics: Designing an in vitro system to study enzymes and metabolism. In *Biochemistry and Molecular Biology of Fishes*, eds. P. W. Hochachka and T. P. Mommsen, Vol. 3 Analytical Techniques, 417–430. Amsterdam: Elsevier Science.

Kennedy, C. J. and A. P. Farrell. 2005. Ion homeostasis and interrenal stress responses in juvenile Pacific herring, *Clupea pallasi*, exposed to the water-soluble fraction of oil. *Journal of Experimental Marine Biology and Ecology* 323:43–56.

Kennedy, C. J. and A. P. Farrell. 2006. Effects of exposure to the water-soluble fraction of crude oil on the swimming performance and the metabolic and ionic recovery postexercise in Pacific herring (*Clupea pallasi*). *Environmental Contamination and Toxicology* 25:2715–2724.

Kennedy, C. J. and A. P. Farrell. 2008. Immunological alterations in juvenile Pacific herring, *Clupea pallasi*, exposed to aqueous hydrocarbons derived from oil. *Environmental Pollution* 153:638–648.

Kennedy, C. J., H. Osachoff, and L. Shelley. 2013. Estrogenic endocrine disrupting chemicals in fish. In *Fish Physiology Series*, ed. K. Tierney. Elsevier Science Publishers, Academic Publishing Division (*in press*).

Kime, D. E. 1998. *Endocrine Disruption in Fish*. Norwell: Kluwer Academic Publishers.

Knudsen, F. R. and T. G. Pottinger. 1998. Interaction of endocrine disrupting chemicals, singly and in combination, with estrogen, androgen and corticosteroid-binding sites in rainbow trout (*Oncorhynchus mykiss*). *Aquatic Toxicology* 44:159–170.

Kocan, R., M. Bradley, N. Elder, T. R. Meyers, W. Batts, and J. Winton. 1997. The North American strain of viral hemorrhagic septicemia virus is highly pathogenic for laboratory reared Pacific herring (*Clupea pallasi*). *Journal of Aquatic Animal Health* 9:279–290.

Kocan, R. M., G. D. Marty, M. S. Okihiro, E. D. Brown, and T. T. Baker. 1996. Reproductive success and histopathology of individual Prince William Sound Pacific herring 3 years after the *Exxon Valdez* oil spill. *Canadian Journal of Fisheries and Aquatic Sciences* 53:2388–2393.

Krimsky, S. 2000. *Hormonal Chaos: The Scientific and Social Origins of the Environmental Endocrine Hypothesis*. Baltimore and London: The Johns Hopkins University Press.

Lemaire-Gony, S., P. Lemaire, and A. L. Pulsford. 1995. Effects of cadmium and benzo(a)pyrene on the immune system, gill ATPase and EROD activity of European sea bass *Dicentrarchus labrax*. *Aquatic Toxicology* 31:297–313.

Lizardo-Daudt, H. M., O. S. Bains, C. R. Singh, and C. J. Kennedy. 2007. Biosynthetic capacity of rainbow trout (*Oncorhynchus mykiss*) interrenal tissue after cadmium exposure. *Archives of Environmental Contamination and Toxicology* 52:90–96.

Madany, I. M., A. Al-Haddad, A. Jaffar, and E. S. Al-Shirbini. 1994. Spatial and temporal distribution of aromatic petroleum hydrocarbons in the coastal waters of Bahrain. *Archives of Environmental Contamination and Toxicology* 26:185–190.

Malmgren, R. A., B. E. Bennison, and T. W. McKinley. 1952. Reduced antibody titers in mice treated with carcinogenic and cancer chemotherapeutic agents. *Proceedings of the Society for Experimental Biology and Medicine* 70:484–488.

Marty, G. D., E. F. Freiberg, T. R. Meyers, J. Wilcock, T. B. Farver, D. E. Hinton. 1998. Viral hemorrhagic septicemia virus, *Ichthyophonus hoferi*, and other causes of morbidity in Pacific herring *Clupea pallasi* spawning in Prince William Sound, Alaska, USA. *Diseases of Aquatic Organisms* 32:15–40.

Marty, G. D., J. W. Short, D. M. Dambach, N. H. Willits, R. A. Heintz, S. D. Rice, et al. 1997. Ascites, premature emergence, increased gonadal cell apoptosis, and cytochrome P4501A induction in pink salmon larvae continuously exposed to oil-contaminated gravel during development. *Canadian Journal of Zoology* 75:989–1007.

Maule, A. G., C. B. Schreck, and S. L. Kaattari. 1987. Changes in the immune system of coho salmon (*Oncorhynchus kisutch*) during the parr-to-smolt transformation and after implantation of cortisol. *Canadian Journal of Fisheries and Aquatic Sciences* 44:161–166.

Maule, A. G., R. A. Tripp, S. L. Kaattari, and C. B. Schreck. 1989. Stress alters immune function and disease resistance in chinook salmon (*Oncorhynchus tshawytscha*). *Journal of Endocrinology* 120:135–142.

McGurk, M. D. and E. D. Brown. 1996. Egg-larval mortality of Pacific herring in Prince William Sound, Alaska, after the *Exxon Valdez* oil spill. *Canadian Journal of Fisheries and Aquatic Sciences* 53:2343–2354.

McKim, J., P. Schmieder, and G. Veith. 1985. Absorption dynamics of organic chemical transport across trout gills as related to octanol-water partition coefficient. *Toxicology and Applied Pharmacology* 77:1–10.

Mearns, A. J. and M. J. Sherwood. 1976. Ocean wastewater discharge and tumors in a southern California flatfish. *Progress in Experimental Tumor Research* 20:75–85.

Meyers, T. R., S. Short, K. Lipson, W. N. Batts, J. R. Winton, J. Wilcock, et al. 1994. Association of viral hemorrhagic septicemia virus with epizootic hemorrhages of the skin in Pacific herring *Clupea harengus pallasi* from Prince William Sound and Kodiak Island, Alaska, USA. *Diseases of Aquatic Organisms* 19:27–37.

Milligan, C. L. 1996. Metabolic recovery from exhaustive exercise in rainbow trout. *Comparative Biochemistry and Physiology* Part A. 113:51–60.

Milligan, C. L., G. B. Hooke, and C. Johnson. 2000. Sustained swimming at low velocity following a bout of exhaustive exercise enhances metabolic recovery in rainbow trout. *Journal of Experimental Biology* 203:921–926.

Milligan, C. L. and C. M. Wood. 1986. Tissue intracellular acid-base status and the fate of lactate after exhaustive exercise in the rainbow trout. *Journal of Experimental Biology* 123:123–144.

Mommsen, T. P., M. M. Vijayan, and T. W. Moon. 1999. Cortisol in teleosts: Dynamics, mechanisms of action, and metabolic regulation. *Reviews in Fish Biology and Fisheries* 9:211–268.

Morrow, M., D. Higgs, and C. J. Kennedy. 2004. The effects of diet composition and ration on biotransformation enzymes and stress parameters in rainbow trout, Oncorhynchus mykiss. *Comparative Biochemistry and Physiology* Part C. 137: 143–154.

Neelakanteswar, A. and M. M. Vijayan. 2004. beta-Naphthoflavone disrupts cortisol production and liver glucocorticoid responsiveness in rainbow trout. *Aquatic Toxicology* 67:273–285.

Neff, J. M. 1990. Composition and fate of petroleum and spill-treating agents in the marine environment. In *Sea Mammals and Oil: Confronting the Risks*, eds. J. R. Geraci and D. J. St Aubin, 1–32. San Diego: Academic Press.

Neff, J. M. and J. W. Anderson. 1981. *Response of Marine Animals to Petroleum and Specific Petroleum Hydrocarbons*. Essex, United Kingdom: Applied Science Publishers.

Noga, E. J. 1996. *Fish Disease: Diagnosis and Treatment*. St. Louis: Mosby.

Norcross, B. L., J. E. Hose, M. Frandsen, and E. D. Brown. 1996. Distribution, abundance, morphological condition, and cytogenetic abnormalities of larval herring in Prince William Sound, Alaska, following the *Exxon Valdez* oil spill. *Canadian Journal of Fisheries and Aquatic Sciences* 53:2376–2387.

Ortuno, J., M. A. Esteban, and J. Meseguer. 2001. Effects of short-term crowding stress on the gilthead seabream (*Sparus aurata* L.) innate immune response. *Fish and Shellfish Immunology* 11:187–197.

Pagnotta, A., L. Brooks, and C. L. Milligan. 1994. The potential regulatory role of cortisol in the recovery from exhaustive exercise in rainbow trout. *Canadian Journal of Fisheries and Aquatic Sciences* 72:2136–2146.

Plaut, I. 2001. Critical swimming speed: Its ecological relevance. *Comparative Biochemistry and Physiology* Part A 131:41–50.

Reddy, C. M., T. I. Eglinton, A. Hounshell, H. K. White, L. Xu, R. B. Gaines, et al. 2002. The West Falmouth oil spill after thirty years: The persistence of petroleum hydrocarbons in marsh sediments. *Environmental Science and Technology* 36:4754–4760.

Reidy, S. P., S. R. Kerr, and J. A. Nelson. 2000. Aerobic and anaerobic swimming performance of individual Atlantic cod. *Journal of Experimental Biology* 203:347–357.

Ruis, M. A. W. and C. J. Bayne. 1997. Effects of acute stress on blood clotting and yeast killing by phagocytes of rainbow trout. *Journal of Aquatic Animal Health* 9:190–195.

Schreck, C. B. and H. W. Lorz. 1978. Stress response of coho salmon (*Oncorhynchus kisutch*) elicited by cadmium and copper and potential use of cortisol as an indicator of stress. *Journal of the Fisheries Research Board of Canada* 35:1124–1129.

Schulte, P. M., C. D. Moyes, and P. W. Hochachka. 1992. Integrating metabolic pathways in post-exercise recovery of white muscle. *Journal of Experimental Biology* 166:181–195.

Seeley, K. R. and B. A. Weeks-Perkins. 1997. Suppression of natural cytotoxic cell and macrophage phagocytic function in oyster toadfish exposed to 7,12-dimethyl-benz[a]anthracene. *Fish and Shellfish Immunology* 7:115–121.

Shelley, L. K., P. S. Ross, and C. J. Kennedy. 2012. The effects of an in vitro exposure to 17ß-estradiol and nonylphenol on rainbow trout (*Oncorhynchus mykiss*) peripheral blood leukocytes. *Comparative Biochemistry and Physiology* Part C. 155:440–446.

Stocco, D. M. 2000. The role of the StAR protein in steroidogenesis: Challenges for the future. *Journal of Endocrinology* 164:247–253.

Thomas, P., B. R. Woodin, and J. M. Neff. 1980. Biochemical responses of the striped mullet *Mugil cephalus* to oil exposure: I. Acute responses—Interrenal activations and secondary stress responses. *Marine Biology* 59:114–141.

Thomas, R. E. and S. D. Rice. 1987. Effect of water soluble fraction of Cook Inlet crude oil on swimming performance and plasma cortisol in juvenile coho salmon (*Oncorhynchus kisutch*). *Comparative Biochemistry and Physiology* Part C. 87:177–180.

Tort, L., J. O. Sunyer, E. Gome, A. Molinero. 1996. Crowding stress induces changes in serum haemolytic and agglutinating activity in the gilthead sea bream (*Sparus aurata*). *Veterinary Immunology and Immunopathology* 51:179–188.

Tripp, R. A., A. G. Maule, C. B. Schreck, and S. L. Kaattari. 1987. Cortisol mediated suppression of salmonid lymphocyte responses in vitro. *Developmental and Comparative Immunology* 11:565–576.

Walsh, L. P., C. McCormick, C. Martin, and D. M. Stocco. 2000. Roundup inhibits steroidogenesis by disrupting steroidogenic acute regulatory (StAR) protein expression. *Environmental Health Perspectives* 108:769–776.

Waterman, B. and H. Kranz. 1992. Pollution and fish diseases in the North Sea. Some historical perspectives. *Marine Pollution Bulletin* 24:131–137.

Weeks, B. A., J. E. Warinner, P. L. Mason, and D. S. McGinnis. 1986. Influence of toxic chemicals on the chemotactic response of fish macrophages. *Journal of Fish Biology* 28:653–658.

Wendelaar Bonga, S. E. 1997. The stress response in fish. *Physiological Reviews* 77:591–625.

Wester, P. W., A. D. Vethaak, and W. B. van Muiswinkel. 1994. Fish as biomarkers in immunotoxicology. *Toxicology* 86:213–232.

White, K. L., T. T. Kawabata, and G. S. Ladics. 1994. Mechanisms of polycyclic aromatic hydrocarbon immunotoxicity. In *Immunotoxicology and Immunopharmacology, 2nd ed.*, eds. J. H. Deqan, M. I. Luster, A. E. Munson, and I. Kimber, 123–142. New York: Raven Press.

Wilson, J. M., M. M. Vijayan, C. J. Kennedy, and G. K. Iwama. 1998. beta-Naphthoflavone abolishes interrenal sensitivity to ACTH stimulation in rainbow trout. *Journal of Endocrinology* 157:63–70.

Zelikoff, J. T. 1993. Metal pollution-induced immunomodulation in fish. *Annual Review of Fish Diseases* 2:305–325.

Zelikoff, J. T. 1994. Fish immunotoxicology. In *Immunotoxicology and Immunopharmacology, 2nd ed.*, eds. J. H. Deqan, M. I. Luster, A. E. Munson, and I. Kimber, 71–95. New York: Raven Press.

4

Proper Handling of Animal Tissues from the Field to the Laboratory Supports Reliable Biomarker Endpoints

Heather M. Olivier and Jill A. Jenkins

CONTENTS

Introduction

In the endeavor to assess potential effects to the Gulf of Mexico ecosystem from the Mississippi Canyon 252 incident, referred to as the *Deepwater Horizon* oil spill, various environmental data have been collected. Whereas initial efforts have included satellite tracking and sediment and water sampling to estimate the geographical scope of oiling, research on biological samples can provide insights into potential physiological responses to oil if it was present in the food web, sediment, or water column. Fish species are ideal model organisms for studying responses to water- and sediment-borne contaminants due to their life history (Jenkins et al. 2014), and several Gulf of Mexico fish species were studied by scientists after this incident. Typical field data collected on fish reflect organism condition and include observations such as fish length, weight, gonad condition, condition factor (weight in relation to length), parasite load, and color of organs (Schmitt and Dethloff 2000). However, if physiological responses occurred due to oil exposure, effects would not be immediately visible using organism-level observations alone. Changes occur first at the organ, tissue, cell, or molecular levels, and these responses can be measured by using biomarker assays (van der Oost et al. 2003).

Biomarkers are early signals of exposure, effect, or susceptibility (NRC 1987) that are generated at several levels of biological organization (Bayne et al. 1985). Alterations associated with health impairments or diseases (e.g., cancer) are biomarkers of effect. Biomarkers of exposure are measures of an exogenous substance, or metabolite, or the product of an interaction between a xenobiotic agent and a target molecule or cell that is measured in a compartment within an organism (van der Oost et al. 2003). Oil does not adhere to the body surface of fish, as it can with birds and mammals, disallowing visual documentation of exposure, but cell and molecular biomarkers in fish can help document oil exposure as well as potential damage (McCain et al. 1978, Aas et al. 2000).

Crude oil is a mixture of hydrocarbons, organics, and nonhydrocarbon compounds. Over 75% of crude oil is composed of hydrocarbons, of which 0.2%–7% are polycyclic aromatic hydrocarbons (PAHs) (Neff 1985, Alber 1995, Eisler 1999). Of the total number of PAHs, 15 are carcinogens (Rice and Arkoosh 2002, U.S. Dept. HHS 2002). The toxicity of some of these chemical constituents may increase once metabolized by the organism (van der Oost et al. 2003), heightening the production of reactive oxygen species and the consequent oxidation of molecular structures, such as DNA (Shen and Ong 2001). Because PAHs show little tendency to biomagnify in food chains and are rapidly metabolized (Eisler 1999), testing for biomarkers of PAH exposure, rather than PAH congeners, is a more informative approach for oil toxicology investigations. This can be done in a variety of ways. The detection of DNA–PAH adducts has been found to be preferable to the detection of PAH levels in tissues in assessing exposure (van der Oost et al. 1994, van der Oost et al. 1997, Aas et al. 2000). Measuring cytochrome P450 (CYP1A)-dependent enzymatic activities involved in the detoxification process (ethoxyresorufin-O-deethylase [EROD] activity) in response to PAH exposure is a proven, reliable biomarker used by several field investigations (Whyte et al. 2000, Jewett et al. 2002). Furthermore, this research on mechanisms of CYP1A-induced toxicity suggests that EROD activity may not only indicate chemical exposure but may also precede effects at various levels of biological organization (Whyte et al. 2000, Whitehead et al. 2012). The metabolites of PAHs can be detected in bile via fixed wavelength fluorescence analysis (Aas et al. 2000). Lastly, measurement of the level of DNA damage in fish exposed to major pollutants (e.g., PAHs, PCBs, dioxins, heavy metals) is another powerful way to monitor fish response to pollution effects (Larno et al. 2001).

Designing experiments to assess potential contaminant effects necessitates the careful selection of species, tissue types, biomarkers, and analytical methods (Holland et al. 2003). Fish in the field are exposed to complex chemical mixtures as well as varying ecological parameters (such as temperature and oxygen level), so the ability to identify physiological responses that clearly point to contaminant effects is strengthened by studying more than one biological endpoint (Depledge 1996). Measuring a suite of biomarkers within the organism or within a single tissue of the organism can serve as a cross-check for exposure or damage (Table 4.1). Flow cytometry is well-suited to this task because the technology allows data on multiple cellular and molecular parameters to be collected simultaneously on the same sample (Rieseberg et al. 2001), and is a well-established clinical research tool used in ecotoxicology studies (Dallas and Evans 1990). Flow cytometry requires the analysis of individual cells, thus the free-flowing nature of both milt and blood makes these tissues good choices for study. In addition, blood and milt can be collected comparatively noninvasively, which is necessary when sampling federally listed species (Fossi and

TABLE 4.1

Examples of Biomarker Options from Various Taxa for Use with Studies on the Effects of Environmental Contamination in Aquatic Ecosystems

	Biomarkers								
	EROD[a]	DNA Integrity/ Genotoxicity	Apoptosis	Gamete Quality	Heat Shock Proteins	Metallo-thioneins	Fluorescent Bile	Histopa-thology	Hema-tology
Birds	X	X	X	X	X	X	X	X	X
Fish	X	X	X	X	X	X	X	X	X
Bivalves		X	X	X	X	X		X	X

[a] Ethoxyresorufin-O-deethylase (EROD) activity.

Leonzio 1994, UFR Committee 2013). Biomarkers from the hematological and reproductive systems can provide insight into reproductive potential, genetic condition, and disease susceptibility (Jirtle and Skinner 2007, Jenkins 2011b).

In order for blood and milt samples to be suitable for biomarker studies, the logistics of sample collection, handling, and transport from remote locations need to be taken into account within the study design. Because field sampling is typically both labor-intensive and expensive, taking precautions to maintain good sample integrity will assure that these initial efforts are not wasted. The requirements for sustaining biological samples in good condition during shipment are dependent on the chosen biological endpoints to be measured (Holland et al. 2003). Cells that are intact, live, and stable can provide a broader range of cellular and molecular parameters for analysis with flow cytometry than cells preserved with fixatives. In this chapter, we discuss various sample handling issues for fish blood and milt, highlight some experiments and results that addressed sample condition, and offer approaches to consider for maintaining blood and sperm quality for biomarker studies.

Considerations for Shipping and Handling Samples

Temperature

Temperature fluctuations during shipment can negatively influence live tissue integrity. If the sample is subjected to high temperatures during transit, cellular enzyme activity increases, speeding decomposition (Holland et al. 2003). Placing samples immediately into a chilled environment, such as a cooler, can lessen degradation processes (Schlenk et al. 2008). Cells may freeze, however, if the sample is overly insulated with ice or icy water, or is in direct contact with ice. Freezing causes the formation of intracellular ice crystals that burst cell membranes, rendering cells nonviable. Milt has been successfully frozen after dilution with cryoprotectants (Tiersch et al. 1994); however, cryopreservation is generally impractical during field collections.

An effective method for transporting live tissues is packing them with frozen gel packs within leakproof shipping containers (hard plastic, not polystyrene) and providing an insulating barrier, such as paper towels or plastic bags, separating the cold packs from samples. To maintain a chilled environment for the length of time in transit, surround blood samples on all sides with cold gel packs, and fill any empty space with insulating material. However for milt in particular, the temperature should remain cool (~8°C), but not near freezing, with cold gel packs placed only below the samples (Tiersch 2011). If procuring frozen gel packs before shipment is not possible, using ice secured within two plastic bags (to prevent leaking) is an acceptable alternative. Choosing a vial, jar, or test tube of an appropriate size for the sample volume will assure that the tissue is not tossed within the container during transport, thereby lessening potential cell damage. If the sample is small relative to the container, adding buffer to mostly fill it, leaving some air space for tissue aeration, helps decrease fluid sloshing.

Foreign Material Inclusions

Preventing the unintended introduction of foreign materials during sample collection and shipment may be problematic in the field, but proactive measures can help maintain sample integrity. Materials such as sediment, mucus, urine, or water can be inadvertently introduced into sample containers. Wiping clean the skin surface of the fish before sample collection and using aseptic tools and techniques can help keep the sample clear of debris.

An example of the difference in quality of field-collected blood due to foreign object inclusion is illustrated in Figure 4.1. Images (a) and (b) depict pallid sturgeon (*Scaphirhynchus albus*) blood cells collected and shipped satisfactorily to the laboratory, while images (c) and (d) show Gulf sturgeon (*Acipenser oxyrinchus desotoi*) blood collected and shipped inadequately. In the latter case, rather than being taped to the outside, a sample identification label was placed within the tube at the time of blood collection. Over the course of transport the blood cells likely reacted to the presence of this foreign material and enzymatic activity increased. Although the blood was collected with anticoagulant and in accord with the other instructions provided to the field crew, the sample was coagulated on arrival and nuclear degradation was apparent. In contrast to the blood sample without paper (Figure 4.1a and Figure 4.1b), nuclear size was decreased (Figure 4.1c), and no clear primary DNA histogram peak was present (Figure 4.1d). Samples of inferior quality due to inadequate shipping and handling are excluded from analyses. The importance of clear communication about sample handling is underscored by this example. Faithful implementation of clear and detailed standard operating procedures, for sample handling and shipping by the field teams, as well as for quality control, quality assurance, and sample analyses by the laboratory team, can help to assure high quality biological endpoints.

Enzymatic Degradation

Enzymatic degradation in blood samples affects many biochemical markers, nucleic acids, and proteins (Vaught 2006). Although proteases and nucleases are integral to natural processes such as apoptosis (programmed cell death), they can become detrimental to otherwise healthy cells during processing, handling, and shipment (Holland et al. 2003). Degradation is intensified by the release of nucleases and other enzymes from cellular compartments as they lose membrane integrity (Williams and Little 1974). Nucleases cleave the molecular structure of DNA, a process that can mimic effects induced by genotoxins, resulting in false positives. Such enzyme activity can be prevented either through cold storage of samples or the addition of preservatives such as ethanol or salt solutions (Seutin et al. 1991, Hobson et al. 1997). Commercial solutions for stabilizing cells and cellular

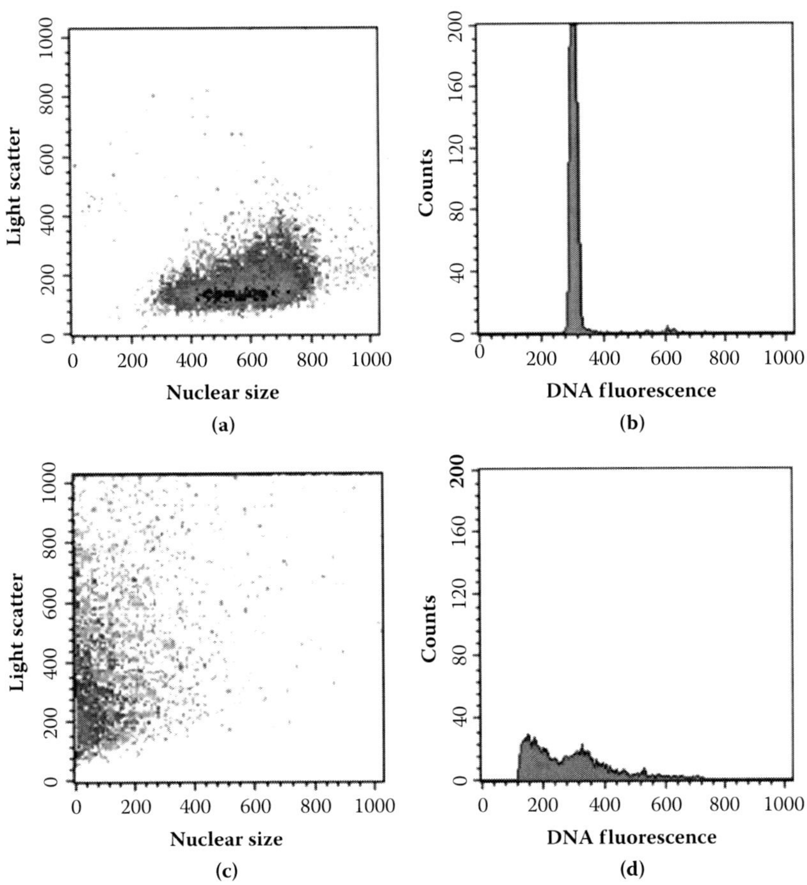

FIGURE 4.1
(See color insert.) Blood cell nuclei from pallid sturgeon (*Scaphirhynchus albus*) (a and b) and Gulf sturgeon (*Acipenser oxyrinchus desotoi*) (c and d) that were stained with propidium iodide and analyzed by flow cytometry. Cytograms from Fish 1 (a + b) represent a blood sample that was shipped and handled correctly from field to lab, cytograms from Fish 2 (c + d) represent a sample that was handled improperly (paper label was placed inside the sample tube). All cytogram axes (except counts) display a linear scale of wavelengths.

components, such as RNA, are also available (Vaught 2006). Fixatives such as paraformaldehyde or buffered formalin stop any biological activity, but the possibility for analyses on living cells is then eliminated.

Microbial Organisms

Microbial organisms can affect the generation of accurate biomarker results (Jenkins 2011a), and can contaminate a sample from the environment, or from the normal bacterial flora. When samples are collected using clean and/or sterile equipment and buffers, and when they are transported at the appropriate temperature (~8°C) and in a timely fashion, microbial contamination may not affect the quality of data in biomarker assays. Preventing contamination can be difficult when collecting milt from freshwater fish because bacteria can infiltrate the sample from the skin surface of the fish or the water. Also, some fish species harbor native bacterial flora within the testis that becomes detrimental to spermatozoa once collected (Jenkins and Tiersch 1997). If no preventative steps are taken, bacteria can

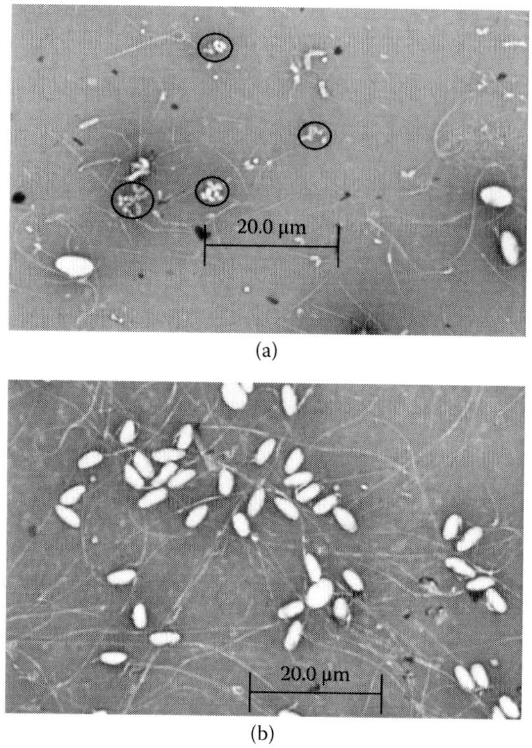

FIGURE 4.2
(See color insert.) Microscopic images of largescale sucker (*Catostomus macrocheilus*) spermatozoa stained with eosin and nigrosin at a total magnification of 1000× (oil immersion). Bacteria (circled) are more prevalent in a than in b, which was relatively free of bacteria.

proliferate, consume oxygen, and produce enzymes that can reduce the motility and quality of otherwise healthy, viable sperm (Moretti et al. 2009). If the fish species of interest is federally listed, then testis removal for milt extraction in the laboratory cannot be performed. Instead, milt is manually expressed into a collection tube containing an extender, such as Hanks' Balanced Salt Solution (HBSS) (Jenkins et al. 2011). Figure 4.2 illustrates a milt sample with a high bacterial load (Figure 4.2a) and one that is essentially free of bacteria (Figure 4.2b). Milt from contaminated collections yields inconsistent and inferior results among individuals within a sampling location, and such samples are excluded from analyses. If the shipping buffers are not prepared with sterile water, microbes will be inherent. To minimize microbial growth, milt and testes can be transported and stored in buffers containing antibiotics that are identified, through preliminary experiments, to be noncytotoxic to the sperm of that species (Saad et al. 1988, Segovia et al. 2000, Jenkins et al. 2014).

Osmolality

Cellular homeostasis is controlled by the plasma membrane, which permits proper chemical functions to be compartmentally maintained. Nutrients, ions, chemicals, and water can cross membranes; therefore, knowledge of the species' normal isosmotic level facilitates preparation of a buffer suitable for maintaining cell quality. Osmometers measure osmolality, or the osmoles (Osm) of solute per kilogram of solvent (osmol/kg or Osm/kg). For blood and milt, generally, the osmolality of the fluid component of either of these tissue types should be measured before buffer preparation. If a buffer's osmolality is too low (hypotonic), too much water will enter the cell causing it to swell; conversely, if the osmolality is too high (hypertonic), too much water will leave the cell causing it to shrink. Shipping milt or testes in slightly hypertonic buffers will maintain cell quality and prevent activation of the spermatozoa (Yang and Tiersch 2009). There are many reasons for shipping milt in slightly hypertonic buffer. Typically, the spermatozoa of fish that spawn in freshwater and practice external fertilization become activated once exposed to low osmolality conditions (Billard and Cosson 1992), such as the waters of their spawning habitat. Activation induces motility, expending ATP reserves and disallowing reliable cell quality assessments later. Because the fish cloaca passes both urine and milt, a milt sample could mix with urine during collection, lowering the osmolality of the collecting buffer, hence the utility of hyperosmotic buffers in such cases (Jenkins et al. 2011). Rinsing the cloaca with the shipment buffer, and waiting until waste has been expelled to collect milt may assist with maintaining osmolality. Adjusting the osmolality of solutions and buffers meant for storing and shipping live tissues of the particular study species before sampling begins will help to maintain cell quality for biomarker analyses.

As part of a breeding program, experiments were performed with razorback sucker (*Xyrauchen texanus*) milt to optimize buffer osmolality when storing or transporting milt (Jenkins et al. 2011). Samples were collected and placed in either iso-osmotic or hyperosmotic HBSS. When the results from both the live cells and cells that had been cryopreserved were compared, the hyperosmotic HBSS was found to better sustain sperm quality parameters (Jenkins et al. 2011). Again, carefully investigating shipping buffer osmolality is beneficial for maintaining parameters reflecting live cell quality.

Experiments Related to Sample Handling

Experiment 1: Temperature

We tested the hypothesis that stored blood would be differentially affected by temperature. Blood from a laboratory-maintained channel catfish (*Ictalurus punctatus*) was collected from the caudal vein with a sterile syringe containing acid citrate dextrose anticoagulant (ACD, final dilution one part ACD to nine parts blood, by volume). This sample was divided into three treatment groups: (1) one part blood: one part ACD (control), (2) one part blood: one part 4% paraformaldehyde, and (3) one part blood: one part Streck Cell Preservative™ (Streck, Omaha, Nebraska). Paraformaldehyde is a polymerized form of formaldehyde that crosslinks protein and DNA, killing the cell. Streck is a live-cell stabilizer designed for maintaining human white blood cells for up to 7 days before analysis in a clinical setting. The solution is purported to maintain cell integrity and antigenic sites without refrigeration. The utility of this stabilizer in the field is three-fold: it is nontoxic, it functions at ambient temperature, and it allows extension of field-to-laboratory transit time for live blood cell samples.

These three treatment groups were divided further into two groups stored at either 8°C or 24°C in 2 mL microcentrifuge tubes for 7 days. To measure the biomarker endpoint of nuclear DNA integrity, an aliquot of catfish blood was removed from each treatment tube daily and analyzed by flow cytometry using a FACSCalibur™ (Becton Dickinson Immunocytometry Systems [BDIS], San Jose, California). Cells were stained with propidium iodide (PI) solution (PI at 0.05 mg/mL, DNase-free RNase A at 1 mg/mL, sodium citrate at 1.12 mg/mL, and 0.1% (v/v) Triton X-100 (Sigma–Aldrich, St. Louis, Missouri)) for 30 minutes at 24°C in the dark (Crissman and Steinkamp 1973). Nuclei were analyzed at 1×10^6 cells/mL at a rate of fewer than 300 per second, and 10,000 events per sample in duplicate were collected using a 1024-channel fluorescence parameter (FL2) measured at 450 nm. Cytograms were generated using CellQuest (BDIS) software upon linear analysis of nuclei with FL2 as the threshold parameter. Aggregates were gated out from analysis using doublet discrimination mode with FL2. The percent nuclei outside the main population (NOMP) (measured with FlowJo software (Tree Star Inc, Ashland, Oregon, USA) was the metric used to determine DNA integrity, for which higher values indicated lower levels of DNA integrity (Jenkins et al. 2011) (see Figure 4.3). Higher NOMP values also reflect higher coefficient of variation (CV) values. Greater DNA dispersion is indicated by higher flow cytometric CV values, a measure of the wideness of the main histogram peak (Shapiro 1995). The NOMP data were expressed as percentages (proportions) and were analyzed using parametric and nonparametric analysis of variance (ANOVA), with multiple comparisons (Tukey's Studentized range test). On the basis of the binomial distribution of proportions, when deviation from normality is great (<30% and >70%), the square root (sqrt) of the proportion is arcsine-transformed to achieve approximate normality (Zar 2010). The NOMP data were transformed in this manner.

Temperature did not significantly affect DNA integrity over time in channel catfish blood cell nuclei stored in the stabilizing solution ($p = .4524$), but paraformaldehyde-stored blood, as well as blood stored in ACD, showed significantly lower levels of intact DNA over time when stored at 24°C ($p = .0002$ and $p = .0014$, respectively) compared to 8°C. Because temperatures can fluctuate from chilled to ambient air temperature during sample packaging and transport from the field, and because the stabilizer helped to maintain nuclear integrity at both temperatures in this experiment, the stabilizer was a useful reagent in sustaining cell condition during transit. Alternatively, samples fixed in the field with paraformaldehyde can be used for analysis of DNA, however the cells will be dead (Jenkins et al. 2011).

Experiment 2: Enzymatic Degradation

The hypothesis tested was that the quality of blood stored in different dilutions of Streck stabilizer over time would change. Blood samples were collected from five laboratory-maintained channel catfish as above and aliquoted into 2 mL microcentrifuge tubes containing three dilutions of the stabilizer: (1) one part blood: one part stabilizer, (2) two parts blood: one part stabilizer, and (3) three parts blood: one part stabilizer. Suspensions were stored at 4°C for 14 days. Blood from four of the catfish was immediately placed into treatment solutions. Blood from a fifth fish was inadvertantly held at 24°C for 5 minutes before adding it to the stabilizer treatments.

To measure biomarker endpoints, aliquots from each treatment were removed on days 0, 1, 2, 3, 4, 7, 9, 10, and 14 and assessed by flow cytometry. Blood was stained and processed as before to determine DNA integrity using NOMP. To determine cell membrane viability, aliquots of catfish blood from each treatment were stained by using a dual fluorescent technique, staining with SYBR 14 and PI (Segovia et al. 2000). SYBR 14 is a membrane-permeant stain (indicating viable cells), but PI enters cells with membranes

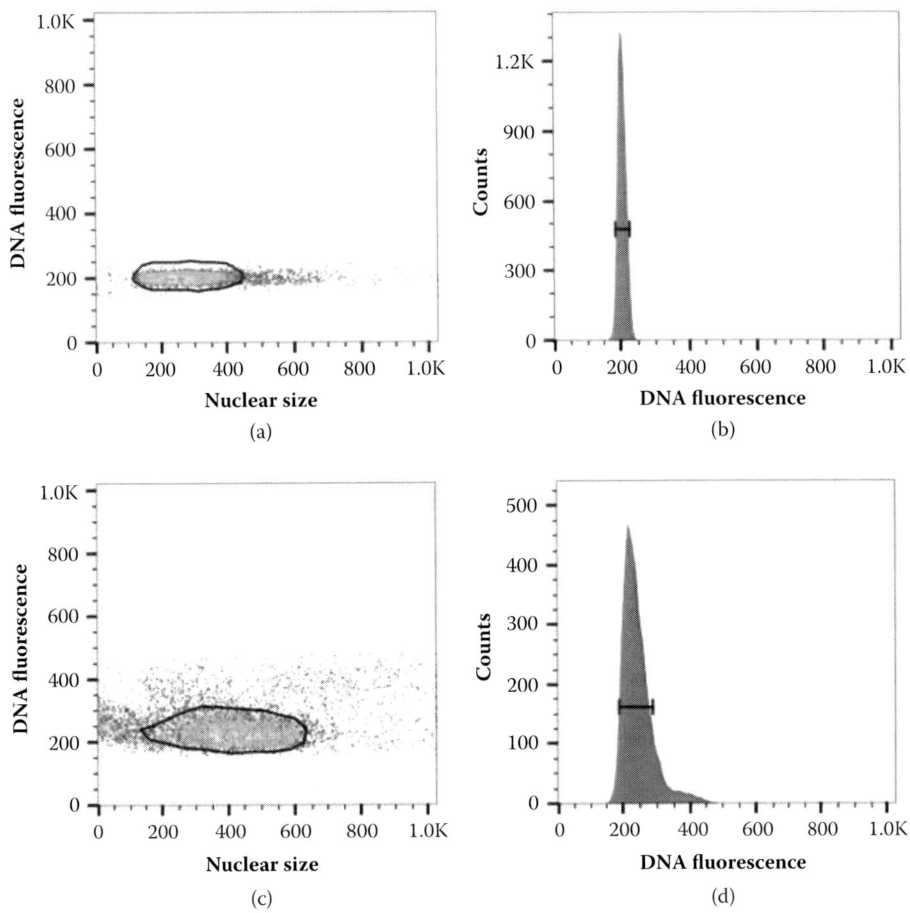

FIGURE 4.3
(See color insert.) Blood from channel catfish (*Ictalurus punctatus*) collected with acid citrate dextrose anticoagulant (ACD) then stained with propidium iodide solution and analyzed immediately (a and b) or after 7 days of storage in ACD at 4°C (c and d). Flow cytometric cytograms reflect nuclear DNA integrity, in which higher percentages of nuclei outside the main population (NOMP) gate (the circles) indicates less intact DNA (c has more nuclei outside the gate than a). The corresponding CV histogram peak widths (bracketed bar) indicate greater levels of DNA dispersion in d as opposed to b.

that are not intact (indicating nonviability). Samples were analyzed in duplicate, collecting 10K data events per replicate, and CellQuest (BDIS) and FlowJo (Tree Star) were used for collection and sample analysis, respectively. NOMP and viability values were arcsin (sqrt) transformed, and analyzed using repeated measures MANOVA, and two-way ANOVA.

DNA integrity and cell membrane viability decreased over time ($p < .0001$) in all samples. The catfish blood sample that was not immediately placed in stabilizer consistently showed the lowest levels of viability (Figure 4.4). The DNA was more intact in the 2:1 and 3:1 blood:stabilizer treatments than the in the 1:1 treatment ($p < .0001$); no significant difference was noted in viability. However at day 14, the viability of catfish blood cells stored in the 3:1 treatment was 85.6% (5.4% SE), as opposed to 83.9% (5.1% SE) and 80.8% (3.5% SE) for 2:1 and 1:1, respectively (Figure 4.4).

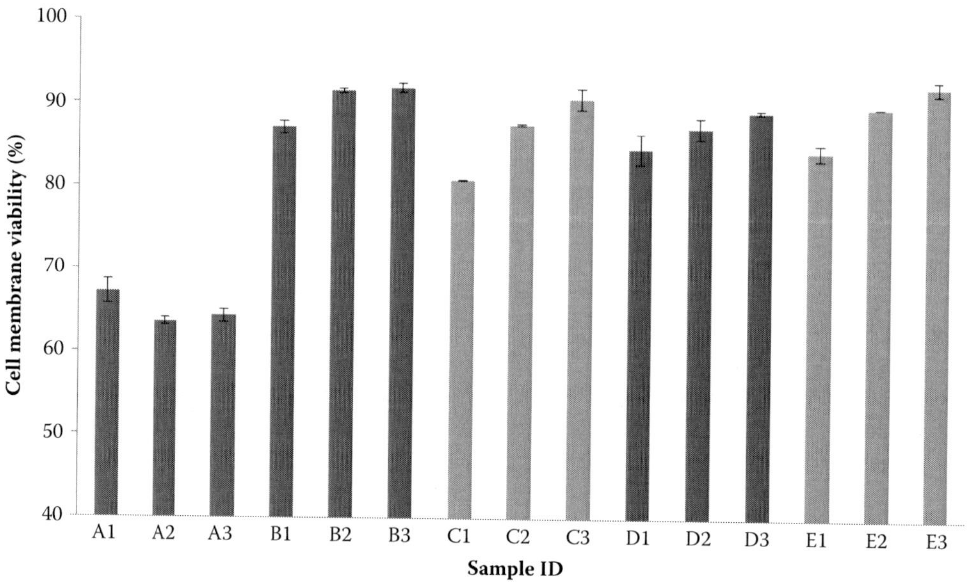

FIGURE 4.4
(See color insert.) Viability of blood cells from channel catfish (*Ictalurus punctatus*) ($n = 5$, Fish A–E) analyzed by flow cytometry after 14 days of storage at 4°C. Sample codes indicate the individual fish (letters A–E) and cell stabilizing treatment (numbers 1–3). Sample treatments were one part blood to one part cell stabilizer (1), two parts blood to one part stabilizer (2), and three parts blood to one part stabilizer (3). Blood from Fish B–E was added to the treatments immediately. Blood from Fish A was added after 5 minutes at 24°C.

In a follow-up experiment, catfish blood was collected as before and stored in either stabilizer (1:1) or ACD (1:1) at 4°C for 7 days. Cell membrane viability was analyzed in duplicate on days 0, 1, 3, 4, and 7, and at the start of the experiment membrane viability was 96.5% (0.1% SE) and 96.3% (0.1% SE) in ACD and stabilizer respectively. By the seventh day, cell viability for the ACD treatment was only 0.6% (0.1% SE), yet it was 87.4% (1.3% SE) in the stabilizer. Thus, the addition of a cell stabilizer extended the number of days that cells remained viable for biomarker analysis.

Experiment 3: Bacteria

Bacterial growth in fish milt and cytotoxic effects on spermatozoa were hypothesized to vary over time depending on the antibiotic solution used during storage (additional details can be found in Jenkins et al. 2014). The antibiotic treatments in HBSS included two concentrations each of streptomycin, penicillin, and a penicillin/streptomycin mixture. Milt was palpated from a laboratory-maintained Koi carp (*Cyprinus carpio*), assessed as 100% motile on visual observation (Jenkins and Tiersch 1997), and then aliquoted into 1 mL of each of the antibiotic treatments in sterile 2 mL microcentrifuge tubes, and one control tube containing only HBSS (7 total). These treatments were stored at 4°C for 6 days, and periodically analyzed for DNA integrity, mitochondrial function, and percent visual motility. DNA integrity differed across treatments ($p = .0014$), with the 100 µg/mL Streptomycin solution showing the highest level of intact DNA. Therefore, this latter treatment was chosen for testis shipments from the field.

Lessons Learned

Because cellular and molecular biomarkers can reflect physiological effects in fish before the effects are noticed at the population, individual, or tissue level, adding laboratory analyses of biomarkers can enhance ecological research. Environmental exposures to chemical factors can alter gene expression, causing epigenetic alterations that can be inherited (Jirtle and Skinner 2007). Biomarkers can be used to detect such epigenetic alterations to the genome, but good quality samples are a prerequisite.

Cell condition can be affected not only by high temperatures that increase decomposition but also by extremely low temperatures that freeze cells and rupture cell membranes. The introduction of foreign materials during sampling is to be prevented, so maintaining debris-free sample collection is a proactive approach. Cells can be affected by inherent enzymes released from lysosomal compartments, but solutions such as protease inhibitors can be added to cell suspensions to deactivate these proteolytic enzymes and to stabilize cells. Microbial presence in samples can influence cell quality, but contamination can be reduced by using asceptic sampling techniques and antibiotics. In addition, both blood and spermatozoal quality is dependent on solution osmolality, which can be proactively adjusted. Establishing clear communications between field and laboratory crews helps ensure that samples reach the laboratory in the best condition possible. Field samples satisfactorily transported have a better chance of yielding accurate results in the laboratory, thus reflecting cell and molecular responses to environmental effects.

Acknowledgments

We thank Meghan Holder and Payton Dupré for technical assistance in the laboratory, and Tiffany Smoak for help in formatting this document. Also, we thank Rassa O. Draugelis-Dale for help with statistical analyses. Finally, we thank Fiona Hollinshead of Matamata Veterinary Services, New Zealand, and Diana Papoulias and Rebecca Howard from USGS for reviewing this manuscript. Any use of trade, firm, or product names is for descriptive purposes only and does not imply endorsement by the U.S. Government.

References

Aas, E., T. Baussant, L. Balk et al. 2000. PAH metabolites in bile, cytochrome P4501A and DNA adducts as environmental risk parameters for chronic oil exposure: A laboratory experiment with Atlantic cod. *Aquatic Toxicology* 51:241–258.

Alber, P. H. 1995. Petroleum and individual polycyclic aromatic hydrocarbons. In *Handbook of Ecotoxicology*, eds. D. J. Hoffman, B. A. Rattner, G. A. Burton, and J. Cairns, 330–355. Boca Raton, FL: Lewis Publishers.

Bayne, B. L., D. A. Brown, K. Burns et al. 1985. *The Effects of Stress and Pollution on Marine Animals.* New York: Praeger.

Billard, R. and M. P. Cosson. 1992. Some problems related to the assessment of sperm motility in freshwater fish. *Journal of Experimental Zoology* 261:122–131.

Crissman, H. A., and J. A. Steinkamp. 1973. Rapid simultaneous measurement of DNA, protein and cell volume in single cells from large mammalian cell populations. *Journal of Cell Biology* 59:766–771.

Dallas, C. E. and D. L. Evans. 1990. Flow cytometry in toxicity analysis. *Nature* 345:557–558.

Depledge, M. H. 1996. Genetic ecotoxicology: An overview. *Journal of Experimental Marine Biology and Ecology* 200:57–66.

Eisler, R. 1999. Polycyclic aromatic hydrocarbon hazards to fish, wildlife, and invertebrates: A synoptic review. U.S. Fish and Wildlife Service, Biological Report 85 (1.11).

Fossi, M. C., and C. Leonzio. 1994. *Nondestructive Biomarkers in Vertebrates*. Boca Raton, FL: CRC Press.

Hobson, K. A., H. L. Gibbs, and M. L. Gloutney. 1997. Preservation of blood and tissue samples for stable-carbon and stable-nitrogen isotope analysis. *Canadian Journal of Zoology* 75:1720–1723.

Holland, N. T., M. T. Smith, B. Escanazi et al. 2003. Biological sample collection and processing for molecular epidemiology studies. *Mutation Research* 543:217–234.

Jenkins, J. A. 2011a. Infectious disease and quality assurance considerations for the transfer of cryopreserved fish gametes. In *Cryopreservation in Aquatic Species*, eds. T. R. Tiersch, and C. C. Green, 939–959. Baton Rouge, LA: World Aquaculture Society.

Jenkins, J. A. 2011b. Male germplasm in relation to environmental conditions: Synoptic focus on DNA. In *Cryopreservation in Aquatic Species*, eds. T. R. Tiersch, and C. C. Green, 227–239. Baton Rouge, LA: World Aquaculture Society.

Jenkins, J. A., B. E. Eilts, A. M. Gautreau et al. 2011. Sperm quality assessments for endangered razorback suckers (*Xyrauchen texanus*). *Reproduction* 141:55–65.

Jenkins, J. A., H. M. Olivier, R. O. Draugelis-Dale et al. 2014. Assessing reproductive and endocrine parameters in male largescale suckers (*Catostomus macrocheilus*) along a contaminants gradient in the Lower Columbia River, USA. *Science of the Total Environment*, 484:365–378.

Jenkins, J. A. and T. R. Tiersch. 1997. A preliminary bacteriological study of refrigerated channel catfish sperm. *Journal of the World Aquaculture Society* 28:282–288.

Jewett, S. C., T. A. Dean, B. R. Wooden et al. 2002. Exposure to hydrocarbons 10 years after the *Exxon Valdez* oil spill: Evidence from cytochrome P4501A expression and biliary FACs in nearshore dimersal fishes. *Marine Environmental Research* 54:21–48.

Jirtle, R. L. and M. K. Skinner. 2007. Environmental epigenomics and disease susceptibility. *Nature Reviews Genetics* 8:253–262.

Larno, V., J. Laroche, S. Launey et al. 2001. Responses of chub (*Leuciscus cephalus*) populations to chemical stress, assessed by genetic markers, DNA damage and cytochrome P4501A induction. *Ecotoxicology* 10:145–158.

McCain, B. B., H. O. Hodgins, W. D. Gronlund et al. 1978. Bioavailability of crude oil from experimentally oiled sediments to English sole (*Parophrys vetulus*), and pathological consequences. *Journal of the Fisheries Research Board of Canada* 35:657–664.

Moretti, E., S. Capitani, N. Figura et al. 2009. The presence of bacteria species in semen and sperm quality. *Journal of Assisted Reproduction and Genetics* 26:47–56.

Neff, J. M. 1985. Polycyclic aromatic hydrocarbons. In *Fundamentals of Aquatic Toxicology*, eds. G. M. Rand and S. R. Petrocilli, 416–454. New York: Hemisphere.

NRC: Committee on Biological Markers of the National Research Council. 1987. Biological markers in environmental health research. *Environmental Health Perspectives* 74:3–9.

Rice, C. D., and M. R. Arkoosh. 2002. Immunological indicators of environmental stress and disease susceptibility in fishes. In *Biological Indicators of Aquatic Ecosystem Stress*, ed. S. M. Adams, 187–220. Bethesda, MD: American Fisheries Society.

Rieseberg, M., C. Kasper, K. F. Reardon et al. 2001. Flow cytometry in biotechnology. *Applied Microbiology and Biotechnology*. 56:350–360.

Saad, A., R. Billard, M. C. Theron et al. 1988. Short-term preservation of carp (*Cyprinus carpio*) semen. *Aquaculture*. 71:133–150.

Schlenk, D., R. Handy, S. Steinert et al. 2008. Biomarkers. In *The Toxicology of Fishes*, eds. R. T. Di Giulio and D. E. Hinton, 684–731. Boca Raton, FL: CRC Press, Taylor & Francis Group.

Schmitt, C. J. and G. M. Dethloff, eds. 2000. Biomonitoring of Environmental Status and Trends (BEST) Program: Selected methods for monitoring chemical contaminants and their effects in aquatic ecosystems. U.S. Geological Survey, Biological Resources Division, Columbia, MO: Information and Technology Report USGS/BRD-2000-0005. 81 pp.

Segovia, M., J. A. Jenkins, C. Paniagua-Chavez, et al. 2000. Flow cytometric evaluation of antibiotic effects on viability and mitochondrial function of stored sperm of Nile tilapia. *Theriogenology* 53:1489-1499.

Seutin, G., B. N. White, and P. T. Boag. 1991. Preservation of avian blood and tissue samples for DNA analyses. *Canadian Journal of Zoology* 69:82–90.

Shapiro, H. M. 1995. *Practical Flow Cytometry*. New York: Wiley-Liss.

Shen, H. and C. Ong. 2001. Detection of oxidative DNA damage in human sperm and its association with sperm function and male infertility. In *Bioassays for Oxidative Stress Status*, ed. W. A. Pryor, 89–96. B. V., Amsterdam: Elsevier Science.

Tiersch, T. R. 2011. Shipping of Refrigerated Samples. In *Cryopreservation in Aquatic Species, 2nd Edition*, eds. T. R. Tiersch and C. C. Green, 690–691. Baton Rouge, LA: World Aquaculture Society.

Tiersch, T. R., C. A. Goudie, and G. J. Carmichael. 1994. Cryopreservation of channel catfish sperm: Storage in cryoprotectants, fertilization trials, and growth of channel catfish produced with cryopreserved sperm. *Transactions of the American Fisheries Society* 123:580–586.

UFR (Use of Fishes in Research) Committee. 2013. *Guidelines for the use of fishes in research*. American Fisheries Society, Bethesda, MD. http://fisheries.org/docs/policy_useoffishes.pdf [last accessed 9.7.13].

U.S. Department of Health and Human Services PHS, National Toxicology Program. 2002. *Report on Carcinogens, Tenth Edition: Carcinogen Profiles*. Vol 1. Research Triangle Park, NC.

van der Oost, R., J. Beyer, and N. P. E. Vermeulen. 2003. Fish bioaccumulation and biomarkers in environmental risk assessment: A review. *Environmental Toxicology and Pharmacology* 13:57–149.

van der Oost, R., F. J. van Schooten, F. Ariese et al. 1994. Bioaccumulation, biotransformation and DNA binding of PAHs in feral eel (*Anguilla anguilla*) exposed to polluted sediments: A field survey. *Environmental Toxicology and Chemistry* 13:859–870.

van der Oost, R., E. Vindimian, P. J. van den Brink et al. 1997. Biomonitoring aquatic pollution with feral eel (*Anguilla anguilla*) III. Statistical analyses of relationships between contaminant exposure and biomarkers. *Aquatic Toxicology* 39:45–75.

Vaught, J. B. 2006. Blood collection, shipment, processing, and storage. *Cancer Epidemiology, Biomarkers, and Prevention* 15:1582–1584.

Whitehead, A., B. Dubansky, C. Bodinier et al. 2012. Genomic and physiological footprint of the *Deepwater Horizon* oil spill on resident marsh fishes. *Proceedings of the National. Academy of Sciences of the United States of America* 109:1–5.

Whyte, J. J., R. E. Jung, C. J. Schmitt et al. 2000. *Ethoxyresorufin-O-deethylase (EROD) Activity in Fish as a Biomarker of Chemical Exposure*. Vol 30. Boca Raton, FL: CRC Press.

Williams, J. R. and J. B. Little. 1974. Association of mammalian cell death with specific endonucleolytic degradation of DNA. *Nature* 252:754–755.

Yang, H. and T. R. Tiersch. 2009. Sperm motility initiation and duration in a euryhaline fish, medaka (*Oryzias latipes*). *Theriogenology* 72:386–392.

Zar, J. H. 2010. *Biostatistical Analysis*. New Jersey: Prentice Hall.

Section II

Oil Impacts to Physical Habitat in Coastal Ecosystems

5

Early Review of Potential Impacts of the Deepwater Horizon *Oil Spill on Gulf of Mexico Wetlands and Their Associated Fisheries*

Matthew E. Andersen

CONTENTS

Introduction

Coastal marsh wetlands and associated estuaries are highly productive habitats found around the world where rivers and streams meet the sea. Wetlands provide many ecosystem services including filtering sediments and nutrients, dissipating energy from storms as they come ashore, atmospheric carbon removal, and providing important marine life habitat and nutrition (Engle 2011; Needles et al. 2013). The 2010 *Deepwater Horizon* (DWH) oil release impacted wetland estuaries along the northern Gulf of Mexico (GOM) coast with the heaviest oiling occurring in Louisiana marshes (National Commission 2011; Sammarco et al. 2013).

Natural wetland habitats, in particular salt marshes, serve as integral components of coastal ecosystems. Wetlands consist of a complex mix of interacting biotic and abiotic components, so a comprehensive evaluation of an anthropogenic perturbation to wetlands should consider multiple factors. Wetlands and the organisms that inhabit them experience population increases and decreases in response to various physical and biotic drivers over long time periods depending on the species examined. Because this chapter is being prepared in 2013, the full story of the DWH event impacts and habitat and species responses cannot be fully known. It is also difficult to separate any DWH impacts from other contemporaneous impacts. For example, coastal Louisiana wetlands are being lost to open water quickly but large-scale changes are generally being observed on a decadal time scale (Couvillion et al. 2011). Impacts to fisheries may be immediate but may also manifest themselves as long-term population level changes (Peterson et al. 2003). Given that petrochemicals and habitat impacts can persist in the environment for many years following an oil spill (Gundlach et al. 1983; Peterson et al. 2003; Short et al. 2004; Castege et al. 2007; Bejarano and Michel 2010; Mendelssohn et al. 2012; Zuijdgeest and Huettel 2012; Perrons 2013; Zhou et al. 2013), and that substrate oiling can harm wetland plants by causing lethal reductions in gas exchange rates (DeLaune and Wright 2011; Wu et al. 2012) and physical breakage (Pezeshki et al. 2001; Lin and Mendelssohn 2012), it is important to continue monitoring affected habitats to evaluate how the largest ever unintentional oil spill in the marine environment (Crone and Tolstoy 2010; Pioch et al. 2010; National Commission 2011) fully impacted the northern GOM ecosystem. The 1991 oil spill in the Arabian Gulf was larger than the DWH spill, but the former was an intentional release (Bejarano and Michel 2010). Thus, a rigorous evaluation of the biological impacts of the DWH oil released into the GOM and washed ashore should consider at a minimum (1) spatial and temporal distribution of the oil and dispersant that came ashore onto wetlands, (2) how harmful this oil and associated dispersant is to wetland plants and associated biota, and (3) how important these habitats are for marine life. This chapter uses some of the literature available in August 2013 to initially address these considerations. The author's brief review of initially available studies allows for personal insights informed by this literature and may provide initial guidance for ecosystem restoration. In particular, the author focused on initial review of studies on wetland plants and studies of fisheries that rely on wetland habitat.

Study Results from Early Available Literature

Oil Provenance

Naturally occurring oil seeps on the coast and offshore in the GOM have been known for centuries and are the basis of an economically viable extraction industry that began in the United States in the 1930s (Burroughs 2011; National Commission 2011). Because of the natural presence of oil in the GOM an evaluation of impacts to wetlands begins with determining whether or not oil or its weathered compounds from the DWH failure approached or inundated the shore. Geochemical analysis of samples from multiple sites in coastal Texas, Louisiana, Mississippi, Alabama, and Florida determined that oil-derived compounds from the Macondo-1 well came ashore in all GOM states (Figure 5.1) (Rosenbauer et al. 2010, 2011; Wong et al. 2011; Sammarco et al. 2013). Rosenbauer et al. (2010, 2011) and Wong et al. (2011) analyzed sediments and tarballs from 49 sites taken from Texas to Florida. Sammarco et al. (2013) analyzed a total of 227 samples from Texas to Florida, which included samples of biota (n = 32), seafood (n = 36), sediment (n = 74), sediment and water (n = 18), and seawater (n = 67). Oiling of the Mississippi study sites of Wu et al. (2012) was consistent with the timing of the initial DWH release and subsequent meteorologic events. Petrochemical materials washed up on the Alabama shoreline in 2011

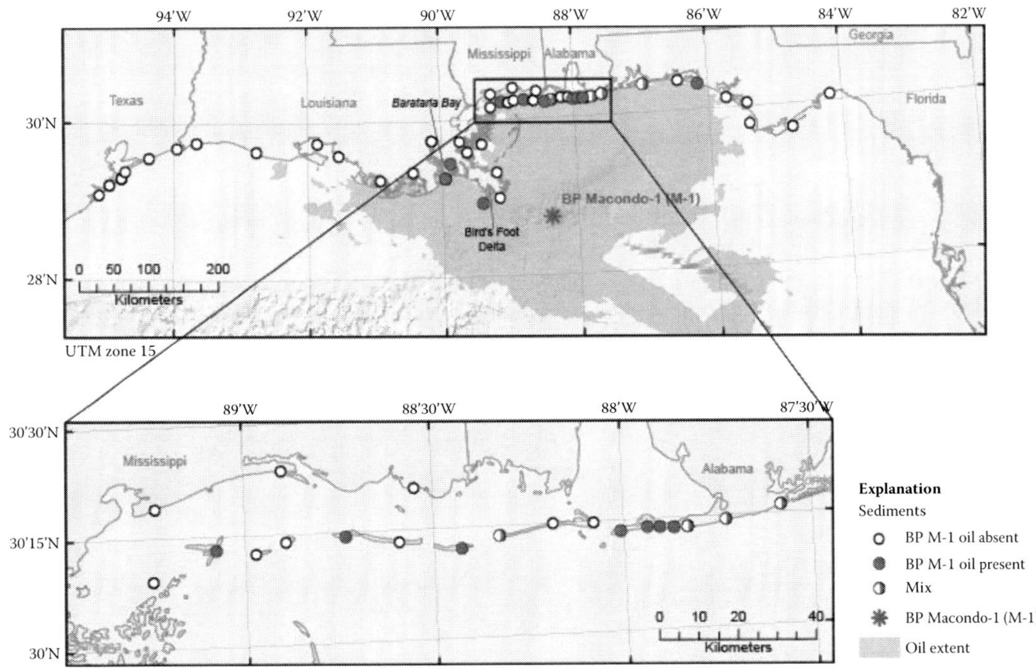

FIGURE 5.1
Macondo-1 well oil released by the failure of the *Deepwater Horizon* came ashore on the northern Gulf of Mexico coast in Louisiana, Mississippi, Alabama, and Florida. In Louisiana, the Macondo-1 well oil was found on the Bird's Foot Delta of the Mississippi River and also in Barataria Bay. (Modified from Wong et al., *Macondo-1 Well Oil in Sediment and Tarballs from the Northern Gulf of Mexico Shoreline,* U.S. Geological Survey Open-File Report 2011-1164, Poster, Reston, VA, 2011.)

and 2012 were subsequently shown to originate from DWH (Mulabagal et al. 2013). In Louisiana, where most of the moderately to heavily oiled marshes were located (National Commission 2011; Moody et al. 2013), Macondo-1 oil was found in Barataria Bay and on the Bird's Foot Delta terminus of the Mississippi River (Kokaly et al. 2011; Wong et al. 2011). The geochemical results and the timing of the contemporary northern GOM field studies strongly suggest that coastal wetlands were exposed to the oil released from the Macondo-1 well by DWH and therefore a discussion of potential impacts is warranted. The extent of oil dispersant was also investigated with other methods as described in the Aerial Imagery and Site Investigations sections of this chapter.

Aerial Imagery

The northern GOM coast is a dynamic habitat affected by multiple drivers. In some locations it is also geographically complex, particularly in Louisiana, between the mouths of the Sabine and Pearl Rivers (Couvillion et al. 2011). After oil and dispersants are washed on to this coast, wind, waves, and tides interact with freshwater inflows and geographic complexity to make specific tracking of pollutants challenging. The absence of topographic relief along the coast means that observers in boats or on foot can only see for limited distances, especially in marshes. Aerial imagery, then, is a potentially important tool for assessing broad geographic areas in a relatively short amount of time. The volume of the DWH release raised the potential for negative impacts to wetlands and associated biota including dispersal inland, so multiple researchers deployed using multiple aerial imagery tools (Ramsey et al. 2011; Mishra et al. 2012; Kokaly et al. 2013) for potential application in assessing the extent of dispersal of oil and dispersants in addition to direct inspection of shorelines (Kokaly et al. 2011; Mendelssohn et al. 2012; Wu et al. 2012; Biber et al., Chapter 7, this volume). Where it can be shown to be accurate, aerial imagery could be a valuable tool for supporting rapid assessment of restoration at broad scales and so should be considered in support of restoration and monitoring plans.

Investigators studying the use of Uninhabited Aerial Vehicles Synthetic Aperture Radar (UAVSAR) (Leifer et al. 2012) began overflights of the Louisiana coastline in 2009 including Barataria Bay (Ramsey et al. 2011) (Figure 5.2). This created baseline images of the habitat before any possible impacts by the oil release. UAVSAR provides many advantages over traditional aerial photography, especially the ability to produce polarimetric signature of the backscatter images from terrestrial habitats even when cloud cover is present. When the DWH failure occurred in 2010, the response strategy for UAVSAR collections incorporated many of the same flight lines that had been used in 2009. Investigators were able to compare backscatter from wetlands before and after the oil spill, especially within Barataria Bay and the Mississippi River Delta. Sampling in June was selected in all years to avoid the potential for phenological changes, active growth or decomposition which could potentially confound observations in the spring and fall, respectively. There were high flooding flows on the Mississippi River in 2009, which potentially carried fresh water

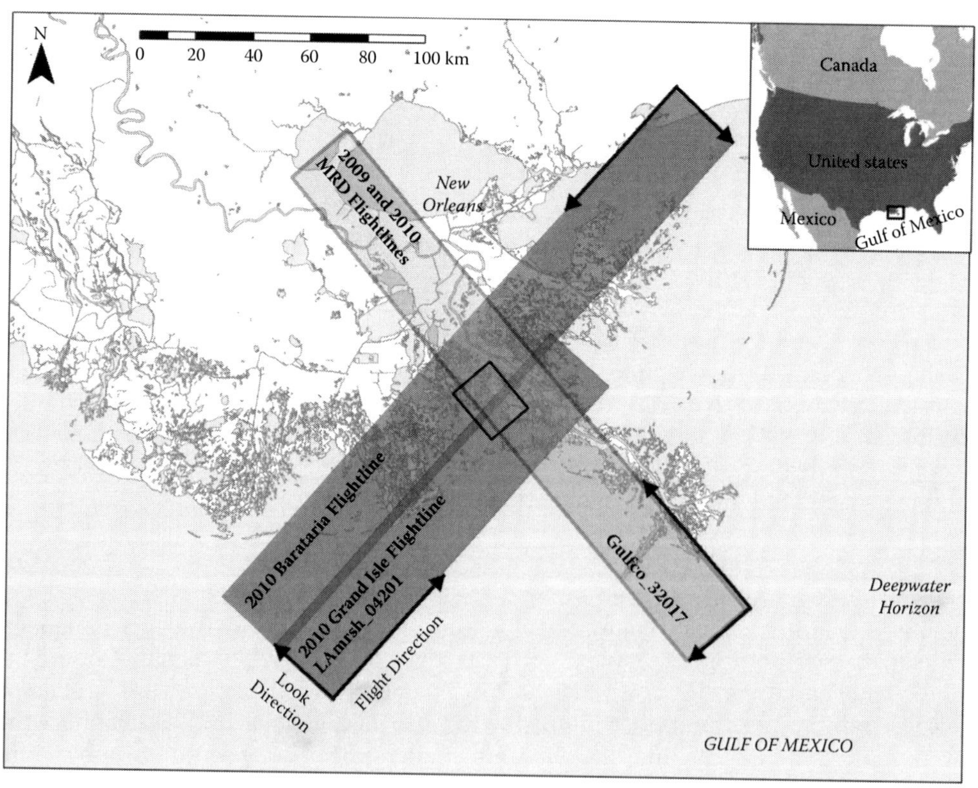

FIGURE 5.2
Uninhabited Aerial Vehicle Synthetic Aperture Radar used by Ramsey et al. (2011) in their study of the *Deepwater Horizon* oil spill. The black box in the region of overlapping data in Barataria Bay was the focus of their study. (Used with permission of the senior author.)

and sediments into the habitats investigated by Ramsey et al. (2011). However, the presence of oil on the shorelines adjacent to the areas of altered backscatter argues that oil was the cause of backscatter changes observed (Ramsey et al. 2011). These results and the methods used to obtain them are still being developed, and include uncertainties that are still being fully quantified and are subject to additional investigation, but still they may offer useful insights and identify locations where additional investigation is warranted.

Ramsey et al. (2011) found that the polarimetric signature of the backscatter had changed when comparing 2009 to 2010 data in select locations, suggesting that oil had come ashore into wetland vegetation. Ground-based visual observations confirmed that oil was observed washed onto a few near shoreline habitats up to 4 m from open water (Kokaly et al. 2011; Mendelssohn et al. 2012; Silliman et al. 2012). Some habitats from which remote images were collected were rendered inaccessible by dense vegetation and unstable substrates, so not all habitats could be directly observed. Where heavy oiling had caused marsh structural damage, however, Ramsey et al. (2011) observed that dramatic changes in backscatter had also occurred. Kokaly et al. (2013) found that in areas of heavy oiling, the average extent of oil into the marshes was 11 m with a documented maximum extent observed of 21 m. Biber et al. (Chapter 7, this volume) also observed oiling 10 m inshore from the water's edge. Ramsey et al. (2011) observed that the same type of backscatter changes extended locally well into the nearshore marsh in some locations up to 40 m from shorelines, suggesting that oil had been brought far ashore by tidal action beyond where it could be observed by crews on boats. These findings are contrary to the expectations of Lin and Mendelssohn (2012) and Mendelssohn et al. (2012), who stated that oiling was limited to shoreline areas and did not extend into marshes. On the basis of limited direct observations, the nearshore marshes were associated with the occurrence of moderate concentrations of oil in the marsh subcanopy, although without the structural damage observed at shorelines. Where backscatter changes were observed beyond 40 m from open water and the nearshore marshes, it is most likely that these areas experienced lower oil exposures compared to shorelines and the nearshore marshes (Ramsey et al. 2011). Limited direct observations on the water and from helicopters were consistent with the assertion that oiling had moved farther inland beyond zones of structural changes to marsh vegetation, leaving petrochemical residues on soils, but investigations of the extent of oil distribution are continuing as of this writing.

Site Investigations

Kokaly et al. (2011) examined Louisiana wetland sites around the Bird's Foot Delta of the Mississippi River and also in Barataria Bay in the summer of 2010 and found extensive shoreline oiling in these locations as did Ramsey et al. (2011), Mendelssohn et al. (2012), and Silliman et al. (2012). Kokaly et al. (2011) documented extensive damage to *Spartina alterniflora* and *Juncus romerianus* along shorelines but found that *Phragmites australis* plants were more robust resisting physical and chemical insults (Figure 5.3). Mendelssohn et al. (2012) noted that the relatively light South Louisiana crude released from the Macondo-1 well allowed some plants that it contacted to survive and eventually produce new growth consistent with the responses observed in earlier years by Pezeshki and DeLaune (1993) and DeLaune and Wright (2011), although Pezeshki et al. (2000) noted that there are many variables affecting marsh vegetation response to oiling, including type of hydrocarbons, plants species, and season of impact. In addition to these variables, Biber et al. (Chapter 7, this volume) also noted that the amount of tidal energy dissipated had a measurable effect on oiling where exposed beaches and marshes recovered more quickly than protected shorelines that receive less tidal impact. Wu et al. (2012) documented how increases

FIGURE 5.3
(See color insert.) DWO-3-BAT-08 photo 3 taken on August 13, 2010. Site vegetation included *Spartina alterniflora* and *Distichlis spicata*. Oiled vegetation and oil-damaged canopy were observed at this site. (From Kokaly et al., *Shoreline Surveys of Oil-Impacted Marsh in Southern Louisiana July to August 2010*, U.S. Geological Survey Open-File Report 2011-1022, Reston, VA, 2011.)

in either or both air temperature and photosynthetically active radiation will accelerate recovery of wetland plants. Much of the mortality caused to *S. alterniflora* and *J. romerianus* was due to oil building up on shoots raising their center of gravity and leading to physical breakage that caused plants to fall over and die. On meter-wide shoreline platforms, plant mortality accelerated land loss in many locations (Mendelssohn et al. 2012; Silliman et al. 2012) consequently increasing the risk of land loss in the future (Kokaly et al. 2013).

Literature

Oil on Wetlands

Although estimates vary (Crone and Tolstoy 2010), the initial federal government estimate concludes that 4.9 million barrels of oil were released by the failure of the DWH and that about one-quarter of that amount came ashore mostly in Louisiana (National Commission 2011). This resulted in more than 1083 km of U.S. Gulf shoreline being oiled with 217 km of the shoreline were moderately to heavily oiled. About 33 km of coastline in Mississippi, Alabama, and Florida were moderately to heavily oiled; Louisiana shorelines were most heavily impacted (National Commission 2011; Moody et al. 2013). This is highly productive habitat of great importance to commercial and recreational fishing interests in the GOM (Boesch and Turner 1984; Cowan et al. 2008) and it is being lost at a relatively rapid rate

(Couvillion et al. 2011) making pollution impacts, such as the 2010 oil release, very important for long-term habitat sustainability (Kokaly et al. 2013).

Lin and Mendelssohn (2012) concluded that because chemical dispersant had been used at low concentrations of around 1:200 relative to oil, its application to oil on the surface and at the wellhead was not of environmental concern. Therefore, they did not study dispersant impacts further but even at low concentrations relative to the oil the volumes of oil released were still large (Zuijdgeest and Huettel 2012). Studies have concluded that the dispersant will lengthen the GOM habitat recovery time following the DWH incident (Fodrie and Heck 2011; Zuijdgeest and Huettel 2012; Campo et al. 2013). The dispersant used, Corexit 9500, has been shown to be toxic to a range of marine animals (Almeda et al. 2012; Goodbody-Gringley et al. 2012) and has the potential to remain in the GOM food web. Zhou et al. (2013) conclude that the dispersant will keep petrochemical products in the water column for at least many months. The dispersant may also have accelerated organismal uptake of petroleum by-products (Graham et al. 2010), although Fodrie and Heck (2011) concluded that the oil release did not immediately affect fish trawl catch per unit effort from the Chandeleur Islands of Louisiana to peninsular Florida north and east of the most heavily oiled marshes. They also noted that large quantities of dispersant applied at the leaking Macando-1 wellhead kept oil degradation products at least temporarily in the water column and more than 1.6 km below the ocean's surface; how those products will behave over time in open water and as they come ashore is not known. With regard to the impacts of dispersant on the coast, Zuijdgeest and Huettel (2012) observed that the polycyclic aromatic hydrocarbons (PAHs), compounds of high toxicity to fishes (Dubansky et al. 2013), were made more mobile by the dispersant. The increased mobility of PAHs exposed to dispersant results in the compounds moving deeper into sands, allowing them to persist years longer than if they had not been treated with dispersant (Zuijdgeest and Huettel 2012). Dispersant application may have limited or no impact on oiled wetland plant community long-term persistence, though if the application breaks plant shoots, reduced marsh survival may be observed (Pezeshki et al. 2001). Another important factor in the activity of the dispersant and oil at great depth is temperature. Campo et al. (2013) documented significantly slower degradation of dispersant at 5°C as compared to 25°C. The environmental persistence of dispersant and potential mechanical breakage suggests careful consideration of potential outcomes in advance of application in future incidents.

Oiled marshes in Louisiana were significantly and immediately impacted by the DWH oil release (Kokaly et al. 2011; Lin and Mendelssohn 2012; Silliman et al. 2012). Oil can have chemical and/or physical impacts on wetland vegetation. Oil from different sources has variable physical properties that contribute to how it will impact vegetation. Heavier crudes will adhere to and topple vegetation, while lighter crudes are less likely to physically pull plants down but are more likely to be capable of entering plant stomata and causing direct toxicity and mortality in addition to interfering with gas exchange and photosynthesis (Pezeshki et al. 2000; Lin and Mendelssohn 2012; Mendelssohn et al. 2012; Silliman et al. 2012). Plants are more likely to withstand light oiling especially if the oil can be washed away. Persistent oiling will begin to interfere with photosynthesis and gas exchange, thereby blocking plant growth and affecting survival (Wu et al. 2012; Biber et al., Chapter 7, this volume). Mature plants are likely more resistant than young plants to oiling (Pezeshki et al. 2000), but if mature plants are killed, there is evidence that young plants can regenerate from root masses provided soils have not been contaminated (DeLaune and Wright 2011; Lin and Mendelssohn 2012). Therefore, it is important to monitor the condition of the physical, chemical, and biological components of soils and belowground biomass following exposure to oil.

Gulf Coast Wetlands and Fish

The State of Louisiana has a large area of coastal wetlands and estuaries, but the wetland habitats are fragile and are being lost, changing over to open water. Coastal Louisiana wetlands converted to open water at a rate of about 43 km^2/year between 1985 and 2010 (Couvillion et al. 2011). Gulf Coast wetland losses are about 80% of the wetland loss observed in the continental United States (Cowan et al. 2008). Wetland losses continue to occur even as scientists and managers strive to better understand the specific roles wetland habitats play in the life histories of fishes, other wetland-dependent organisms, and ecosystem services.

Wetlands are important habitat for multiple life stages of invertebrates and fish (Vidal-Hernandez and Pauly 2004; Rountree and Able 2007; Quan et al. 2011; Boys et al. 2012; Corrêa et al. 2012; Green et al. 2012). Northern GOM wetlands associated with the current and historic mouths of the Mississippi River–like wetlands found elsewhere around the world support diverse assemblages of marine life out of proportion to the limited distribution of these habitats (Chesney et al. 2000; Gelwick et al. 2001; Boswell et al. 2010) including economically important fish and invertebrates (Boesch and Turner 1984; Cowan et al. 2008). The GOM coastal wetland marshes, like those elsewhere, are especially important for early life stages of fishes (Boesch and Turner 1984; Cowan et al. 2008; Moody et al. 2013). Fishes of the northern GOM may be able to move offshore and utilize soft-bottomed, estuary-like habitats along the continental shelf, as for example at the Mississippi River outflow when available estuaries are reduced (Cowan et al. 2008). However, in general, they are heavily dependent on coastal wetlands at some stage in their life cycle, especially for feeding, reproduction, and rearing (nurseries).

Discussion and Conclusions

Potential Impacts

Oil released from the DWH failure washed up onto a highly important area of the northern GOM coastline in the United States mostly in Louisiana. In sampled areas, the provenance of the oil on wetlands and beaches in 2010 and later has been established as originating from the Macondo-1 well drilled by DWH (Wong et al. 2011; Mulabagal et al. 2013; Sammarco et al. 2013). Where it came ashore emulsified with seawater and commercial dispersants, it caused mortality in wetland plants. This was most noticeable at the wetland/open water interface of shorelines, but remote sensing revealed that the oil also travelled farther into marsh interiors (Ramsey et al. 2011; Mishra et al. 2012; Kokaly et al. 2013) than could be observed with the naked eye. Developing the capacity to detect oil on substrates within the interior wetland canopy with remote sensing tools was an important technological advance, because it documented previously undetectable oil spill impacts. Although *S. alterniflora* and *J. roemerianus* were negatively affected by the DWH oil (Kokaly et al. 2011; Mendelssohn et al. 2012; Silliman et al. 2012), it is likely that many of these areas are capable of rapid regeneration and recovery especially given that they were exposed to relatively light South Louisiana crude (Smith et al. 1984; DeLaune and Wright 2011; Lin and Mendelssohn 2012; Wu et al. 2012; Biber et al., Chapter 7, this volume). Oil extending into interior northern GOM marshes is of concern because oil on soils and at depth may limit respiration and reduce plant survival and reproduction (DeLaune and Wright 2011).

Because the largest GOM fish catches in the United States are observed in Louisiana (Adams et al. 2012; NOAA 2013b; Petrolia et al. 2013), the commercial fish harvest from the state provides a reference point for evaluating Gulf fishery productivity, and because wetlands support fish reproduction (Vidal-Hernandez and Pauly 2004) these data are an indirect evaluation of habitat health and productivity. Between 1960 and 2010, for example, the annual mean mass of fish harvested commercially was over 1.182 million pounds (over 536,000 metric tons [t]). Commercial fish landings in Louisiana were greatest during the 1980s, peaking at more than 1.942 million pounds (over 880,000 t) landed in 1984. Commercial fish landings in Louisiana were about 1.172 million pounds (over 531,000 t) in 2009, about 1.007 million pounds (over 456,000 t) in 2010, and 1.286 million pounds (over 583,000 t) in 2011. Commercial landings generally increased between 1960 and 1984. They have been lower than the mean for the period since then (NOAA 2013b), although there are many factors that can affect catch in addition to habitat impacts, including harvest pressure. Harvest pressure was reduced in 2010 because authorities closed most commercial and recreation fisheries in response to health concerns (Levy and Gopalakrishnan 2010; NOAA 2013a). Catch reductions are temporally correlated with the loss of coastal Louisiana wetlands that have generally been decreasing throughout the period 1960–2010 (Couvillion et al. 2011). Care must be taken when using catch data, especially of only a single species low in the trophic web, when inferring linkages to ecosystems (de Mutsert et al. 2008), thus the linkage between catch and habitat degradation deserves additional investigation. In their sampling to the north and east of the most heavily oiled marshes, Fodrie and Heck (2011) and Moody et al. (2013) did not observe dramatic declines in trawled fish catch. Their findings may have been limited because they were sampling offshore far from the heavily oiled Louisiana wetlands and in the absence of most of the harvest pressure exerted in a typical year.

The available evidence suggests that direct contact with oil from the DWH failure had a negative impact on both wetlands (Mendelssohn et al. 2012) and the health of individual resident fishes (Whitehead et al. 2011; Dubansky et al 2013). Caveats to this conclusion are based on the potential for rapid recovery in GOM wetlands. Wu et al. (2012) noted that high temperatures and high solar radiation accelerated recovery, and Biber et al. (Chapter 7, this volume) noted that shorelines subjected to more wind and wave energy recovered more quickly than more protected shorelines. As noted earlier, some evidence of low impact to fish populations has been published from sampling sites distant from the heavily oiled Louisiana coast (Fodrie and Heck 2011; Moody et al. 2013). Although coastal fishes likely have adapted to shifting habitat availability in Louisiana (Cowan et al. 2008), no one can say when wetland losses, no matter the cause, may have an irrevocable negative impact on fish production. The trends of both resources as of June 2013 are negative. All available evidence found that oil immediately decreased habitat health (DeLaune and Wright 2011; Silliman et al. 2012; Wu et al. 2012; Biber et al., Chapter 7, this volume) and that dispersant was neutral to negative (Graham et al. 2010; Almeda et al. 2012; Goodbody-Gringley et al. 2012; Zuijdgeest and Huettel 2013), further decreasing habitat quality condition. Where marsh plants were not entirely killed regeneration at the root mass was noted in the field (Lin and Mendelssohn 2012; Silliman et al. 2012) suggesting at least some level of salt marsh resilience. Continued observations will be necessary to evaluate long-term impacts if any. The full description of the environmental impacts of the DWH incident cannot yet be complete, because the available literature reports that Macando-1 oil and dispersant remain in the marine and terrestrial environment. Continued monitoring of multiple natural resources (Castege et al. 2007), including sediment plants and animals, will be necessary to complete the description because environmental impacts of oil releases are events that potentially take years to completely unfold (Peterson et al. 2003; Bejarano and Michel 2010).

The Mississippi River has changed its course and outlet to the GOM at least six times in the last 8000 years (Coleman et al. 1998; Blum and Roberts 2009). Native fish and other marine life appear to have adapted (Cowan et al. 2008), foregoing habitat stability for many other advantages such as complex heterogeneous aquatic habitats offering cover for rearing and feeding, and high nutrients to support a rich and complex trophic web. However, wetland changes by humans began to increase in the twentieth century and continue to the present day. How much anthropogenic change can the habitats and the animals they support tolerate? The fish landings to date suggest that the breaking point has not yet been reached (Cowan et al. 2008), but because the habitat loss continues (Couvillion et al. 2011) it is sobering to consider that deteriorating coastal marshes may eventually have irrevocable negative impacts on fish and invertebrate populations. Because ecosystems are complex, consisting of physical and biological elements, all experiencing a range of forcing factors, it is hard to predict how many cumulative insults the habitats may absorb before dramatic negative events may occur (Hilborn and Walters 1981). Chronic effects may take years or even decades to be observable (Peterson et al. 2003; Perrons 2013). Ecosystem services and values associated with wetlands are at serious risk. Bjerstedt (2011) suggests that impacts to wetlands from acute events such as oil spills are dwarfed by larger continental-scale chronic processes, such as subsidence and sea-level rise. As additional wetland loss occurs due to subsidence and sea-level rise (Twilley et al. 2001), wetlands and the important ecosystem services they provide humans (Needles et al. 2013) and the natural environment should be conserved and restored where feasible. Because of the disproportionate benefits offered by wetlands such as ecosystem services and species conservation and unknown long-term responses of wetlands to multiple stressors (Silliman et al. 2012), current efforts by agencies to conserve coastal wetland habitats (State of Louisiana 2012) are vital for supporting and maintaining these benefits.

There are known factors that support recovery of the affected wetlands and known factors that will accelerate their decline. When compared, there are direct conflicts between the assumptions of recovery or decline based on laboratory and field observations and multiple compounding physical and biological factors. Which factor or factors will ultimately be most influential on wetland survivorship cannot be known at this time but we know many of the factors that must be monitored. To illustrate the conflicting nature of these factors some of them are presented as contrasting pairs in Table 5.1.

Despite the limited geographical extent of direct oiling in the northern GOM, this event was only the most recent perturbation to a heavily stressed and disproportionately important environment. Therefore, it would be prudent to proceed with both long-term monitoring and restoration actions that possess a reasonable probability of success.

Restoration and Monitoring

The extent of oiled wetlands has been documented initially and these estimates continue to be refined but Oiled locations, including interior marshes, need to remain under surveillance. Given that this was the second worst marine oil spill known (Bejarano and Michel 2010), the habitat observations associated with its aftermath provide valuable parameterization of habitat response to such a perturbation. Observation and documentation of oiled shorelines and the physical and biological processes occurring there should proceed including monitoring of soil cores, soil surfaces, and the plant species. If soils are sufficiently clean to support plantings, then restoration is warranted, but care must be taken to not disturb fragile marsh substrates in the conduct of restoration.

TABLE 5.1

Potential Factors That May Accelerate or Slow Recovery of Gulf of Mexico Wetlands Following Exposure to Oil and Dispersant in the Summer of 2010

Accelerates Recovery	Reference(s)	Slows Recovery	Reference(s)
River distributaries naturally tend to prograde seaward	Coleman et al. (1998)	Natural processes in Mississippi River delta have been interrupted leading to land loss	Coleman et al. (1998), Couvillion et al. (2011)
Gulf of Mexico vegetation recovers quickly from exposure to South Louisiana crude	DeLaune and Wright (2011), Mendelssohn et al. (2012), Silliman et al. (2012), Wu et al. (2012)	Applications of Corexit dispersant will extend the presence of PAHs in the environment for months or years	Fodrie and Heck (2011), Zuijdgeest and Huettel (2012), Campo et al. (2013), Zhou et al. (2013)
Quick recovery of *Spartina* when lightly or moderately oiled	Pezeshki and DeLaune (1993), DeLaune et al. (2003)	Localized high concentrations of oil in the largest industrial accidental oil release ever	Crone and Tolstoy (2010), Pioch et al. (2010), National Commission (2011), Kokaly et al. (2013)
Quick recovery of *Spartina* when lightly or moderately oiled	Pezeshki and DeLaune (1993)	Signs of oil contamination decrease with time	Ramsey et al. (2011)
Perennial marsh plants can generate new growth when conditions favorable	Pezeshki et al. (2001), DeLaune and Wright (2011)	Louisiana coastal wetlands rapidly subsiding	Blum and Roberts (2009), Couvillion et al. (2011), Bjerstedt (2011)
Perennial marsh plants can generate new growth when conditions favorable	Pezeshki et al. (2001), DeLaune and Wright (2011)	Oiling of benthos changes nutrient cycling	DeLaune et al. (1979), Mendelssohn et al. (2012)
Marshes only oiled along shorelines	Mendelssohn et al. (2012), Silliman et al. (2012)	Localized penetration of oil meters into marshes	Ramsey et al. (2011), Mishra et al. (2012), Kokaly et al. (2013)
Oysters can help restore nutrient balance	DeLaune and Wright (2011), Mendelssohn et al. (2012)	Oysters were negatively impacted by direct oil exposure and freshwater inputs	Mendelssohn et al. (2012)
Dispersant can reduce plant exposure	Pezeshki et al. (2001), DeLaune and Wright (2011)	Dispersant application can physically damage plants slowing recovery	Pezeshki et al. (2001)
Acute short-term impacts have already occurred	Crone and Tolstoy (2010)	Chronic and long-term impacts may be the most serious and damaging	Peterson et al. (2003), Short et al. (2004)
Louisiana fish production currently high despite wetland subsidence and sea-level rise	Cowan et al. (2008)	Ecosystems are complex and may reach latent tipping points followed by dramatic changes such as fish stock loss	Hilborn and Walters (1981), Cowan et al. (2008)

What is less well known is the occurrence of oil degradation products and dispersant that may be hard to quantify either because they are buried or because they are inland away from shore. Efforts under development and deployment by Ramsey et al. (2011), Mishra et al. (2012), Kokaly et al. (2013), and others should be considered during surveys to help confirm that oiling may have moved tens of meters inland from the obviously oiled physically damaged shorelines to monitor recovery rates of impacted wetlands and to support monitoring of restoration. The initial efforts of scientists to use remote sensing technologies including the UAVSAR and Landsat imagery should be further supported so that increasing levels of confidence can be associated with such technologies making them useful now and in the future.

Environmental response times are long and to fully understand how perturbation affects a habitat and associated populations can often take years, if not decades. Understanding how the DWH oil release will affect ecologically and economically important species of fish and invertebrates in the GOM will not be achieved with short-term monitoring. The negative impacts on the physical and mental health of human communities, especially those associated with the fishing profession, have also now been documented (Cope et al. 2013). The oil was released into important nursery habitat on which fish species depend for feeding, reproduction, and recruitment of young into the adult population. Adult fish may be able to tolerate exposure to pollution, but surviving to reproduce and producing offspring capable of robust reproduction are critically important responses that must be evaluated with continued monitoring data because of the impact on long-term population survival and success and the potential for continued release of contaminants.

Acknowledgments

H. Wang provided thoughtful, useful discussion. A. M. Dausman, M. S. Peterson, J. Powell, E. W. Ramsey III, C. M. Swarzenski, S. E. Finger, and two anonymous reviewers all provided constructive criticism that improved the manuscript. L. C. Broussard provided critically important research assistance that is much appreciated.

References

Adams, C. M., E. Hernandez, and J. C. Cato. 2012. The economic significance of the Gulf of Mexico related to population income employment minerals fisheries and shipping. *Ocean and Coastal Management* 47:565–580.

Almeda, R., Z. Wambaugh, Z. Wang, C. Hyatt, Z. Liu, and E. J. Buskey. 2012. Interactions between zooplankton and crude oil: Toxic effects and bioaccumulation of polycyclic aromatic hydrocarbons. *PLOS ONE* 8(6):e67212.

Bejarano, A. C. and J. Michel. 2010. Large-scale risk assessment of polycyclic aromatic hydrocarbons in shoreline sediments from Saudi Arabia: Environmental legacy after twelve years of the Gulf war oil spill. *Environmental Pollution* 158:1561–1569.

Bjerstedt, T. W. 2011. Impacting factors and cumulative impacts by midcentury on wetlands in the Louisiana coastal area. *Journal of Coastal Research* 27(6):1029–1051.

Blum, M. D. and H. H. Roberts. 2009. Drowning of the Mississippi Delta due to insufficient sediment supply and global sea-level rise. *Nature Geoscience* 2:488–491. doi:10.1038/NGEO553.

Boesch, D. F. and R. E. Turner. 1984. Dependence of fishery species on salt marshes: The role of food and refuge. *Estuaries* 7(4):460–468.

Boswell, K. M., M. P. Wilson, P. S. D. MacRae, C. A. Wilson, and J. H. Cowan Jr. 2010. Seasonal estimates of fish biomass and length distributions using acoustics and traditional nets to identify estuarine habitat preferences in Barataria Bay, Louisiana. *Marine and Coastal Fisheries: Dynamics Management and Ecosystem Science* 2(1):83–97.

Boys, C. A., F. J. Kroon, T. M. Glasby, and K. Wilkinson. 2012. Improved fish and crustacean passage in tidal creeks following floodgate remediation. *Journal of Applied Ecology* 49(1):223–233.

Burroughs, R. 2011. Oil, pp. 43–65. In *Coastal Governance*. Washington, DC: Springer and Island. doi:10.5822/978-1-61091-016-3_4.

Campo, P., A. D. Venosa, and M. T. Suidan. 2013. Biodegradability of Corexit 9500 and dispersed South Louisiana crude oil at 5 and 25°C. *Environmental Science and Technology* 47:1960–1967. dx.doi:org/10.1021/es30388lh.

Castege, I., Y. Lalanne, V. Gouriou, G. Hemery, M. Girin, F. D'Amico et al. 2007. Estimating actual seabirds mortality at sea and relationship with oil spills: Lessons from the "Prestige" oil spill in Aquitaine (France). *Ardeola* 54(2):289–307.

Chesney, E. J., D. M. Baltz, and R. G. Thomas. 2000. Louisiana estuarine and coastal fisheries and habitats: Perspectives from a fish's eye view. *Ecological Applications* 10(2):350–366.

Coleman, J. M., H. H. Roberts, and G. W. Stone. 1998. Mississippi River delta: An overview. *Journal of Coastal Research* 14(3):698–716.

Cope, M. R., T. Slack, T. C. Blanchard, and M. R. Lee. 2013. Does time heal all wounds? Community attachment, natural resource employment, and health impacts in the wake of the *BP Deepwater Horizon* disaster. *Social Science Research* 42:872–881.

Corrêa, F., M. C. Claudino, R. F. Bastos, S. Huckembeck, and A. M. Garcia. 2012. Feeding ecology and prey preferences of a piscivorous fish in the Lagoa do Peixe National Park a Biosphere Reserve in southern Brazil. *Environmental Biology of Fishes* 93:1–12.

Couvillion, B. R., J. A. Barras, G. D. Steyer, W. Sleavin, M. Fishcer, H. Beck et al. 2011. *Land Area Change in Coastal Louisiana from 1932 to 2010*. Reston, VA: U.S. Geological Survey Scientific Investigations Map 3164 scale 1:265000, 12 page pamphlet.

Cowan, J. H., C. B. Grimes, and R. F. Shaw. 2008. Life history hysteresis and habitat changes in Louisiana's coastal ecosystem. *Bulletin of Marine Science* 83(1):197–215.

Crone, T. J. and M. Tolstoy. 2010. Magnitude of the 2010 Gulf of Mexico oil leak. *Science* 330:634.

DeLaune, R. D., W. H. Patrick Jr., and R. J. Buresh. 1979. Effect of crude oil on a Louisiana *Spartina alterniflora* salt marsh. *Environmental Pollution* 20:21–31.

DeLaune, R. D., S. R. Pezeshki, A. Jugsujinda, and C. W. Lindau. 2003. Sensitivity of US Gulf of Mexico coastal marsh vegetation to crude oil: Comparison of greenhouse and field response. *Aquatic Ecology* 37:351–360.

DeLaune, R. D. and A. L. Wright. 2011. Projected impact of *Deepwater Horizon* oil spill on U.S. Gulf Coast wetlands. *Soil Science Society of America Journal* 75(5):1602–1612.

de Mutsert, K., J. H. Cowan Jr., T. E. Essington, and R. Hilborn. 2008. Reanalyses of Gulf of Mexico fisheries data: Landings can be misleading in assessments of fisheries and fisheries ecosystems. *Proceedings of the National Academy of Sciences of the United States of America* 105(7):2740–2744. Accessed June 7, 2014, www.pnas.org/cgi/doi/10.1073/pnas.0704354105.

Dubansky, B., A. Whitehead, J. T. Miller, C. D. Rice, and F. Galvez. 2013. Multitissue molecular genomic and developmental effects of the *Deepwater Horizon* oil spill on resident Gulf killifish (*Fundulus grandis*). *Environmental Science and Technology* 47:5074–5082.

Engle, V. D. 2011. Estimating the provision of ecosystem services by Gulf of Mexico coastal wetlands. *Wetlands* 31:179–193.

Fodrie, F. J. and K. L. Heck Jr. 2011. Response of coastal fishes to the Gulf of Mexico oil disaster. *PLOS ONE* 6(7):1–8. doi:10.1371/journal.pone.0021609.

Gelwick, F. P., S. Akin, D. A. Arrington, and K. O. Winemiller. 2001. Fish assemblage structure in relation to environmental variation in a Texas Gulf coastal wetland. *Estuaries* 24(1):285–296.

Goodbody-Gringley, G., D. L. Wetzel, D. Gillon, E. Pulster, A. Miller, and K. B. Ritchie. 2012. Toxicity of *Deepwater Horizon* source oil and the chemical dispersant Corexit 9500 to coral larvae. *PLOS ONE* 8(1):e45574 (10 p.).

Graham, W. M., R. H. Condon, R. H. Carmichael, I. D'Ambra, H. K. Patterson, L. J. Linn et al. 2010. Oil carbon entered the coastal planktonic food web during the *Deepwater Horizon* oil spill. *Environmental Research Letters* 5:045301 (6 p.).

Green, B. C., D. J. Smith, J. Grey, and C. J. C. Underwood. 2012. High site fidelity and low site connectivity in temperate salt marsh fish populations: A stable isotope approach. *Oecologia* 168:245–255.

Gundlach, E. R., P. D. Boehm, M. Marchand, R. M. Atlas, D. M. Ward, and D. A. Wolfe. 1983. The fate of *Amoco Cadiz* oil. *Science* 221(4606):122–129.

Hilborn, R. and C. J. Walters. 1981. Pitfalls of environmental baseline and process studies. *Environmental Impact Assessment Review* 2(3):265–278.

Kokaly, R. F., B. R. Couvillion, J. M. Holloway, D. A. Roberts, S. L. Ustin, S. H. Peterson et al. 2013. Spectroscopic remote sensing of the distribution and persistence of oil from the *Deepwater Horizon* spill in Barataria Bay marshes. *Remote Sensing of Environment* 129:210–230.

Kokaly, R. F., D. Heckman, J. Holloway, S. Piazza, B. Couvillion, G. D. Steyer et al. 2011. *Shoreline Surveys of Oil-Impacted Marsh in Southern Louisiana July to August 2010*. Reston, VA: U.S. Geological Survey Open-File Report 2011-1022. Accessed June 7, 2014, http://pubs.er.usgs.gov/publication/ofr20111022.

Leifer, I., W. J. Lehr, D. Simecek-Beatty, E. Bradley, R. Clark, P. Dennison et al. 2012. State of the art satellite and airborne marine oil spill remote sensing: Application to the BP *Deepwater Horizon* oil spill. *Remote Sensing of Environment* 124:185–209.

Levy, J. K. and C. Gopalakrishnan. 2010. Promoting ecological sustainability and community resilience in the U.S. Gulf Coast after the 2010 *Deepwater Horizon* oil spill. *Journal of Natural Resources Policy Research* 2(3):297–315.

Lin, Q. and I. A. Mendelssohn. 2012. Impacts and recovery of the *Deepwater Horizon* oil spill on vegetation structure and function of coastal salt marshes in the northern Gulf of Mexico. *Environmental Science and Technology* 46:3737–3743.

Mendelssohn, I. A., G. L. Andersen, D. M. Baltz, R. H. Carrey, K. R. Carman, J. W. Fleeger et al. 2012. Oil impacts on coastal wetlands: Implications for the Mississippi River Delta ecosystem after the *Deepwater Horizon* oil spill. *BioScience* 62(6):562–574.

Mishra, D. R., H. J. Cho, S. Ghosh, A. Fox, C. Downs, P. B. T. Merani et al. 2012. Post-spill state of the marsh: Remote estimation of the ecological impact of the Gulf of Mexico oil spill on Louisiana salt marshes. *Remote Sensing of Environment* 118:176–185.

Moody, R. M., J. Cebrian, and K. L. Heck Jr. 2013. Interannual recruitment dynamics for resident and transient marsh species: Evidence for a lack of impact by the Macondo oil spill. *PLOS ONE* 8(3):1–11. doi:10.1371/journal.pone.0058376.

Mulabagal, V., F. Yin, G. F. John, J. S. Hayworth, and T. P. Clement. 2013. Chemical fingerprinting of petroleum biomarkers in *Deepwater Horizon* oil spill samples collected from Alabama shoreline. *Marine Pollution Bulletin* 70:147–154.

National Commission (National Commission on the *BP Deepwater Horizon* Oil Spill and Offshore Drilling). 2011. *Deep Water: The Gulf Oil Disaster and the Future of Offshore Drilling*. Washington DC: U.S. Government Printing Office.

Needles, L. A., S. E. Lester, R. Ambrose, A. Andren, M. Beyeler, M. Connor et al. 2013. Managing bay and estuarine ecosystems for multiple services. *Estuaries and Coasts*, ISSN 1559-2723. doi:10.1007/s12237-013-9602-7.

NOAA (National Oceanic and Atmospheric Administration). 2013a. *Deepwater Horizon/BP* Oil Spill: Size and Percent Coverage of Fishing Area Closures Due to *BP* Oil Spill. Available at http://sero.nmfs.noaa.gov/deepwater_horizon/size_percent_closure/index.html. Accessed August 2013.

NOAA (National Oceanic and Atmospheric Administration). 2013b. Gulf of Mexico Commercial Fish Landings Data. Available at http://www.st.nmfs.noaa.gov/commercial-fisheries/. Accessed June 2013.

Perrons, R. K. 2013. Assessing the damage caused by *Deepwater Horizon*: Not just another *Exxon Valdez*. *Marine Pollution Bulletin* 71(1–2):20–22. Accessed May 25, 2013, http://dx.doi.org/10.1016/j.marpolbul.2013.03.016.

Peterson, C. H., S. D. Rice, J. W. Short, D. Esler, J. L. Bodkin, B. E. Ballachey, and D. B. Irons. 2003. Long-term ecosystem response to the *Exxon Valdez* oil spill. *Science* 302(5653):2082–2086.

Petrolia, D. R., M. G. Interis, J. Hwang, M. K. Hidrue, R. G. Moore, and G. Kim. 2013. *America's Wetland? A National Survey of Willingness to Pay for Restoration of Louisiana's Coastal Wetlands*. Final Project Report, Department of Agricultural Economics, Mississippi State University, Starkville. NOAA Office of Ocean and Atmospheric Research award NA06OAR320264 06111039.

Pezeshki, S. R. and R. D. DeLaune. 1993. Effects of crude oil on gas exchange functions of *Juncus roemerianus* and *Spartina alterniflora*. *Water Air and Soil Pollution* 68:461–468.

Pezeshki, S. R., R. D. DeLaune, and A. Jugsujinda. 2001. The effects of crude oil and the effectiveness of cleaner application following oiling on US Gulf of Mexico coastal marsh plants. *Environmental Pollution* 112:483–489.

Pezeshki, S. R., M. W. Hester, Q. Lin, and J. A. Nyman. 2000. The effects of oil spill and clean-up on dominant US Gulf coast marsh macrophytes: A review. *Environmental Pollution* 108:129–139.

Pioch, S., J. Hay, and H. Levrel. 2010. Faraway, so close: les enjeux de la mare noire *Deepwater Horizon* vus depuis la France. *Natures Sciences Sociétés* 18:305–308.

Quan, W., L. Shi, and Y. Chen. 2011. Comparison of nekton use for cordgrass *Spartina alterniflora* and bulrush *Scirpus marqueter* marshes in the Yangtze River estuary China. *Estuaries and Coasts* 34(2):405–416.

Ramsey III, E., A. Rangoonwala, Y. Suzuoki, and C. E. Jones. 2011. Oil detection in a coastal marsh with polarimetric synthetic aperture radar (SAR). *Remote Sensing* 3:2630–2662. doi:10.3390/rw3122630.

Rosenbauer, R. J., P. L. Campbell, A. Lam, T. D. Lorenson, F. D. Hostetller, B. Thomas et al. 2010. *Reconnaissance of Macondo-1 Well Oil in Sediment and Tarballs from the Northern Gulf of Mexico Shoreline Texas to Florida*. Reston, VA: U.S. Geological Survey Open-File Report 2010-1290. 22 pages. Accessed July 25, 2013, http://pubs.usgs.gov/of/2010/1290/.

Rosenbauer, R. J., P. L. Campbell, A. Lam, T. D. Lorenson, F. D. Hostetller, B. Thomas et al. 2011. *Petroleum Hydrocarbons in Sediment from the Northern Gulf of Mexico Shoreline Texas to Florida*. Reston, VA: U.S. Geological Survey Open-File Report 2011-1014. Accessed July 25, 2013, http://pubs.usgs.gov/of/2011/1014/.

Rountree R. A. and K. W. Able. 2007. Spatial and temporal habitat use patterns for salt marsh nekton: Implications for ecological functions. *Aquatic Ecology* 41:25–45.

Sammarco, P. W., S. R. Kolian, R. A. F. Warby, J. L. Bouldin, W. A. Subra, and S. A. Porter. 2013. Distribution and concentrations of petroleum hydrocarbons associated with the *BP/Deepwater Horizon*, Oil Spill, Gulf of Mexico. *Marine Pollution Bulletin* 73:129–143. Accessed June 7, 2014, http://dx.doi.org/10.1016/j.marpolbul.2013.05.029.

Short, J. W., M. R. Lindeberg, P. M. Harris, J. M. Maselko, J. J. Pella, and S. D. Rice. 2004. Estimate of oil persisting on the beaches of Prince William Sound 12 years after the *Exxon Valdez* oil spill. *Environmental Science and Technology* 38(1):19–25.

Silliman, B. R., J. van de Kippel, M. W. McCoy, J. Diller, G. N. Kasozi, K. Earl et al. 2012. Degradation and resilience in Louisiana salt marshes after the *BP-Deepwater Horizon* oil spill. Proceedings of the National Academy of Sciences Early Edition. Available at www.pnas.org/cgi/doi/10.1073/pnas.1204922109. Accessed July 2012.

Smith, C. J., R. D. DeLaune, W. H. Patrick Jr., and J. W. Fleeger. 1984. Impact of dispersed and undispersed oil entering a Gulf Coast salt marsh. *Environmental Toxicology and Chemistry* 3:609–616.

State of Louisiana. 2012. *Louisiana's Comprehensive Master Plan for a Sustainable Coast*. Coastal Protection and Restoration Authority of Louisiana, Baton Rouge.

Twilley, R. R., E. J. Barron, H. L. Gholz, M. A. Harwell, R. L. Miller, D. J. Reed et al. 2001. *Confronting Climate Change in the Gulf Coast Region: Prospects for Sustaining Our Ecological Heritage*. Washington, DC: Union of Concerned Scientists Cambridge Massachusetts and Ecological Society of America. Accessed June 7, 2014, http://www.ucsusa.org/assets/documents/global_warming/gulfcoast.pdf, http://www.ucsusa.org/gulf/gcchallengereport.html.

Vidal-Hernandez, L. and D. Pauly. 2004. Integration of subsystem models as a tool toward describing feeding interactions and fisheries impacts in a large marine ecosystem the Gulf of Mexico. *Ocean and Coastal Management* 47:709–725.

Whitehead, A., B. Dubansky, C. Bodinier, T. I. Garcia, S. Miles, C. Pilley et al. 2011. Genomic and physiological footprint of the *Deepwater Horizon* oil spill on resident marsh fishes. *Proceedings of the National Academy of Sciences of the United States of America* 109(50):20298–20302. Available at www.pnas.org/cgi/doi/10.1073/pnas.1109545108. Accessed July 2012.

Wong, F. L., R. J. Rosenbauer, P. L. Campbell, A. Lam, T. D. Lorenson, F. D. Hostettler et al. 2011. *Macondo-1 Well Oil in Sediment and Tarballs from the Northern Gulf of Mexico Shoreline*. Reston, VA: U.S. Geological Survey Open-File Report 2011-1164, Poster. Accessed June 7, 2014, http://pubs.usgs.gov/of/2011/1164/.

Wu, W., P. Biber, M. S. Peterson, and C. Gong. 2012. Modeling photosynthesis of *Spartina alterniflora* (smooth cordgrass) impacted by the *Deepwater Horizon* oil spill using Bayesian inference. *Environmental Research Letters* 7:045302. doi:10.1088/1748-9326/7/4/045302.

Zhou, Z., Z. Liu, and L. Guo. 2013. Chemical evolution of Macondo crude oil during laboratory degradation as characterized by fluorescence EEMs and hydrocarbon composition. *Marine Pollution Bulletin* 66:164–175.

Zuijdgeest, A. and M. Huettel. 2012. Dispersants as used in response to the MC252-spill lead to higher mobility of polycyclic aromatic hydrocarbons in oil-contaminated Gulf of Mexico sand. *PLOS ONE* 7(11):1–13. doi:10.1371/journal.pone.0050549.

6

A Brief Review of the Effects of Oil and Dispersed Oil on Coastal Wetlands Including Suggestions for Future Research

John A. Nyman and Christopher G. Green

CONTENTS

Introduction

Coastal Louisiana contains vast wetlands that have a range of plant associations that reflect differences in freshwater inflow (Figure 6.1). Louisiana's coastal marshes account for 44% of all coastal wetlands in the conterminous United States, including 39% of all coastal salt marshes (Field et al. 1988). Coastal ecosystems vary from place to place primarily because of differences in riverine inflow. For example, the Mississippi–Atchafalaya River system discharges 610 km^3 freshwater into the Gulf of Mexico each year, representing approximately 80% of its freshwater inflow. The river system also carries a large quantity of sediment, nutrients, and carbon that affect physical and chemical conditions in those wetlands and adjacent coastal waters (Coleman 1989). On the other hand, the Florida panhandle receives much less freshwater, sediment, nutrients, and carbon; river deltas are smaller and beaches prevail there. Thus, spatial differences among areas within ecosystems (e.g., a salt or freshwater marsh) are a result of differences in freshwater inflow among these areas. The remainder of this chapter focuses on vegetation and nekton in coastal marshes with most examples coming from coastal Louisiana.

It is widely assumed that some ecosystems are more sensitive to crude oil pollution than others; for example, wetlands are assumed to be more sensitive than beaches but the relative sensitivity of all the coastal ecosystems remains unclear. Crude oil pollution can be viewed ecologically as a disturbance, which is an event that reduces the abundance of

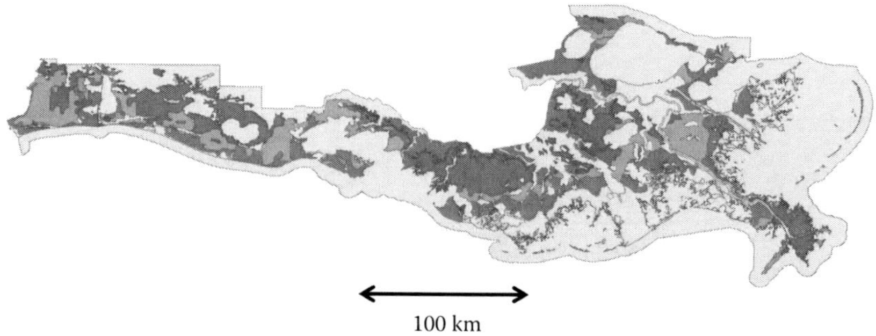

FIGURE 6.1
(See color insert.) This map of coastal Louisiana shows some coastal wetland forests and the most common marsh classification system in coastal Louisiana. From inland to the Gulf of Mexico: pink areas are bald cypress swamp (dominated by *Taxodium distichum* and *Nyssa aquatic*), dark green areas are fresh marsh (dominated by *Panicum hemitomon, Sagittaria lancifolia*, or *Typha* spp.), light green areas are intermediate marsh (dominated by *Spartina patens* and supporting many other species), orange areas are brackish marsh (dominated by *Spartina patens* and supporting few other species), and yellow areas are saline marsh (dominated by *Spartina alterniflora*). The different plant associations also support different communities of fish and wildlife. The data were collected in 1997 and made available by Louisiana Department of Wildlife and Fisheries (LDWF [2001]); the data are described in Visser et al. (1998, 2000).

dominant species, allowing new species to establish and less common, existing species to increase in abundance (Figure 6.2). Disturbance often leads to a partially predictable change in species abundance and eventually to a return in dominance by the original species (van der Valk 1981). These fairly predictable changes in species composition are known as succession; ecosystems that are easily disturbed are said to be more sensitive or to have less resistance whereas ecosystems that recover quickly from a disturbance are said to be more resilient (Halpern 1988). Classic examples of disturbance and succession effects generally include timber harvesting that initially leads to a progression of grasses, then to shrubs, then to shade-intolerant trees, and then to shade-tolerant trees. The concepts of disturbance and succession equally apply to crude oil pollution of coastal marsh ecosystems, because it allows for the initial domination by petroleum-degrading microbes (Panov 2013) and early successional plant species to increase in coastal marshes following plant death (Pahl et al. 2003), and because such changes generally alter production of invertebrate, fish, and wildlife communities (Nyman and Chabreck 1995).

Ecosystems that differ in how much they are initially affected by disturbance are said to differ in sensitivity (or resistance). Ecosystems that differ in the time required to recover from disturbance are said to differ in resiliency. Spill-response planners unconsciously incorporate concepts of sensitivity and resiliency into spill-response actions when they choose to protect some ecosystems (coastal marshes and beaches) at the expense of others (shallow and deep open water areas). In coastal ecosystems, such as a salt marsh or barrier beach, long-term trends in riverine inputs determine which of these ecosystems will come to dominate an area, thus hydrology ultimately regulates the geography of these sensitive and resilient ecosystems. Furthermore, even within an ecosystem, its sensitivity and resiliency to a disturbance can be affected by climatic stresses (e.g., drought) and resource limitation (e.g., nutrient availability) that may be operating concurrently or prior to the disturbance event. Short-term differences in riverine inputs govern the sensitivity and resiliency of coastal ecosystems to disturbance by crude oil pollution. For example, we speculate that crude oil biodegradation and the recovery of preexisting microbial and

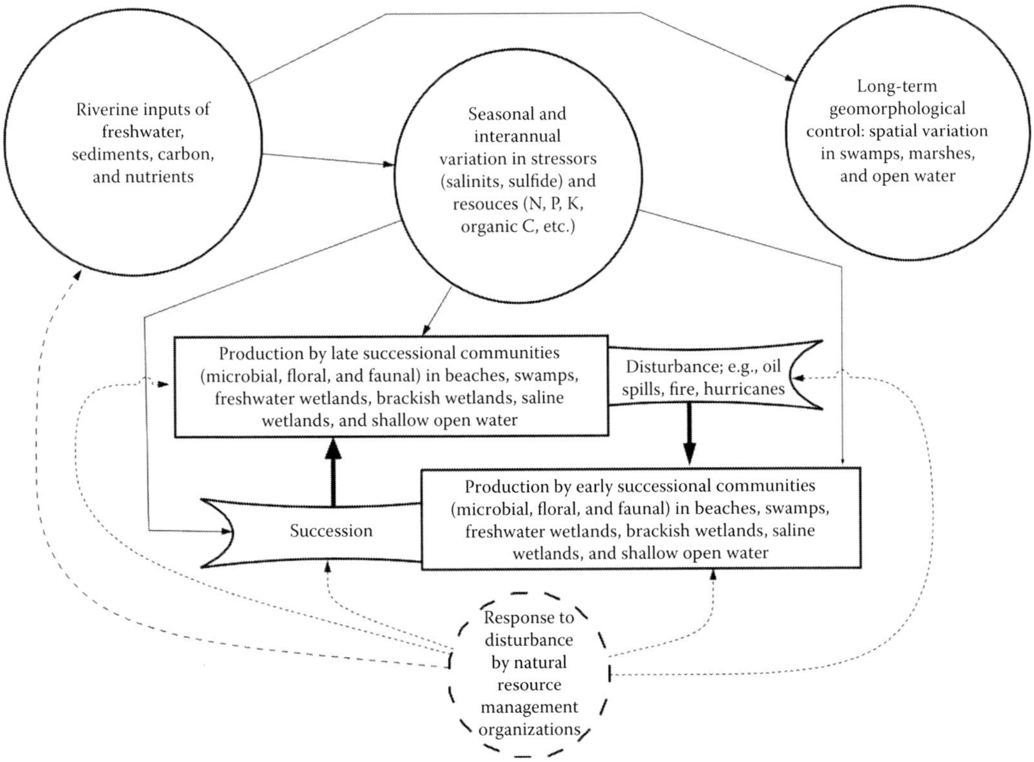

FIGURE 6.2
Diagram of long-term and short-term control by rivers in creating different coastal ecosystems and in governing the availability of resources such as freshwater and nutrients within different ecosystems.

plant communities might differ among beaches that may have been contaminated by the same oiling event if the beaches also differ in salinity and nutrient availability.

Response planners can improve the efficiency of allocating limited resources to prevent crude oil pollution and to speed the recovery of polluted ecosystems as they increase understanding of the effects of sediment transport, salinity fluctuations, and nutrient availability on the magnitude of disturbance to coastal ecosystems. As a consequence, recovery efforts may be fine-tuned to facilitate the speed of succession following crude oil pollution. In this chapter, we present the results of our review of the effects of oil spills and response options, such as oil spill dispersants, on coastal ecosystems. The purpose of this review was to identify effects that are well understood and effects where understanding is lacking.

Factors Affecting Biodegradation of Oil in Coastal Wetlands

The biodegradation of crude oil in marsh soils appears to be slowed by high levels of organic matter. The initial step of toxic biodegradation of saturated alkanes begins on the terminal methyl group (Stroud et al. 2007); we assume that biodegradation continues with additional attacks on the new terminal methyl group. We speculate that a consequence of the continual removal of methyl groups is that hydrocarbons that initially are longer chained, and hence less water soluble, toxic, and labile, become shorter chained and hence more water soluble, toxic, and labile as biodegradation proceeds. Thus, biodegradation might gradually produce a small amount of shorter chained hydrocarbons that are abundant only in fresh crude oils.

Biodegradation of oil in wetlands generally is limited by oxygen availability. Oils persist in anoxic soils and sediments because anaerobic degradation, although well documented at least for alkanes (Stroud et al. 2007), is assumed to be so much slower than aerobic degradation. Thus, oils can be buried in anoxic sediments for months or years but appear relatively unweathered when storm events expose them. The aerobic pathway of aliphatic hydrocarbon metabolism has been well described for terrestrial soils (Stroud et al. 2007), and that information probably has much relevance to the surface layer of marine sediments and wetland soils, which are toxic. Hydrocarbons are considered hydrophobic, but solubility among hydrocarbons is highly variable. For example, the aromatic polycyclic aromatic hydrocarbons (PAHs) naphthalene and phenanthrene are 33,000 and over 1,200 times more water soluble than the nonaromatic, aliphatic hexadecane (Stroud et al. 2007). The lower solubility of aliphatic relative to aromatic hydrocarbons might account somewhat for the lower degree of biodegradation of aliphatic compounds compared to aromatic compounds. At least two fertilization experiments in seawater have shown that fertilization enhances biodegradation of alkanes to a greater extent than PAHs (Bachoon et al. 2001a; Coulon et al. 2004). Most coastal wetland soils have porewaters that contain abundant nutrients and export inorganic nitrogen (Mendelssohn and Morris 2000). Given the general lack of oxygen and the general abundance of nutrients, it therefore is not surprising that nutrient additions generally fail to increase hydrocarbon biodegradation in coastal wetlands (Tate et al. 2012). Another factor that might slow biodegradation of crude oil in marsh soils is the preference for plant-based compounds over petroleum-based compounds. In general, marsh soils on the Gulf of Mexico coastline are highly organic and rich in lignin and humics. Some microorganisms that can degrade oil appear to prefer those plant-based compounds over petroleum-based compounds (Bachoon et al. 2001b), which might explain observations of biodegradation being slower in marsh soils that contain high levels of organic matter (see Dowty et al. 2001).

Studies of the biodegradation of crude oil generally measure petroleum hydrocarbon concentrations over time. A review of these studies quickly illustrates that there are a variety of ways to measure petroleum hydrocarbons: (1) total petroleum hydrocarbons (TPHs)—fluorometry, (2) TPH—gas chromatography (GC)/flame ionization detector (FID), (3) TPH—GC/mass spectrometry (GC/MS), (4) total aromatic hydrocarbons—GC/MS, (5) total polycyclic aromatic hydrocarbons—GC/MS, and (6) target aromatic hydrocarbons—GC/MS. The first five techniques all measure "total" petroleum hydrocarbons; those techniques were reviewed by the TPH Criteria Working Group (TPHCWG 1998). The TPHCWG concluded that no TPH method adequately describes contamination (TPHCWG 1998). They recommended an indicator and surrogate approach to estimating hazards because of a lack of toxicity data regarding the mixture of thousands of compounds in crude oil (TPHCWG 1997:4). This approach, driven by a lack of toxicity data on natural mixtures, was adopted and modified by many entities, including the Louisiana Department of Environmental Quality (LDEQ), to assess human and environmental risks from petroleum hydrocarbons (LDEQ 2003). The sixth technique listed, that is, target aromatic hydrocarbons, often is referred to as "fingerprinting." The list of hydrocarbons used to "fingerprint" crude oil has evolved over time, such that during the 1990s, the U.S. Environmental Protection Agency (EPA) list of target PAHs contained 24 PAHs. Overton et al. (2004) recommended using 40 PAHs and biomarkers, and we obtained measurements of 79 PAHs and biomarkers from a commercial laboratory during recent research (J. A. Nyman, unpublished data). McKenna et al. (2013) recently characterized over 30,000 unique elemental compositions in crude oil. A result is that some entities measure multiple metrics (see EPA 2010) and, even when focusing on specific compounds measured via GC/MS, different entities focus on different lists of hydrocarbons (see LDEQ 2003; EPA 2010; U.S. Food and Drug Administration [USFDA] 2010). It thus is important to determine which measurement is best.

We believe that there is no single best measurement of crude oil component concentrations, because information needs differ among professionals responsible for exploration, refining, and spill response. We also believe that spill responders are most concerned about toxicity to organisms contaminated by oil products. Thus, from the point of view of spill response, the best chemical measurement of crude oil component concentrations would be the one that most accurately reflects toxicity. Although it is true that virtually all of the compounds measured when crude oil is fingerprinted are toxic, it also is true that there are tens of thousands of compounds in crude oil (McKenna 2010a,b). Thus, it is important to design research that quantifies the concentration of specific hydrocarbons as well as toxicity on common samples followed by statistical analyses to identify hydrocarbons most related to measured toxicity. For example, Bhattacharyya et al. (2003) noted that water exposed to dispersed South Louisiana crude oil remained more toxic to a fish and a crustacean than water exposed to undispersed South Louisiana crude oil even after 6 months of biodegradation in laboratory microcosms. Working with those same samples, Nyman et al. (2007) failed to detect a difference in chemical measurements of oil in water between water exposed to dispersed South Louisiana crude oil and water exposed to undispersed South Louisiana crude oil. When Klerks et al. (2004) combined the toxicity data and the chemical data, they concluded that to accurately rank the toxicity of samples, chemical measures could not substitute for actual toxicity determinations. Wong et al. (1999) and Mao et al. (2009) also illustrate the difficulty of attempting to relate chemical measurements of crude oil to toxicity. Wong et al. (1999) studied toxicity of crude oil contaminated soils from eight sites across the United States to earthworms and four plant species. Wong et al. (1999) concluded that TPH measured via GC was a better predictor of toxicity than measurement of target polyaromatic hydrocarbons. Conversely, Mao et al. (2009) showed that TPH measurements were not indicative of the toxicity of a diesel-contaminated soil to *Vibrio fischeri*, daphnia, plant growth, and seed germination. Mao et al. (2009) also showed that a novel HPLC-GCxGC/FID analytical technique measured hydrocarbons that were related to toxicity. Presumably, the inability of ranked chemical measurements to match the rank of toxicity measurements is a consequence of the toxicity resulting from chemicals in oil that are not being measured via GC/MS when fingerprinting oil. Aeppli et al. (2013) recently noted that a class of compounds called oxyhydrocarbons are not detectable with traditional crude oil fingerprinting and also increase during biodegradation. Such compounds may contribute to the inability of chemical tests to reflect toxicity.

For the remainder of this review, we use biological activity to describe the effects of crude oil on coastal ecosystems because, as noted earlier, chemical measurements of oil in water and soil do not necessarily correspond to effects of oil on biota in water and soil. We consider three biotic communities: (1) soil microbial communities, (2) plant communities, and (3) fish. It appears to us that crude oil is more toxic to wetland fish and wildlife than to wetland plants because of the greater exposure needed to stress and/or kill vegetation (Pezeshki et al. 1995) than on fish (Bhattacharyya et al. 2003). We also believe that the soil microbial community is the least sensitive of all three communities to crude oil, because activity of the soil microbial community generally increases following the addition of crude oil (Nyman 1999). Crude oil is presumably toxic to many bacteria, but apparently oil-tolerant bacteria quickly replace oil-intolerant bacteria, and so forth. Such rapid replacement apparently is not possible in plant or animal communities. Such replacement and temporarily accelerated microbial activity does not mean that oil has a beneficial effect on wetland ecosystems; increased soil microbial activity in wetland soils might slow or reverse the accumulation of soil organic matter, which is needed by many coastal marshes to offset local subsidence and global sea-level rise (Nyman et al. 2006; Neubauer et al. 2008). It is therefore possible for a small loss in elevation to induce a vicious cycle of flooding stress and inadequate plant production that ultimately ends in emergent wetlands converting to shallow open water (Nyman et al. 1993).

Effects of Crude Oil on Wetland Soil and Microbial Activity

Even though consortia of bacterial species in marsh soils degrade oil and are necessary for nutrient cycling, plant growth, and other ecosystem functions, understanding the consequences of crude oil on soil microbe succession and community dynamics lags far behind our understanding of crude oil effects on marsh vegetation. The effects of oil, and response options such as dispersants, on the soil microbial community should be understood because the microbial community regulates the flow of energy from plants to consumers in food webs (Knox 1986). Upland soils generally are oxidized throughout their depth. In contrast, wetland soils have only a thin oxidized layer, because the presence of water in wetland soils slows the diffusion of oxygen into the soil sediments. Nonetheless, the soil microbial community continues metabolism, but by using alternative electron acceptors such as ferric iron (Fe^{+3}) or sulfate (SO_4^{2-}).

The lack of oxygen and subsequent reduction of alternate electron acceptors produces reduced substances that accumulate in wetland soils, such as Fe^{+2} or S^{2-}. The degree of microbial reduction under the oxidized layer, which generally is measured via soil Eh (redox potential), directly controls plant growth in marshes (Chalmers 1982; Good et al. 1982). Changes in soil water chemistry coincide with changes in soil Eh (Feijtel et al. 1988), and soil Eh has been used as indicator of stress on fresh, brackish, and saline marsh vegetation (DeLaune et al. 1983; Burdick et al. 1989; McKee and Mendelssohn 1989). Nyman and McGinnis (1999) showed that with the addition of Alaskan and South Louisiana crude oils at the surface of marsh soil in microcosms, Eh remained lower than controls (no added oil) for up to 5 months (Figure 6.3). Dispersed oil also can reduce the Eh for up to 5 months at the surface and at least up to 3 cm deep (Figure 6.4). The soil microbial community also regulates the release of nutrients from marshes to adjacent water. Continual nutrient release by microbes appears to maintain a rapid transfer of energy when demand is high, regardless of the degree of primary productivity (Knox 1986). The recovery of oiled vegetation may therefore depend partly on biodegradation rates by microbes and the degree to which oil affects soil conditions. Consequently, oil may have long-term effects on wetland function via microbial inhibition, even after plant growth resumes. Thus, whereas short-term effects of oil spills in marshes are dominated by effects on plants, the long-term functioning of the wetland depends on how oil affects soil microbial processes. Burns and Teal (1979) found oil in marsh soil 7 years after one spill, which indicates the potential for long-term effects.

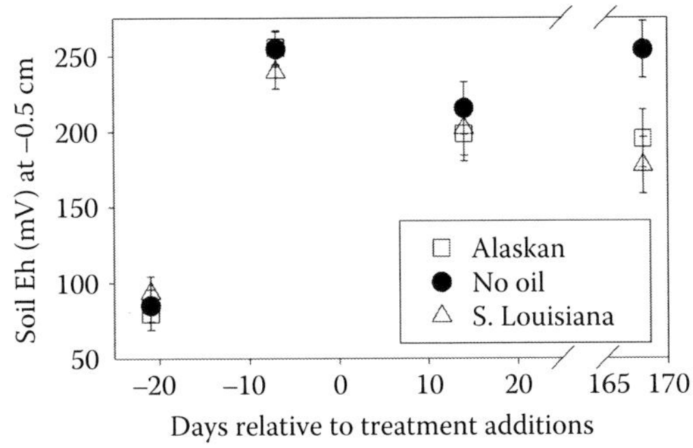

FIGURE 6.3
From data in Nyman and McGinnis (1999): soil Eh in soils from salt marshes, held for several months to allow labile carbon to degrade, and then treated with a crude oil (no oil, South Louisiana crude oil, or Alaskan crude oil) and a response option (no response, fertilize the oiled marsh, oiled with dispersed oil [COREXIT 9550], or oiled with cleaned oil [COREXIT 9580]). Soil Eh was measured as 5 months following treatment additions. The oil by term was statistically significant only at 0.5 cm below the surface and is graphed here.

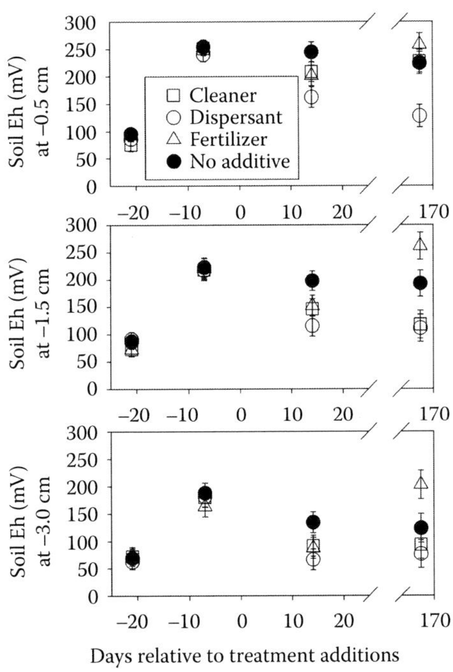

FIGURE 6.4
From data in Nyman and McGinnis (1999): soil Eh in soils from salt marshes, held for several months at three different depths (3.0 cm, 1.5 cm, and 0.5 cm) to allow labile carbon to degrade, and then treated with a crude oil (no oil, South Louisiana crude oil, or Alaskan crude oil) and a response option (no response, fertilize the oiled marsh, oiled with dispersed oil [COREXIT 9550], or oiled with cleaned oil [COREXIT 9580]). Soil Eh was measured for 5 months following treatment additions. The oil by response was statistically significant at all three depth measures; all three depths are graphed here.

Studies involving crude oils and soil microbes have generally focused on the presence or absence of soil. Some workers found no or little effects of oil on abundance in the soil microbial community (DeLaune et al. 1979, 1984). Other studies have suggested that oil adversely affects abundance of soil organisms (Bender et al. 1977; Alexander and Shwarz

1980; Sanders et al. 1980). In contrast, positive effects of oil on soil microbes have been found by others (Bachoon et al. 2001a,b, Wright and Weaver 2004). One possible reason for these inconsistencies is that there are different ways to measure microbial abundance and diversity. In a comparison of standard methods for detecting/estimating abundances of microbes, Bachoon et al. (2001b) found that most probable number and DNA analyses with domain-specific probes failed to detect changes evident with whole bacterial community rDNA hybridization in salt marsh sediments. This would suggest that the techniques typically used to assess microbial communities in most other environments may not work as well for marsh soils.

Studies designed to determine the effects of crude oil, and response options such as dispersants, on soil microbial activity are especially useful because the microbial consumption of soil organic matter is directly tied to processes that influence ecosystem productivity and greenhouse gas emissions. These processes include (1) making nutrients available to vegetation by releasing them during decomposition of soil organic matter remineralization, (2) consuming soil oxygen while decomposing soil organic matter, which creates anoxic conditions that allow wetland plants to outcompete upland plants, but may lower their productivity due to stress, and (3) converting soil organic matter into CO_2 and CH_4.

South Louisiana crude oil stimulated soil microbial activity in laboratory microcosms for several months following the addition of oil in fresh marsh soils dominated by *Sagittaria lancifolia*, in fresh marsh soils dominated by *Panicum hemitomon* (Nyman 1999), in saline marsh soils dominated by *Spartina alterniflora* (Figure 6.5), and in saline marsh soils dominated by *Juncus romerianus* (Figure 6.6). Combining the results of those studies suggest that soil with inherently faster microbial activity also recover faster after being oiled, making nutrients available to plant roots. If true, then soil from marshes dominated by *Spartina patens*, which dominates intermediate and brackish marshes, would be the least resilient in

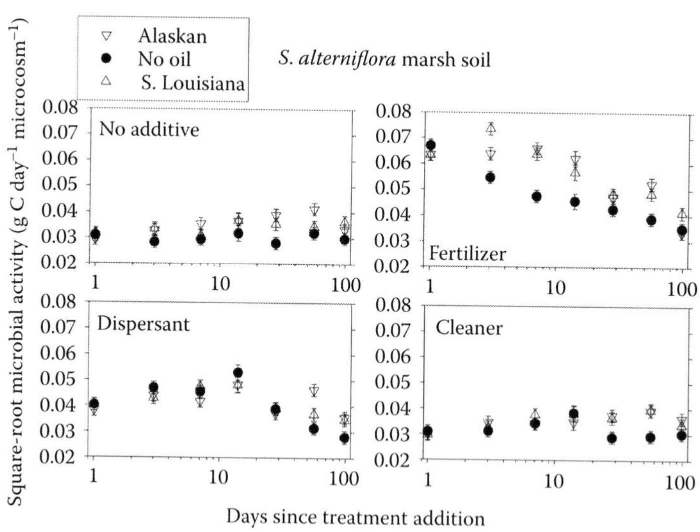

FIGURE 6.5
From data in Nyman and McGinnis (1999): microbial activity in soils from marshes dominated by *Spartina alterniflora*, held for several months to allow labile carbon to degrade, and then treated with a crude oil (no oil, South Louisiana crude oil, or Alaskan crude oil) and a response option (no response, fertilize the oiled marsh, oiled with dispersed oil [COREXIT 9550], or oiled with cleaned oil [COREXIT 9580]). Soil microbial activity was measured as the sum of CO_2 and CH_4 emission for 5 months following treatment additions. The oil by response interaction term was statistically significant and is graphed here.

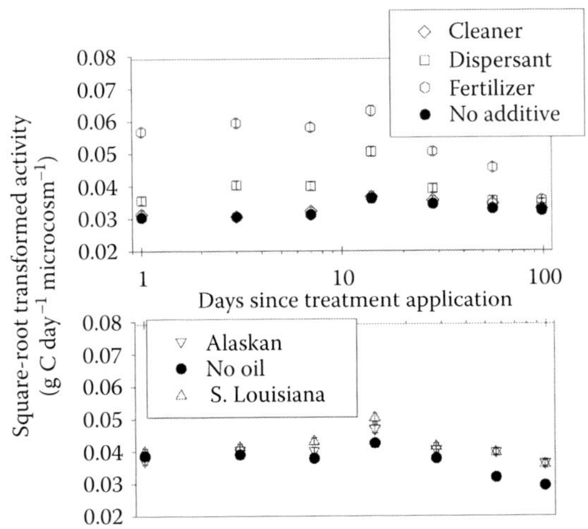

FIGURE 6.6
From data in Nyman and McGinnis (1999): microbial activity in soils from marshes dominated by *Juncus roemerianus*, held for several months to allow labile carbon to degrade, and then treated with a crude oil (no oil, South Louisiana crude oil, or Alaskan crude oil) and a response option (no response, fertilize the oiled marsh, oiled with dispersed oil [COREXIT 9550], or oiled with cleaned oil [COREXIT 9580]). Soil microbial activity was measured as the sum of CO_2 and CH_4 emission for 5 months following treatment additions. The oil by response interaction term was statistically insignificant; the oil term and the response terms were statistically significant and are graphed here.

coastal Louisiana because it has the slowest soil microbial activity (Nyman and DeLaune 1991). This hypothesis agrees with observations from sediments in Charleston Harbor where sites with the highest bacterial production also had the highest rates of PAH mineralization (Montgomery et al. 2010).

It is possible that the wind, waves, and currents that carried undispersed oil to Louisiana's coastal marshes also carried dispersed oil to the coastal marshes because the effects of dispersion on biodegradation cannot be predicted despite numerous studies (National Research Council [NRC] 2005). Chemical tests showing the presence of oil in wetland soils (Beazley et al. 2012) cannot determine if the oil was dispersed or not because chemical tests cannot differentiate between dispersed and undispersed oil (Nyman et al. 2007). Laboratory studies indicated that dispersed oil had the same effect as undispersed oil in fresh marsh soils (Nyman 1999) and saline marsh soils (Nyman and McGinnis 1999). In both of those marsh types, the dispersant alone had less effect on microbial activity than dispersed oil did (Nyman 1999; Nyman and McGinnis 1999).

Effects of Crude Oil on Wetland Vegetation

The effect that crude oil has on wetland vegetation depends more on the portion of the soil that is covered by oil than the portion of the plant covered by oil, probably because plants have a greater capacity to replace all their shoots and leaves than to replace all of their roots. Plants have been observed to produce new leaves following oiling in the greenhouse (Pezeshki et al. 1995) and in the field (Figure 6.7) (Nyman and McGinnis 1999). Oil that covers the soil surface prevents oxygen from entering surface roots and the underlying soil. Underlying roots that are not oiled apparently cannot replace the oxygen supplied by surface roots. Thus, in areas where wetland soils are completely covered in oil, plant mortality can be total. In uplands, such areas might remain barren for years, but in coastal wetlands such areas can convert to permanent open water within a few months (Silliman et al. 2012). Soil strength depends upon live roots (McGinnis 1997).

The effect of hydrocarbons on vegetation also depends partly upon the season of the spill, upon the plant species being fouled, and upon whether or not the petroleum has been refined, with crude oils being less toxic than refined products. For a review of that information published before 2000, see Pezeshki et al. (2000); examples since then include Hester and Mendelssohn (2000), Dowty et al. (2001), Lin et al. (2002), DeLaune et al. (2003), Pahl et al. (2003), Lin et al. (2005a,b), and Silliman et al. (2012). Those studies rightfully focused on emergent plants that dominate most coastal wetlands on the northern Gulf of Mexico (e.g., *Spartina alterniflora, Spartina patens, Juncus romerianus, Phragmites australis, Typha dominensis, Panicum hemitomon, and Sagittaria lancifolia*) because dominant plants produce roots that provide most of soil strength (McGinnis 1997), produce roots that provide most of the vertical accretion needed to offset subsidence and global sea-level rise (Nyman et al. 2006), produce most of the biomass that enters coastal wetland food webs, and provide most of the cover for fish and wildlife.

The relative sensitivity of the dominant plant species to crude oil can be determined by comparing the results of the previous research. Examination of those rankings shows that the species at the mouth of the Mississippi River, where marshes are fresh and intermediate, are among the most sensitive (Table 6.1).

The rankings provided by previous research (Table 6.1) indicate that the first marshes to be oiled during the 2010 spill in the Gulf of Mexico, which were *Phragmites australis* marshes in the Bird's Foot Delta of the Mississippi River, are some of the most sensitive to crude oil. It is possible that the wind, waves, and currents that carried undispersed oil to Louisiana's

FIGURE 6.7
(See color insert.) Photograph by the authors of *Phragmites australis* at the edge of the Gulf of Mexico in the Bird's Foot Delta of the Mississippi River on July 10, 2010. The death of unoiled leaves suggests that toxins in the oil were transported from the lower stems that were oiled to the upper leaves that are assumed to be unoiled. The production of new leaves suggests that the plant may survive if its soil is not eroded before recovery is completed.

TABLE 6.1
Relative Sensitivity of Five Dominant Wetland Plant Species Inferred Primarily from Previous Research[a]

Marsh Type	Species	Sensitivity Class	Sensitivity Rank
Saline	*Spartina alterniflora*	Intermediate tolerance	3.5
Brackish and intermediate	*Spartina patens*	Sensitive	2
Fresh	*Sagittaria lancifolia*	Most tolerant	5
Fresh	*Panicum hemitomon*	Intermediate tolerance	3.5
Fresh	*Phragmites australis*	Most sensitive	1

[a] Studies most responsible for these rankings are Lin et al., *Environ. Sci. Technol.* 39, 1848–1854, 2005a; Dowty et al., *Mar. Environ. Res.* 52, 195–211, 2001; DeLaune et al., *Aquat. Ecol.* 37, 351–360, 2003.

coastal marshes also carried dispersed oil to the coastal marshes because the effects of dispersion on biodegradation cannot be predicted despite numerous studies (NRC 2005) and because chemical tests cannot differentiate between dispersed and undispersed oil (Wong et al. 1999; Nyman et al. 2007; Mao et al. 2009). We are unaware of any studies that compared the effects of dispersed oil and undispersed oil on marsh vegetation.

Effects of Abiotic Estuarine Conditions on the Composition of Oil and Dispersed Oil

Crude oil is a mixture of thousands of different compounds with extreme variations in composition (McKenna et al. 2010a,b). Principle compounds within crude oil that receive the largest amount of attention with respect to both chemical and biological investigations include the following: aliphatic hydrocarbons, naphthenic acids, monocyclic aromatic hydrocarbons, and PAHs. Many highly soluble and toxic constituents (low molecular weight alkanes and monocyclic hydrocarbons) are also highly volatile resulting in oil that is enriched in PAHs including parent compounds and alkylated homologues (NRC 2005). The majority of the remaining compounds are polar hydrocarbons (including the resins and asphaltenes); however, there is limited evidence for bioavailability (Uraizee et al. 1998) or toxicity (Vrabie et al. 2012). Various hydrocarbon measurements (including several PAH analyses) are conducted in routine monitoring programs and for assessing the impact of oil spills but researchers have not yet identified a chemical measurement that accurately predicts toxicity under a wide range of conditions (Wong et al. 1999; Klerks et al. 2004; Mao et al. 2009).

Estuaries are extremely dynamic habitats in both space and time, where physical parameters such as temperature, salinity, and hypoxia can fluctuate widely both periodically and episodically. In addition to affecting the physiology of aquatic animals inhabiting these coastal marshes, variations in salinity and oxygen may alter hydrocarbon toxicity by influencing its physical dispersion, chemical dispersion, and/or biodegradation. The acute toxicity of primary oil dispersants used for the past 25 years is well established (Singer et al. 2000; Khan and Payne 2005), but synergistic effects of weathering and salinity on acute and sublethal toxicity are not well known. Data regarding the toxicity of weathered and biodegraded products are especially limited (TPHCWG 1998). The most significant abiotic factors attributed to PAH degradation in aquatic sediments and water are temperature and salinity (Chen et al. 2008). Thus, an important line of research quantifies chemistry as well as toxicity on common samples followed by statistical analyses to identify chemical parameters most closely related to observed toxicity on a variety of biological levels of organization (Wong et al. 1999; Klerks et al. 2004; Mao et al. 2009).

During the initial response to the 2010 oil spill in the Gulf of Mexico, much media attention was paid to reports by the EPA that dispersed oil was no more toxic to fish than undispersed oil (Hemmer et al. 2010, 2011). However, we believe that the dilution amounts used by Hemmer et al. (2010, 2011) (1 g of oil to 32 g of water) made their study unable to detect real differences between dispersed and undispersed oil in acute toxicity to an aquatic invertebrate (*Americamysis bahia*) and an estuarine fish (*Menidia beryllina*). For example, we detected differences in acute toxicity to a fish (*Oryzias latipes*), a crustacean (*Daphnia pulex*), and a benthic invertebrate (*Chironomus tentans*) with dilutions of approximately 1 g of oil to 7 g of water (Bhattacharrya et al. 2003) and differences in sublethal toxicity to a small estuarine fish (*Fundulus grandis*) with dilutions of approximately 1 g of oil to 125 g of water (C. G. Green, unpublished data). Likewise, Milinkovitch et al. (2011) observed more acute toxicity in dispersed oil than undispersed oil. Bhattacharrya et al. (2003) also observed that the toxicity of dispersed oil disappeared slower with time than the toxicity of undispersed oil. Thus, dispersed oil not only can be more toxic under some conditions, it may increase the persistence of acutely toxic conditions for a longer time under some conditions.

Chemically dispersed oil is a complex mixture of dispersant, petroleum hydrocarbons, oil/dispersant droplets, and nondispersed oil. Environmental factors can affect both the solubility of crude oil constituents and the performance of the dispersant in forming oil/dispersant droplets. Hydrocarbon solubility and dispersant efficacy is directly influenced by salinity. A study of 12 aromatic hydrocarbons showed a decrease in PAH solubility moving from a gradient of freshwater to seawater (Sutton and Calder 1975). As a result, oil spilled in estuarine environments would likely have greater concentrations of hydrocarbons in the water column then oil spills in the open ocean. Low salinities, however, also increase the solubility of the *dispersant* into the water phase, which lessens dispersant efficacy by reducing the interaction of the dispersant with the oil (MacKay et al. 1984). Modulations in both acute and chronic toxicity could result from variations across a salinity gradient within the estuarine ecosystem. While decreased dispersant solubility may decrease dispersion of hydrocarbons into the water column, it may also decrease biodegradation and sequestration rates (Nyman and McGinnis 1999; Nyman et al. 2007) resulting in greater persistence in the environment for chronic exposures. Studies have demonstrated a three- to fourfold increase in PAH solubility as temperatures increased from 5°C to 30°C (Neff 1979) while solubility decreased 68 ± 4.4% in saltwater as compared to freshwater (Sutton and Calder 1975 as reviewed in McAuliffe 1987).

Suggestions for Future Research

We believe that one of the greatest challenges facing researchers who want to improve oil spill response is increasing the accuracy of chemical tests at predicting the toxicity of crude oil. Such research probably will require reiterations of chemists improving the ability to quantify the thousands of compounds in crude oil and of toxicologists improving understanding of how different classes and/or functional groups of compounds in crude oil cause toxicity. Until such advances in technology and knowledge have been developed, spill response, restoration, and mitigation efforts will be forced to rely upon measurements that, although chemically precise, are toxicologically imprecise (Wong et al. 1999; Klerks et al. 2004; Mao et al. 2009).

We believe that numerous but more manageable challenges are related to understanding how salinity and nutrient availability alters biodegradation and toxicity of oil in wetlands. The effects of crude oil pollution on dominant plants that are important to wildlife for cover, such as *Spartina* and *Typha*, have been studied enough to rank their sensitivity to crude oil, but additional research is needed to determine how salinity and nutrient availability alter their sensitivity and resilience. The next generation of research regarding the effects of crude oil pollution on marsh vegetation also probably will focus on plant species important at providing food for fish and wildlife. These species generally dominate the emergent communities only after disturbances such as fire, such as annual plants and some species of *Schoenoplectus* (Nyman and Chabreck 1995), or they are submersed aquatic vegetation (Kanouse et al. 2006). Currently there is little to no information that could be used to rank their sensitivity to disturbance by crude oil or to judge how salinity and nutrient availability alter their sensitivity and resilience. The edge of marsh ponds, where emergent vegetation transitions to shallow open water is especially valuable fish and wildlife habitat (see LaPeyre et al. 2007; O'Connell and Nyman 2010 and articles cited therein) but has only begun to be studied regarding impacts from crude oil pollution (Roth and Baltz 2009; Silliman et al. 2012).

Appendix

Methods used by Nyman and McGinnis (1999) to generate data presented and discussed in this chapter are provided in this Appendix, because the *Louisiana Applied Oil Spill Research and Development Program* no longer provides web access to that publication.

Microcosms are a valid and efficient way of examining factors affecting soil bacteria (Bolton et al. 1991; Kroer and Coffin 1992), thus, this experiment used microcosms consisting of marsh soil (approximately 300 mL) in 500 mL Erlenmeyer flasks. Soils from three sites were collected to make the microcosms. The three sites occupied different geologic settings, were separated by at least 50 km, and were intended to represent the range in conditions among saline marsh soils. Site One soils were collected on July 24, 1997 from the Chenier Plain at Rockefeller Wildlife Refuge. Site Two soils were collected on August 20, 1997 from the Inactive Deltaic Plain in Cocodrie at the Louisiana Universities Marine Consortium. Site Three soils were collected on November 21, 1997 from the Deltaic Plain at Four League Bay. Although in the Inactive Delta Plain, Site Two receives fresh water and sediment from the Atchafalaya River each spring.

Microcosms were created with bulk samples. Within each site, separate bulk samples were collected from *Spartina alterniflora*- and *Juncus roemerianus*-dominated areas. Bulk soil was collected from the upper 30 cm, placed in covered plastic tubs, and returned to the lab. Each bulk sample was homogenized by manually stirring and cutting with a serrated knife. All soils were converted from firm peat to pastes following destruction of the living-root network; deionized water was added as needed to create pastes. After allowing several months of storage and occasional stirring to facilitate the decomposition of labile organic carbon, approximately 300 mL of soil were placed in clean, preweighed, and numbered 500 mL Erlenmeyer flasks. Thus, these soils were dominated by recalcitrant organic matter, which is assumed to control marsh vertical accretion (see Nyman et al. 2006 and articles cited therein). The microcosms were then reweighed to determine the mass of soil paste added. The flasks were wrapped in foil to prevent algae from growing below the soil surface on the sides of the flasks. The flasks were unsealed, which allowed a surface oxidized layer to form on the soils. This layer is a natural characteristic of marsh soils (Mitsch and Gosselink 1984) and where the bulk of biodegradation occurs (DeLaune et al. 1980; Hambrick et al. 1980). Forty-eight flasks from both plant types were prepared from Sites One and Two; 24 flasks from both plant types were prepared from Site Three. In all there were 240 microcosms. Twenty-four of the microcosms from Site One and 24 from Site Two were used to study microbial activity and soil Eh. The remaining soil from the sites was used to characterize hydrocarbon content using GC-FID at the start of the experiment (see Nyman and McGinnis 1999 for results but realize that differences among flasks in chemical measurements of petroleum hydrocarbons may not correspond to differences in toxicity among flasks; see main text for more information).

Three 50-mL subsamples of both soil pastes were placed in preweighed containers, weighed wet, dried at 100°C, and reweighed to determine the bulk density of microcosm soil. The organic matter content of the samples was determined using a subsample, which was dried via combustion at 400°C for 12 hours (Davies 1974). The samples were processed to determine the soil characteristics of the microcosm soils. Microcosm soils had a bulk density of 0.35 $g{\cdot}cm^{-3}$ (SD = 0.07), a moisture content of 35%, and a mineral content of 88% (SD = 0.04). Field conditions in marsh soils also were determined. Within each site, two 15 cm diameter cores were collected from an area dominated by *Spartina alterniflora*, and two from an adjacent area dominated by *Juncus roemerianus*. These cores were returned to the lab and the upper 36 cm were used to characterize field soils. Field soils had a bulk density of 0.30 $g{\cdot}cm^{-3}$ (SD = 0.19), a moisture content of 71%, and a mineral content of 58% (SD = 32). The lower organic content in the mesocosm soils is assumed to have resulted from losses of labile carbon caused by the intentional long storage and occasional stirring.

Treatment solutions were mixed in 1000 mL glass beakers. There were nine treatments: (1) no oil, no response, (2) South Louisiana crude oil, no response, (3) Alaskan North Slope crude oil, no response, (4) no oil, dispersant, (5) no oil, cleaner, (6) dispersed South Louisiana crude oil, (7) dispersed Alaskan North Slope crude oil, (8) cleaner and South Louisiana crude oil, and (9) cleaner and Alaskan North Slope crude oil. Louisiana crude (100 mL), Alaskan North Slope crude (100 mL), dispersed oil (100 mL oil and 20 mL COREXIT 9500), and cleaned oil (100 mL oil and 20 mL COREXIT 9580) were weathered in 1000 mL beakers containing 300 mL of deionized water. Beakers were continuously stirred under a fume hood overnight (16 hours). This degree of weathering was mild relative to that used by some other researchers, which can involve ultraviolet light and heat. Approximately 13% of the crude oil was lost during weathering when no chemicals were used or when the cleaner was used; dispersants greatly increased the weathering losses to 22% and 30% for South Louisiana crude oil and Alaskan crude oil, respectively. After weathering, oil and water fractions were separated with a separatory funnel and stored separately in amber glass jars until they were applied to the microcosms. Water and oil fractions were separated so that the oil to water ratio added to microcosms could be precisely replicated.

Treatments were applied over a 2-month period during the summer of 1998 because of the large number of microcosms. Two millimeters of oil fraction (1.8 g) and 6 mL (6 g) of the water fraction were added to the appropriate microcosm. This oil to water ratio maintained the 1:3 ratio used during weathering. In addition, this volume of oil was added because preliminary trials indicated that it covered approximately 75% of the surface area in similar flasks containing similar amounts of soil. Expressed on an aerial basis, the oiling rate was approximately 3.5 L·m^2 (320 g·m^{-2}). Complete coverage of water with 4 mm of oil resulted from applying 4 L·m^{-2}; higher rates simply increased the thickness of the oil layer. The fertilizer solution was Shultz 10-15-10 All Purpose Plant Food Plus. Fertilizer was applied at a rate equivalent to 40 g N·m^{-2}, 60 g P·m^{-2}, and 40 g K·m^{-2}. Fe, Mn, and Zn were also present in trace amounts. This rate was chosen because the nitrogen-loading rate was within the range recommended for commercial rice fields in south Louisiana. For microcosms receiving cleaner or dispersant but no oil, 0.4 mL of chemical in 10 mL of deionized water were added.

The best and most easily measured index of gross metabolic activity of mixed microbial populations in soil is CO_2 emissions (Stotzky 1997). However, CH_4 can also be an important avenue of carbon emissions in wetland soils, thus both were measured. The detectors on the GC were calibrated for measuring CO_2 (thermal conductivity detector) and CH_4 (FID). Activity was measured on or about days 1, 3, 7, 14, and 28, and 2, 3.5, 5, and 6 months after treatments began. Microcosms were sealed with airtight caps equipped with septa on measurement days. Immediately after sealing, the time was recorded and a 0.25 mL air sample was withdrawn for determination of CO_2 and CH_4 concentrations in the headspace gas. After 1–2 hours, another 0.25 mL sample was withdrawn from the microcosm, and the exact elapsed time was recorded. CO_2 and CH_4 emissions were determined from changes in CO_2 and CH_4 concentrations, the elapsed time, and the headspace volume. Microcosms were unsealed after the second air sample was analyzed.

Soil Eh was measured with platinum electrodes (Faulkner et al. 1989). We had initially planned to measure Eh on days 1, 3, 7, 14, and 28, and 2, 3.5, and 5 months after treatments began, but we concluded that too much oil was sticking to the electrodes and being lost from the microcosms. Therefore, Eh was measured 2 weeks before applying oil, and 1, 3, and 24 weeks following treatment additions. Soil Eh was measured in microcosms at 0.5 cm, 1.5 cm, and 3 cm below the soil surface. Electrodes were inserted into the soil and allowed to equilibrate >15 minutes prior to both measurements (de la Cruz and Hackney 1989). Since measuring Eh disturbs the soil, this measurement was taken after CO_2 and CH_4 samples were collected.

All glassware used during the hydrocarbon extractions was washed with hot soapy water, rinsed in a 5% acid bath, rinsed in deionized water, and then rinsed with pesticide-grade dichloromethane. Hydrocarbon extraction began by adding enough distilled deionized water to create a soil slurry. Approximately 100 g of slurry were added to preweighed 250 mL Teflon-lined, wide-mouth centrifuge bottles. The mass of the sample was then recorded.

Metabolic activity and Eh data were analyzed as an analysis of variance with a 3 × 4 factorial (three oils: no oil, Louisiana crude, Alaska crude, four additives: no additive, fertilizer, dispersant, cleaner) with repeated measures over time (Steele and Torrie 1980). Metabolic activity data were square-root transformed to improve compliance with assumptions of parametric statistics. Eh data did not require transformation and were analyzed as above but separately for both depths. Cumulative carbon emissions did not require transformation and were analyzed as 3 × 4 factorial (three oils: no oil, Louisiana crude, Alaska crude, four additives: no additive, fertilizer, dispersant, cleaner). A transformation to normalize TPH data could not be found, but all three variables analyzed (raw data, log transformed, and square-root transformed) indicated the same conclusions. Untransformed data were reported.

References

Aeppli, C., C. A. Carmichael, R. K. Nelson, K. L. Lemkau, W. M. Graham, M. C. Redmond, et al. 2013. Oil weathering after the *Deepwater Horizon* disaster led to the formation of oxygenated residues. *Environmental Science and Technology* 46:8799–8807. doi: 10.1021/es3015138.

Alexander, S. K. and J. R. Schwarz. 1980. Short-term effects of South Louisiana and Kuwait crude oils on glucose utilization by marine bacterial populations. *Applied and Environmental Microbiology* 40:341–345.

Bachoon, D. S., R. Araujo, M. Molina, and R. E. Hodson. 2001a. Microbial community dynamics and evaluation of bioremediation strategies in oil-impacted salt marsh microcosms. *Journal of Industrial Microbiology and Biotechnology* 27:72–79.

Bachoon, D. S., R. E. Hodson, and R. Araujo. 2001b. Microbial community assessment in oil-impacted salt marsh sediment microcosms by traditional and nucleic acid-based indices. *Journal of Microbiological Methods* 46:37–49.

Beazley, M. J., R. J. Martinez, J. Rajan, J. Powell, Y. M. Piceno, et al. 2012. Microbial community analyses of a coastal salt marsh affected by the *Deepwater Horizon* oil spill. *PLoS One* 7:e-41305. doi: 10.1371/journal.pone.0041305.

Bender, M. E., E. A. Shearls, and R. P. Ayres. 1977. Ecological effects of experimental oil spills on eastern coastal plain estuarine ecosystems. *Proceedings 1977 International Oil Spill Conference*, 505–509. Washington, DC: American Petroleum Institute.

Bhattacharyya, S., P. L. Klerks, and J. A. Nyman. 2003. Toxicity to freshwater organisms from oils and oil spill chemical treatments in laboratory microcosms. *Environmental Pollution* 122:205–215. doi: 10.1016/S0269–7491(02)00294-4.

Bolton, H. Jr., J. K. Fredrickson, S. A. Bentjen, D. J. Workman, S. W. Li, and J. M. Thomas. 1991. Field calibration of soil-core microcosms: Fate of a genetically altered rhizobacterium. *Microbial Ecology* 21:163–173.

Burdick, D. M., I. A. Mendelssohn, and K. L. McKee. 1989. Live standing crop and metabolism of the marsh grass *Spartina patens* as related to edaphic factors in a brackish, mixed marsh community in Louisiana. *Estuaries* 12:195–204.

Burns, K. A. and J. M. Teal. 1979. The West Falmouth oil spill: Hydrocarbons in the salt marsh ecosystem. *Estuarine and Coastal Marine Science*. 8:349–360.

Chalmers, A. G. 1982. Soil dynamics and the productivity of *Spartina alterniflora*. In *Estuarine Comparisons*, ed. V. S. Kennedy, 231–242. New York: Academic Press.

Chen, J., M. H. Wong, Y. S. Wong, and. N. F. Tam. 2008. Multi-factors on biodegradation kinetics of polycyclic aromatic hydrocarbons (PAHs) by *Sphingomonas sp.* a bacterial strain isolated from mangrove sediment. *Marine Pollution Bulletin* 57:695–702.

Coleman, J. M. 1988. Dynamic changes and the processes in the Mississippi river delta. *Geological Society of America Bulletin* 100:999–1015. doi: 10.1130/0016-7606(1988)100<0999:DCAPIT>2.3.CO,2.

Coulon, F., E. Pelletier, R. St. Louis, L. Gourhant, and D. Delille. 2004. Degradation of petroleum hydrocarbons in two sub-arctic soils: influence of an oleophilic fertilizer. *Environmental Toxicology and Chemistry* 23:1893–1901.

Davies, B. E. 1974. Loss-on-ignition as an estimate of soil organic matter. *Proceedings of the Soil Science of America* 38:150–151.

de la Cruz, A. A. and C. T. Hackney. 1989. Temporal and spatial patterns of redox potential (Eh) in three tidal marsh communities. *Wetlands* 9(2):181–190.

DeLaune, R. D., G. A. Hambrick, and W. H. Patrick, Jr. 1980. Degradation of hydrocarbons in oxidized and reduced sediments. *Marine Pollution Bulletin* 11:103–106.

DeLaune, R. D., W. H. Patrick, Jr., and R. J. Buresh. 1979. Effect of crude oil on a Louisiana *Spartina alterniflora* salt marsh. *Environmental Pollution* 1979:21–31.

DeLaune, R. D., C. J. Smith, and W. H. Patrick, Jr. 1983. Relationship of marsh elevation, redox potential, and sulfide to *Spartina alterniflora* productivity. *Soil Science of America Journal* 47:930–935.

DeLaune, R. D., C. J. Smith, and W. H. Patrick, Jr., J. W. Fleeger, and M. D. Tolley. 1984. Effect of oil on salt marsh biota: Methods for restoration. *Environmental Pollution (A)* 36:207–227.

DeLaune, R. D., S. R. Pezehski, A. Jugsujindam, and C. W. Lindau. 2003. Sensitivity of US Gulf of Mexico coastal marsh vegetation to crude oil: Comparison of greenhouse and field responses. *Aquatic Ecology* 37:351–360.

Dowty, R. A., G. P. Shaffer, M. W. Hester, G. W. Childers, F. M. Campo, and M. C. Greene. 2001. Phytoremediaton of small-scale oil spills in fresh marsh environments: A mesocosm simulation. *Marine Environmental Research* 52:195–211.

EPA. 2010. BP spill quality assurance sampling plan to evaluate the effects to water and sediment from oil and dispersant to shoreline, nearshore and far off-shore areas. Available at: http://epa.gov/bpspill/samplingplanjune1/BP_Spill_QASP_May_30_v2.pdf (last accessed on July 15, 2013).

Faulkner, S. P., W. H. Patrick, Jr., and R. P. Gambrell. 1989. Field techniques for measuring wetland soil parameters. *Soil Science Society of America Journal* 53:883–890.

Feijtel, T. S., R. D. DeLaune, and W. H. Patrick, Jr. 1988. Seasonal porewater dynamics in marshes of Barataria Basin, Louisiana. *Soil Science of America Journal* 52:59–67.

Field, D. W., C. E. Alexander, and M. Broutman. 1988. Toward developing an inventory of U.S. coastal wetlands. *Marine Fisheries Review* 50:40–46.

Good, R. E., N. F. Good, and B. R. Frasco. 1982. A review of primary production and decomposition dynamics of the belowground marsh component. In *Estuarine Comparisons*, ed. V. S. Kennedy, 139–157. New York: Academic Press.

Halpern, C. B. 1988. Early successional pathways and the resistance and resilience of forest communities. *Ecology* 69:1703–1715.

Hambrick, G. A., R. D. DeLaune, and W. H. Patrick, Jr. 1980. Effect of estuarine sediment pH and oxidation-reduction potential on microbial hydrocarbon degradation. *Applied and Environmental Microbiology* 40:365–369.

Hemmer, M. J., M. G. Barron, and R. M Greene. 2010. Comparative toxicity of Louisiana sweet crude oil (LSC) and chemically dispersed LSC to two Gulf of Mexico aquatic test species. U.S. Environmental Protection Agency, Office of Research and Development, National Health and Environment Effects Research Laboratory. Dated July 31, 2010. Available at: http://www.epa.gov/bpspill/reports/phase2dispersant-toxtest.pdf (last accessed on July 15, 2013).

Hemmer, M. J., M. G. Barron, and R. M. Greene. 2011. Comparative toxicity of eight oil dispersants, Louisiana sweet crude oil (LSC), and chemically dispersed LSC to two aquatic test species. *Environmental Toxicology and Chemistry* 30:2244–2252. doi: 10.1002/etc.619.

Hester, M. W. and I. A. Mendelssohn. 2000. Long-term recovery of a Louisiana brackish marsh plant community from oil-spill impact: Vegetation response and mitigation effects of marsh surface elevation. *Marine Environmental Research* 49:233–254.

Kanouse, S., M. K. La Peyre, and J. A. Nyman. 2006. Nekton use of *Ruppia maritima* and non-vegetated bottom habitat types within brackish marsh ponds. *Marine Ecology Progress Series* 327:61–69.

Khan, R. A., and J. F. Payne. 2005. Influence of a crude oil dispersant, Corexit 9527, and dispersed oil on a capelin, Atlantic cod, longhorn sculpin, and cunner. *Bulletin of Environmental Contamination and Toxicology* 75:50–56.

Klerks, P. L., J. A. Nyman, and S. Bhattacharyya. 2004. Relationship between hydrocarbon measurements and toxicity to a chironomid, fish larvae, and daphnid for oils and oil spill chemical treatments in laboratory freshwater marsh microcosms. *Environmental Pollution* 129:345–353. doi: 10.1016/j.envpol.2003.12.001.

Knox, G. A. 1986. *Estuarine Ecosystems: A Systems Approach*, Vol. II. Boca Raton, FL: CRC Press.

Kroer, N. and R. B. Coffin. 1992. Microbial trophic interactions in aquatic mesocosms designed for testing genetically engineered microorganisms: A field comparison. *Microbial Ecology* 23:143–157.

La Peyre, M. K., B. Gossman, and J. A. Nyman. 2007. Assessing functional equivalency of nekton habitat in enhanced habitats: Comparison of terraced and unterraced marsh ponds. *Estuaries and Coasts* 33(3): 526–536.

LDEQ. 2003. Appendix D: Guidelines for assessing petroleum hydrocarbons, polycyclic aromatic hydrocarbons, lead, polychlorinated dibenzodioxins and polychlorinated dibenzofurans, and non-traditional parameters. Risk Evaluation Correction Action Program (RECAP), Louisiana Department of Environmental Quality, Baton Rouge, LA.

LDWF. 2001. Louisiana Coastal Marsh Vegetative Type (poly), Geographic NAD83, LDWF (2001) [marsh_veg_type_poly_LDWF_2001]: Louisiana Department of Wildlife and Fisheries, Fur and Refuge Division, and the U.S. Geological Survey's National Wetlands Research Center, Lafayette, LA. Available at: http://lagic.lsu.edu/data/losco/marsh_veg_type_poly_LDWF_2001.zip.

Lin, Q., I. A. Mendelssohn, K. Carney, N. P. Bryner, and W. D. Walton. 2002. Salt marsh recover and oil spill remediation after in-situ burning: Effects of water depth and burn duration. *Environmental Science and Technology* 36:576–581.

Lin, Q., I. A. Mendelssohn, N. P. Bryner, and W. D. Walton. 2005a. In-situ burning of oil in coastal marshes. 1. Vegetation recovery and soil temperature as a function of water depth, oil type, and marsh type. *Environmental Science and Technology* 39:1848–1854.

Lin, Q., I. A. Mendelssohn, N. P. Bryner, and W. D. Walton. 2005b. In-situ burning of oil in coastal marshes. 2. Oil spill cleanup efficiency as a function of oil type, marsh type, and water depth. *Environmental Science and Technology* 39:1855–1860.

Mackay, D. C. A., K. Hossain, and M. Bobra. 1984. Measurement and prediction of the effectiveness of oil spill chemical dispersants. In *Oil Spill Chemical Dispersants*. Philadelphia, PA: American Society for Testing and Materials.

Mao, D., R. Lookman, H. Van de Wegh, R. Weltans, G. Vanermen, N, de Brucker, et al. 2009. Estimation of ecotoxicity of petroleum hydrocarbon mixtures in soil based on HPLC-GCXGC analysis. *Chemosphere* 77:1508–1513.

McAuliffe, C. D. 1987. Measuring hydrocarbons in water. *Chemical Engineering Progress* 83:40–45.

McGinnis II, T. E. 1997. Factors of soil strength and shoreline movement in a Louisiana coastal marsh. Master's Thesis. University of Southwestern Louisiana, Lafayette, LA.

McKee, K. L., and I. A. Mendelssohn. 1989. Response of a freshwater marsh plant community to increased salinity and water level. *Aquatic Botany* 34:301–316.

McKenna, A. M., J. M. Purcell, R. P. Rodgers, and A. G. Marshall. 2010a. Heavy petroleum composition. 1. Exhaustive compositional analysis of athabasca bitumen HVGO distillates by Fourier transform ion cyclotron resonance mass spectrometry: A definitive test of the Boduszynski model. *Energy and Fuels* 24:2929–2938.

McKenna, A. M., J. M. Purcell, R. P. Rodgers, and A. G. Marshall. 2010b. Heavy petroleum composition. 2. Progression of the Boduszynski model to the limit of distillation by ultrahigh-resolution FT-ICR mass spectrometry. *Energy and Fuels* 24:2939–2946.

McKenna, R. K. C. M. Nelson, J. J. Reddy, N. K. Savory, J. E. Kaiser, A. G. Fitzsimmons, et al. 2013. Expansion of the analytical window for oil spill characterization by ultrahigh resolution mass spectrometry: Beyond gas chromatography. *Environmental Science and Technology* 47:7530–7539. doi: 10.1021/es305284t.

Mendelssohn, I. A. and J. T. Morris. 2000. Eco-physioligical controls on the productivity of *Spartina alterniflora* Loisel. In *Concepts and Controversies in Tidal Marsh Ecology*, eds. M. P. Weinstein and D. A. Kreeger, 59–80. The Netherlands: Kluwer Academic Publishers.

Milinkovitch, T., R. Kanan, H. Thomas-Guyon, and S. Le Floch. 2011. Effects of dispersed oil exposure on the bioaccumulation of polycyclic aromatic hydrocarbons and the mortality of juvenile *Liza ramada*. *Science of the Total Environment* 409:1643–1650.

Mitsch, W. J. and J. G. Gosselink. 1984. *Wetlands*. New York: Van Nostrand Reinhold Co.

Montgomery, M. T., T. J. Boyd, C. L. Osburn, and D. C. Smith. 2010. PAH mineralization and bacterial organotolerance in surface sediments of the Charleston Harbor estuary. *Biodegradation* 21:257–266.

Neff, J. M. 1979. *Polycyclic Aromatic Hydrocarbons in the Aquatic Environment: Sources, Fates and Biological Effects*. Essex, United Kingdom: Applied Science Publishers Ltd.

Neubauer, S. C. 2008. Contribution of mineral and organic components to tidal freshwater marsh accretion. *Estuarine Coastal and Shelf Science* 78:78–88.

NRC. 2005. Understanding oil spill dispersants: Efficacy and effects. Committee on understanding oil spill dispersants, National Research Council. ISBN 0-309-54793-8.

Nyman, J. A. 1999. Effects of crude oil and chemical additives on metabolic activity of mixed microbial populations in fresh marsh soils. *Microbial Ecology* 37:152–162.

Nyman, J. A., and R. H. Chabreck. 1995. Fire in coastal marshes: history and recent concerns. In *Proceedings 19th Tall Timbers Fire Ecology Conference—Fire in wetlands: A management perspective*, eds. S. I. Cerulean and R. T. Engstrom, 135–141. Tallahassee, FL: Tall Timbers Research, Inc.

Nyman, J. A. and R. D. DeLaune. 1991. CO_2 emission and soil Eh responses to different hydrological conditions in fresh, brackish, and saline marsh soils. *Limnology and Oceanography* 36:1406–1414.

Nyman, J. A., R. D. DeLaune, H. H. Roberts, and W. H. Patrick, Jr. 1993. Relationship between vegetation and soil formation in a rapidly submerging coastal marsh. *Marine Ecology Progress Series* 96:269–279. doi: 10.3354/meps096269.

Nyman, J. A., P. L. Klerks, and S. Bhattacharyya. 2007. Effects of chemical additives on hydrocarbon disappearance and biodegradation in freshwater marsh microcosms. *Environmental Pollution* 149:227–238. doi: 10.1016/j.envpol.2006.12.028.

Nyman, J. A and T. E. McGinnis. 1999. Louisiana Applied and Educational Oil Spill Research and Development Program, *OSRADP Technical Report Series* 99-007.

Nyman, J. A., R. J. Walters, R. D. DeLaune, and W. H. Patrick, Jr. 2006. Marsh vertical accretion via vegetative growth. *Estuarine Coastal and Shelf Science* 69:370–380.

O'Connell, J. L., and J. A. Nyman. 2010. Marsh terraces in coastal Louisiana increase marsh edge and densities of waterbirds. *Wetlands*. 30:125–135.

Overton, E. G., B. M. Ashton, and M. S. Miles, 2004. Historical polycyclic aromatic and petrogenic hydrocarbon loading in Northern Central Gulf of Mexico shelf sediments. *Marine Pollution Bulletin* 49:557–563.

Pahl, J. W., I. A. Mendelssohn, C. Henry, and T. J. Hess. 2003. Recovery trajectories after in situ burning of an oiled wetland in coastal Louisiana, USA. *Environmental Management* 31:236–251.

Panova, A. V., T. Z. Esikovab, S. L. Sokolovb, I. A. Koshelevaa, and A. M. Boronina. 2013. Influence of soil pollution on the composition of a microbial community. *Microbiology* 82:241–248. doi: 10.1134/S0026261713010116.

Pezeshki, S. R., R. D. DeLaune, J. A. Nyman, R. R. Lessard, and G. P. Canevari. 1995. Removing oil and saving oiled marsh grass using a shoreline cleaner. *Proceedings of the 1995 Oil Spill Conference*, 203–209. Washington, DC: American Petroleum Institute.

Pezeshki, S. R., M. W. Hester, Q. Lin, and J. A. Nyman, 2000. The effects of oil spill and clean-up on dominant US Gulf Coast marsh macrophytes: A review. *Environmental Pollution* 108:129–139.

Roth, A. M. F. and D. M. Baltz. 2009. Short-term effects of an oil spill on marsh-edge fishes and decapods crustaceans. *Estuaries and Coasts* 32:565–572.

Sanders, H. L., J. F. Grassle, G. R. Hampson, L. S. Morse, S. Garner-Price, and C. C. Jones. 1980. Anatomy of an oil spill: Long-term effects from the grounding of the barge Florida off West Falmouth, Massachusetts. *Journal of Marine Research* 38(2):265–380.

Silliman, B. R., J.van de Koppel, M. W. McCoy, J. Diller, G. N. Kasozi, K. Earl, et al. 2012. Degradation and resilience in Louisiana salt marshes after the *BP-Deepwater Horizon* oil spill. *Proceedings of the National Academy of Sciences of the United States of America*. 109:11234–11239. doi: 10.1073/pnas.1204922109.

Singer, M. M., S. George, D. Benner, S. Jacobson, R. S. Tjeerdema, and M. L. Sowby. 1993. Comparative toxicity of two oil dispersants to the early life stages of two marine species. *Environmental Toxicology and Chemistry* 12:1855–1863.

Steele, R.G.D., and J. H. Torrie. 1980. *Principles and Procedures of Statistics: A Biometrical Approach*. Second Edition, 633 pp. New York: McGraw-Hill.

Stotzky, F. 1997. Quantifying the metabolic activity of microbes in soil. In *Manual of Environmental Microbiology*, eds. C. J. Hurst, G. R. Knudsen, M. J. McInerney, L. D. Stetzenback, and W. V. Walter WV, 453–458. Washington, DC: American Society for Microbiology.

Stroud, J. L., G. I. Paton, and K. T. Semple. 2007. Microbe-aliphatic hydrocarbon interactions in soil: Implications for biodegradation and bioremediation. *Journal of Applied Microbiology* 102:1239–1253.

Sutton C. and J. A. Calder. 1975. Solubility of alkylbenzenes in distilled water and seawater at 25.0 degrees C. *Journal of Chemical Engineering Data* 20:320–322.

Tate, P. T., W. S. Shin, J. H. Pardue, and W. A. Jackson. 2012. Bioremediation of an experimental oil spill in a coastal Louisiana salt marsh. *Water Air Soil Pollution* 223:1115–1123. doi: 10.1007/s11270-011-0929-z.

TPHCWG. 1997. Development of fraction specific references doses (RfDs) and reference concentrations (RfCs) for Total Petroleum Hydrocarbons (TPH). *Total Petroleum Hydrocarbon Criteria Working Group Series*, Vol. 4. Amherst, MA: Amherst Scientific Publishers.

TPHCWG. 1998. Analysis of petroleum hydrocarbons in environmental media. *Total Petroleum Hydrocarbon Criteria Working Group Series*, Vol. 1. Amherst, MA: Amherst Scientific Publishers.

Uraizee F. A., A. D Venosa, and M. T. Suidan. 1998. A model for diffusion controlled bioavailability of crude oil components. *Biodegradation* 8:287–296.

USFDA. 2010. Protocol for interpretation and use of sensory testing and analytical chemistry results for re-opening oil-impacted areas closed to seafood harvesting. Available at: http://www.fda.gov/downloads/Food/RecallsOutbreaksEmergencies/Emergencies/UCM233818.pdf (last accessed on July 15, 2013).

van der Valk, A. G. 1981. Succession in wetlands: A Gleasonian approach. *Ecology* 62:688–696. doi: 10.2307/1937737.

Visser, J. M., C. E. Sasser, R. H. Chabreck, and R. G. Linscombe. 1998. Marsh vegetation types of the Mississippi River Deltaic Plain. *Estuaries* 21:818–828. doi: 10.2307/1353283.

Visser, J. M., R. H. Chabreck, C. E. Sasser, and R. G. Linscombe. 2000. Marsh vegetation types of the Chenier Plain, Louisiana, USA. *Estuaries* 23:318–327. doi: 10.2307/1353324.

Vrabie C. M., T. L. Sinnige, A. J. Murk, and M. T. O. Jonker. 2012. Effect-directed assessment of the bioaccumulation potential and chemical nature of Ah receptor agonists in crude and refined oils. *Environmental science and technology* 46:1572–1580.

Wong, D. C., E. Y. Chai, K. K. Chuand, and P. B. Dorn. 1999. Prediction of ecotoxicity of hydrocarbon-contaminated soils using physiochemical parameters. *Environmental Toxicology and Chemistry* 18:2611–2621.

Wright, A. L. and R. W. Weaver. 2004. Fertilization and bioaugmentation for oil biodegradation in salt marsh microcosms. *Water, Air, and Soil Pollution* 156:229–2004.

7

Oil Contamination in Mississippi Salt Marsh Habitats and the Impacts to Spartina alterniflora *Photosynthesis*

Patrick D. Biber, Wei Wu, Mark S. Peterson, Zhanfei Liu, and Linh Thuy Pham

CONTENTS

Introduction

The *Deepwater Horizon* (DWH) explosion and subsequent oil spill in the northern Gulf of Mexico (GOM), from April 20 to July 15, 2010, is the largest accidental marine oil spill in the history of the U.S. petrochemical industry (Read 2011). This accident released about 4.9 million barrels (7.0×10^6 m^3) of crude oil into the open ocean from the leaking Macondo (MC252) well at 1.5 km depth (Crone and Tolstoy 2010) and at approximately 170 km from the Mississippi mainland coast. Salt marshes are important coastal ecosystems because they are highly productive. They also provide valuable ecosystem services, such as the provision of food and shelter for many organisms (Turner 1977; Boesch and Turner 1984; Phillips 1987; Cai et al. 2000; Beck et al. 2001), carbon sequestration (Chmura et al. 2003), shoreline stabilization and storm protection (King and Lester 1995; Moeller et al. 1996), filtration of excess nutrients (Valiela et al. 2000; Tobias et al. 2001a,b; Valiela and Cole 2002), and valued recreational and aesthetic opportunities (Lee et al. 1992; Engle 2011; Jordan and Peterson 2012). Coastal wetlands are not only threatened by stressors from the terrestrial side (*sensu* Peterson and Lowe 2009), but are also at risk from ocean-side stressors that include pollutants such as oil.

Oil spills in the marine environment have a long history, with numerous studies following the acute and chronic impacts of oil and how they affect the recovery of various coastal habitats and species (Baca et al. 1987; Kenworthy et al. 1993; Duke et al. 1997; Peterson et al. 2001; Graham et al. 2010; Fodrie and Heck 2011; Ortmann et al. 2012; Kolian et al. 2013). From this literature it is clear that the rate of recovery varies by habitat type largely as a function of coastline energetics (NOAA 2012), degree of direct exposure to oil experienced by subtidal versus intertidal organisms (Kenworthy et al. 1993; Peterson et al. 2001; Venosa and Zhu 2003), and mean annual temperature (Baca et al. 1987; Duke et al. 1997). In combination with these factors, the overall toxicity of oil decreases rapidly over time, with warmer conditions promoting more rapid "weathering" (Lee and Page 1997; Venosa and Zhu 2003; Aeppli et al. 2012; Mendelsohn et al. 2012).

The degradation of oil in the marine environment varies substantially as a function of temperature, oxygen content, nutrients, pH, and salinity (Venosa and Zhu 2003). Crude oil in water is broken down by important weathering processes, including spreading, evaporation, dissolution, emulsification, sedimentation, biodegradation (microbial oxidation), and photooxidation (Hunt 1996; Fingas 1999; Plata et al. 2008; Aeppli et al. 2012; Liu et al. 2012; Mendelssohn et al. 2012), all of which reduce the toxicity of the oil over time. In addition, the GOM is widely recognized as having a native microbial biota that is adapted to consuming hydrocarbons as an energy source, in part coming from extensive natural seeps on the seafloor, and in part a response to nearly a century of oil and gas extraction activities (Lin and Mendelssohn 1998; Venosa and Zhu 2003; Hazen et al. 2010; Orcutt et al. 2010; Edwards et al. 2011; Bik et al. 2012).

Oil impacts along the northern GOM coast in 2010 were documented extensively (e.g., daily shoreline cleanup assessment technique team maps and daily Administration-Wide Response to DWH oil spill press releases) and where feasible (e.g., sand beaches) oil was removed soon after arrival. In less accessible locations, such as salt marsh shorelines, oil was documented but generally left in place or received minimal cleanup actions to prevent further damage to the habitat. Silliman et al. (2012) reported that about 75 km of Louisiana salt marsh shoreline was affected by medium to heavy oiling, the most of any of the affected states. In Mississippi, small oil slicks (<100 m^2) arrived sporadically and intermittently from early June 2010 through the remainder of the year; in fact, British

Petroleum (BP) contractors continued to monitor the arrival of heavily weathered oil during summer 2012 along barrier island beaches. The initial 2010 oil arrived onshore more frequently and as larger patches around the time tropical systems were active in the northern GOM, primarily because of storm surge and onshore winds. Oil impacts to salt marshes were generally restricted to a narrow ribbon (5–15 m wide) right along the coastline, with little to no penetration of oil into the marsh interior (Mendelsohn et al. 2012; Silliman et al. 2012). This fringing zone is typically vegetated by smooth cordgrass (*Spartina alterniflora*) in the northern GOM, although in portions of Louisiana and Texas the black mangrove (*Avicennia germinans*) may also occur (McMillan and Sherrod 1986; Lloyd and Tracy 1901; Patterson et al. 1993). All these factors contributed to an emerging pattern of heavily weathered oil arriving sporadically in small quantities, impacting localized areas of shoreline, and not resulting in an extensive, simultaneous oiling event. This is atypical of previously studied large-scale oil spill disasters (Lee and Page 1997) that included massive volumes of oil released in a short time and near to shore from either shipwrecks, such as the *Torrey Canyon* (Southward and Southward 1978), *Amoco Cadiz* (Baca et al. 1987), and *Exxon Valdez* (Wolfe et al. 1994), or coastal oil processing facilities such as in Panama (Duke at al. 1997) and Kuwait (Kenworthy et al. 1993).

The response of salt marsh vegetation to weathered crude oil is complex and variable, ranging from short-term reductions in photosynthesis and rapid subsequent recovery to complete mortality and long-term wetland loss when the roots are killed (Pezeshki et al. 2000, 2001; Roth and Baltz 2009; Engle 2011). Oil can affect the plants directly by coating the leaves and blocking gas exchange through the stomata, chemical toxicity can disrupt plant–water relations or directly kill living cells, and thick oil on the sediment can reduce oxygen exchange with the atmosphere causing negative consequences for root health (Baker 1970; Pezeshki et al. 2000; Ko and Day 2004). Oil that smothers plants can also increase temperature stress, especially during summer, and in combination with reduced photosynthetic gas exchange, it will cause rapid and acute mortality in leaves and stems (Baker 1970; Lin and Mendelssohn 1996; Ko and Day 2004). The extent of the acute impacts to salt marshes varies with the amount and type of oil, the weather and hydrologic conditions at the time, the species of plant, season, soil composition, and any physical disturbance caused by cleanup activities (Lin and Mendelssohn 1996; Hester and Mendelssohn 2000; Pezeshki et al. 2000; Mendelssohn et al. 2012). For instance, in a study that compared oil impacts on photosynthetic gas exchange of two important U.S. Gulf Coast plant species, smooth cordgrass and black needlerush (*Juncus roemerianus*), smooth cordgrass was shown to be more sensitive to partial oil coating than black needlerush (Lytle and Lytle 1987; Pezeshki and DeLaune 1993). In contrast, moderate oiling from the DWH oil spill had no significant effect on smooth cordgrass, but it significantly lowered live aboveground biomass and stem density of black needlerush in the Bay Jimmy area of northern Barataria Bay, Louisiana (Lin and Mendelssohn 2012). Furthermore, smooth cordgrass was shown to be more sensitive to oiling during the spring/summer growing season than during the predormancy or dormant season in winter (Pezeshki et al. 2000; Mishra et al. 2012). These factors notwithstanding marshes in the northern GOM region can be resilient to oil effects in the long term (2 years or longer), and some studies show complete recovery is attainable 4 years after a serious oil spill event (Mendelssohn et al. 1993; Silliman et al. 2012).

Oil can cause long-term chronic consequences in coastal wetlands when it becomes incorporated into the fine anaerobic sediments and persists for many years (Krebs and Tanner 1981; Alexander and Webb 1987; Lin and Mendelssohn 1998; Mendelssohn et al. 2012). The severity of these chronic impacts will be influenced by the factors aforementioned for acute impacts, as well as the amount of oil penetrating into the sediments, the physical nature of the coastline (i.e., high- or low-wave energy), and the composition of the plant community (Dicks and Hartley 1982; Peterson et al. 2003; Ko and Day 2004; Mendelssohn et al. 2012). Once oil penetrates into the sediments, recovery to reference conditions may take 3–4 years (Hester and Mendelssohn 2000) or longer (Bergen et al. 2000; Michel et al. 2009; Mendelssohn et al. 2012). Under extreme circumstances, recovery may never occur due to sediment removal (Baca et al. 1987; Gilfillan et al. 1995) or accelerated erosion after vegetation morality (Mendelssohn et al. 2012). In one study following a crude oil spill in Galveston Bay, oil concentrations in the sediment ranged between 5 and 51 mg/g, which significantly reduced growth of smooth cordgrass over 18 months. This reduction in the plants led to shoreline erosion that did not become evident until 16–32 months after the oil spill (Alexander and Webb 1987). In another study, Silliman et al. (2012) reported that heavily oiled (>100 mg/kg polyaromatic hydrocarbons, PAH) wetlands in Louisiana resulted in plant mortality within less than 5 months and the subsequent loss of plant roots increased the rate of coastline erosion over 18 months compared to adjacent control sites that remained vegetated. The interplay of vegetation and geomorphic processes requires better understanding to properly assess the acute and chronic impacts of oil on salt marshes and determine their long-term resilience.

One approach commonly used to evaluate the impacts to an organism, population, or habitat from an injury (physical or chemical) is the impact-recovery model (Kirsch et al. 2005), which has its underpinnings in ecological succession theory. This conceptual model outlines an acute stress response immediately after an impact (e.g., an oil spill) as a reduction in the function of a process (e.g., photosynthesis), with gradual recovery back to preimpact function over time (Figure 7.1). In some instances recovery may be protracted, or may remain depressed compared to adjacent control sites, resulting in a chronic stress effect. In oil spills that impact soft-sedimentary shorelines, chronic impacts are often observed when oil is incorporated into subsurface sediments causing a persistent source of toxic chemicals that are slow to degrade under anaerobic conditions, resulting in ongoing exposure to the affected plants and animals. In the case of plants, this incessant exposure to oil at the roots can result in chronic depression of photosynthesis and a lack of recovery, compared to rates observed at unimpacted control sites. The impact-recovery model has been commonly employed in Before After Control Impact (BACI) designs and is also increasingly being applied in habitat equivalency analysis (HEA) as part of the valuation of lost ecosystem services (Underwood 1994; Unsworth and Bishop 1994; Dunford et al. 2004; Cacela et al. 2005). As HEA will be included in the development of compensatory mitigation projects arising from DWH impacts, the question of how much photosynthesis was reduced due to oil-induced stress is an important part of the overall evaluation.

Determining acute or chronic stress in plants can be accomplished by measuring photosynthesis (Schulze and Caldwell 1990; Krause and Weis 1991). As net primary productivity is also a measure of energy available to support ecosystem structure and function (Cardoch et al. 2002), photosynthesis can be used to relate the impacts of oil exposure to ecosystem health. However, photosynthesis is affected by factors other than oil stress, such as seasonal changes in temperature and light with greater photosynthesis during the summer growing season and less activity during the dormant winter season. There are two main techniques for measuring photosynthesis in emergent plants: chlorophyll fluorescence and gas-exchange rate (Schulze and Caldwell 1990; Rohacek and Bartak 1999; Maricle et al. 2007).

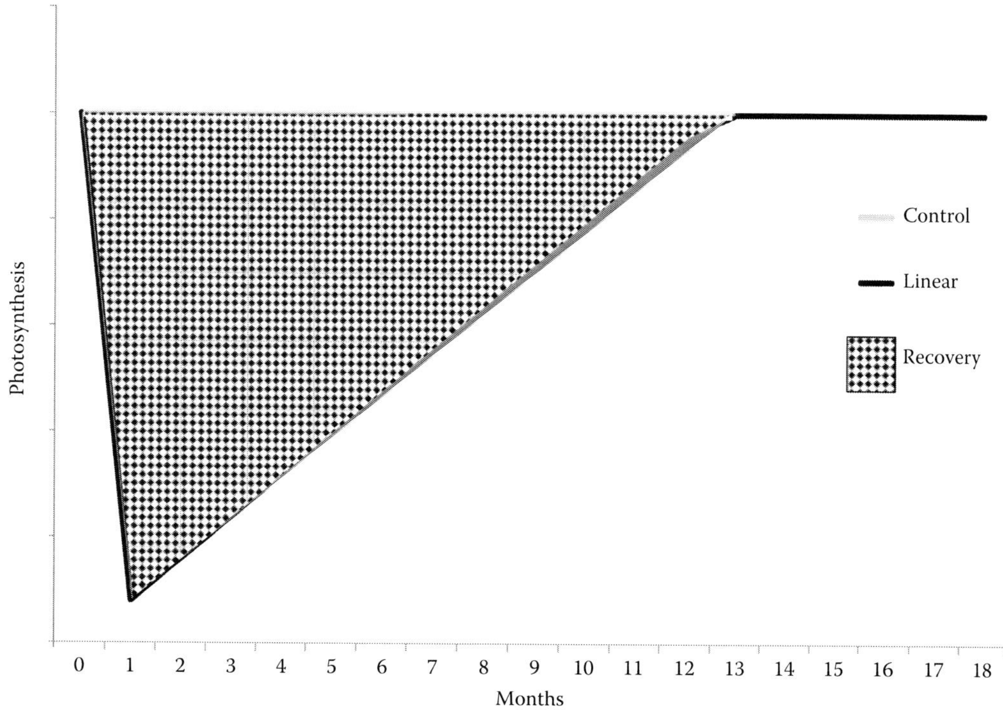

FIGURE 7.1
Conceptual diagram of recovery following an impact relative to control levels for photosynthesis. A linear recovery is assumed over 12 months, although nonlinear responses could also be used.

These two approaches target different steps in the photosynthesis reaction, with chlorophyll fluorescence measuring conversion efficiency of photon energy to biochemical energy occurring in the electron transport chain of the light reactions at the very beginning of the photosynthesis process (Figure 7.2). In contrast, the incorporation of CO_2 measured during gas exchange is indicative of the carbon fixation (Calvin) cycle occurring in the dark reactions at the end of the photosynthesis process. Because of the difference in the process being measured, the two approaches should present similar patterns in response to stress, but they are not directly comparable. Furthermore, stress measured by the chlorophyll fluorescence technique may show more rapid recovery than stress measured by gas exchange, as plants tend to optimize light reaction chemistry more quickly than the downstream dark reactions (Figure 7.2).

The goals of this study were to determine the effect of oil contamination in Mississippi salt marsh plants and how oil may have affected photosynthesis of smooth cordgrass, the main species affected, over 1 year. More specifically, the objectives of this study were to (1) determine the location and spatial extent of oil contamination in Mississippi salt marsh habitats; (2) quantify the concentration of oil that was present in sediments, plant tissues, and animal tissues, and how it changed over the course of 1 year; (3) measure the effects of oiling on plant photosynthesis (measured by F_v/F_m, CO_2 flux) to determine acute and chronic responses; and (4) assess the time to recovery of photosynthesis at two contrasting locations—an eroding coastline exposed to high wave and tide energy and a protected, depositional coastline with low energy.

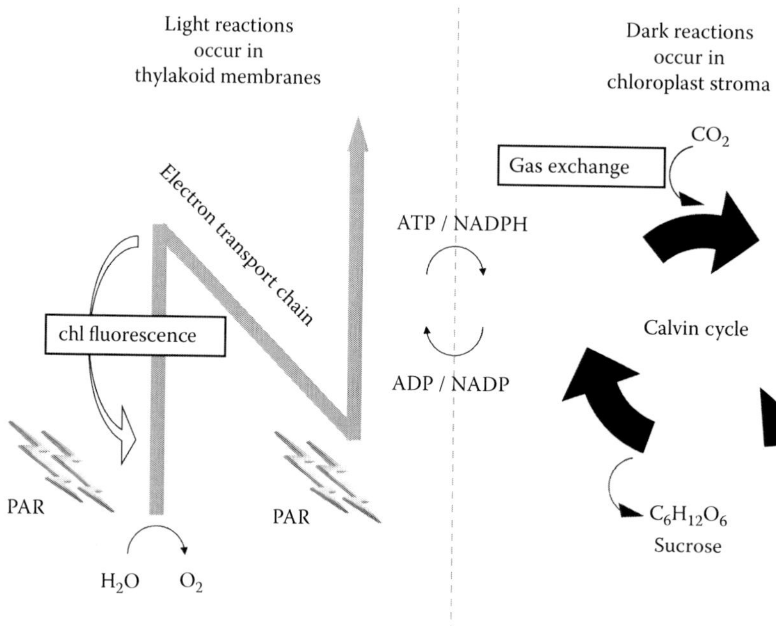

FIGURE 7.2
Conceptual diagram of the two different reaction steps in photosynthesis. The light reactions occur in the thylakoid membranes within the chloroplast, which is responsible for using photosynthetically active radiation and chlorophyll to split water into oxygen and hydrogen ions. These ions are then used to generate the energy molecules ATP and NADPH. The products of the light reaction fuel the subsequent dark reaction steps of the Calvin cycle, located in the fluid-filled stroma of the chloroplast where carbon dioxide is "fixed" into sucrose. The two measurements of photosynthesis we report occur at two different places in this reaction sequence: chlorophyll fluorescence is measured at the very beginning of the light reaction using a Pulse Amplitude Modulated fluorometer, whereas the carbon flux is measured at the later dark reaction step using the gas exchange system in the LI-COR 6400XT portable photosynthesis system.

Methods

We contacted local, state, and federal agency personnel, as well as private citizens, to ascertain locations in Mississippi where oil had been documented to come ashore during the summer months of 2010. Only locations with documented impacts (photos, GPS coordinates) were selected for the oil contaminant sampling. We had two different goals when sampling for oil contamination. Our first goal was to determine the spatial distribution of residual oil contamination in salt marsh habitats when we sampled in fall (October–November 2010) along the Mississippi coastline. Our second goal was to determine the temporal change in contamination over 1 year (July 2010–July 2011) at selected locations.

Study Sites

To determine the spatial distribution of oil contaminants along the Mississippi coastline, we visited four locations: Grand Bay National Estuarine Research Reserve (GBNERR), Marsh Point (MP), a lagoon on Horn Island (HI), and Sand Bayou (SB) near Waveland, Mississippi (Figure 7.3). These locations covered the east to west extent of oil in Mississippi salt marshes, as well as mainland and barrier island coastlines. In each location, we sampled

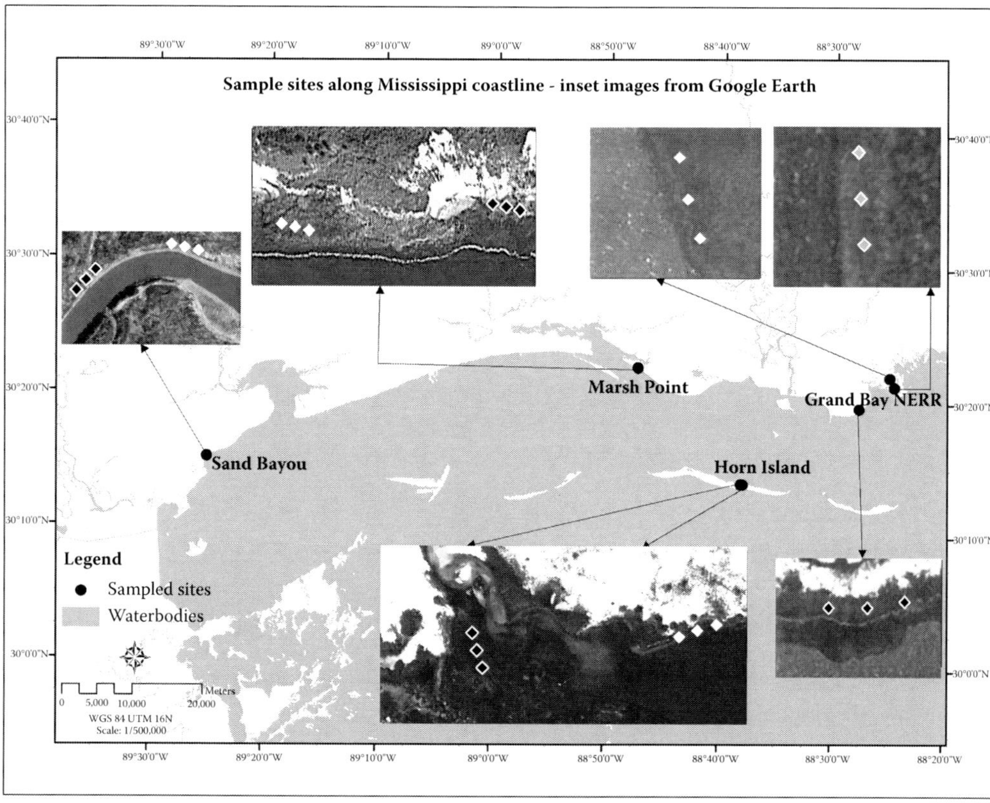

FIGURE 7.3
(**See color insert.**) Map of the four study locations: Grand Bay National Estuarine Research Reserve, Marsh Point, a lagoon on Horn Island, and Sand Bayou near Waveland, Mississippi with control and heavily oiled treatments indicated. White symbols are control quadrats (no oil) and dark symbols are impacted quadrats (medium oiling, gray; heavy oiling, black). Within each treatment, three 1-m^2 quadrats were placed 5 m apart and parallel to the shoreline in the smooth cordgrass fringe zone.

replicate quadrats (for contaminants) or replicate plants (for photosynthesis) at one control and at least one oil-impacted marsh treatment. At one location (GBNERR), oil was no longer visible in the oiled treatment at the time of sampling, highlighting the need for prior documentation during the acute phase of this event. A base map of the Mississippi coastline was obtained from Mississippi Geospatial Clearinghouse (MGC 2011) in NAD83 and then converted to WGS84 UTM 16N projection and the sampling locations are indicated thereon. Inset images were created to the same scale (1:800) from Google Earth and the three replicate quadrats sampled for contamination within each treatment at each location are indicted (Figure 7.3). Plant photosynthesis sampling occurred in the same general area within each treatment, but was not necessarily exactly within the quadrats.

To determine the temporal change in oil contamination, we sampled two locations with different energy coastlines: a high energy, exposed, and eroding coastline at MP and a low energy, protected, and depositional coastline inside a lagoon at HI. Sampling at MP occurred less than 2 weeks after the initial oiling in July 2010, in fall (October 2010), in spring (March 2011), and 1 year (July 2011) after the initial oiling had occurred. Sampling at HI was only done twice, November 2010 and August 2011, due to logistic issues prohibiting us from access to the location during the "cleanup" phase.

Sampling of Sediments, Plants, and Animals for Oil Contamination

At each treatment within a location, we haphazardly selected 10 m of salt marsh dominated by smooth cordgrass. Three replicate quadrats (1 m^2) located no more than 5 m apart were placed along this 10-m transect. As smooth cordgrass is a fringing species in northern GOM salt marshes, quadrats were always within 1–3 m of open water, ensuring that we sampled where oil impacts were most likely to have affected vegetation. In each quadrat, all sediment and plant biomass samples were collected in duplicate, and immediately placed in labeled glass jars provided by the analytical laboratory. We always sampled the control treatment first and the heavily oiled treatment last, to avoid potential contamination of tools and samples. Control treatments were sampled as close as possible to oiled treatments within each location to minimize differences due to spatial position.

In addition to sample collection, the following data were gathered at each treatment within a location as part of the chain of custody documentation: GPS coordinates of each quadrat, photos from the boat of the shoreline, and at least one overview and one close-up photo of each quadrat to document the condition of the plants and sediments, air temperature, and the time of collection (U.S. Central Standard Time). All samples were placed on ice immediately after collection then transferred to the analytical laboratory later that same day.

To obtain a representative sample of plant tissues within each quadrat, we harvested 20–30 smooth cordgrass stems near the base, cut them into pieces that fit into 1000-mL glass jars, and divided the sample equally between two duplicate jars. After removing the plant sample, it was easy to access the sediments in the quadrat. Using a stainless steel gardening trowel, we collected sediments to a depth of no more than 8–12 cm and filled two 250-mL glass jars about two-thirds full. Preference was given for shallow sediments with visible oil contaminant over deeper sediments, where possible. All hardware (trowel, plant cutters) were cleaned with acetone until all traces of oil were removed and then wiped down twice more with fresh paper towels. Afterward, a triple rinse with deionized water was used to minimize potential contamination of samples between quadrats and treatments. Only one of the duplicate samples was submitted for analysis; the second was retained as an archival sample.

We collected animal tissue samples from small resident killifishes (*Fundulus* spp.) and grass shrimp (*Palaemonetes* spp.). To obtain these species, a bag seine net (3.05 m long × 1.83 m tall, mesh size 3.17 mm) or a dipnet (0.46 m × 0.36 m, mesh size 3.17 mm) was run along the 10 m of shoreline in front of the quadrats at low tide, and all specimens were hand picked out of the catch. Target animals were placed in a sealed plastic bag and immediately kept on ice. We attempted to collect a minimum of 50 shrimp and 10 fish from the treatments within each location, but were not always successful despite up to 1 hour of intensive sampling at each treatment within location. Nekton samples were processed by removing the lateral muscle tissue of the fish and the tail muscle of the shrimp after removal of the carapace. All animal tissues from individuals collected from a treatment within a location were pooled for a minimum weight of 3–5 g required for analysis of hydrocarbons. Each sample was frozen at –20°C in 7-mL glass scintillation tubes until analysis.

Analysis of oil range organics (ORO, C_{19}–C_{36}) in sediments and tissues was performed by gas chromatography–flame ionization detection (GC–FID) following EPA 8015C protocol, and analysis of PAH was performed by gas chromatography–mass spectrometry (GC–MS) following EPA 8270C protocol at Micro-Methods Laboratory Inc. (Ocean Springs, Mississippi). Results were reported in milligrams per kilogram (i.e., ppm) for each sample provided or "ND" if not detectable. Because of interference with plant structural compounds, varying dilutions were required to stay within instrument detection limits that otherwise may have affected the minimum reporting limits (MRL). Using these methods, detection limits were 10 ppm for ORO and 0.030 ppm for PAH (Douglas et al. 2007). Hydrocarbon concentration data and MRL were transcribed from the analytical reports into a Microsoft Excel© v. 2007 spreadsheet for each replicate quadrat within treatment at each of the four locations.

Comparison of Marsh Oil with Macondo Oil

Samples collected at MP in July 2010 were also sent to the University of Texas, Marine Science Institute to compare oil from a heavily impacted marsh treatment with the oil originating from the MC252 well. The oil smothering the smooth cordgrass leaves was scraped carefully into a glass vial using a Teflon knife and frozen at −20°C until analysis. MC252 reference oil was acquired from BP. The oil obtained from the surface of the MP plants was normalized to total solvent-extractable materials (TSEM); the TSEM was 750 mg/g. The oil extraction followed the protocols of Wang et al. (2004). Briefly, about 1 g of sample was weighed and extracted five times with Dichloromethane (DCM) using sonication. The extracts were combined and dried by filtering through a glass column packed with anhydrous sodium sulfate. The DCM extract was concentrated by a rotovap and exchanged with hexane. An aliquot of the concentrated extract was transferred into a chromatographic column packed with activated silica gel and topped with anhydrous sodium sulfate. After the columns were conditioned with hexane, the concentrated extract was loaded into the column and eluted with hexane as the fraction for measuring saturated *n*-alkanes, and subsequently eluted with benzene in hexane (50% v/v) as the fraction for measuring aromatic hydrocarbons. For the MC252 oil, the crude oil was diluted with hexane and filtered through a sodium sulfate column. The procedures for cleanup and fractionation were the same as for the MP plant oil. Analysis for *n*-alkanes (C_8–C_{40}), pristane (Pr), and phytane (Ph) was performed on GC–FID, whereas PAHs were analyzed with GC–MS in a selective ion mode.

Plant Stress Responses

To ascertain the impact of oil contamination on plant photosynthesis and time to recovery to preoiled performance, we sampled plants repeatedly at MP over 1 year (July 2010–July 2011). In particular, we were interested in determining recovery of photosynthesis from acute effects of oiling versus potential chronic depression of photosynthesis lasting longer than a few months after oiling. To compare the responses obtained from plants at the high energy, exposed, and eroding coastline at MP, we also measured plant photosynthesis at the low energy, protected, and depositional coastline in the lagoon at HI after 1 year (August 2011). At MP we sampled three treatment areas about 300 m^2 each (Figure 7.3): one treatment was not impacted by crude oil (control), the second treatment experienced medium impact (some oil observed on the sediment and plants), and the third treatment was heavily impacted (plants covered by crude oil extensively) in July 2010. It is important to note that the spatial extent of the heavy impact area was not larger than about 5 m wide × 10 m long (50 m^2). At HI we sampled only at the control treatment and the heavily oiled treatment (Figure 7.3); there was no medium-oiled treatment. Within each treatment at both locations, we randomly sampled up to 10 individual plants, taking measurements on one leaf per plant (see below), at each visit.

Photosynthesis was measured using two instruments. First, chlorophyll fluorescence was measured with a pulsed amplitude modulation (PAM) fluorometer, the Walz Mini-PAM (www.walz.com), to obtain plant stress (F_v/F_m) and electron transport rate (ETR). Second, CO_2 flux and chlorophyll fluorescence were measured with a LI-6400XT portable photosynthesis and fluorescence system (www.LICOR.com/env) to obtain carbon fixation rates (μmol $CO_2/m^2/s$) and plant stress (F_v/F_m). The PAM is able to measure fluorescence very quickly and all measurements were obtained in a few hours during the morning, whereas the LI-COR relies on slower gas-exchange processes requiring repeated visits over the course of 3 days to obtain the same number of replicate measurements. For this reason, the results between instruments were not compared directly.

Pulsed Amplitude Modulation Chlorophyll Fluorescence

We used the PAM to measure the chlorophyll fluorescence yield on 10 randomly selected individuals of smooth cordgrass in each of the three treatments at MP during each time point. Plants were not tagged, so each visit likely resulted in us selecting different individuals. Measurements were made at MP on July 21, 2010 (13 days after oiling); October 4, 2010 (88 days); January 19, 2011 (195 days); March 10, 2011 (245 days); and July 7, 2011 (351 days). In addition, we had already made chlorophyll fluorescence yield measurements on smooth cordgrass plants on May 9, 2010 in the lower Pascagoula River delta 18 km away from the east in anticipation of future oil spill impacts, so this data were included in our analysis as a before-impact sample for the control treatment.

The yield measurement was always taken from the middle portion of the second leaf from the top of the plant; leaves used for PAM measurements were always green. To obtain the measurements, the instrument-supplied leaf clip (DLC-8) was attached to the midpoint of the leaf and immediately the chlorophyll fluorescence under ambient sunlight (effective quantum yield [EQY]) was recorded. Subsequently, the shutter on the leaf clip was closed for a minimum of 15 minutes and the dark-adapted measurement (potential quantum yield [PQY]) was then taken from the same leaf. As there were 10 leaf clips, all 10 individuals were measured within a few minutes of each other.

In addition to these yield measurements, three additional individuals were randomly selected and a rapid light curve (RLC) was measured using the same leaf selection criteria outlined for the yield measurements. The RLC was obtained by measuring the yield over nine incremental light levels (0, 55, 81, 122, 183, 262, 367, 616, and 1115 μmol photons/m^2/s) generated by the halogen lamp inside the PAM. From these measurements the instrument calculates the ETR by the following equation:

$$\text{ETR} = \text{EQY} \times \text{PAR} \times \text{AF} \times 0.5$$

where EQY, effective quantum yield; PAR, light level generated by halogen lamp inside the PAM; AF, absorption factor set to instrument default setting of 0.84; and 0.5, 50% of incident light absorbed at photosystem II (PSII) (Walz 1999). The RLC so obtained can be considered as the equivalent of a photosynthesis–irradiance (P–I) curve for the light reactions of photosynthesis. The PAM measurements were typically conducted between 0900 and 1400 hours (CST), with the exact time for each measurement recorded in the instrument memory. Measurements were conducted on sunny days when possible.

In summer 2011, we revisited the HI location and measured chlorophyll fluorescence on plants in the control and the heavily oiled treatments using the same methods outlined above for MP. From this 1-year post-oiling time point we wanted to determine whether the chronic exposure to residual oil at the HI location was evident as depressed photosynthesis measurements (EQY, PQY, and RLC).

LI-COR Chlorophyll Fluorescence and Gas Exchange

We used a LI-6400 XT portable photosynthetic gas exchange and chlorophyll fluorescence system to measure the photosynthesis rate (μmol $CO_2/m^2/s$) and dark-adapted chlorophyll fluorescence (PQY) at each time point. Measurements at MP were conducted monthly between July 2010 and June 2011. No photosynthesis data were able to be collected in December 2010 due to lack of boat availability, nor at the 1-year time point in July 2011 because of a broken instrument that was accidentally exposed to saltwater.

At each visit, we randomly chose up to 15 individuals for measurement of the photosynthesis rate on the middle portion of the second leaf from the top of the plant. Leaves used for LI-COR measurements were not always green, and included a range of healthy to senescing/unhealthy leaves. The photosynthetic rate for each leaf was measured by CO_2 exchange under five photosynthetically active radiation (PAR) intensities: approximately 400, 800, 1200, 1600, and 2000 μmol photons/m^2/s. In addition, for each leaf we measured PQY after at least 30-minute dark adaption and used this as an indicator of leaf and individual stress.

To assess the recovery of photosynthesis at the two oiled treatments (medium, heavy), we assumed that their close proximity to the control treatment provided similar environmental conditions. Time series of monthly PQY for each of the three treatments were compared against only the light saturated (1200–2000 μmol photons/m^2/s) gas-exchange measurements (P_{max}) from the same individuals, and then the time until the mean response at an impacted treatment was not significantly different from the control response was determined. In a second approach to assess time to recovery, we used all the gas-exchange measurements obtained by the LI-COR to construct the mean P–I curve for each season. Summer was defined as June–September, fall was October–November, winter was December–February, and spring was March–May, based on mean monthly air temperatures. Seasonal mean P–I curves for the heavily oiled and the control treatments were plotted, and recovery determined when the 95% confidence intervals overlapped at all irradiance levels.

Statistical Analyses

All plant photosynthesis data were imported from the instruments into Excel (Microsoft Office© v. 2007) and processed prior to analyses, which were performed in JMP Intro v.5 (SAS Cary, NC). The replicate was an individual plant (one leaf measured per plant), with ≤10 plants measured in each treatment (control, oiled) within each location (MP, HI) at each time step (month). Data were tested for normality and homoscedasticity using the Shapiro-Wilk test and Bartlett's test, respectively. Data were transformed ($\log_{10}[Y + 1]$), where appropriate, to meet assumptions for subsequent statistical models. The EQY and PQY data (collected with PAM) and F_v/F_m and P_{max} data (collected with LI-COR) were each analyzed separately by a two-way mixed model analysis of variance (ANOVA) with treatment (i.e., control vs. medium vs. heavy oil) designated as a fixed effect and month designated as a random effect. Resulting significant treatment × time interaction effects for EQY, PQY ($n = 10$ plants) and F_v/F_m, P_{max} ($n = 2$–10 plants) were determined post hoc by separate one-way ANOVA (treatment) tests for each month sampled, with the alpha-level adjusted (Dunn-Sidak) to compensate for multiple pair-wise comparisons (PAM = 5 months,

LI-COR = 10 months). Finally, EQY and PQY data collected with the PAM 1-year post impact at MP and HI were tested with a two-way fixed factor ANOVA with treatment (i.e., control vs. oiled) and location considered as fixed effects. For all test results with significantly different means, Tukey's HSD was used to determine means that were not significantly different. Data for RLC and P–I curves were assessed for differences among means either by comparing the 95% confidence intervals (~2σ) to the curves, or from the slopes and R^2 coefficients of the linear regression between control and oiled variables.

Results

Spatial Contamination

During fall 2010, ORO were found in both sediment and plant tissue samples, but not in animal muscle (both fish and shrimp) samples. Not all locations yielded sufficient animal tissues for analysis (minimum 3 g required), hence the low number of results for these samples (Table 7.1). Only at MP were all three species sampled found to co-occur at the time of collection (Table 7.1). No ORO were detected in sediments or plant tissues from control treatments at GBNERR and MP, both relatively energetic shorelines, or from SB, a more protected shoreline but with strong tidal currents. In contrast, in the low energy lagoon on HI, ORO were detected at the control treatment in both sediment and plant samples despite being almost 300 m away from the oiled treatment, indicating chemical contamination had spread throughout the enclosed lagoon.

ORO were detected in all oiled treatments, but not always in both sediment and plant fractions. Detected concentrations ranged from 101 to 67,500 ppm in the sediments and from 465 to 31,300 ppm in the smooth cordgrass tissues (Table 7.1). There tended to be as much variability among quadrats within the impacted treatment as there was between oiled locations; for example, MP sediments in October 2010 ranged between 0, 6,470, and 67,500 ppm in the three quadrats each located 5 m apart (Table 7.1). This was confirmed by visual observation at the time of collection, with heavy oil coating the sediment surface to a depth of up to 3 cm in the quadrat with the heaviest ORO concentration, whereas no visible oil was seen in the quadrat that gave a result of 0 ppm. This pattern was consistently encountered in oiled treatments that had visible oil at the time of collection (MP, HI, SB), and indicates the degree of patchiness in both the initial oiling event(s) and the subsequent breakdown or removal of oil contaminants by natural processes.

PAHs were generally not detected in samples analyzed from fall 2010. The only PAH detected was chrysene, found in one plant tissue sample each at the oiled treatment from SB (0.47 ppm) and HI (0.68 ppm). As chrysene was found in the absence of other byproducts of combustion, such as fluoranthene, pyrene, or benzo[a]anthracene, this result suggests crude oil contamination was subsequently taken up or sorbed into the plant tissues. This conclusion is supported by the dominance of chrysene in the weathered oil collected at MP (see the Time-Series Contamination section of this chapter).

Time-Series Contamination

The oil on the plants at MP in July 2010 (Figure 7.4) was analyzed further to determine the hydrocarbon composition, and compared to a reference oil sample obtained from the

TABLE 7.1

Oil Contamination in Each of Three 1-m² Quadrats (Q1–Q3) Sampled in Fall 2010, 3 Months after Initial Oiling, from Four Locations along the Mississippi Coastline

Sample	Location	Oiling	Q1	MRL	Q2	MRL	Q3	MRL
Sediment	GBNERR	Control	ND	50	ND	50	ND	50
		Heavy	ND	50	ND	50	ND	50
	Marsh Point	Control	ND	50	ND	50	ND	50
		Heavy	6,740	2,500	ND	50	67,500	2,500
	Horn Island	Control	101	50	67	50	ND	250
		Heavy	173	100	109	50	101	50
	Waveland	Control	ND	50	ND	50	ND	50
		Heavy	3,950	1,500	ND	500	130	50
Plants	GBNERR	Control	ND	930	ND	735	ND	1,280
		Heavy	1,070	600	ND	1,340	963	690
	Marsh Point	Control	ND	300	ND	300	ND	750
		Heavy	ND	750	ND	600	ND	1,500
	Horn Island	Control	ND	150	192	150	170	150
		Heavy	547	360	1,740	1,500	465	300
	Waveland	Control	ND	1,130	ND	1,140	ND	1,050
		Heavy	3,360	1,020	31,300	11,300	13,100	9,000
			P	**MRL**	**Fs**	**MRL**	**Fg**	**MRL**
Animals	GBNERR	Control					ND	268
		Heavy						
	Marsh Point	Control	ND	485	ND	517	ND	600
		Heavy	ND	500				
	Horn Island	Control					ND	526
		Heavy			ND	441	ND	455
	Waveland	Control						
		Heavy	ND	455				

Note: Grand Bay National Estuarine Research Reserve (GBNERR) and Marsh Point are both eroding shorelines, whereas Horn Island Garden Pond and Waveland are both protected, depositional shorelines. Sediments, plant tissues, and animal tissues (P = *Palaeomonetes* spp., Fs = *Fundulus similis*, Fg = *Fundulus grandis*) were analyzed for the concentration (mg/kg) of oil range organics (C_{19}–C_{36}) using EPA method 8015C, MRL is minimum reporting limit that varies depending on the sample dilution, ND indicates ORO was below detection for that sample.

MC252 well. The Pr/Ph ratio of the MP oil (1.0) is close to that of the MC252 well (0.9), suggesting that MP oil originated from the BP oil spill. The overall hydrocarbon compositions of the MP oil, including alkanes, PAHs, and alkylated PAHs, are similar to the oil collected on the sea surface of northern GOM during the oil spill, a further indication of MC252 origin (Liu et al. 2012). The MC252 oil shows a predominance of *n*-alkanes with short carbon chains (C_9–C_{18}) at concentrations exceeding 3 mg/mL (~3000 ppm) and a lower concentration of longer carbon chains (C_{20}–C_{38}) present (Figure 7.5a). In comparison, the oil

FIGURE 7.4
(**See color insert.**) Time-series photos of oil contamination and smooth cordgrass plant responses at Marsh Point from July 2010 (2–13 days after oiling) until July 2011 (351 days after oiling). The left panels show the same general area of the heavily oiled treatment, marked by either two tall bamboo marker posts or a short, red-painted wooden stake. The right panels show close-ups of the 1-m^2 quadrats that were sampled for contamination and show the condition of the weathered oil over time, as well as the recovery of the initially heavily oiled plants through the seasons.

from plants at MP was dominated by intermediate *n*-alkanes (C_{19}–C_{26} at a concentration ≥0.6 mg/g TSEM), and longer carbon chains (C_{27}–C_{38} at a concentration ≤0.6 mg/g TSEM) (Figure 7.5b). There was a very low concentration of short carbon chains (C_{10}–C_{18}), unlike what was observed in the reference MC252 oil sample.

Naphthalene dominates PAHs in the MC252 oil, accounting for 64% of the total PAHs, followed by phenanthrene and fluorene (Figure 7.5c). Chrysene and other PAHs are minor components. These distribution patterns of alkanes and PAHs are typical of fresh crude oil samples. For the PAHs in the MP oil, naphthalene accounted for only 9% of the total, and other PAHs with 2–3 aromatic rings, including fluorene and phenanthrene, were minor components relative to the reference MC252 oil (Figure 7.5d). Chrysene became the dominant (53%) PAH in the MP oil. Other PAHs with a high number of aromatic rings (>4), from benzo[b]fluoranthene (BbF) to benzo[ghi]perylene (BgP), became more concentrated relative to the reference oil (Figure 7.5). These results indicate that the oil present at MP in July 2010 was substantially weathered, as the low-molecular-weight *n*-alkanes and PAHs were preferentially lost. This agrees with field observations at the time of a rust-red colored, very sticky, tar-like consistency oil coating on the plants (Figure 7.4). It is also evident that PAHs overall had been more weathered than the alkanes, as the ratio of total alkanes/PAHs was 70 in the crude oil, but was as high as 1400 in the MP oil. Thus, by the time the oil had reached the salt marsh it had weathered to an extent that it was less toxic to smooth cordgrass relative to the potential toxicity of oil at the well, based on the PAHs.

At MP (the high-energy coastline), the initial sampling in July 2010 (13 days post-impact) was from a single quadrat in each of the control and heavily oiled treatments (Table 7.2). ORO concentrations at this time were orders of magnitude higher in the plant tissues (2,660–137,000 ppm) than in the sediments (32–68 ppm), concurring with observations that the oil had arrived during a storm-driven high tide and coated the plants within 3–5 m of the shoreline. Little oil was deposited on the sediments at the time of initial oiling (Figure 7.4). In later months, the pattern of oiling changed with substantially more ORO found in sediments in October 2010 (88 days postimpact), ranging from 6,740 to 67,500 ppm, compared to the plant tissues where no oil was detected in any of the three quadrats sampled (Table 7.2). This concurs with observations made at this sampling event, which indicated that regrowth of new plant shoots had occurred since the initial oiling and the oil-coated shoots had died, falling over and leaving the sediment prone to oil transfer from the plant detritus (Figure 7.4). During the spring (March 2011), ORO were detected in sediments (96 ppm) and plant tissues of one quadrat in the control treatment, but at substantially lower concentrations than those found in the quadrats (≥3600 ppm) in the heavily oiled treatment (Table 7.2). Finally, 1 year after oiling, samples from July 2011 (Figure 7.4, Table 7.2) indicate minimal sediment contamination (164 ppm), but some residual contamination occurred in plant tissues at both the control treatment (563–909 ppm) and the heavily oiled treatment (1270 ppm). These final results suggested that oil contamination may have spread out from the original impact area and resulted in low-level contamination over a broader area, including the plants at the control treatment.

At HI (the low-energy coastline) sampling occurred only twice, in fall 2010 and 1-year postimpact in summer 2011 (Table 7.2). In November 2010, we detected oil contamination in both sediments and plant tissues at both the control and heavily oiled treatments, with greater ORO concentrations at the oiled treatment than the control (Table 7.2). As the HI location is in a lagoon, we anticipated that oil would persist in the sediments for longer than at the MP location. Results from August 2011 confirm this with persistent low-level contamination in all three quadrats sampled at the oil-impacted treatment (59–86 ppm),

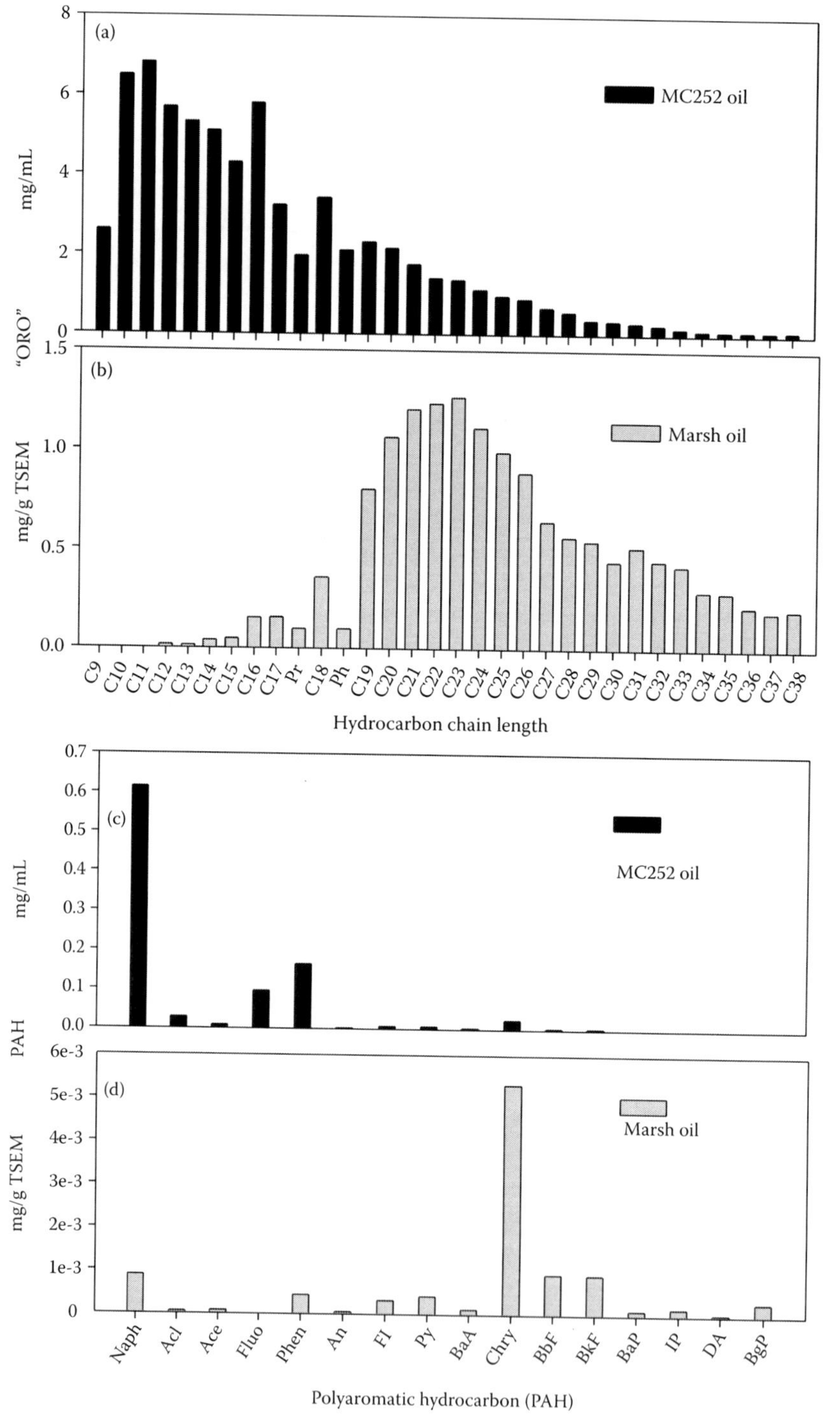

FIGURE 7.5
Concentrations of hydrocarbons in crude oil from the Macondo Canyon well (MC252), and weathered oil attached to smooth cordgrass tissues at Marsh Point in July 2010. (Modified from Liu et al., *Environ. Res. Lett.* 7, e035302, 2012.) The measured hydrocarbons include (a, b) *n*-alkanes (C_9–C_{38}), pristane (Pr), and phytane (Ph), and (c, d) PAHs. The hydrocarbon concentrations in MP oil were normalized to total solvent-extractable materials. These chemical fingerprints indicate substantial weathering of the oil had occurred by the time it washed ashore in the marsh.

TABLE 7.2

Oil Contamination in Each of Three 1-m² Quadrats (Q1–Q3) Sampled (a) Approximately Quarterly (Only One Quadrat Was Sampled in July 2010) from Marsh Point, an Erosional Shoreline and (b) Sampled Semiannually from Horn Island Garden Pond, a Depositional Shoreline

Sample	Month	Oiling	Q1	MRL	Q2	MRL	Q3	MRL
Marsh Point—Erosional								
Sediment	July-10	Control	69	20				
		Heavy	32	20				
	October-10	Control	ND	50	ND	50	ND	50
		Heavy	6,740	2,500	ND	50	67,500	2,500
	March-11	Control	ND	50	ND	50	97	50
		Heavy	7,040	500	ND	50	3,600	500
	July-11	Control	ND	50	ND	50	ND	50
		Heavy	ND	50	164	50	ND	50
Plants	July-10	Control	2,660	2,030				
		Heavy	137,000	54,000				
	October-10	Control	ND	300	ND	300	ND	750
		Heavy	ND	750	ND	600	ND	1,500
	March-11	Control	ND	600	ND	1,350	953	750
		Heavy	375	300	ND	900	ND	3,000
	July-11	Control	678	660	909	570	563	450
		Heavy	ND	810	ND	705	1,270	705
Horn Island—Depositional								
Sediment	November-10	Control	101	50	67	50	ND	250
		Heavy	173	100	109	50	101	50
	August-11	Control	ND	50	ND	50	ND	50
		Heavy	65	50	87	50	59	50
Plants	November-10	Control	ND	150	192	150	170	150
		Heavy	547	360	1,740	1,500	465	300
	August-11	Control	362	300	ND	1,020	ND	510
		Heavy	ND	1,800	ND	690	604	420

Note: Sediments and plant tissues were analyzed for the concentration (mg/kg) of oil range organics (C_{19}–C_{36}) using EPA method 8015C, MRL is minimum reporting limit that varies depending on the sample dilution, ND indicates ORO was below detection for that sample.

whereas no ORO were found in sediments at the control treatment (Table 7.2). Results from plant tissue samples also suggest that ORO contamination 1 year after initial oiling was greater at the oiled treatment (604 ppm) than at the control treatment (362 ppm). Like the situation at MP, these final plant tissue results suggested that oil contamination may have spread out from the original impact area and resulted in low-level contamination over a broader area, including the previously unoiled control treatment within the lagoon.

In addition to the results reported in Table 7.2 from the heavily oiled treatment at HI, one extra sediment and two extra plant tissue samples were collected in August 2011 from an area with visibly oiled dead plant stems that had spent inflorescences on them, indicating they were standing detritus leftover from the 2010 growing season. The sediment sample collected from immediately adjacent to the dead stems was found to have 1250 ppm (MRL = 100), and the 2010 plant detritus samples had 8430 ppm (MRL = 6480) of ORO, whereas the plant tissues that were green and from new growth in 2011 had ND (MRL = 3300) contamination. These three extra samples indicate that at the heavily oiled treatment 1 year later, there was still visible oil contamination on the previous year's plant detritus and contamination in the adjacent sediments, but that new plant growth in the same sediments was not detectably contaminated. This result highlights the very patchy nature of the oil impacts and the difficulty in determining oil impacts to plant function based on contamination data alone.

Photosynthesis Responses to Oil Contamination

Pulsed Amplitude Modulation Chlorophyll Fluorescence at Marsh Point

Optimal values of PQY range from 0.75 to 0.83 in field measurements (Krause and Weis 1991; Maxwell and Johnson 2000), with EQY often slightly depressed compared to PQY due to photochemical quenching (i.e., active electron transport). Both the EQY and PQY of the plants at the control treatment at MP remained above 0.6, even during the winter when this species underwent senescence (Figure 7.6). In contrast, both EQY ($F_{2,27} = 179$, $p < .0001$) and PQY ($F_{2,27} = 80.9$, $p < .0001$) were significantly depressed immediately (13 days, July 2010) after the oil impact compared to the control plants, and yields in the heavily oiled plants were significantly lower than those in the medium impact treatment. Quantum yields show rapid recovery of photosynthesis in plants at the medium and heavily oiled treatments to values not significantly different from plants at the control treatment by 88 days (October 2010) post-spill (Figure 7.6). Quantum yields remained depressed in plants at all three treatments during the winter months (88–195 days), but with no significant difference among treatments. The yields recovered to preimpact values by March 2011 (day 245), however, PQY was significantly lower ($F_{2,27} = 11.7$, $p < .0002$) in plants from the two oil-impacted treatments than plants at the control treatment. In July 2011 (351 days), there was no longer any indication of chronic stress in plants at the two oiled treatments compared to the control at MP, suggesting that a complete recovery of photosynthesis had occurred (Figure 7.6).

The RLC results support this same trajectory of rapid recovery when comparing plants from the heavily oiled treatment to plants from the control treatment at MP (Figure 7.7a). Data obtained from the medium-impacted plants are not plotted, but were aligned between the responses shown for the control and the heavily oiled plants. The P–I curves on day 13 show that plants at the heavily oiled treatment have a very depressed response (ETR < 5 μmol electrons/m^2/s) compared to the plants at the control treatment (ETR > 10 μmol electrons/m^2/s). In all subsequent sampling dates (days 88–351), the P–I curves for plants at the two treatments are not substantially different (Figure 7.7a). When we plotted the data in control versus impact space and fit a straight line with intercept of zero, the slope of the fit for days 88–351 ranged from 0.89 ($R^2 = 0.957$) to 1.17 ($R^2 = 0.992$), with a slope of 1.0 being a perfect match. Also, 95% confidence intervals for the means included the 1:1 line indicating no significant difference between the ETR obtained from the heavily impacted plants compared to the controls. However, immediately after the initial oiling in July 2010, the slope of the fit was 0.18 ($R^2 = 0.187$) and was substantially less than on the subsequent days. The 95% confidence intervals also did not overlap the 1:1 line, indicating a significant depression of the ETR in plants at the heavily oiled treatment compared to the control at MP.

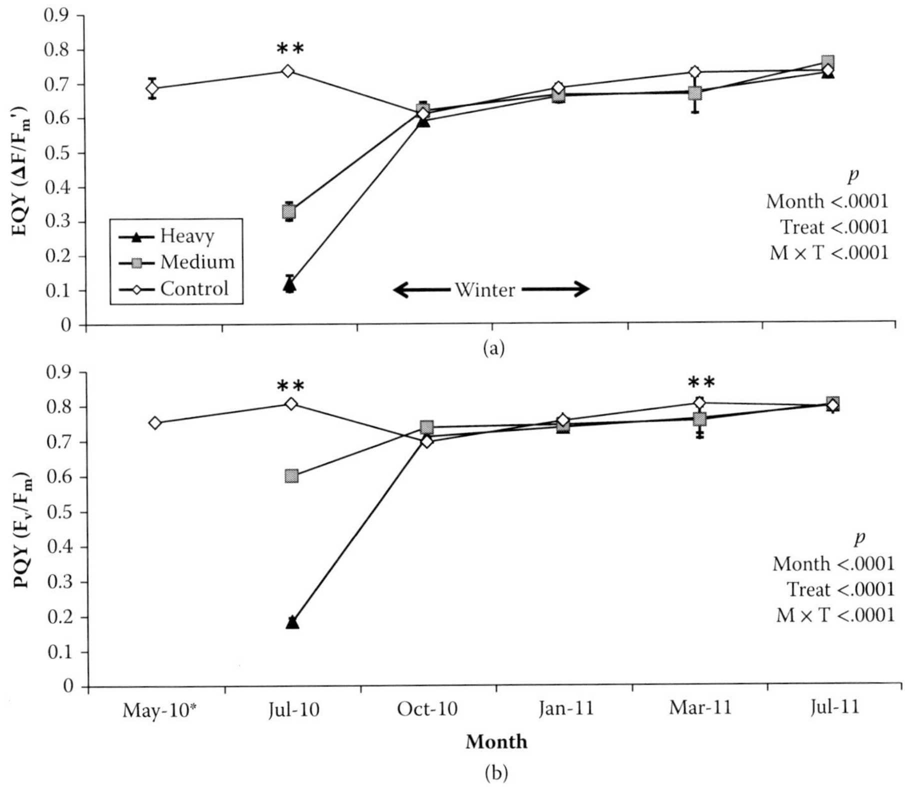

FIGURE 7.6
Time series of quantum yields of photosynthesis measured on smooth cordgrass at three locations (control, medium oil, and heavily oiled) at Marsh Point over 1 year. (a) Effective quantum yield (EQY) is the response obtained from leaf samples that are actively photosynthesizing under ambient light conditions. (b) Potential quantum yield (PQY) is the response measured in dark-adapted leaf samples where photosynthesis has been effectively "shut down," and this is often used as a proxy for plant stress. May 2010 measurements were obtained from plants located in the lower Pascagoula River delta, about 18 km east of Marsh Point and 60 days prior to oil impacts. ** Indicates a significant ($p < .05$) difference due to oiling for that month.

The chlorophyll fluorescence data obtained with the PAM indicate a rapid recovery of the light reactions of photosynthesis in the plants at both the medium and the heavily oiled treatments to levels observed in plants at the control treatment by October 2010, 88 days after the impact (Figures 7.6 and 7.7). The results obtained during winter months tended to be depressed compared to spring and summer months, reflecting winter senescence known to occur in this species. The rapid recovery of photosynthesis, as measured by chlorophyll fluorescence, indicates a short acute impact and minimal chronic impact of the oil to smooth cordgrass at MP.

LI-COR Fluorescence and Gas Exchange at Marsh Point

Chlorophyll fluorescence (PQY) measurements collected with the LI-COR at MP showed a recovery of photosynthesis during the fall 2010 months to control levels (Figure 7.8), whereas CO_2 flux (P_{max}) remained depressed in the control and medium impact treatments. Statistical analyses of the recovery were complicated by the lack of control data for August, September, and December 2010, so these months had to be dropped from the dataset. The

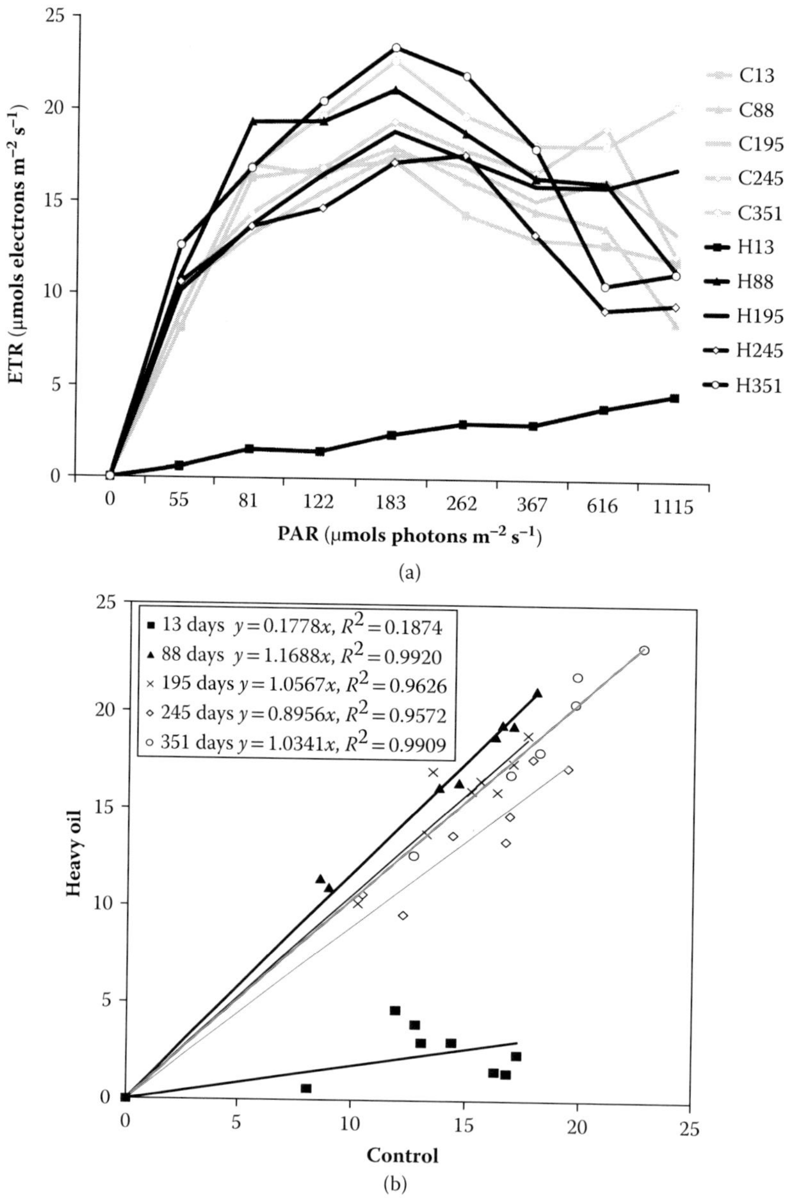

FIGURE 7.7
(a) Rapid light curves (RLC) of photosynthesis measured on smooth cordgrass in two treatments (control and heavily oiled) at Marsh Point over 1 year showing the rapid recovery of the light reaction of photosynthesis in oil-impacted plants (within 88 days) to performance levels similar to the adjacent control plants. Legend: C is the control, H is the heavily oiled treatment, and the number is days postimpact. For example, H13 = heavily oiled treatment 13 days after impact and C351 = control treatment at 351 days (1 year) after impact. (b) Scatterplot of observed (control treatment) versus predicted (heavily oiled) for the RLC data. Linear regression ($y = mx$) and R^2 values for each date are presented; 95% confidence intervals are not shown to improve clarity.

PQY results showed a significant ($F_{2,27} = 122$, $p < .0001$) difference in July 2010 among the three treatments at MP. Plants at the medium impact treatment were more stressed than the control plants and those at the heavily oiled treatment were more stressed than the

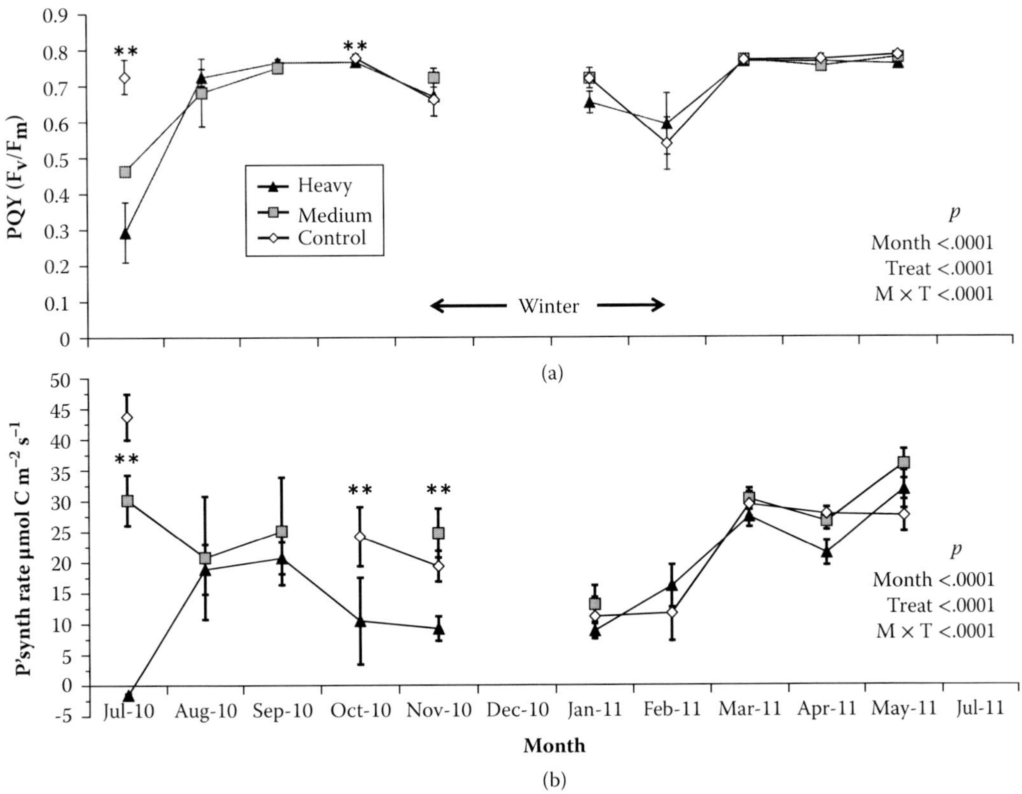

FIGURE 7.8
(a) Dark-adapted chlorophyll fluorescence yield and (b) light-saturated (1200–2000 µmol photons/m^2/s) rate of P_{max} measured monthly in smooth cordgrass at three treatments (control, medium, and heavily oiled) at Marsh Point over 1 year. ** Indicates a significant ($p < .05$) difference due to oiling for that month.

medium-impacted plants (Figure 7.8). Monthly measurements from August 2010 until June 2011 at MP showed a very rapid recovery of PQY to control values (≥0.70) within 60–90 days (Figure 7.8a). During October 2010, only the heavily oiled plants had a mean PQY significantly lower ($F_{1,18} = 15.4$, $p < .001$) than that measured in the control plants. PQY declined in all three treatments in November and remained depressed below 0.7 until March 2011. Smooth cordgrass has a seasonal dieback in winter resulting in depressed photosynthesis compared to summer (Figure 7.8). From March until June 2011, when measurements stopped, PQY remained high (>0.75) and steady at all three treatments. The chlorophyll fluorescence data indicate there was no chronic stress from the 2010 oil impacts in the plants measured at all three treatments in June 2011 at MP.

During July 2010, P_{max} measurements taken at saturating irradiances (1200–2000 µmol photons/m^2/s) also showed a significant ($F_{2,27} = 114$, $p < .0001$) difference among the three treatments, with control > medium > heavily oiled plants (Figure 7.8b). During the late summer and fall of 2010, the plants at the control and medium impact treatment exhibited a decline in P_{max}, whereas plants in the heavily oiled treatment initially showed an increase in P_{max} in August and September, before declining again in October–November. By October 2010, the heavily oiled plants had a mean P_{max} significantly lower ($F_{1,18} = 23.3$, $p < .0001$) than that measured in the control plants, whereas in November 2010, the heavily oiled plants were significantly lower ($F_{2,27} = 9.1$, $p < .0009$) than both the medium impact and control plants (Figure 7.8b). By November 2010, P_{max} declined at all three treatments and then remained below 20 μmol $CO_2/m^2/s$ during the winter. In March 2011, P_{max} began to increase with the spring regrowth and returned to a range of 20–30 μmol $CO_2/m^2/s$ (Figure 7.8b). The P_{max} measurements also indicate there was no chronic stress from the 2010 oil impacts in the plants measured at all three treatments in June 2011 at MP. The LI-COR data confirm our hypothesis that chlorophyll fluorescence recovers very rapidly from acute stress (within less than 2 months after the impact), whereas the gas-exchange measurements show a slower recovery with some chronic depression evident in the heavily oiled plants for up to 4 months after the impact.

Seasonal mean P–I curves measured in smooth cordgrass at the control and heavily oiled treatments at MP also indicated that recovery of the heavily oiled plants took between 4 and 6 months (Figure 7.9). Initially in summer 2010, immediately following oiling of the plants, the P–I curve for the heavily oiled plants was essentially flat near zero, whereas the P–I curve of the control plants had a P_{max} around 45 μmol $CO_2/m^2/s$ (Figure 7.9). Over the course of the following year, the P–I curve of the heavily oiled plants recovered back to levels measured in the same season in the adjacent control plants, with convergence of the P–I curves by winter (December–February). The depression of the P–I curve in control plants during the autumn (P_{max} = 20) and winter (P_{max} = 10) compared to spring (P_{max} = 25) and summer (P_{max} = 45 μmol $CO_2/m^2/s$) is an inherent natural response of this species to cold air temperatures and reduced day-lengths. The P–I curve data further strengthens

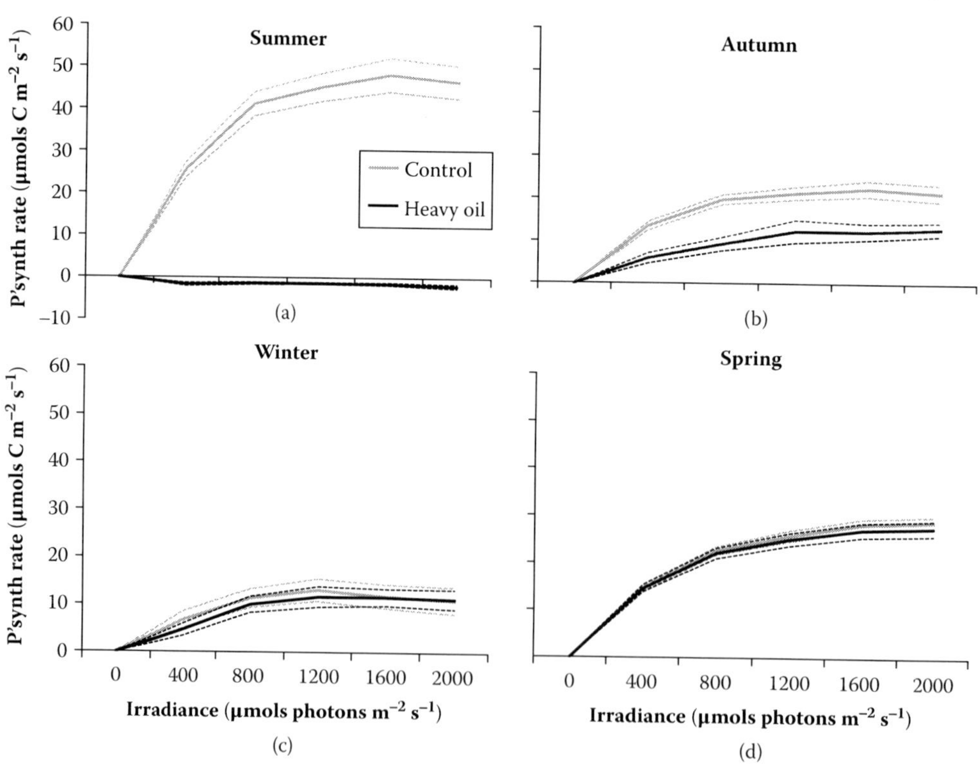

FIGURE 7.9
Seasonal mean (±1 se) photosynthetic gas exchange rates measured as photosynthesis–irradiance curves in smooth cordgrass in two treatments (control and heavily oiled) at Marsh Point over 1 year. Note the depression of photosynthesis in control plants during the fall and winter compared to spring and summer, an inherent natural response of this species to cold air temperatures and reduced day-lengths.

the previous findings that plants in the heavily oiled MP treatment required until winter/spring to fully recover photosynthesis to rates seen in the control plants, with no evidence of chronic effects by 1 year (July 2011) after the oil impacts.

One-Year Assessment

In July 2011, the quantum yields of plants in the control treatment and the heavily oiled treatment at the two locations, MP and HI, were compared. There were significant differences due to location for both EQY ($F_{1,35} = 9.05, p < .005$) and PQY ($F_{1,35} = 22.1, p < .0001$), and significant interaction effects (location × treatment) for EQY ($F_{1,35} = 7.91, p < .008$) and PQY ($F_{1,35} = 8.37, p < .007$). At the high-energy MP location, the quantum yields were not significantly different between the plants at the control and heavily oiled treatments (Figure 7.10). In contrast, at the low energy HI location, the mean EQY ($t_{1,18} = 2.47, p < .011$) and PQY ($t_{1,18} = 1.84, p < .041$) were both significantly lower in the plants at the heavily oiled treatment compared to the plants at the control treatment (Figure 7.10). Variances were also much greater in the heavily oiled treatments at both MP and HI compared to the control, reflecting a range of plants from those that were very stressed (EQY < 0.55) to those that exhibited less stress (PQY > 0.8).

The RLCs support this finding with the 95% confidence intervals of the mean ETR overlapping at all irradiances for the control and heavily oiled treatments in the MP location (Figure 7.11a). The slope of the linear fit between control and heavily oiled data points was 1.034 with an R^2 of 0.99. As expected from the previous quantum yield results, at HI the RLC confidence intervals did not overlap for the majority of the curve; exceptions were 55 and 367 μmol photons/m²/s (Figure 7.11b). The slope of the linear fit between control and heavily oiled data points was 0.788 with an R^2 of 0.86. These results indicate that the plants at the heavily oiled treatment continue to exhibit chronic depression of photosynthesis, as measured by chlorophyll fluorescence, up to 1 year after the initial oil impacts.

Plant stress in the heavily oiled treatment at MP was less than at HI as indicated by the quantum yields and RLC results (Figure 7.11). Combining these data with the previous results from the contaminant analyses (Table 7.2), the summer 2011 findings suggest that there were major differences in recovery after 1 year between MP and HI. At the depositional location (HI) there was prolonged and ongoing exposure to oil residues in the sediments that possibly continue to cause a chronic depression of photosynthesis in the smooth cordgrass plants. In comparison, at the erosional location (MP), there were fewer oil residues in the sediments, and the plants appeared to have recovered from the oiling impacts.

Discussion

Contamination

The DWH disaster was the largest accidental marine oil spill in U.S. history and released about 4.9 million barrels (7.0×10^6 m³) of crude oil into the open ocean (Crone and Tolstoy 2010). Oil gradually dispersed through the water column as its buoyancy caused it to rise toward the surface of the ocean from the 1544 m deep source at the MC252 wellhead (Camilli et al. 2010; McNutt et al. 2011; Oil Spill Commission 2011). Oil in the water was

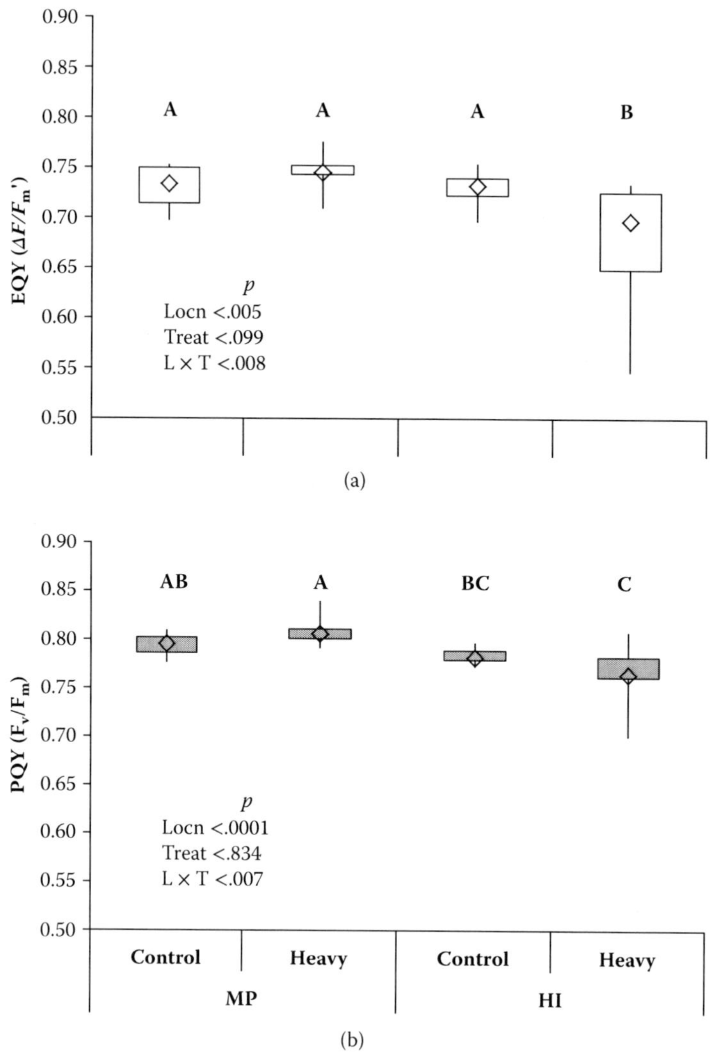

FIGURE 7.10
Boxplots comparing (a) effective quantum yield (EQY) and (b) potential quantum yield (PQY) responses at Marsh Point ([MP] erosional, exposed shoreline) versus Horn Island ([HI] depositional, protected shoreline) in July 2011, 1 year after initial oil impacts. Significant effects due to treatment and location are indicated by the different letter groups.

visible from April 24, 2010 onward (Oil Spill Commission 2011) throughout much of the summer in 2010 and was extensively monitored and mapped using satellite and aerial imagery (U.S. Government 2010; White House 2010; McNutt et al. 2011). Despite the leaking wellhead being contained by July 15, 2010, oil from the extensive offshore slicks that cumulatively covered up to 180,000 km^2 continued to disperse and eventually contaminated more than 1050 km of shoreline (Skytruth 2010). Oil periodically came ashore in Mississippi salt marshes, with peak frequency of oiling from early June until late September 2010. Timing of heavy oiling in the marsh was observed to coincide with tropical storm systems, including Hurricane Alex (June 30– July 1), Tropical Storm Bonnie (July 24–25), and Tropical Depression number 5 (August 12–13), which all brought strong southerly winds

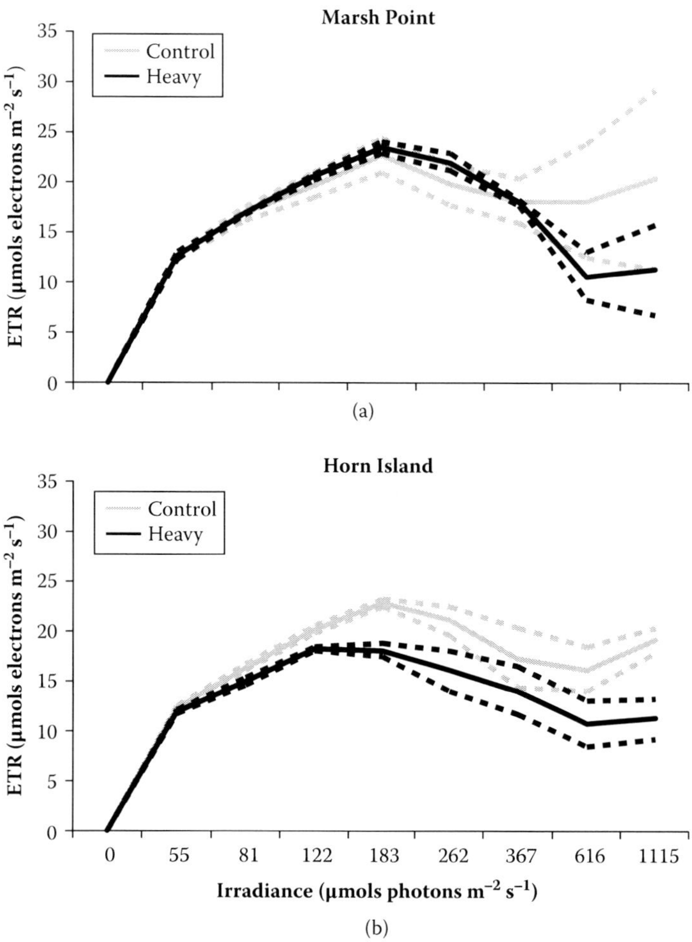

FIGURE 7.11
Rapid light curves (RLC) of photosynthesis measured on smooth cordgrass at two treatments (control and heavily oiled) within (a) Marsh Point (erosional, exposed shoreline) and (b) Horn Island (depositional, protected shoreline) locations in July 2011, 1 year after initial oil impacts. 95% confidence intervals of the mean are plotted as dashed lines.

and a storm surge that pushed ashore water contaminated with oil (NOAA 2010). The elevated tides, measured at the Gulf Coast Research Laboratory pier, associated with the storm surge (+0.6 m above MHHW) contributed to further inland penetration of the floating oil slicks than would otherwise have occurred with normal tidal amplitudes (0.47 m). Despite this, most oil contamination was primarily to smooth cordgrass within <10 m from open water in most areas, affecting the coastal fringe disproportionately.

The first part of our study assessed crude oil impacts in sediments, plant tissues, and selected nekton in salt marsh habitats along the Mississippi coastline. This information will be invaluable in determining the long-term ecological impacts and potential food web implications of the oil, which ended up being trapped in the delicate coastal marsh habitats of Mississippi and the northern GOM. Our contaminant sampling began after funds were obtained, and occurred about 90 days after the wellhead had been effectively sealed. This allowed sufficient time for any oil remaining in the water column to make it to the shoreline and begin to breakdown into less harmful constituents at the time of our synoptic coast-wide contaminant sampling in fall 2010. The areas we chose to study in detail were locations with prior documentation (photos, GPS coordinates) of oil impacts throughout summer 2010. The four areas selected for our study encompass marsh areas along the Mississippi mainland and barrier island coastlines and were among the most heavily contaminated locations. Despite this, the total area of heavily impacted marsh remained less than a few hundred square meters, and within this area, oil was spatially highly variable even at the submeter scale. Some of the heavily oiled areas were $<5 \times 10$ m^2 in total, and substantial patchiness of the oil within this size area was also observed. Many of the oil patches that were washed ashore were much smaller than this, on the order of <0.25 m^2, making it hard to detect without significant sampling effort on a fine spatial scale (1 m^2 or finer). Because of this highly localized impact, it was important to use a hierarchical sampling approach, based on prior knowledge of impact locations, to capture both within treatment (replicate quadrats) and among treatment (control vs. oil impacted) differences within a single location. Much of the shoreline remained clear of heavy oiling, making it difficult to extrapolate our findings to a Gulf-wide scale of contamination.

Our results suggest that the MC252 oil that reached Mississippi's salt marshes was highly weathered, which lessened the toxic impacts that potentially could have occurred with a more proximate spill location. In general, PAHs were not detected in the sediment or plant samples with the exception of two samples that contained low (<1 ppm) concentrations of chrysene. Sediment sampling by the U.S. EPA (EPA 2010) also found chrysene and other PAH contaminants at very low concentrations (2.9–39 ppb) in beach sediments off Bay St. Louis (report number BCH01-SD-201008) and Pascagoula (report number BCH04-SD-201008) in late summer 2010, whereas no PAH contamination was reported from a subtidal sediment sample near Ocean Springs (report number R4-10-B-SD-09152010). Earlier sampling prior to oil impacts at these locations in May 2010 found no PAH contaminants in the sediments (EPA 2010). These findings are further supported by the extensive "fingerprinting" of a single sample of weathered oil from MP immediately following the initial oiling in July 2010, where chrysene was the dominant PAH present in the sample. As PAHs are the more toxic fraction compared to long-chain alkanes, the lack of substantial PAH contamination in the samples from fall 2010 is a positive finding, indicating a less toxic form of weathered oil contamination was found in the salt marsh. The contamination detected in patches where oil was generally still visible at the time of sampling was composed of mainly longer carbon-chain *n*-alkanes (C_{19}–C_{38}), which have lower acute toxicity (Lin et al. 2012). ORO were always detected analytically when oil was visible at the time of sampling, but were also found in areas where oil was no longer visible (e.g., GBNERR, HI). When oil was not readily visible at the time of sample collection, the *n*-alkanes were mainly found in plant tissues. Without further petroleum biomarker analysis, it is difficult to evaluate the degree of contamination, as *n*-alkanes are also sourced from plant tissues such as cuticles. Some contamination of plant samples was found in the adjacent control treatment in summer 2011 at both MP and HI, suggesting that oil may have become dispersed along the coastline from the original impact site and now affected a wider area, albeit at a much reduced concentration.

No oil was found in any of the animal muscle tissue samples analyzed, suggesting no or only limited trophic transfer had occurred. However, the limited number of samples that satisfied the requirement for the minimum amount of sample needed (3 g) to detect contamination affects the robustness of this conclusion. An extensive and better designed sampling strategy is required to determine whether habitat contamination might be affecting the lower trophic-level populations in salt marshes (Kahn et al. 2005; Roth and

Baltz 2009; Whitehead et al. 2011). Some scientific studies conducted on offshore planktonic and larval populations detected potential contamination of the lower trophic guilds with entrained oil (Graham et al. 2010; Chanton et al. 2012; Ortmann et al. 2012; Vestheim et al. 2012); however, other studies indicated nondetectable contamination of commercial seafood or no discernible population response at higher trophic levels (Fodrie and Heck 2011; NOAA 2011; Carmichael et al. 2012). At the time of publication, no studies have conclusively linked oil-contaminated plant material as a source of ingested pollutants in coastal marine biota along the northern GOM.

Location-Specific Oil Impacts

The second part of our study assessed plant responses to crude oil impacts in the sediments and on plant tissues at two salt marsh locations with contrasting coastlines: a high energy, exposed, and eroding coastline at MP and a low energy, protected, and depositional coastline in a lagoon at HI. Salt marshes are highly sensitive to oil contamination (Gundlach and Hayes 1978; Jensen et al. 1998), and are ranked by the NOAA Office of Response and Restoration as 10 out of 10 in terms of vulnerability on their Environmental Sensitivity Index (NOAA 2012). For this reason, cleanup crews that were effective at removing oil that washed ashore on beaches were not able to remove oil in the salt marshes effectively, and in many instances were tasked not to do so in an effort to minimize further damage (Shogren and Gonyea 2010).

During the DWH oil spill and after the well was closed, more than 1050 km of shorelines along the GOM were exposed to weathered crude oil, with 210 km of those designated as moderately to heavily oiled, and 75 km of these occurred in Louisiana (Oil Spill Commission 2011; Silliman et al. 2012). Previous studies on oil spill impacts to GOM marsh plants have focused on Louisiana because spill events occur more frequently there, and there are larger marsh areas (Pezeshki and DeLaune 1993; Hester and Mendelssohn 2000; Pezeshki et al. 2000; Ko and Day 2004; Mendelssohn et al. 2012; Mishra et al. 2012). Some studies concerning the impact of the DWH oil spill on vegetation have been published recently. Mishra et al. (2012) assessed the ecological impact on the salt marshes along the southeastern Louisiana coast using photosynthetic capacity and physiological status through satellite and ground truth data, and found extensive reduction in photosynthetic activity during the peak of the growing season in 2010. Lin and Mendelssohn (2012) documented variable impacts depending on oiling intensity in the Bay Jimmy area of northern Barataria Bay, Louisiana. As of the fall of 2011, many of the most heavily oiled shorelines at that location had minimal to no recovery (Mendelssohn et al. 2012). In Mississippi, some salt marshes experienced crude oil impacts in summer 2010 and our four study locations represent the high-end of oil contamination for the documented impact locations in the state. Despite these and other ongoing studies, data on the impact and recovery of the GOM plants are still sparse. Nevertheless, this chapter enhances our understanding of the impacts of the DWH oil spill on photosynthesis of smooth cordgrass, the dominant marsh plant in this ecosystem.

As most of the time-series research documenting acute and chronic impacts to photosynthesis was done at the MP location, we will discuss these results first. In July 2010, almost 2 weeks after the initial arrival of the oil during Tropical Depression number 2 on July 8–9 (NOAA 2010), we were able to obtain initial data on oil contamination and plant stress using the PAM and LI-COR. The thick oil coating immediately affected leaves within the heavily oiled treatment and appears to have resulted in acute leaf effects by blocking transpiration and gas exchange through the stomata (Ferrell et al. 1984; Snedaker et al.

2001). This oiling also increased leaf temperatures, which caused them to absorb large quantities of infrared solar radiation and resulted in leaf mortality during the first month. Temperature over the heavily oiled plants was discernibly higher than at the adjacent control treatment when collecting the chlorophyll fluorescence data during the midday hours of an already hot summer day. At the beginning of the crude oil impact (July 2010), the concentration of ORO was high on the plants and low in the sediment. Despite the acute oiling stress and loss of existing leaves, the plants at MP were able to recover by growing new leaves and shoots from intact, unoiled roots within 2–3 months after the initial contamination. The oil concentration on the plants decreased over time but increased on the sediment because of gravitational migration of the oil down to the sediment surface. New growth was evident in both the medium and heavily oiled treatments by the October 2010 sampling event, 88 days after initial oiling, with the oiled dead leaves and stems laying prone on the sediment surface (Figure 7.4). This rapid regrowth of new leaves and shoots from the intact root–rhizome complex afforded the smooth cordgrass plants a short window of opportunity for additional carbon fixation and energy storage prior to the onset of winter senescence, and contributed to a successful recovery the following spring. It is likely that an oiling event later in the year may have resulted in higher mortalities of individuals, as the plants would not have been able to store sufficient reserves to survive dormancy for a prolonged period of many months (Baker 1971; Webb et al. 1985).

A second potential reason for the quick natural degradation of the remaining oil residues on the plants and on the sediments at the MP location over the course of the year-long study is the physical break down of the tar-mats and their removal by waves and tides. Natural degradation processes of the oil in the marsh not only include weathering (temperature, sunlight) but also physical removal by waves and tides (Mendelssohn et al. 2012). In particular, locations such as MP and GBNERR, which are exposed to the predominant southeasterly winds that are common during summer, experienced substantial tidal "cleaning." Oil was removed during high tides and during periods of strong wave action created by tropical storms in the GOM. Warm temperatures, often exceeding 35°C during the day, are also common in the region from June through September promoting rapid volatilization of lighter carbon fractions. Oil that was observed to persist over time at MP quickly became more asphalt-like in its appearance and consistency, which correlates to reduced toxicity (Irwin et al. 1997). At least one location that was sampled is on a well-documented eroding shoreline, Point Aux Chenes in GBNERR, having lost 2.5 m of coastline per year, on average (Hilbert 2006; Otvos and Carter 2008). This means that the oil-impacted sediments and plants could be lost to open coastal waters within 1–3 years. Similar erosion rates have occurred in some Louisiana marshes (Day et al. 2000; Silliman et al. 2012), suggesting similar fates to oiled salt marsh habitats in the Mississippi River Delta region.

In contrast to the rapid removal of oil from sediments at MP, the sediment inside the protected lagoon at HI was still contaminated (59–87 ppm) by oil in August 2011. At this location, plants exhibited significantly reduced photosynthesis compared to the HI control treatment and previously oiled plants at MP. The difference in residence times between the erosional edge location (MP) and lagoon location (HI), in part, helps to explain why oil persisted in HI sediments. Oiled and standing dead shoots from 2010 were also still present at HI, and sampling of these confirmed there was still heavy oil contamination (ORO = 8340 ppm), further indicating the lack of removal of the contaminants from this low-energy environment. A location with similar characteristics to HI is Bay Jimmy in northern Barataria Bay, Louisiana, which is also a low-energy shoreline with chronic DWH oil pollution problems (Schleifstein 2012; Silliman et al. 2012). Despite cleanup efforts, oil persists over 1 year later and is causing ongoing plant mortality and loss of coastline (Silliman et al. 2012). Oil trapped in waterlogged fine sediments is known to pose a long-term threat (Teal et al. 1992; Burns et al. 1994) in low-energy coastlines, explaining why salt marshes and mangroves rank as the most susceptible coastal habitats to oil spills (Michel et al. 1978; Jensen et al. 1998). This is because the anaerobic conditions in these sediments inhibit the complete breakdown of oil, resulting in pools of organic and heavy metal pollutants that are extremely slow to degrade (Lin et al. 1999; Andrade et al. 2004). The eventual result is plant mortality from root die-off (Proffitt et al. 1997; Ko and Day 2004), and salt marsh recovery can be further inhibited by new recruits that also fail to establish in the contaminated sediments. In both salt marsh and mangrove habitats, this is the key reason that oil pollution poses a threat of chronic stress in low-energy locations (Proffitt et al. 1997; Snedaker et al. 2001).

These examples from Mississippi and Louisiana demonstrate the importance of coastal energetics in the persistence of oil pollutants and their chronic toxicity to the vegetation. Salt marshes in the GOM are often found along eroding edges and are constantly migrating shoreward in response to coastline erosion (DeLaune et al. 1983; Turner 1990). In sections of coastline that are more exposed and are eroding, oil was removed rapidly by tidal and wave energy and the plants are recovering with new growth. If the plant roots are killed by oil contamination, however, the resulting erosion of the sediments will convert the previously oiled coastal fringe into open-water subtidal habitat (Silliman et al. 2012), diluting and dispersing the remaining hydrocarbons back into the water column.

Acute and Chronic Photosynthesis Responses

Chlorophyll fluorescence is a well-established technique to rapidly and noninvasively measure photophysiological processes in vivo and has been used successfully to demonstrate physiological stress in a wide variety of plant species (Critchley and Smilie 1981; Havaux and Lannoye 1983; Bowyer et al. 1991; Filiault and Stier 1999; Maxwell and Johnson 2000). Chlorophyll fluorescence of PSII can provide an instantaneous measure of the EQY under prevailing ambient light conditions (Genty et al. 1989). Alternatively, more standardized differences among leaves can also be determined by measuring the PQY of dark-adapted samples. Generally, the maximum possible proportion of the solar energy absorbed into photosynthesis is around 83%, equivalent to a quantum yield of 0.830 (Maxwell and Johnson 2000). As plants become stressed, reductions in the quantum yield (either EQY or PQY) indicate a reduction in the efficiency with which light is converted to photosynthetic product and subsequently plant growth or reproductive output (Schulze and Caldwell 1990; Krause and Weis 1991; Rohacek and Bartak 1999).

Another technique commonly used to quantify photosynthesis is by measuring leaf gas exchange (Šesták et al. 1971; Farquhar et al. 1980; Collatz et al. 1992; Long et al. 1996; Long and Bernacchi 2003). With the gas-exchange technique, the rate of CO_2 assimilation is determined from the change in the CO_2 mole fraction in a chamber or by the eddy covariance method (Long et al. 1996). It is nearly instantaneous and nondestructive, and allows measurement of the total carbon gain by a plant leaf, which can then be extrapolated to the whole plant or stand (Long et al. 1996). As environmental stressors generally reduce CO_2 assimilation, the gas-exchange technique has been applied extensively to study plant responses under stressed conditions (Ciompi et al. 1996; Lima et al. 1999; Pereira et al. 2000; Mielke et al. 2003; Huang et al. 2004). Some studies have also shown that a pronounced decline in gas exchange was the first response to acute stress due to stomatal closure, and that the damage to PSII, reflected in reduced quantum yields, occurred later

(Eastman and Camm 1995; Guidi et al. 1997; Souza et al. 2004). Thus, measurements of leaf gas-exchange complement data obtained by the chlorophyll fluorescence method for quantifying photosynthesis. In our study, by measuring photosynthesis under field conditions at oil-impacted and control treatments simultaneously, one can determine whether the oiled plants are under additional stress compared to uninjured controls.

The photosynthesis data collected with the PAM/LI-COR instruments and field observations suggested a faster recovery in plants at MP, where oil was removed by physical (waves/erosion) processes, than in plants at HI, where there was little physical removal of oil from the quiescent lagoon. The photosynthesis results from these two locations indicate a different recovery trajectory after oiling, with plants at MP showing recovery to control rates within less than 6 months. In contrast, the plants at HI still exhibited signs of depressed photosynthesis even 1 year after oiling, and contaminant sampling indicated that oil persisted in the sediments and on standing detritus at this location. It is likely that plant photosynthesis at HI may take up to 2 years or longer to fully recover to pre-spill rates, which requires further monitoring.

One of the potential difficulties with the method used to measure photosynthesis is the problem of scaling up from leaf-level measurements to whole-plant stress (Krause and Weis 1991; Long et al. 1996; Maricle et al. 2007). One solution to this problem is to collect many data points on an individual and determine whole-plant stress from the resulting distribution of the data. To ascertain the stress condition of a population of individuals, many similar measurements are required before the underlying pattern can emerge. Luckily, the chlorophyll fluorescence measurements are very rapid and it is easy to obtain many measurements at a site in a short period, dark-adaptation time required for PQY measurements notwithstanding. This is not the case for the gas-exchange measurements with the LI-COR, where a single P–I curve took as many as 20–30 minutes to complete. Therefore, the gas-exchange technique requires much more time to collect a statistically robust data set, reducing the investigator's ability to obtain sufficient samples to determine population-level condition with a high degree of confidence. As photosynthesis is the initial process that drives the subsequent increase in biomass and shoot density over time, it can be used as an early indicator of recovery. In this study, we found that smooth cordgrass photosynthesis recovered rapidly on a high-energy eroding shoreline, even after heavy oiling, suggesting a high potential for recovery of shoot density and biomass, two variables that are more routinely collected (Lin and Mendelssohn 1996, 1998; Michel et al. 2009). Compared with other studies of DWH oiling on smooth cordgrass habitats, our findings from MP are in agreement with results reported from exposed coastlines by Mendelssohn et al. (2012) and Mishra et al. (2012); however, in heavily oiled and wave-protected areas recovery was proceeding at a slower rate (Silliman et al. 2012) analogous to our findings at HI. Monitoring for recovery from oil impacts should ideally involve both photosynthesis and growth measurements, but growth is slower to respond than photosynthesis.

An additional caveat that is important to consider in any photosynthesis monitoring is that seasonal changes in plant photosynthesis rates may be misinterpreted as chronic oil stress effects. In the data collected over the course of 1 year, the seasonal dormancy of smooth cordgrass was clearly evident as depressed photosynthesis during the winter months (Figure 7.12). It was important, therefore, to have data from adjacent unimpacted control treatments at each location to properly interpret whether depressed photosynthesis was a function of oil pollution or whether it was a response to the low temperatures and reduced day-lengths that induce winter dormancy in this species. Failing to recognize this important natural process could easily have been misinterpreted as a chronic depression due to oiling, which would have been incorrect (compare Figures 7.1 and 7.12).

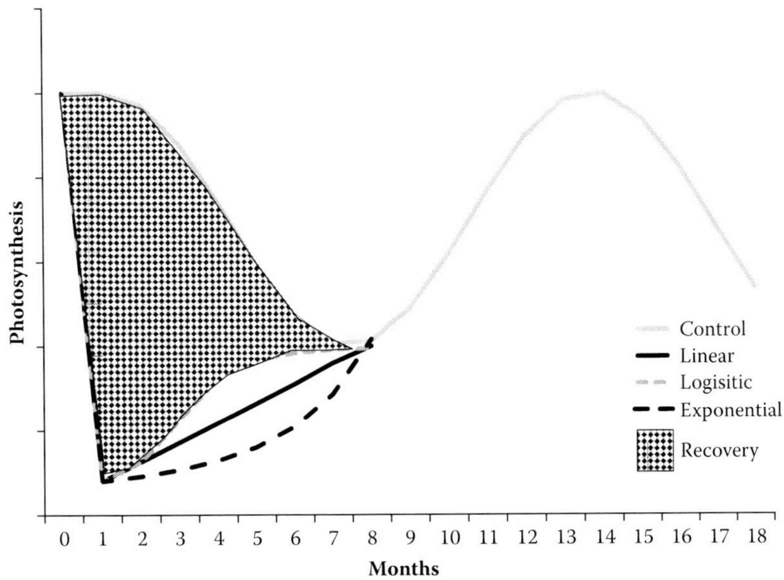

FIGURE 7.12
Refined conceptual diagram of time it took to recover photosynthesis in smooth cordgrass back to control levels following the oil spill impacts. In this figure seasonal differences in photosynthesis in the control plants exhibit a sine wave response. Three different recovery trajectories are shown (months 1–8; linear, logistic, exponential). The shaded impact area corresponds to the loss of photosynthesis due to the oil spill in summer and fall 2010, with recovery to control levels by the winter following a logistic response.

During the winter months, the contrast between the control and the two impacted locations in the field data was not as obvious as during the growing season earlier in 2010, which could be explained either by complete recovery of photosynthesis from oil impacts or by seasonally lower air temperature, which is a limiting factor keeping photosynthesis uniformly low. To better understand the interplay of recovery from oil impacts and seasonality on the photosynthesis dynamics, we developed a Hierarchical Bayesian (HB) model to simulate these two sources of plant stress from the field data obtained at MP in a closely related study (see Wu et al. [2012] for details on the model development and interpretation). The results from this model suggest that the photosynthesis rates at the heavily oiled treatment recovered to the status of the control treatment at MP in about 140 days (4.7 months), which was in early December 2010. On the other hand, the oil impact was never severe enough to make the photosynthesis rates at the medium impact treatment significantly lower than those at the control treatment (Wu et al. 2012). This HB model allowed us to better interpret the dynamic interplay of photosynthesis recovery from oil stress and the opposing seasonal depression of photosynthesis as the plants went into winter dormancy, and then to correctly interpret the time to recovery as being within 6 months, even at the heavily oiled treatment at MP. Other recent studies also show rapid recovery in marshes in Louisiana, especially in high-energy coastlines where oil contamination was rapidly diluted and removed (McCall and Pennings 2012; Silliman et al. 2012). The results of these and our study suggest that salt marshes can be very resilient to even heavy oil contamination and that photosynthesis and growth can recover quickly, within less than 1 year, if oil contamination is rapidly reduced by tidal and wave flushing.

Conclusions

Oil persisted in salt marsh environments for at least 1 year after arrival. High-energy coastlines, such as MP, allowed more rapid removal or degradation of the oil contamination. Plant recovery was more rapid in these locations based on photosynthesis measurements. In contrast, at low-energy locations with long residence times, such as HI, oil was still detected in sediments and on plants 1 year later. Plants in these locations exhibited chronic stress, which depressed photosynthesis, despite a full year of recovery. As many of the salt marsh coastlines in the northern GOM are eroding, the effects of DWH oil contamination are likely to be short term (1–3 years), with oil residues being resuspended back to the shallow coastal waters adjacent to the coastline. The implications of these findings on salt marsh function are that along exposed shorelines, removal of oil contaminants by tidal and wave action helps to speed up recovery of photosynthesis in smooth cordgrass, whereas in protected lagoonal shorelines, oil contamination tends to persist in the sediments and may result in protracted recovery.

Acknowledgments

We thank J. D. Caldwell, Scott Caldwell, Thomas Albaret, Lina Fu, Jennifer Frey, Hailong Huang, and Jiqing Liu for the field sampling and analysis, Gretchen Grammer at GBNERR for transportation, Erik Lang and Bradley Ennis for obtaining and processing the nekton samples, and Z. B. Leist for providing valuable insights during the writing of the manuscript. We would like to acknowledge the Parasitology Laboratory (in particular, Dr. Steve Curren and Eric Pulis), and the Fishery Center at the Gulf Coast Research Laboratory for boat time. This research was funded in part by grants from the National Science Foundation RAPID program (award numbers: DEB-1048342 to WW and PB, OCE-1042908 to ZL), Gulf Research Initiative (DROPPS Consortium), and Northern Gulf Institute Phase I BP Oil Spill Research (Task order number 191001-306811-04/TO 001 to PB, MSP, and WW).

References

Aeppli, C., C. A. Carmichael, R. K. Nelson, K. L. Lemkau, W. M. Graham, M. C. Redmond et al. 2012. Oil weathering after the *Deepwater Horizon* disaster led to the formation of oxygenated residues. *Environmental Science and Technology* 46:8799–8807.

Alexander, S. K. and J. W. Webb. 1987. Relationship of *Spartina alterniflora* growth to sediment oil content following an oil spill. In *Proceedings of the 1987 Oil Spill Conference*, edited by American Petroleum Institute, 445–449. Washington, DC.

Andrade, M. L., E. F. Covelo, F. A. Vega, and P. Marcet. 2004. Effect of the Prestige oil spill on salt marsh soils on the coast of Galicia (northwestern Spain). *Journal of Environmental Quality* 33:2103–2110.

Baca, B. J., T. E. Lankford, and E. R. Gundlach. 1987. Recovery of Brittany coastal marshes in the eight years following the Amoco Cadiz incident. In *Proceedings of the 1987 Oil Spill Conference*, edited by American Petroleum Institute, 459–464. Washington, DC.

Baker, J. M. 1970. The effects of oils on plants. *Environmental Pollution* 1:27–44.

Baker, J. M. 1971. Seasonal effects of oil pollution on salt marsh vegetation. *Oikos* 22:106–110.

Beck, M., K. Heck Jr., K. Able, D. Childers, D. Eggleston, B. Gillanders et al. 2001. The identification, conservation, and management of estuaries and marine nurseries for fish and invertebrates. *BioScience* 51:633–641.

Bergen, A., C. Alderson, R. Bergfors, C. Aquila, and M. A. Matsil. 2000. Restoration of a *Spartina alterniflora* salt marsh following a fuel oil spill, New York City, NY. *Wetlands Ecology and Management* 8:185–195.

Bik, H. M., K. M. Halanych, J. Sharma, and W. K. Thomas. 2012. Dramatic shifts in benthic microbial eukaryote communities following the *Deepwater Horizon* oil spill. *PLoS One* 7:e38550.

Boesch, D. F. and R. E. Turner. 1984. Dependence of fishery species on salt marshes: The role of food and refuge. *Estuaries* 7:460–468.

Bowyer, J. R., P. Camilleri, and W. F. J. Vermaas. 1991. Photosystem II and its interaction with herbicides. In *Herbicides, Topics in Photosynthesis*, edited by N. R. Baker and M. P. Percival, 27–85. Amsterdam, the Netherlands: Elsevier.

Burns, K. A., S. D. Garrity, D. Jorissen, J. MacPherson, M. Stoelting, J. Tierney et al. 1994. The Galeta oil spill II: Unexpected persistence of oil trapped in mangrove sediments. *Estuarine Coastal and Shelf Science* 38:349–364.

Cacela, D., J. Lipton, D. Beltman, J. Hansen, and R. Wolotira. 2005. Associating ecosystem service losses with indicators of toxicity in habitat equivalency analysis. *Environmental Management* 35:343–351.

Cai, W., W. Wiebe, Y. Wang, and J. Sheldon. 2000. Intertidal marsh as a source of dissolved inorganic carbon and a sink of nitrate in the Satilla River-estuarine complex in the southeastern U.S. *Limnology and Oceanography* 45:1743–1752.

Camilli, R., C. M. Reddy, D. R. Yoerger, B. A. S. Van Mooy, M. V. Jakuba, J. C. Kinsey et al. 2010. Tracking hydrocarbon plume transport and biodegradation at *Deepwater Horizon*. *Science* 330:201–204.

Cardoch, L., J. W. Day Jr., and C. Ibàñez. 2002. Net primary productivity as an indicator of sustainability in the Ebro and Mississippi deltas. *Ecological Applications* 12:1044–1055.

Carmichael, R. H., A. L. Jones, H. K. Patterson, W. C. Walton, A. Perez-Huerta, E. B. Overton et al. 2012. Assimilation of oil-derived elements by oysters due to the *Deepwater Horizon* oil spill. *Environmental Science and Technology* 46:12787–12795.

Chanton, J. P., J. Cherrier, R. M. Wilson, J. Sarkodee-Adoo, S. Bosman, A. Mickle et al. 2012. Radiocarbon evidence that carbon from the *Deepwater Horizon* spill entered the planktonic food web of the Gulf of Mexico. *Environmental Research Letters* 7:045303.

Chmura, G. L., S. C. Anisfeld, D. R. Cahoon, and J. C. Lynch. 2003. Global carbon sequestration in tidal, saline wetland soils. *Global Biogeochemical Cycles* 17:1–22.

Ciompi, S., E. Gentili, L. Guidi, and G. F. Soldatini. 1996. The effect of nitrogen deficiency on leaf gas exchange and chlorophyll fluorescence parameters in sunflower. *Plant Science* 118:177–184.

Collatz, G. J., M. Ribas-Carbo, and J. A. Berry. 1992. Coupled photosynthesis-stomatal conductance model for leaves of C_4 plants. *Australian Journal of Plant Physiology* 19:519–538.

Critchley, C. and R. M. Smilie. 1981. Leaf chlorophyll fluorescence as an indicator of photoinhibition in *Cucumis sativus* L. *Australian Journal of Plant Physiology* 8:133–141.

Crone, T. J. and M. Tolstoy. 2010. Magnitude of the 2010 Gulf of Mexico oil leak. *Science* 330:634.

Day, J. W., L. D. Britsch, S. R. Hawes, G. P. Shaffer, D. J. Reed, and D. Cahoon. 2000. Pattern and process of land loss in the Mississippi Delta: A spatial and temporal analysis of wetland habitat change. *Estuaries and Coasts* 23:425–438.

DeLaune, R., D. R. H. Baumann, and J. G. Gosselink. 1983. Relationships among vertical accretion, coastal submergence, and erosion in a Louisiana Gulf Coast marsh. *Journal of Sedimentary Research* 53:147–157.

Dicks, B. and J. P. Hartley. 1982. The effects of repeated small oil spillages and chronic discharges. *Philosophical Transactions of the Royal Society B*. 297:285–307.

Douglas, G. S., S. D. Emsbo-Mattingly, S. A. Stout, A. D. Uhler, and K. J. McCarthy. 2007. Chemical fingerprinting methods. In *Introduction to Environmental Forensics (2nd Edition)*, edited by B. L. Murphy and R. D. Morrison, 311–454. Academic Press, Burlington MA.

Duke, N. C., Z. S. Pinzón, and M. C. Prada. 1997. Large-scale damage to mangrove forests following two large oil spills in Panama. *Biotropica* 29:2–14.

Dunford, R. W., T. C. Ginn, and W. H. Desvousges. 2004. The use of habitat equivalency analysis in natural resource damage assessments. *Ecological Economics* 48:49–70.

Eastman, P. A. K. and E. L. Camm. 1995. Regulation of photosynthesis in interior spruce during water stress: Changes in gas exchange and chlorophyll fluorescence. *Tree Physiology* 15:229–235.

Edwards, B. R., C. M. Reddy, R. Camilli, C. A. Carmichael, K. Longnecker, and B. A. S. Van Mooy. 2011. Rapid microbial respiration of oil from the *Deepwater Horizon* spill in offshore surface waters of the Gulf of Mexico. *Environmental Research Letters* 6:e035301.

Engle, V. D. 2011. Estimating the provision of ecosystem services by Gulf of Mexico coastal wetlands. *Wetlands* 31:179–193.

EPA (U.S. Environmental Protection Agency). 2010. Coastal Sediment Sampling. Available at: http://www.epa.gov/emergency/bpspill/sediment.html (accessed March 2013).

Farquhar, G. D., S. von Caemmerer, and J. A. Berry. 1980. A biochemical model of photosynthetic (CO_2) assimilation in leaves of C_3 species. *Planta* 149:78–90.

Ferrell, R. E., E. D. Seneca, and R. A. Linthurst. 1984. The effects of crude oil on the growth of *Spartina alterniflora* Loisel. and *Spartina cynosuroides* (L.) Roth. *Journal of Experimental Marine Biology and Ecology* 83:27–39.

Filiault, D. L. and J. C. Stier. 1999. The use of chlorophyll fluorescence in assessing the cold tolerance of three turfgrass species. *Wisconsin Turfgrass Research* 16:109–110.

Fingas, M. F. 1999. The evaporation of oil spills: Development and implementation of new prediction methodology, poster number 131. In *1999 International Oil Spill Conference*, edited by U.S. Environmental Protection Agency, 1–11. Seattle, WA.

Fodrie, F. J. and K. L. Heck Jr. 2011. Response of coastal fishes to the Gulf of Mexico oil disaster. *PLoS One* 6:e21609.

Genty, B., J. Briantais, and N. Baker. 1989. The relationship between the quantum yield of photosynthetic electron transport and quenching of chlorophyll fluorescence. *Biochemica et Biophysica Acta* 990:87–92.

Gilfillan, E. S., N. P. Maher, C. M. Krejsa, M. E. Lanphear, C. D. Ball, J. B. Meltzer et al. 1995. Use of remote sensing to document changes in marsh vegetation following the Amoco Cadiz oil spill (Brittany, France, 1978). *Marine Pollution Bulletin* 30:780–787.

Graham, W. M., R. H. Condon, R. H. Carmichael, I. D'Ambra, H. K. Patterson, L. J. Linn et al. 2010. Oil carbon entered the coastal planktonic food web during the *Deepwater Horizon* oil spill. *Environmental Research Letters* 5:e045301.

Guidi, L., C. Nali, S. Ciompi, G. Lorenzini, and G. F. Soldatini. 1997. The use of chlorophyll fluorescence and leaf gas exchange as methods for studying the different responses to ozone of two bean cultivars. *Journal of Experimental Botany* 48:173–179.

Gundlach, E. R. and M. O. Hayes. 1978. Vulnerability of coastal environments to oil spill impacts. *Marine Technology Society Journal* 12:18–27.

Havaux, M. and R. Lannoye. 1983. Chlorophyll fluorescence induction: A sensitive indicator of water stress in maize plants. *Irrigation Science* 4:147–151.

Hazen, T. C., E. A. Dubinsky, T. Z. DeSantis, G. L. Andersen, Y. M. Piceno, N. Singh et al. 2010. Deep-sea oil plume enriches indigenous oil-degrading bacteria. *Science* 330:204–208.

Hester, M. W. and I. A. Mendelssohn. 2000. Long-term recovery of a Louisiana brackish marsh plant community from oil-spill impact: Vegetation response and mitigating effects of marsh surface elevation. *Marine Environmental Research* 49:233–254.

Hilbert, K. W. 2006. Land cover change within the Grand Bay National Estuarine Research Reserve: 1974–2001. *Journal of Coastal Research* 226:1552–1557.

Huang, Z. A., D. A. Jiang, Y. Yang, J. W. Sun, and S. H. Jin. 2004. Effects of nitrogen deficiency on gas exchange, chlorophyll fluorescence, and antioxidant enzymes in leaves of rice plants. *Photosynthetica* 42:357–364.

Hunt, J. M. 1996. *Petroleum Geochemistry and Geology*. New York: W. H. Freeman.

Irwin, R. J., M. Van Mouwerik, L. Stevens, M. D. Seese, and W. Basham. 1997. *Environmental Contaminants Encyclopedia*. National Park Service, Water Resources Division, Fort Collins, CO.

Jensen, J. R., J. N. Halls, and J. Michel. 1998. A systems approach to Environmental Sensitivity Index (ESI) mapping for oil spill contingency planning and response. *Photogrammetric Engineering & Remote Sensing* 64:1003–1014.

Jordan, S. J. and M. S. Peterson. 2012. Contributions of estuarine habitat to major fisheries. In *Estuaries: Classification, Ecology, and Human Impacts*, edited by S. J. Jordan, 75–92. Hauppauge, NY: Nova Science Publishers.

Kenworthy, W., M. Durako, S. Fatemy, H. Valavi, and G. Thayer. 1993. Ecology of seagrasses in northeastern Saudi Arabia one year after the Gulf War oil spill. *Marine Pollution Bulletin* 27:213–222.

Khan, M. A. Q., S. M. Al-Ghais, B. Catalin, and Y. H. Khan. 2005. Effects of petroleum hydrocarbons on aquatic animals. In *Oil Pollution and its Environmental Impact in the Arabian Gulf Region*, edited by M. Al-Azab, W. El-Shorbagy, and S. Al-Ghais, 159–185. Elsevier, Amsterdam, The Netherlands.

King, S. E. and J. N. Lester. 1995. The value of salt marsh as a sea defense. *Marine Pollution Bulletin* 30:180–89.

Kirsch, K. D., K. A. Barry, M. S. Fonseca, P. E. Whitfield, S. R. Meehan, W. J. Kenworthy et al. 2005. The Mini-312 Program: An expedited damage assessment and restoration process for seagrasses in the Florida Keys National Marine Sanctuary. *Journal of Coastal Research* SI 40:109–119.

Ko, J. Y. and J. W. Day. 2004. A review of ecological impacts of oil and gas development on coastal ecosystems in the Mississippi Delta. *Ocean and Coastal Management* 47:597–623.

Kolian, S. R., S. Porter, P. W. Sammarco, and E. W. Cake Jr. 2013. Depuration of Macondo (MC-252) oil found in heterotrophic scleractinian corals (*Tubastrea coccinea* and *Tubastrea micranthus*) on offshore oil/gas platforms in the Gulf. *Gulf and Caribbean Research* 25:99–103.

Krause, G. H. and E. Weis. 1991. Chlorophyll fluorescence and photosynthesis: The basics. *Annual Review of Plant Physiology and Plant Molecular Biology* 42:313–349.

Krebs, C. T. and C. E. Tanner. 1981. Restoration of oiled marshes through sediment stripping and *Spartina* propagation. In *Proceedings of the 1981 Oil Spill Conference*, edited by American Petroleum Institute, 375–385. Washington, DC.

Lee, J. K., R. A. Park, and P. W. Mausel. 1992. Application of geoprocessing and simulation modeling to estimate impacts of sea level rise on the northeast coast of Florida. *Photogrammetric Engineering and Remote Sensing* 58:1579–1586.

Lee, R. F., and D. S. Page. 1997. Petroleum hydrocarbons and their effects in subtidal regions after major oil spills. *Marine Pollution Bulletin* 34:928–-940.

Lima, J. D., P. R. Mosquim, and F. M. Matta. 1999. Leaf gas exchange and chlorophyll fluorescence parameters in *Phaseolus vulgaris* as affected by nitrogen and phosphorus deficiency. *Photosynthetica* 3791:113–121.

Lin, Q. and I. A. Mendelssohn. 1996. A comparative investigation of the effects of south Louisiana crude oil on the vegetation of fresh, brackish and salt marshes. *Marine Pollution Bulletin* 32:202–209.

Lin, Q. and I. A. Mendelssohn. 1998. The combined effects of phytoremediation and biostimulation in enhancing habitat restoration and oil degradation of petroleum contaminated wetlands. *Ecological Engineering* 10:263–74.

Lin, Q. and I. A. Mendelssohn. 2012. Impacts and recovery of the *Deepwater Horizon* oil spill on vegetation structure and function of coastal salt marshes in the Northern Gulf of Mexico. *Environmental Science and Technology* 46:3737–43.

Lin, Q., I. A. Mendelssohn, B. Charles, J. Henry, M. W. Hester, and E. C. Webb. 1999. Effect of oil cleanup methods on ecological recovery and oil degradation of *Phragmites* marshes. In *1999 International Oil Spill Conference*, edited by U.S. Environmental Protection Agency, 1–13. Seattle, WA.

Liu Z., J. Liu, Q. Zhu, and W. Wu. 2012. The weathering of oil after the *Deepwater Horizon* oil spill: Insights from the chemical composition of the oil from the sea surface, salt marshes and sediments. *Environmental Research Letters* 7:e035302.

Lloyd, F. and S. Tracy. 1901. The insular flora of Mississippi and Louisiana. *Bulletin of the Torrey Botanical Club* 28:61–105.

Long, S. P. and C. J. Bernacchi. 2003. Gas exchange measurements, what can they tell us about the underlying limitations to photosynthesis? Procedures and sources of error. *Journal of Experimental Botany* 54:2393–2401.

Long, S. P., P. K. Farage, and R. L. Garcia. 1996. Measurement of leaf and canopy photosynthetic CO_2 exchange in the field. *Journal of Experimental Botany* 47:1629–1642.

Lytle, J. S. and T. F. Lytle. 1987. The role of *Juncus roemerianus* in clean-up of oil-polluted sediments. In *Proceedings of the 1987 Oil Spill Conference*, edited by American Petroleum Institute, 495–501. Washington, DC.

Maricle, B. R., R. W. Lee, C. E. Hellquist, O. Kiirats, and G. E. Edwards. 2007. Effects of salinity on chlorophyll fluorescence and CO_2 fixation in C_4 estuarine grasses. *Photosynthetica* 45:433–440.

Maxwell, K. and G. N. Johnson. 2000. Chlorophyll fluorescence: A practical guide. *Journal of Experimental Botany* 51:659–668.

McCall, B. D. and S. C. Pennings. 2012. Disturbance and recovery of salt marsh arthropod communities following BP *Deepwater Horizon* oil spill. *PLoS One* 7:e32735.

McMillan, C. and C. L. Sherrod. 1986. The chilling tolerance of black mangrove, *Avicennia germinans*, from the Gulf of Mexico coast of Texas, Louisiana and Florida. *UTMSI Contributions in Marine Science* 29:9–16.

McNutt, M., R. Camilli, T. J. Crone, G. D., Guthrie, P. A. Hsieh, T. B. Ryerson et al. 2011. Review of flow rate estimates of the *Deepwater Horizon* oil spill. *Proceedings of the Natural Academy of Sciences* pnas.1112139108.

Mendelssohn, I. A., G. L. Andersen, D. M. Baltz, R. H. Caffey, K. R. Carman, J. W. Fleeger et al. 2012. Oil impacts on coastal wetlands: Implications for the Mississippi River delta ecosystem after the *Deepwater Horizon* Oil Spill. *BioScience* 62:562–574.

Mendelssohn, I. A., M. W. Hester, and J. M. Hill. 1993. *Effects of Oil Spills on Coastal Wetlands and Their Recovery*. OCS Study MMS 93-0045. U.S. Department of the Interior, Minerals Management Service, Gulf of Mexico OCS Regional Office, New Orleans, LA.

MGC (Mississippi Geospatial Clearinghouse). 2011. Available at: www.gis.ms.gov (accessed March 2013).

Michel, J., M. Hayes, and P. Brown, 1978. Application of an oil spill vulnerability index to the shoreline of lower Cook Inlet, Alaska. *Environmental Geology* 2:107–117.

Michel, J., Z. Nixon, J. Dahlin, D. Betenbaugh, M. White, D. Burton et al. 2009. Recovery of interior brackish marshes seven years after the Chalk Point oil spill. *Marine Pollution Bulletin* 58:995–1006.

Mielke, M. S., A. F. de Almeida, F. P. Gomes, M. A. G. Aguilar, and P. A. O. Mangabeira. 2003. Leaf gas exchange, chlorophyll fluorescence and growth responses of *Genipa americana* seedlings to soil flooding. *Environmental and Experimental Botany* 50:221–231.

Mishra, D. R., H. J. Cho, S. Ghosh, A. Fox, C. Downs, P. B. T. Merani et al. 2012. Post-spill state of the marsh: Remote estimation of the ecological impact of the Gulf of Mexico oil spill on Louisiana salt marshes. *Remote Sensing of Environment* 118:176–185.

Moeller, I., T. Spencer, and J. R. French. 1996. Wind wave attenuation over salt marsh surfaces: Preliminary results from Norfolk, England. *Journal of Coastal Research* 12:1009–1016.

NOAA (National Oceanic and Atmospheric Administration). 2010. National Hurricane Center: 2010 Atlantic Hurricane Season. Available at: www.nhc.noaa.gov/2010atlan.shtml (accessed March 2013).

NOAA (National Oceanic and Atmospheric Administration). 2011. National Marine Fisheries Service: *Deepwater Horizon*/BP Oil Spill Information. Available at: sero.nmfs.noaa.gov/deepwater_horizon_oil_spill.html (accessed March 2013).

NOAA (National Oceanic and Atmospheric Administration). 2012. Office of Response and Restoration: Environmental Sensitivity Index (ESI). Available at: response.restoration.noaa.gov/maps-and-spatial-data/shoreline-rankings.html (accessed March 2013).

Oil Spill Commission. 2011. Deep Water: The Gulf Oil Disaster and the Future of Offshore Drilling. Available at: www.gpo.gov/fdsys/pkg/GPO-OILCOMMISSION/content-detail.html (accessed March 2013).

Orcutt, B. N., S. B. Joye, S. Kleindienst, K. Knittel, A. Ramette, A. Reitz et al. 2010. Impact of natural oil and higher hydrocarbons on microbial diversity, distribution, and activity in Gulf of Mexico cold-seep sediments. *Deep Sea Research Part II: Topical Studies in Oceanography* 57:2008–2021.

Ortmann, A. C., J. Anders, N. Shelton, L. Gong, A. G. Moss, and R. H. Condon. 2012. Dispersed oil disrupts microbial pathways in pelagic food webs. *PLoS One* 7:e42548.

Otvos, E. G. and G. A. Carter. 2008. Hurricane degradation: Barrier development cycles, northeastern Gulf of Mexico: Landform evolution and island chain history. *Journal of Coastal Research* 24:463–478.

Patterson, C. S., I. A. Mendelssohn, and E. M. Swenson. 1993. Growth and survival of *Avicennia germinans* seedlings in a mangal/salt marsh community in Louisiana. *Journal of Coastal Research* 9:801–810.

Pereira, W. E., D. L. de Siqueira, C. A. Mart⊠nez, and M. Puiatti. 2000. Gas exchange and chlorophyll fluorescence in four citrus rootstocks under aluminum stress. *Journal of Plant Physiology* 157:513–520.

Peterson, C. H., L. L. McDonald, R. H. Green, and W. P. Erickson. 2001. Sampling design begets conclusions: The statistical basis for detection of injury to and recovery of shore-line communities after the Exxon Valdez oil spill. *Marine Ecology Progress Series* 210:255–283.

Peterson, C. H., S. D. Rice, J. W. Short, D. Esler, J. L. Bodkin, B. E. Ballachey et al. 2003. Long-term ecosystem response to the Exxon Valdez oil spill. *Science* 302:2082–2086.

Peterson, M. S. and M. R. Lowe. 2009. Implications of cumulative impacts to estuarine and marine habitat quality for fish and invertebrate resources. *Reviews in Fisheries Science* 17:505–523.

Pezeshki, S. R. and R. D. DeLaune. 1993. Effect of crude oil on gas exchange functions of *Juncus roemerianus* and *Spartina alterniflora*. *Water Air Soil Pollution* 68:461–468.

Pezeshki, S. R., R. D. DeLaune, and A. Jugsujinda. 2001. The effects of crude oil and the effectiveness of cleaner application following oiling on US Gulf of Mexico coastal marsh plants. *Environmental Pollution* 112:483–489.

Pezeshki, S. R., M. W. Hester, Q. Lin, and J. A. Nyman. 2000. The effects of oil spill and clean-up on dominant US Gulf coast marsh macrophytes: A review. *Environmental Pollution* 108:129–139.

Phillips, J. 1987. Shoreline processes and establishment of *Phragmites australis* in a coastal plain estuary. *Vegetatio* 71:139–144.

Plata, D. L., C. M. Sharpless, and C. M. Reddy. 2008. Photochemical degradation of polycyclic aromatic hydrocarbons in oil films. *Environmental Science and Technology* 42:2432–2438.

Proffitt, C. E. 1997. *Managing Oil Spills in Mangrove Ecosystems: Effects, Remediation, Restoration, and Modeling*. Lake Charles, LA: McNeese University.

Read, C. 2011. *BP and the Macondo Spill: The Complete Story*. New York: Palgrave Macmillan.

Rohacek, K. and M. Bartak. 1999. Technique of the modulated chlorophyll fluorescence: Basic concepts, useful parameters, and some applications. *Photosynthetica* 37:339–363.

Roth, A. M. F. and D. M. Baltz. 2009. Short-term effects of an oil spill on marsh-edge fishes and decapod crustaceans. *Estuaries and Coasts* 32:565–572.

Schleifstein, M. 2012. Photos document BP oil still contaminates "cleaned" Louisiana marshes, state officials say. New Orleans, LA: Times-Picayune.

Schulze, E. D. and M. M. Caldwell. 1990. *Ecophysiology of Photosynthesis*. Berlin, Germany: Springer.

Šesták, A., J. Catský, and P. G. Jarvis. 1971. *Plant Photosynthetic Production: Manual of Methods*. Berlin, Germany: Springer.

Shogren, E. and D. Gonyea. 2010. Determining who's in charge of Gulf cleanup. Available at: http://www.npr.org/templates/story/story.php?storyId = 128839225 (accessed March 2013).

Silliman, B. R., J. Van de Koppel, M. W. McCoy, J. Diller, G. N. Kasozi, K. Earl et al. 2012. Degradation and resilience in Louisiana salt marshes after the BP: *Deepwater Horizon* oil spill. *Proceedings of the Natural Academy of Sciences* 109:11234–11239.

SkyTruth. 2010. BP/Gulf Oil Spill: 68,000 Square Miles of Direct Impact. Available at: blog.skytruth.org/2010/07/bp-gulf-oil-spill-68000-square-miles-of.html (accessed March 2013).

Snedaker, S. C., S. M. Smith, P. Biber, and R. J. Araujo. 2001. Comparative effects of Orimulsion and Fuel Oil No. 6 on floating and stranded propagules, and established seedlings of *Rhizophora mangle*. In *Mangrove Ecosystems: Natural Distribution, Biology and Management*, edited by N. R. Bhat, F. K. Taha, and A. Al-Nasser, 109–126. Kuwait Institute for Scientific Research, Safat, Kuwait.

Southward, A. J. and E. C. Southward. 1978. Recolonization of rocky shores in Cornwall after use of toxic dispersants to clean up the Torrey Canyon spill. *Journal of the Fisheries Research Board of Canada* 35:682–706.

Souza, R. P., E. C. Machado, J. A. B. Silva, A. M. M. A. Lagôa, and J. A. G. Silveira. 2004. Photosynthetic gas exchange, chlorophyll fluorescence and some associated metabolic changes in cowpea (*Vigna unguiculata*) during water stress and recovery. *Environmental and Experimental Botany* 51:45–56.

Teal, J. M., J. W. Farrington, K. A. Burns, J. J. Stegeman, B. W. Tripp, B. Woodin et al. 1992. The West Falmouth oil spill after twenty years: Fate of oil compounds and effects on animals. *Marine Pollution Bulletin* 24:607–614.

Tobias, C. R., I. C. Anderson, E. A. Canuel, and S. A. Macko. 2001a. Nitrogen cycling through a fringing marsh-aquifer ecotone. *Marine Ecology Progress Series* 210:25–39.

Tobias, C. R., S. Macko, I. Anderson, E. Canuel, and J. Harvey. 2001b. Tracking the fate of a high concentration nitrate plume through a fringing marsh: A combined groundwater tracer and in-situ isotope enrichment study. *Limnology and Oceanography* 46:1977–1989.

Turner, R. E. 1977. Intertidal vegetation and commercial yields of penaeid shrimp. *Transactions of the American Fisheries Society* 106:411–416.

Turner, R. E. 1990. Landscape development and coastal wetland losses in the northern Gulf of Mexico. *American Zoologist* 30:89–105.

Underwood, A. J. 1994. On beyond BACI: Sampling designs that might reliably detect environmental disturbances. *Ecological Applications* 4:3–15.

Unsworth, R. E. and R. C. Bishop. 1994. Assessing natural resource damages using environmental annuities. *Ecological Economics* 11:35–41.

US Government. 2010. ERMA Gulf Response. Available at: http://www.restorethegulf.gov/release/2011/10/28/erma-gulf-response (accessed March 2013).

Valiela, I. and M. L. Cole. 2002. Comparative evidence that salt marshes and mangroves may protect seagrass meadows from land-derived nitrogen loads. *Ecosystems* 5:92–102.

Valiela, I., G. Tomasky, J. Hauxwell, M. L. Cole, J. Cebrian, and K. D. Kroeger. 2000. Operationalizing sustainability: Management and risk assessment of land-derived nitrogen loads to estuaries. *Ecological Applications* 10:1006–1023.

Venosa, A. D. and X. Zhu. 2003. Biodegradation of crude oil contaminating marine shorelines and freshwater wetlands. *Spill Science and Technology Bulletin* 8:163–178.

Vestheim, H., K. Langford, and K. Hylland. 2012. Lack of response in a marine pelagic community to short-term oil and contaminant exposure. *Journal of Experimental Marine Biology and Ecology* 416:110–114.

Walz, H. 1999. *Photosynthesis Yield Analyzer Mini-PAM Handbook*. Germany: Effeltrich.

Wang, Z. D., M. Fingas, P. Lambert, G. Zeng, C. Yang, and B. Hollebone. 2004. Characterization and identification of the Detroit river mystery oil spill (2002). *Journal of Chromatography A* 1038:201–214.

Webb, J. W., S. K. Alexander, and J. K. Winters. 1985. Effects of autumn application of oil on *Spartina alterniflora* in a Texas salt marsh. *Environmental Pollution Series A, Ecological and Biological* 38:321–337.

Whitehead, A., B. Dubansky, C. Bodinier, T. I. Garcia, S. Miles, C. Pilley et al. 2011. Genomic and physiological footprint of the *Deepwater Horizon* oil spill on resident marsh fishes. *Proceedings of the Natural Academy of Sciences* 108:pnas.1109545108.

White House. 2010. Ongoing Administration-wide response to the Deepwater BP oil spill: By the numbers to date. Available at: www.restorethegulf.gov/release/2010/08/23/ongoing-administration-wide-response-deepwater-bp-oil-spill (accessed March 2013).

Wolfe, D. A., M. J. Hameedi, J. A. Galt, G. Watabayashi, J. Short, C. O'Claire et al. 1994. The fate of the oil spilled from the Exxon Valdez. *Environmental Science and Technology* 28:560A–568A.

Wu, W., P. D. Biber, M. S. Peterson, and C. Gong. 2012. Modeling photosynthesis of *Spartina alterniflora* (smooth cordgrass) impacted by the *Deepwater Horizon* oil spill using Bayesian inference. *Environmental Research Letters* 7:e045302.

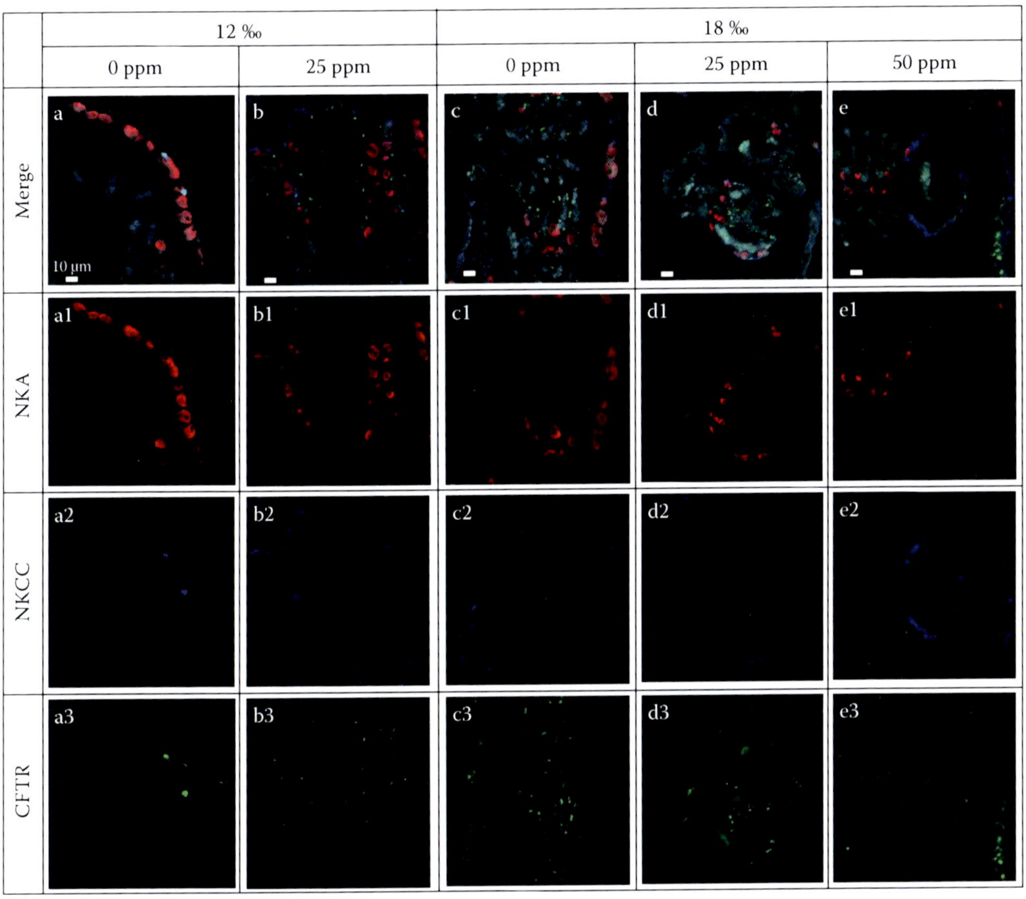

FIGURE 2.3
Triple immunolocalization of Na^+/K^+-ATPase (NKA) (red), $Na^+/K^+/2Cl^-$ cotransporter (NKCC) (blue), and cystic fibrosis transmembrane regulator channel (CFTR) (green) in *Fundulus grandis* larvae after 96-hour exposure to 0 (a through a3; c through c3), 25 (b through b3; d through d3), and 50 ppm (e through e3) of dioctyl sodium sulfosuccinate (DOSS) at salinities of 12‰ (a through a3; b through b3) and 18‰ (c through c3; d through d3; e through e3).

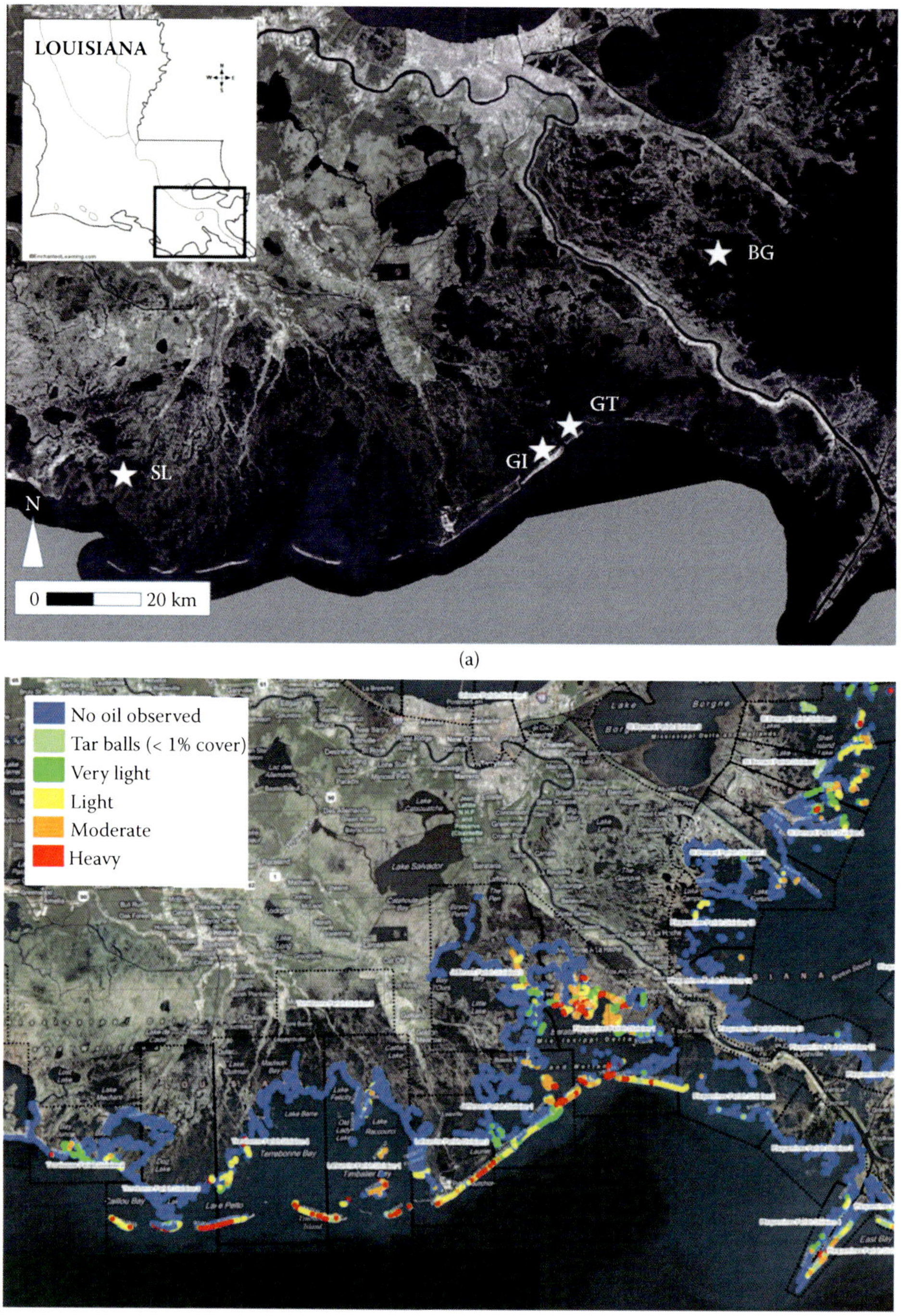

(a)

(b)

FIGURE 14.1
Maps indicating study sites where caged oysters were deployed and wild oysters were collected in Louisiana estuaries (a) and *Deepwater Horizon* shoreline oiling on September 20, 2010 obtained from the Shoreline Cleanup Assessment Teams (SCAT) (b). (From http://www.noaa.gov/deepwaterhorizon/maps/pdf/scat_la/Reduced_MC252_ShorelineCurrentOilingSituation_092010.pdf.)

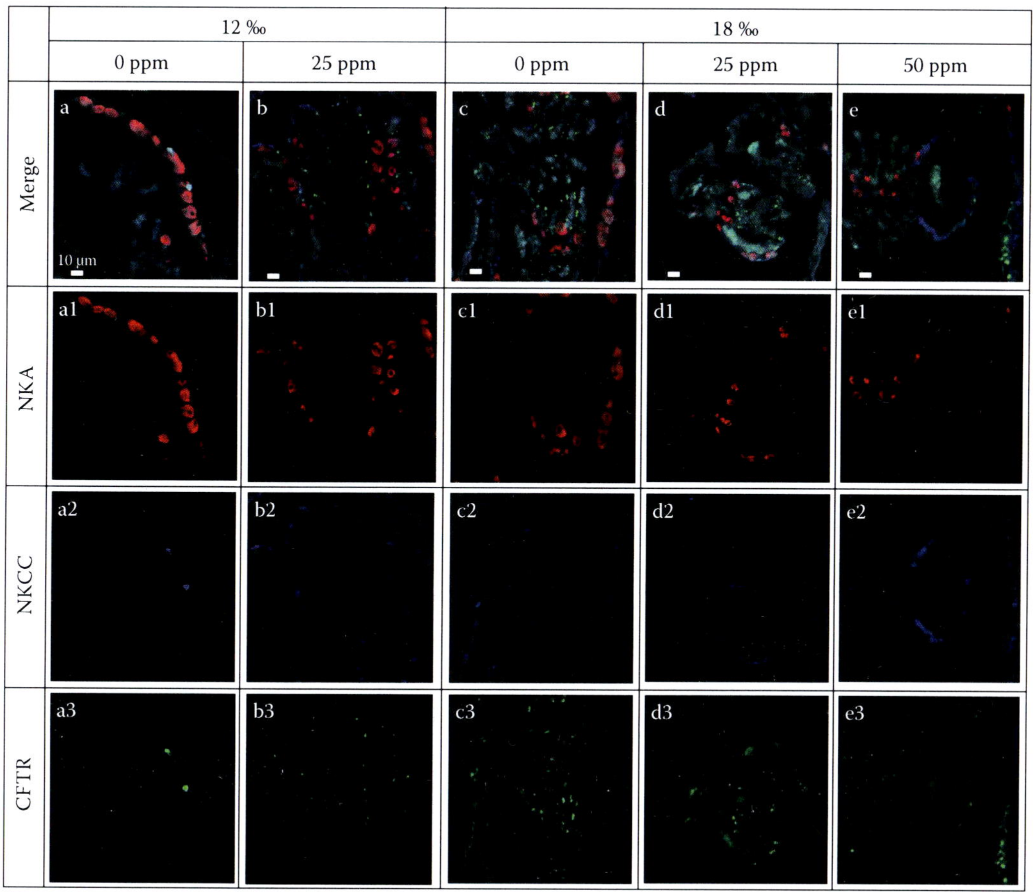

FIGURE 2.3
Triple immunolocalization of Na^+/K^+-ATPase (NKA) (red), $Na^+/K^+/2Cl^-$ cotransporter (NKCC) (blue), and cystic fibrosis transmembrane regulator channel (CFTR) (green) in *Fundulus grandis* larvae after 96-hour exposure to 0 (a through a3; c through c3), 25 (b through b3; d through d3), and 50 ppm (e through e3) of dioctyl sodium sulfosuccinate (DOSS) at salinities of 12‰ (a through a3; b through b3) and 18‰ (c through c3; d through d3; e through e3).

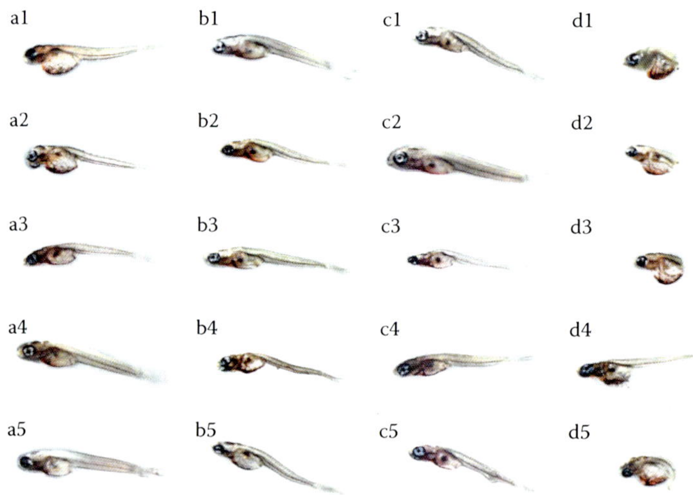

FIGURE 2.5
Representative newly hatched Gulf killifish, *Fundulus grandis*, exposed to 0, 1, 25, and 40 mg·L^{-1} dioctyl sodium sulfosuccinate (DOSS) for 96 hours beginning at stage 19 at a salinity of 12‰.

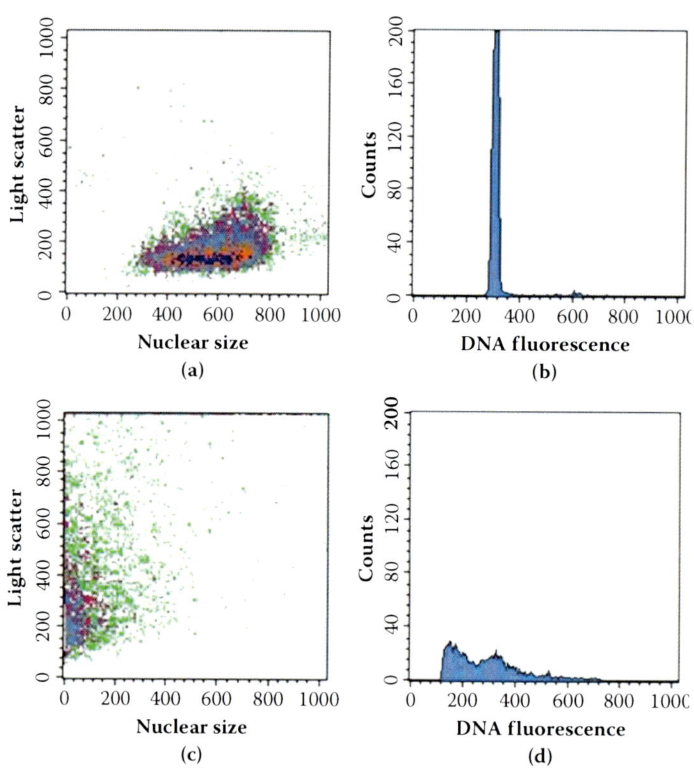

FIGURE 4.1
Blood cell nuclei from pallid sturgeon (*Scaphirhynchus albus*) (a and b) and Gulf sturgeon (*Acipenser oxyrinchus desotoi*) (c and d) that were stained with propidium iodide and analyzed by flow cytometry. Cytograms from Fish 1 (a + b) represent a blood sample that was shipped and handled correctly from field to lab, cytograms from Fish 2 (c + d) represent a sample that was handled improperly (paper label was placed inside the sample tube). All cytogram axes (except counts) display a linear scale of wavelengths.

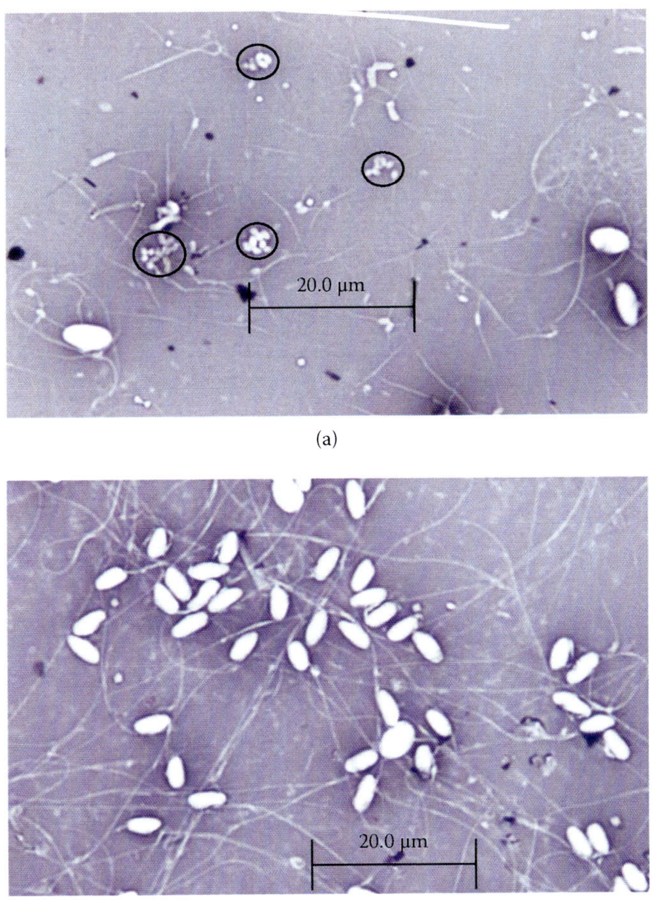

(a)

(b)

FIGURE 4.2
Microscopic images of largescale sucker (*Catostomus macrocheilus*) spermatozoa stained with eosin and nigrosin at a total magnification of 1000× (oil immersion). Bacteria (circled) are more prevalent in a than in b, which was relatively free of bacteria.

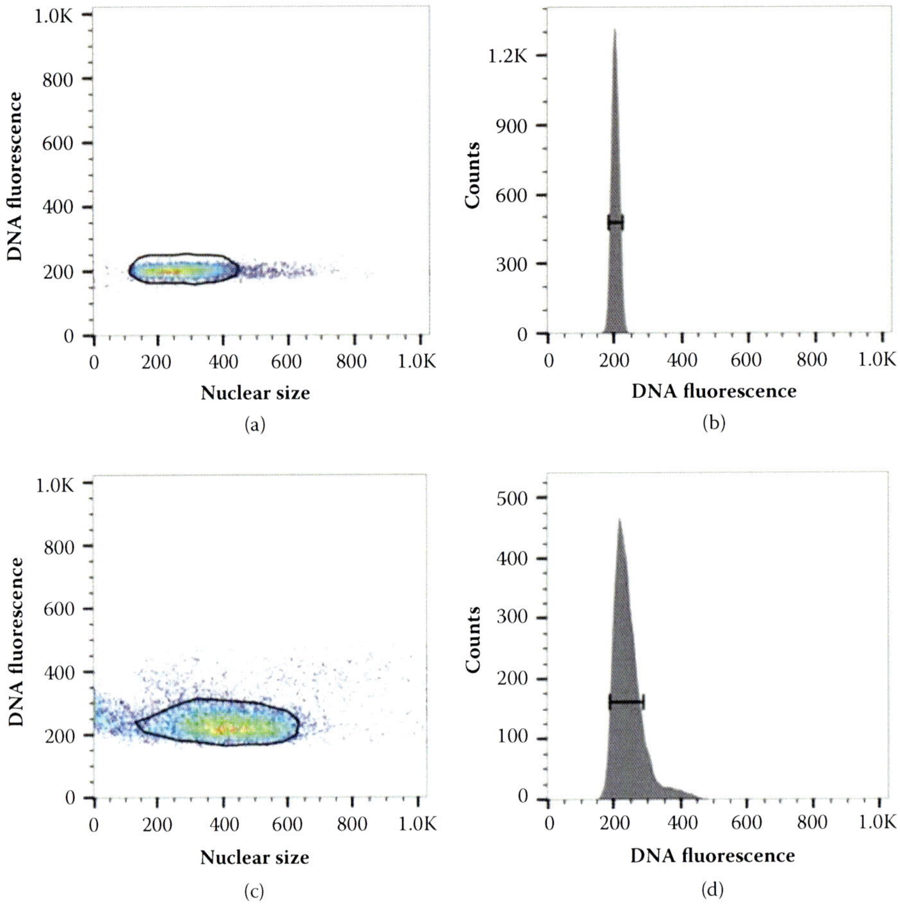

FIGURE 4.3
Blood from channel catfish (*Ictalurus punctatus*) collected with acid citrate dextrose anticoagulant (ACD) then stained with propidium iodide and analyzed immediately (a and b) or after 7 days of storage in ACD at 4°C (c and d). Flow cytometric cytograms reflect nuclear DNA integrity, in which higher percentages of nuclei outside the main population (NOMP) gate (the circles) indicates less intact DNA (c has more nuclei outside the gate than a). The corresponding CV histogram peak widths (bracketed bar) indicate greater levels of DNA dispersion in d as opposed to b.

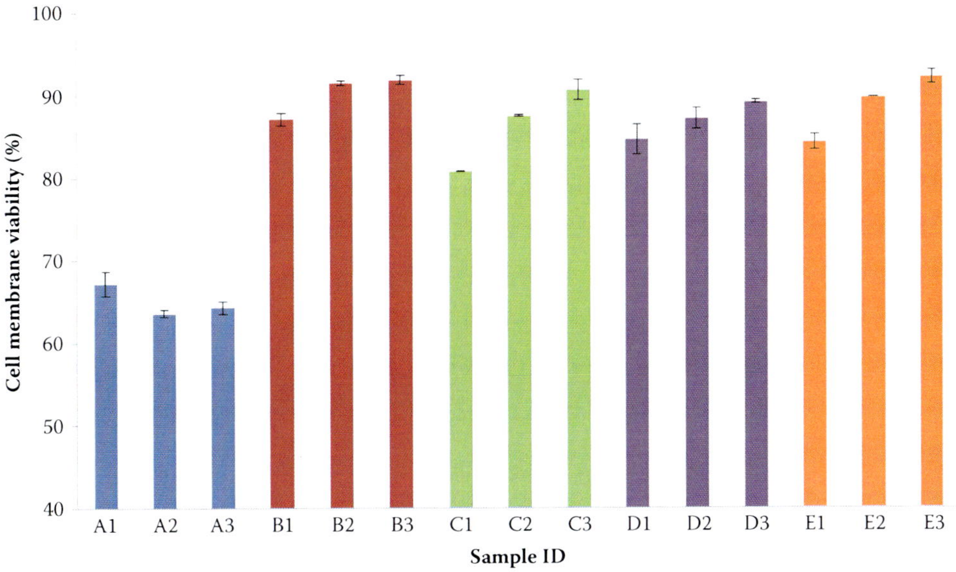

FIGURE 4.4
Viability of blood cells from channel catfish (*Ictalurus punctatus*) ($n = 5$, Fish A–E) analyzed by flow cytometry after 14 days of storage at 4°C. Sample codes indicate the individual fish (letters A–E) and cell stabilizing treatment (numbers 1–3). Sample treatments were one part blood to one part cell stabilizer (1), two parts blood to one part stabilizer (2), and three parts blood to one part stabilizer (3). Blood from Fish B–E was added to the treatments immediately. Blood from Fish A was added after 5 minutes at 24°C.

FIGURE 5.3
DWO-3-BAT-08 photo 3 taken on August 13, 2010. Site vegetation included *Spartina alterniflora* and *Distichlis spicata*. Oiled vegetation and oil-damaged canopy were observed at this site. (From Kokaly et al., *Shoreline Surveys of Oil-Impacted Marsh in Southern Louisiana July to August 2010*, U.S. Geological Survey Open-File Report 2011-1022, Reston, VA, 2011.)

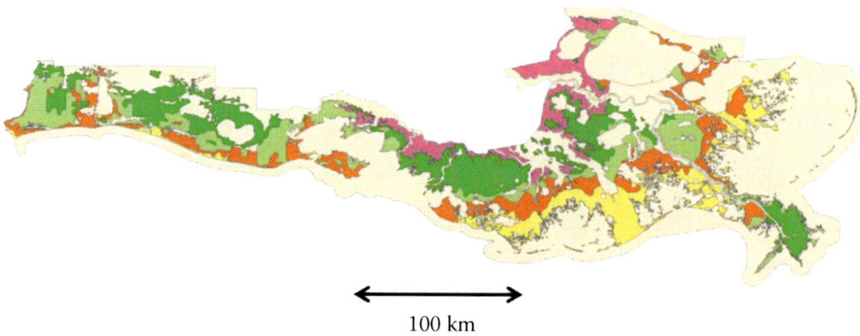

FIGURE 6.1
This map of coastal Louisiana shows some coastal wetland forests and the most common marsh classification system in coastal Louisiana. From inland to the Gulf of Mexico: pink areas are bald cypress swamp (dominated by *Taxodium distichum* and *Nyssa aquatic*), dark green areas are fresh marsh (dominated by *Panicum hemitomon*, *Sagittaria lancifolia*, or *Typha* spp.), light green areas are intermediate marsh (dominated by *Spartina patens* and supporting many other species), orange areas are brackish marsh (dominated by *Spartina patens* and supporting few other species), and yellow areas are saline marsh (dominated by *Spartina alterniflora*). The different plant associations also support different communities of fish and wildlife. The data were collected in 1997 and made available by Louisiana Department of Wildlife and Fisheries (LDWF [2001]); the data are described in Visser et al. (1998, 2000).

FIGURE 6.7
Photograph by the authors of *Phragmites australis* at the edge of the Gulf of Mexico in the Bird's Foot Delta of the Mississippi River on July 10, 2010. The death of unoiled leaves suggests that toxins in the oil were transported from the lower stems that were oiled to the upper leaves that are assumed to be unoiled. The production of new leaves suggests that the plant may survive if its soil is not eroded before recovery is completed.

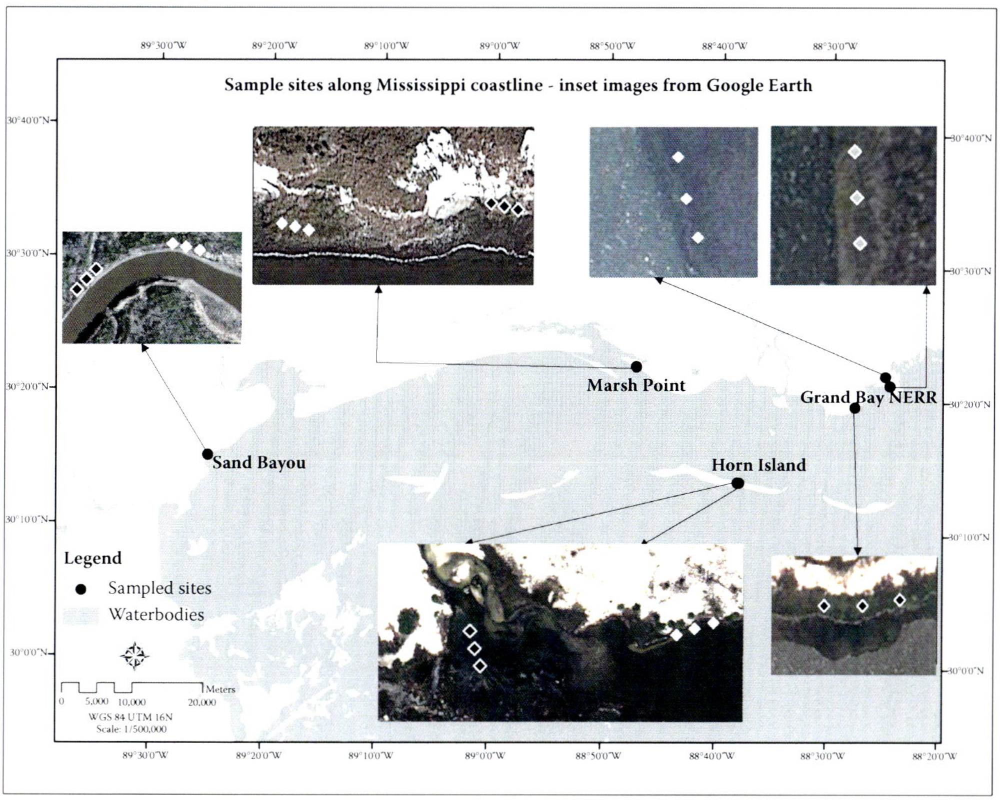

FIGURE 7.3
Map of the four study locations: Grand Bay National Estuarine Research Reserve, Marsh Point, a lagoon on Horn Island, and Sand Bayou near Waveland, Mississippi with control and heavily oiled treatments indicated. White symbols are control quadrats (no oil) and dark symbols are impacted quadrats (medium oiling, gray; heavy oiling, black). Within each treatment, three 1-m^2 quadrats were placed 5 m apart and parallel to the shoreline in the smooth cordgrass fringe zone.

FIGURE 7.4
Time-series photos of oil contamination and smooth cordgrass plant responses at Marsh Point from July 2010 (2–13 days after oiling) until July 2011 (351 days after oiling). The left panels show the same general area of the heavily oiled treatment, marked by either two tall bamboo marker posts or a short, red-painted wooden stake. The right panels show close-ups of the 1-m^2 quadrats that were sampled for contamination and show the condition of the weathered oil over time, as well as the recovery of the initially heavily oiled plants through the seasons.

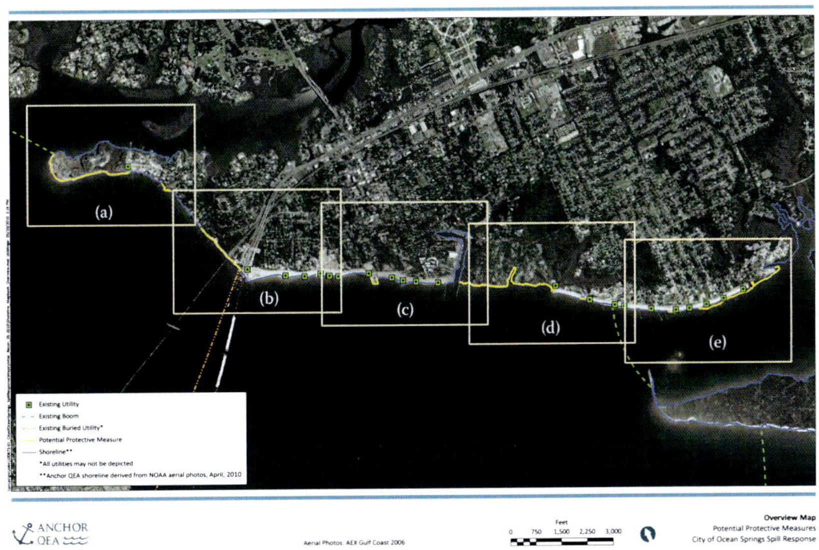

FIGURE 8.1
A map of sensitive marine resources (a through e) that were delineated for protection in Ocean Springs, Mississippi, following the *Deepwater Horizon* oil spill in 2010.

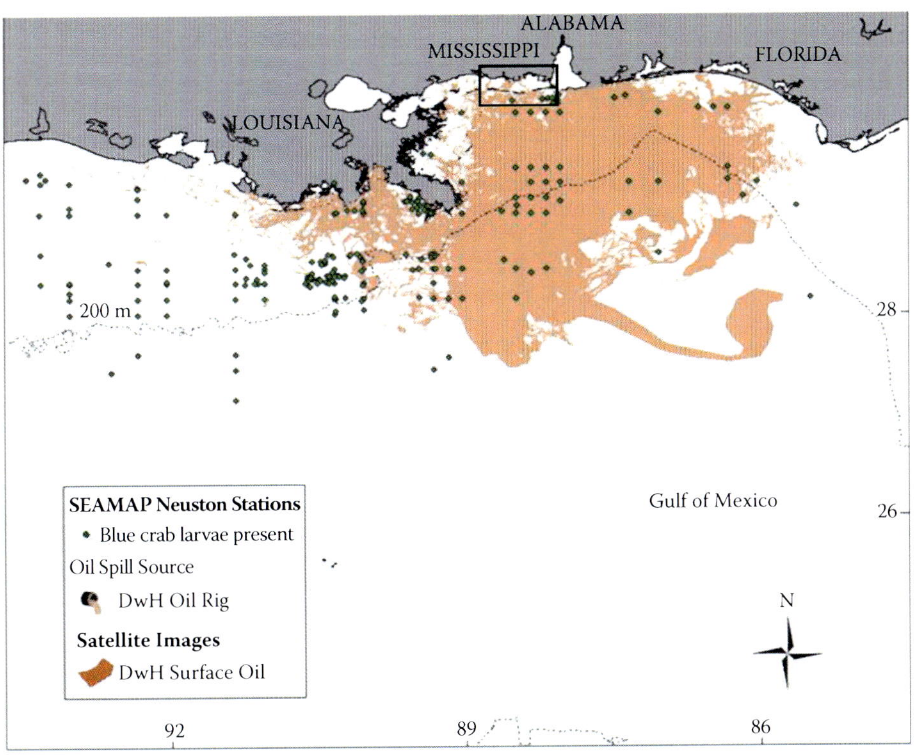

FIGURE 13.1
Map showing largest extent of the *Deepwater Horizon* oil spill April 20-July 15, 2010 (from National Oceanic and Atmospheric Administration [NOAA]) overlaid with distribution of *Callinectes sapidus* larvae in the summer based on SEAMAP sampling (from NOAA). The area used in this study for measurement of daily settlement is indicated by the black rectangle.

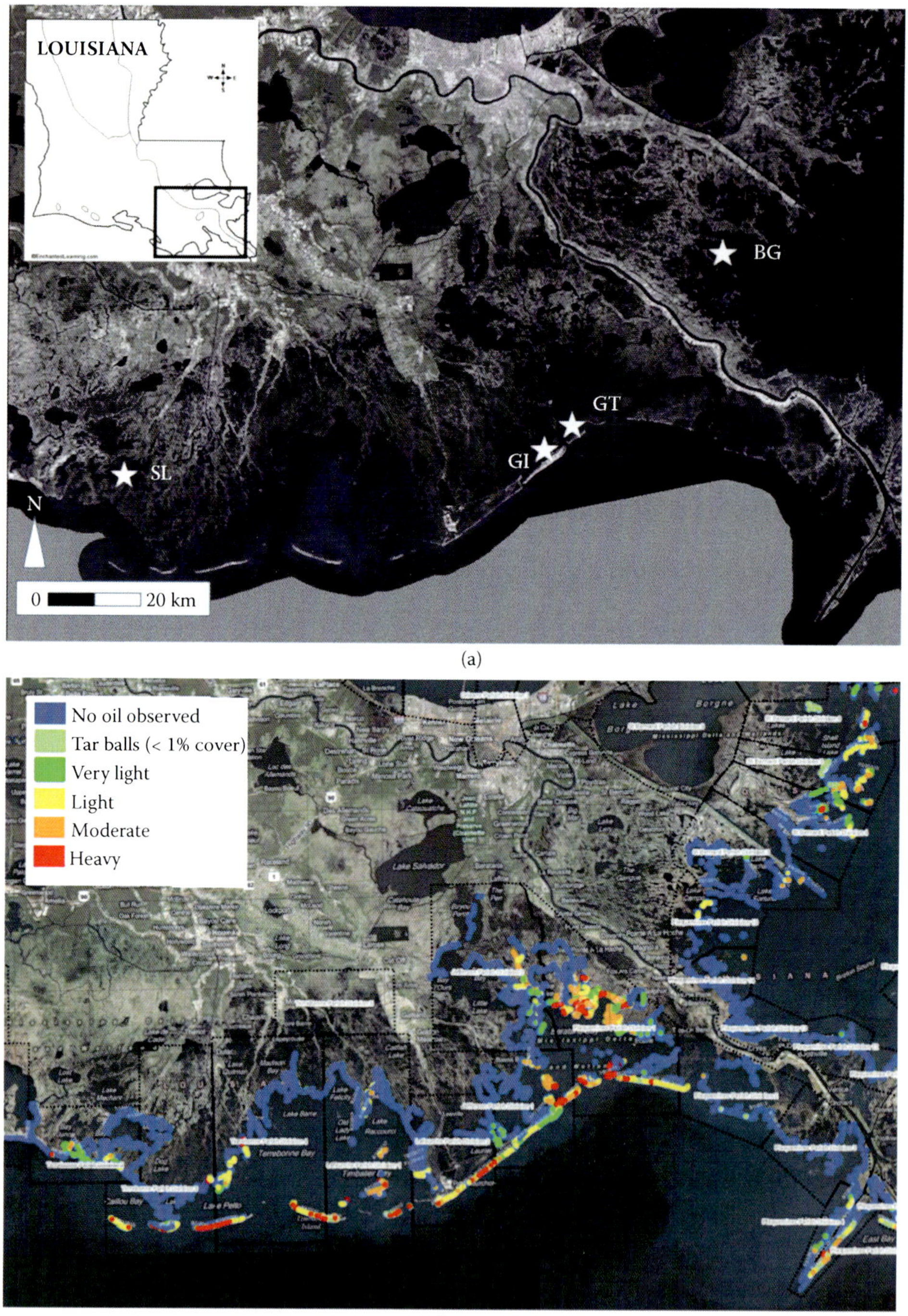

FIGURE 14.1
Maps indicating study sites where caged oysters were deployed and wild oysters were collected in Louisiana estuaries (a) and *Deepwater Horizon* shoreline oiling on September 20, 2010 obtained from the Shoreline Cleanup Assessment Teams (SCAT) (b). (From http://www.noaa.gov/deepwaterhorizon/maps/pdf/scat_la/Reduced_MC252_ShorelineCurrentOilingSituation_092010.pdf.)

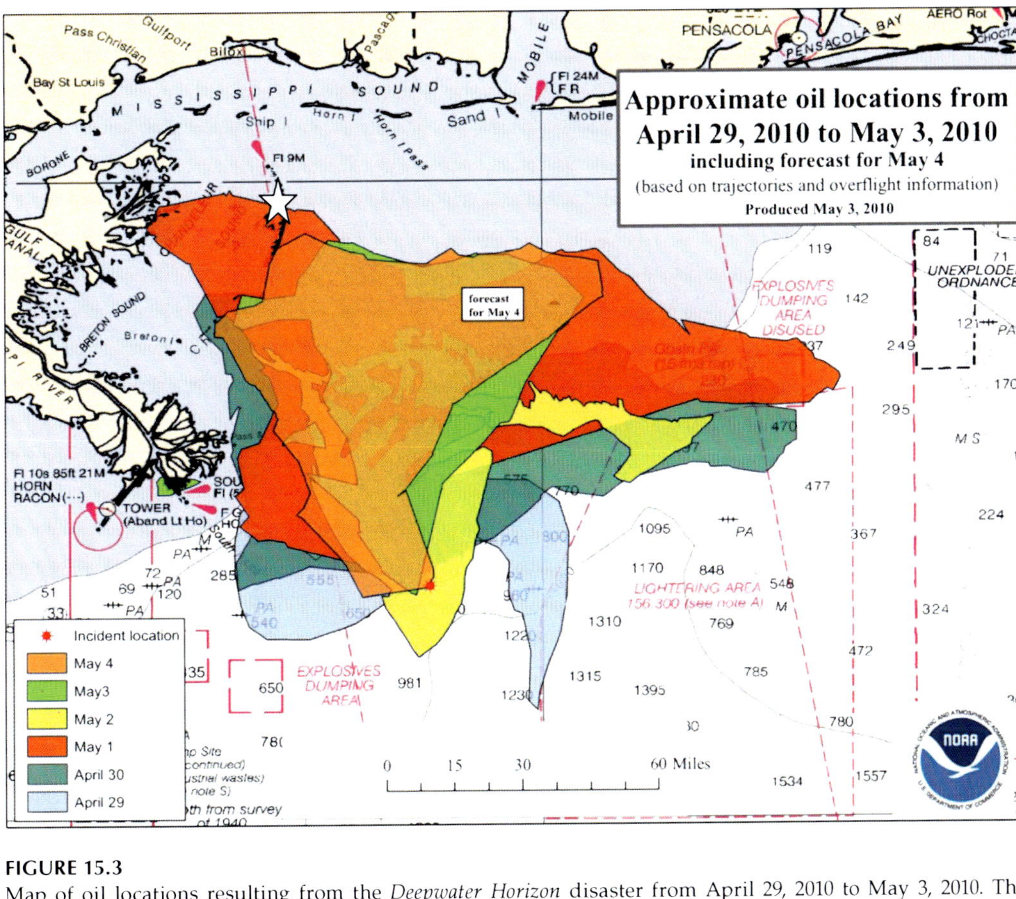

FIGURE 15.3
Map of oil locations resulting from the *Deepwater Horizon* disaster from April 29, 2010 to May 3, 2010. The Chandeleur Islands (upper left corner, starred) were inundated with oil on May 1, 2010. (Courtesy of NOAA.)

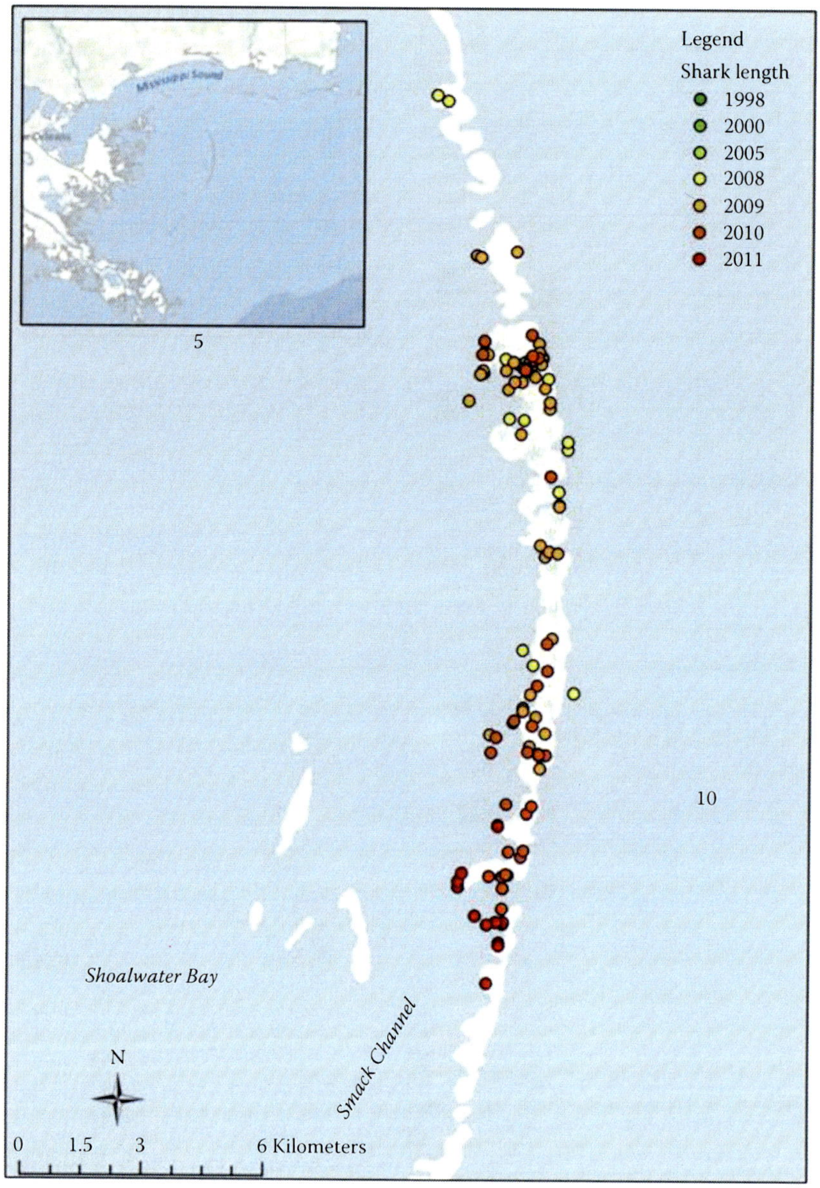

FIGURE 15.8
Map of the Chandeleur Islands showing locations of lemon shark collections from 1998 to 2011. Oil spill occurred in 2010 and the number of lemon sharks collected in back barrier marshes appears to have decreased since the creation of the oil prevention sand berm.

8

City, County, and State Methods to Protect Nearshore Fisheries Habitat in Mississippi during the Deepwater Horizon *Oil Spill*

Joseph R. Pursley and David Keith

CONTENTS

Overview

In the first few days following the explosion of the *Deepwater Horizon* (DWH) drilling rig on April 20, 2010 and its eventual sinking, the citizens and agencies of the Mississippi Gulf Coast were obviously concerned. As the full extent of the tragedy unfolded in the following weeks, the local, county, and state agencies organized to address the impending myriad of environmental, economics, and health concerns they were about to face. With lingering memories of the devastating impacts of Hurricane Katrina in 2005, the coastal communities of Mississippi recognized they would need immediate action—swift political decisions and strong community volunteerism—to protect the Gulf Coast from this new crisis.

Beginning at the local level, the City of Ocean Springs (COS) reached out to BP and the federally established Unified Command to hold community outreach meetings. It was at these meetings that the COS began coordinating with state and federal agencies for guidance regarding protection of their nearshore marine environment. During the development of the "BP MC252 *Deepwater Horizon* Block Grant Application," the COS and Anchor QEA, LLC, designed and implemented methods to protect Mississippi's nearshore freshwater, estuarine, and marine fisheries habitats. Using available aerial photos, geographic information systems, and local knowledge of shorelines, detailed maps of sensitive coastal resources were generated, and priority rankings were assigned to these areas. To support shoreline protection, adaptive designs, in particular reinforced oilphylic fencing, were developed that fit the diverse and dynamic shoreline of COS. The mapping, designs, and implementation methods were presented to the Mississippi Department of Marine Resources (DMR) as part of the Block Grant to BP. Continued coordination with DMR ultimately led to an expedited permit process with the U.S. Army Corps of Engineers to install the protective measures. As a result, implementation of 2 miles of reinforced oilphylic fencing in Ocean Springs began on June 2, 2010, approximately 43 days after the spill began and prior to any active shoreline oiling in the area.

News of the protective fencing design and rapid permitting process quickly spread to the adjacent coastal counties of Mississippi. These adjacent communities and Anchor QEA continued to design and develop innovative approaches such as Mississippi Sound—wide booming with large ocean boom systems and the development of reinforced jetty covers. Ultimately, 30 miles of reinforced oilphylic fencing were permitted for implementation along each of the three counties in Mississippi but the additional protection designs were not finalized or implemented. Ultimately, the innovative methods developed helped protect the bayous, estuaries, and marine fisheries habitats of coastal Mississippi.

Setting the Stage

Coastal Mississippi has a diverse coastline interface with the Gulf of Mexico. There are barrier islands (Gulf Islands National Seashore), the Mississippi Sound, bayous, back bays, marshes, rivers, and other freshwater wetlands. It has a highly productive marine ecosystem that supports multiple finfish and shellfish fisheries. The coastline also has several culturally and historically significant municipalities. One of these, the COS, is a small artisan community of less than 20,000 people. The entire waterfront and much of the surrounding residential neighborhoods were either destroyed or rehabilitated after being flooded by Hurricane Katrina in 2005, but the historic downtown environment was spared and quickly rebounded. It was this tragic event that trained and subsequently set the stage for the citizens and government to take proactive measures to protect themselves from the DWH spill in 2010.

During the first week of the spill, the COS Board of Aldermen activated the Emergency Operations Center. This emergency action provided the COS the ability to mobilize and finance the hiring of contractors and specialists to assist in the local response. This early action enabled the COS to reserve and contract with these specialists before the larger response from the federal government and other Gulf states consumed most of the available personal, equipment, and vessels.

Actions Taken

By May 1, 2010, BP officials had arrived in Biloxi, Mississippi, to hold the first of many public forums. Local residents created a volunteer network and signed up 5000 participants coordinated through a website (www.oilspillvolunteers.com). Anchor QEA was contracted by the COS to help design measures, assess priority locations, and provide response technologies to protect the shoreline of the city. Anchor QEA quickly began a detailed mapping exercise to create a set of field maps outlining COS infrastructure, parks, and sensitive natural resources. Once all the COS shoreline and infrastructure were ground-truthed, a resource matrix was compiled to prioritize the protection of the most sensitive areas (Figure 8.1). To protect these areas several design ideas were entertained, but several design parameters were needed to be met to make the implementation feasible. The protective measure selected was an oilphylic fabric, which enabled water to pass through, while trapping oil and debris in the fabric fibers (Figures 8.2 and 8.3). The fabric was attached to a wire fence with nylon zip-ties, to provide vertical stability, and the entire fence was stabilized by vertical metal t-posts driven into the ground. The fencing was designed and implemented as a secondary protective measure to the small booms and oil-skimming operations being implemented Gulf-wide. The fencing was installed relatively easily near sensitive areas, but the fencing needed to be maintained or replaced if oiled, damaged by waves or tidal fluctuations, or fouled by marine aquatic growth. As manufactured, the fencing fabric had limitations to how much weathered or raw hydrocarbon product it could hold, so the primary function of the fencing was seen as exclusion of oil from designated areas while accepting the inevitable oiling of adjacent shorelines. Minimizing impacts to the most sensitive habitats became the focus of all efforts.

Before the reinforced oilphylic fencing could be installed, the State-identified trustees for the response, which were the Mississippi Department of Environmental Quality (DEQ) and the DMR, had to review the design and implementation locations. Anchor QEA and COS quickly drafted a "Shoreline Protection Plan" (herein, the plan), which was presented to DEQ and DMR. Similarly, the plan was submitted to BP to cover the $4 million cost of

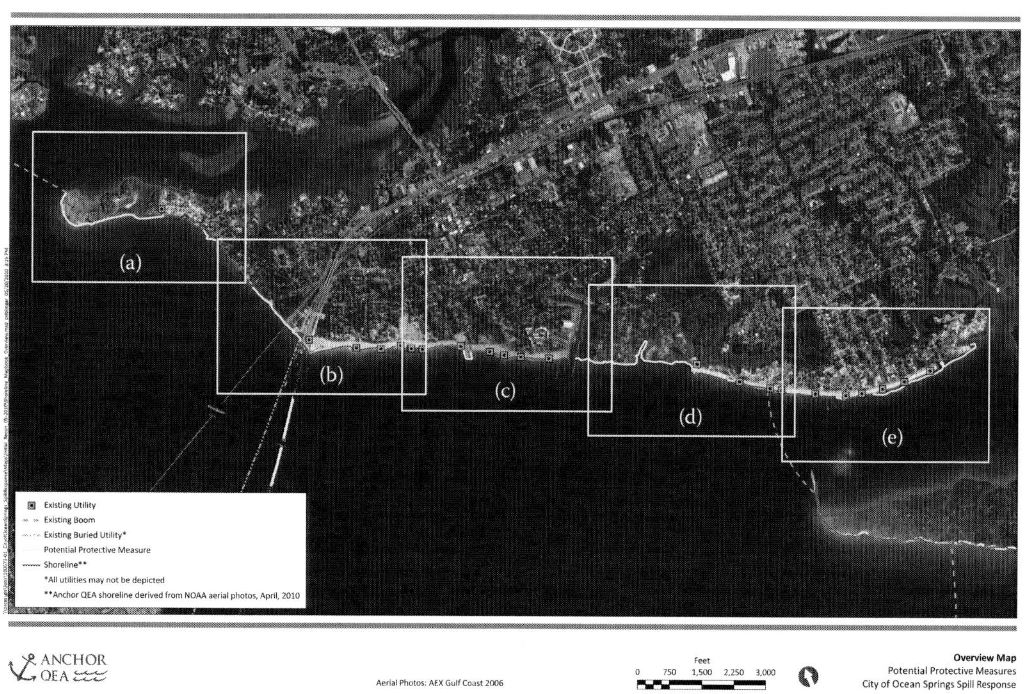

FIGURE 8.1
(See color insert.) A map of sensitive marine resources (a through e) that were delineated for protection in Ocean Springs, Mississippi, following the *Deepwater Horizon* oil spill in 2010.

FIGURE 8.2
Installation of the reinforced oilphylic fencing around a small patch of nearshore vegetation and a city stormwater culvert outfall in Ocean Springs, Mississippi. The fencing was installed as a protective oil spill response to the approaching oil from the *Deepwater Horizon* spill during summer 2010.

materials, equipment, labor, and engineering. Half of the requested amount, $2 million, was awarded. The DMR contacted the U.S. Army Corps (ACOE) to also review the plan. Soon additional federal and state agencies were asked to provide a review of the plan through an Emergency Public Notice. As Public Notice comments, questions, and concerns were received by the ACOE, Anchor QEA and COS responded within minutes or hours to each participant. In late May 2010, a Letter of Permission (LOP) from the ACOE was issued for the plan, and it contained several conditions. Aquatic vegetation surveys had to be conducted in all areas where fencing installation were to be implemented. Within 2 days, a survey plan was developed and a team of biologists was mobilized to conduct the surveys by boat and by foot. Similarly, the LOP also required a design change to how the protective measures were to be installed. To avoid entrapment of fish, turtles, and other marine organisms the fabric fencing had to provide gaps that were large enough to pass the largest of known sea turtles that are known to use the area. To accommodate the LOP condition, the fencing was installed with overlapping areas in the deepest water while still providing protection should surface or subsurface oiling occur. Once the LOP conditions were met, the implementation began. The plan implementation quickly fell under the now organized Unified Command and Incident Command System. A local firm was hired to install the protective measures with the trustees (DMR/DEQ) monitoring the environmental compliance of the LOP.

FIGURE 8.3
Installation of the reinforced oilphylic fencing around shoreline in Ocean Springs, Mississippi. The fencing was installed as a protective oil spill response to the approaching oil from the *Deepwater Horizon* spill during summer 2010.

Lessons Learned

Looking back now on the first few months of the spill, there was a constant feeling of incompleteness while waiting for impending slow motion disaster to occur. However, we observed that the resolve of the people of Mississippi was there, and the state and federal agencies were supportive, but the waiting game of "Where is the oil?" and "When will it get here?" was a constant source of anxiety. The people of the Mississippi Gulf Coast are a tried and tested bunch. The yearly threat of hurricanes and episodic destruction brought to them has taught these people to be proactive in the face of physical and financial ruin. From a technical standpoint, we learned that the fabric used as a preventative measure requires periodic replacement, while the t-posts and wire fencing withstood remained intact.

Section III

Population and Community Dynamics Following Oil Spill Disasters

9

Lessons from the 1989 Exxon Valdez *Oil Spill: A Biological Perspective*

Brenda E. Ballachey, James L. Bodkin, Daniel Esler, and Stanley D. Rice

CONTENTS

Introduction

On March 24, 1989, the tanker vessel *Exxon Valdez* altered its course to avoid floating ice, and ran aground on Bligh Reef in northeastern Prince William Sound (PWS), Alaska (Figure 1). The tanker was carrying about 53 million gallons of Prudhoe Bay crude, a heavy oil, and an estimated 11 million gallons spilled (264,000 barrels or about 42 million liters) in what was, prior to the *Deepwater Horizon* (DWH) spill of 2010, the largest accidental release of oil into U.S. waters (Morris and Loughlin 1994; Spies et al. 1996; Shigenaka 2014). Following the *Exxon Valdez* oil spill (EVOS), a broad range of studies was implemented and 25 years later, monitoring and research efforts to understand the long-term impacts of the spill continue, although now at a lesser intensity. The *Exxon Valdez* and DWH spills differed in many ways (Plater 2010; Atlas and Hazen 2011; Sylves and Comfort 2012), but there are also similarities, and lessons from the EVOS experience may offer valuable insights as research efforts proceed in the wake of the DWH spill. Here we provide an overview of the EVOS, summarize key findings from several long-term biological research programs, and conclude with some considerations of lessons learned after two and a half decades of study.

PWS is located in the north-central Gulf of Alaska, with much of the area considered pristine and uncontaminated prior to the spill (Short and Babcock 1996). The Sound is surrounded on three sides by mountain ranges, and the shoreline is complex, with hundreds of islands, bays and passages, including a high proportion of rocky benches and boulder beaches, areas of unconsolidated sediments in sheltered bays, and finer sediments in estuaries (Lindeberg et al. 2009). The climate is maritime, characterized by cool temperatures (averaging about 20°F in winter and 60°F in summer) and high precipitation

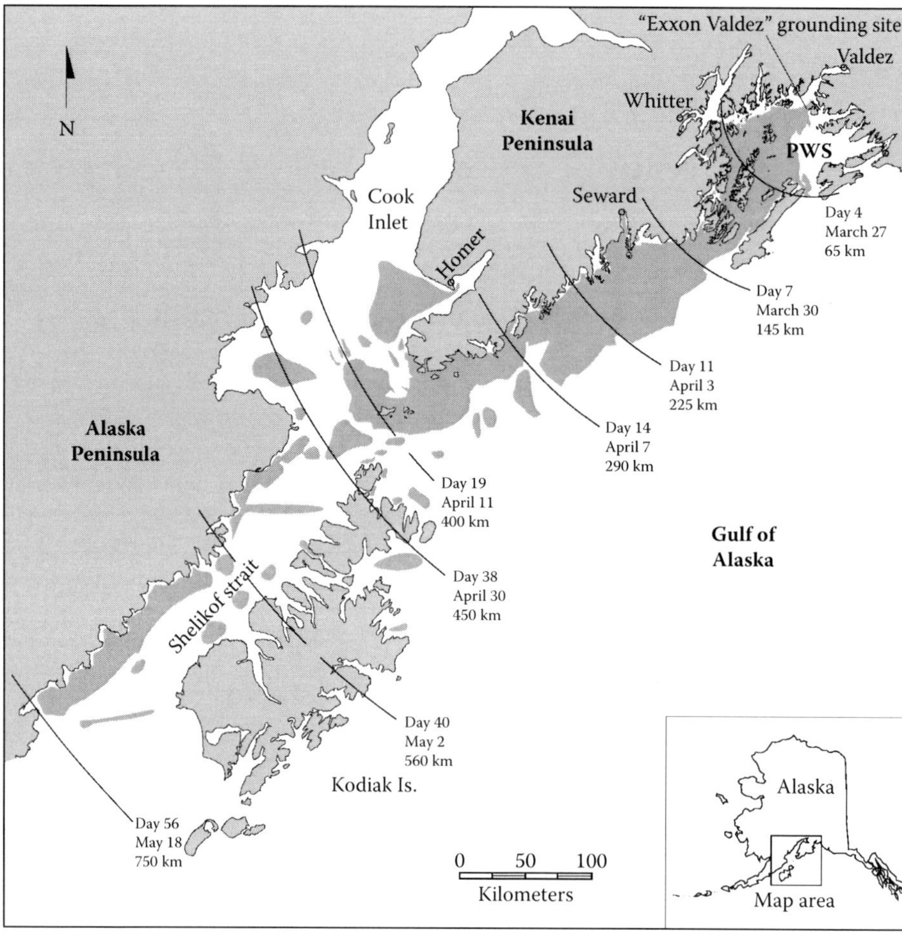

FIGURE 9.1
Map of Prince William Sound and the western Gulf of Alaska, showing site of the grounding of the T/V *Exxon Valdez* and extent of distribution of oil (dark grey shaded area) over the subsequent weeks (*Source*: NOAA Fisheries).

(ranging from about 60 inches to > 300 inches annually, depending on location within the Sound). About 150 glaciers feed into PWS, providing additional freshwater input. Terrestrial vegetation is primarily northern temperate rainforest. The dominant oceanographic feature is the Alaska Coastal current, which transports nutrients and productivity into the Sound from the southeast (Morris and Loughlin 1994). Kelp and seagrass are abundant and supplement phytoplankton production, providing habitat and nutrients for a variety of fishes and an assemblage of nearshore benthic invertebrates including crabs, clams, snails, urchins, and mussels. Unconsolidated sediments constitute much of the benthic habitat, supporting diverse assemblages of infaunal bivalves that serve as prey for higher trophic levels. The marine vertebrate community in PWS and the neighboring Gulf of Alaska is rich and diverse, with hundreds of species of fishes, birds, and marine and terrestrial mammals (Mundy 2005).

Within days of the tanker striking the reef, weather conditions deteriorated, with winds gusting up to 70 knots, severely restricting initial response efforts and rapidly distributing the spilled oil, destroying any illusions that responders may have had of containing the spill (Shigenaka 2014). Oil spread on the ocean surface in a southwesterly direction, leaving a heavy layer on numerous beaches within western PWS before exiting Montague Strait and other passages at the southwest corner of the Sound. Islands in the central portion of PWS were in the direct path of the moving oil and not far from the source, and were heavily contaminated. Oil eventually moved a linear distance of almost 900 km, covered over 26,000 km^2 of water in PWS and the Gulf of Alaska, and directly impacted about 2,000 km of coastline (Figure 1; Bragg et al. 1994; Morris and Loughlin 1994; Spies et al. 1996; Shigenaka 2014).

The greatest impacts from the spill were felt in PWS, although oil moved hundreds of kilometers beyond the Sound and contaminated shorelines on the Kenai and Alaska Peninsulas. In the weeks after the spill, approximately 150,000 barrels of oil washed ashore on beaches in PWS (Wolfe et al. 1994). Cleanup efforts occurred during spring and summer 1989, and continued for several years, involving over 11,000 workers before finishing in 1992 (Mearns 1996; Bohn 2011). Exxon estimated its cleanup costs to be $2.1 billion (Shigenaka 2014). The overall effectiveness and benefit of cleanup methods (including steam cleaning of beaches), and effects on the shorelines, were controversial, with questions as to the relative value of oil removal versus damage to intertidal communities and habitats (Mearns 1996; Shigenaka 2014). The most heavily oiled shorelines were located in the islands of the central western Sound (Figure 1), and accordingly, most studies of biological effects of the spill focused on species and communities in western PWS.

Early Impacts

The initial, acute effects of the spilled oil were dramatic and relatively well documented for birds and marine mammals, through carcass collections and response activities, with acute mortality estimates of 250,000 seabirds comprising over 90 species (Piatt et al. 1990; Piatt and Ford 1996) and several thousand sea otters (*Enhydra lutris*) (DeGange et al. 1994; Loughlin et al. 1996). Birds and sea otters were particularly vulnerable as the oil compromised the insulative properties of feathers and fur, leaving them subject to hypothermia in addition to direct toxic exposure from inhalation or ingestion. Direct toxic effects were apparent, as evidenced by mortality of seals and whales. Three hundred harbor seals (*Phoca vitulina*) and over 20 killer whales (*Orcinus orca*) were estimated to have died (Frost et al. 1994; Matkin et al. 1994, 2008). Although no whale carcasses were recovered (whale carcasses are known to sink), some killer whales were missing from counts (through photo identification) made in the years prior to the spill, and were thought to have died from exposure to the oil, probably from inhalation of toxic fumes, or possibly from ingestion of contaminated prey (Matkin et al. 1994, 2008; Bodkin et al. 2014). Both seals and whales rely on blubber for insulation and thus were not compromised directly by hypothermia when exposed to oil. Little information is available for other marine mammals (e.g., Steller sea lion, *Eumetopias jubatus*, and Dall's porpoise, *Phocoenoides dalli*), but acute effects were not thought to be significant (Loughlin et al. 1996).

In the first 2–3 years post-spill, numerous studies were conducted on a range of vertebrate species (see articles in Rice et al. 1996). In general, it was expected that oil-related effects would abate after several years (Neff et al. 1995). Herring (*Clupea pallasi*), a primary forage fish in the Sound, have been the focus of extensive post-spill studies, as they were commercially harvested in PWS and their numbers plummeted within a few years of the spill. Lack of population recovery persisted through the second decade (reviewed by Rice 2009) and continues, as demonstrated by the absence of a herring fishery through 2014, 25 years after the spill. However, despite numerous studies, the role of the spill in the decline and lack of recovery of herring, relative to other factors including disease, predation, and recruitment, has not been fully clarified and it is not likely that it ever will be completely understood (Carls et al. 2002; Rice 2009). Effort also was directed toward understanding effects and recovery among the algae, seagrasses, invertebrates, and intertidal fishes, which were affected by direct oiling as well as cleanup efforts that included destructive high pressure washing in intertidal habitats (Lees et al. 1996; Stekoll et al. 1996; Hoff and Shigenaka 1999; Shigenaka 2014).

Subsequent to the spill, both the *Exxon Valdez* Oil Spill Trustee Council ("Trustees", comprised of state and federal representatives) and the Exxon Corporation conducted multiple studies of acute and chronic effects, to support the pending litigation (Rice 2009; Bohn 2011). In 1991, an initial settlement of close to one billion dollars was reached. As of 2014, additional litigation is still possible, under a "Reopener for Unknown Injury" provision. The provision requires proof of specific conditions that include continuing loss or decline of a population, habitat or species, attributable to the spill, which could not have been known or reasonably anticipated by the federal or state Trustees at the time of settlement in 1991, and for which restoration projects exist whose costs are not grossly disproportionate to the magnitude of their benefits. For Trustee scientists, who were tasked with assessing damages to species for which their respective agencies were responsible, the legal guidance initially focused on documenting damage. After the 1991 settlement, the direction of the research evolved to restoration, which included understanding spill effects and documenting the long-term recovery of species and habitats. Differing perspectives among government and industry scientists resulted in studies on similar topics that frequently varied in concept, design, and statistical rigor. In many cases, conclusions in response to the same questions were inconsistent, and communications and collaborative efforts were limited or precluded because of litigation concerns (Paine et al. 1996; Rice 2009; Bohn 2011).

Persistence of Oil on Shorelines

Based on loss rates of oil during the first few years post-spill, it was assumed that lingering oil would soon be negligible and spill effects on fish and wildlife would rapidly diminish, with recovery of affected populations anticipated within a few years (Neff et al. 1995). However, over the decade following the spill, evidence mounted that oil was persisting in unconsolidated sediments of the intertidal zone longer than anticipated (Hayes and Michel 1999), that it was contaminating invertebrates (Fukuyama et al. 2000; Carls et al. 2001), and that some vertebrate consumers (including fishes and seabirds) were still being exposed to toxic oil (Trust et al. 2000; Golet et al. 2002; Jewett et al. 2002). Continued contamination of subtidal areas was not thought to be a concern. In 2001, Short et al. (2004, 2006) conducted an extensive survey of shorelines in western PWS. They conservatively estimated that over 55,000 kg of *Exxon Valdez* oil remained in intertidal habitats, and predicted that subsurface oil might persist at some sites for several decades (Short et al. 2007). Further, the remaining oil was largely unweathered and remained toxic, as it occurred in patches below the surface where little or no oxygen was available to degrade it. In 2014, 25 years after the spill, oil still can be found in the subsurface of some shorelines. Recent studies have examined factors contributing to the persistence of oil on PWS shorelines (Li and Boufadel 2010; Nixon et al. 2013) as well as potential approaches to bioremediation (Venosa et al. 2010; Boufadel and Bobo 2011; Boufadel et al. 2011). Oil also was documented to persist outside of PWS, on shorelines on the Alaska Peninsula, for more than two decades post-spill (Irvine et al. 2006, 2014).

Long-term Effects – The Unexpected

By the mid-1990s, evidence was beginning to accumulate that recovery was not progressing as anticipated for some species, and lingering effects of oil were suspected as a factor constraining recovery. Studies initiated at that point were designed to have a multispecies, multidisciplinary approach to evaluating population and ecosystem recovery. Based on a recognition of the lack of population recovery from the spill, or on evidence of continuing exposure to oil persisting in the environment, some studies continued for a subset of species recognized as not recovered and vulnerable to lingering oil. Long-term research was directed at sea otters (Bodkin et al. 2002, 2012; Ballachey et al. 2014), river otters (*Lutra canadensis*; Bowyer et al. 2003), harlequin ducks (*Histrionicus histrionicus*; Esler et al. 2002, 2010), Barrow's goldeneyes (*Bucephala islandica*; Esler et al. 2011), pigeon guillemots (*Cepphus columba*; Golet et al. 2002; Bixler et al. 2010), killer whales (Matkin et al. 2008, 2012, 2014), pink salmon (*Oncorhynchus gorbuscha*; Rice et al. 2001), and herring (Carls et al. 2002; Thorne and Thomas 2008; Rice 2009). Below we present a synopsis of long-term efforts to understand the effects of oil on three of those species: pink salmon, harlequin ducks, and sea otters. All three species have life histories that closely link them to intertidal habitats, and thus were at risk of prolonged effects from exposure to lingering oil in intertidal sediments. For pink salmon, studies were done on fish in the wild, comparing oiled and unoiled groups, and there was also a series of laboratory studies with controlled exposures to known quantities of oil to discern long-term effects. For harlequin ducks and sea otters, studies were done at the individual and population levels, and generally involved comparisons of oiled and unoiled areas.

Pink Salmon

In fall, pink salmon adults return to their natal streams to spawn, and embryos/larvae develop in the gravels over the winter. In spring, young of the year "fry" emerge and are swept downstream to estuarine marine waters where they feed along shore for several weeks as they migrate to the open ocean. Two years later, they return as adults. There are over 2,000 spawning streams in PWS (most less than one mile long), and it was estimated that the mouths of about one-third of the streams in the southwest portion of the Sound were oiled to some extent. The most productive spawning habitat is the intertidal portion of a stream, and after the spill, spawning in intertidal areas placed developing embryos at risk in the tidal zone where oil initially was stranded and subsequently persisted. Thus after the spill, there was concern over the potential both for embryos to be exposed to oil and out-migrating fry to be exposed as they dispersed and foraged along contaminated shorelines.

In late March 1989, when the spill occurred, fry were beginning to emerge from the spawning gravels, feed along the shoreline, and migrate out of PWS. Acute mortalities of out-migrating fry were never detected, but immediate effects on growth and subsequent survival of fry from oiled areas were identified in 1989. Wild pink salmon fry collected from nearshore waters in oiled areas grew at half the rate of fry from reference areas (Wertheimer and Celewycz 1996) and Willette (1996) found lower growth in fry released from hatcheries and collected near oiled shorelines compared to unoiled sites. Evidence of oil exposure included tissue polycyclic aromatic hydrocarbons (PAHs) and induction of the biomarker enzyme cytochrome P4501A (Carls et al. 1996; Willette 1996; Wiedmer et al. 1996), and oil observed in stomachs and intestines (Sturdevant et al. 1996). By 1990, growth of fry was similar in oiled and unoiled reference areas of the Sound, and there was no further evidence of increased P4501A induction or tissue hydrocarbons. Detrimental effects on growth in 1989 were not surprising as oil was in the marine environment was widely evident, nor was the recovery in 1990, when oil was no longer visible at the water surface, even in inshore areas.

Elevated pink salmon embryo mortalities were detected in fall 1989 in spawning gravels of oiled beaches, which was not unexpected as oil was still obvious on many beaches. However, additional sampling over the subsequent 4- year period (through 1993) continued to detect elevated embryo mortalities, although differences between oiled and unoiled reference streams declined over time (Bue et al. 1994, 1996, 1998). The observation of higher mortality several years after the spill was unprecedented and raised questions about the mechanism of exposure.

Initially, it was anticipated that exposure of pink salmon embryos would be negligible as oil floats and the spawning gravels were underwater, and thus were not subject to direct contamination. However, the observation of continued elevated mortalities of embryos led to investigation of possible exposure mechanisms. Murphy et al. (1999) confirmed that lingering oil adjacent to streams was associated with elevated mortalities, and Carls et al. (2003) demonstrated that embryo mortality was consistent with areas of interstitial drainage of oil-contaminated water into spawning gravels from surrounding stream banks. Concentrations of PAHs in gravels exposed to oil were measured and found to be very low, increasing concerns about the effects of low-level, chronic PAH loads on developing embryos.

A series of controlled laboratory exposure tests on pink salmon embryos to low parts per billion (ppb) levels of PAHs demonstrated significant effects on survival, indicating that embryos were particularly vulnerable to oil exposure. Prior to the spill, studies of acute PAH toxicity (4-10 day studies) suggested that concentrations of several parts per million were required to affect survival (Moles and Rice 1983). The field observations in PWS after the spill, however, suggested that chronic exposure concentrations in the ppb range were sufficient to affect survival. The first in a series of controlled laboratory tests confirmed that long term, low level exposures of embryos to PAH concentrations of 20–50 ppb resulted in increased deformities, slower development, histopathological damage, and lower survival (Marty et al. 1997; Heintz et al. 1999, 2000). Analyses of hydrocarbons in tissues and induction of cytochrome P4501A in embryonic tissues indicated the PAHs were permeating the outer egg membranes, resulting in lower growth and survival (Carls et al. 2005). In later experiments that combined controlled laboratory exposures with releases of fry to the marine environment, impacts on returning adult success were detected after long-term exposure of embryos to 5-20 ppb PAHs (Heintz et al. 1999, 2000).

For these studies, embryos were exposed, and resulting fry were tagged in the spring before release to the marine environment to migrate. Success was measured when adults returned in the following year, after surviving the stresses of life in the wild (foraging, avoiding predators, growing, migrating). These low level exposures were found to have affected growth, which in turn affected survival. Fitness impacts were proportionate to the exposure dose: returns of adults more than 1 year post-exposure were shown to decrease by 16% (at PAH exposure of 5 ppb) to 36% (at PAH exposure of 19 ppb). To put this in perspective, Rice et al. (2001) calculated that a run producing 10 million eggs would result in 46,000 fewer adult fish if those eggs had been exposed to 19 ppb total PAHs. Concentrations of this magnitude were still present in the interstitial waters of a number of salmon streams in 1995, 6 years after the EVOS (Murphy et al. 1999).

Overall, there were three major research efforts that provided compelling evidence of chronic impacts to pink salmon from oil persisting in intertidal habitats: (1) observations of elevated embryo mortality in the 5 years following the spill, (2) identification of the exposure mechanism from contaminated beaches to spawning gravels, and (3) measured effects on fitness following embryonic exposures under controlled conditions. Collectively, this group of pink salmon studies provided unprecedented evidence that exposure of embryonic life stage to low level PAHs (in ppb) from persistent oil can have a population level effect. These findings influenced damage assessment of future spills, such as Cosco Busan and Deep Water Horizon spills (Incardona et al. 2012, 2014).

Harlequin Ducks

Harlequin ducks are a common winter resident of PWS. Much of their annual life cycle is spent in nearshore marine habitats, where they show high fidelity to relatively small areas (Iverson et al. 2004). As a consequence, they may remain in an area even if there is an event that degrades the quality of their habitat (Iverson and Esler 2006). They forage exclusively on marine invertebrates in intertidal and shallow subtidal habitats, and thus were at risk of exposure to oil from the EVOS that persisted in intertidal areas (Short et al. 2006). Harlequin ducks have relatively long life spans, up to 20 years. The generally high rates of adult survival, combined with a low reproductive potential, lead to population dynamics that are sensitive to even slight changes in adult survival rates (Saether and Bakke 2000). In northern areas, harlequin ducks exist close to their energetic threshold during the winter, and have little margin for increasing caloric intake to meet any elevated metabolic demands (Esler et al. 2002), as may result from chronic exposure to PAHs (Holmes et al. 1979; Jenssen 1994).

Several lines of evidence suggested that recovery of harlequin ducks was constrained during the first decade after the oil spill (Esler et al. 2002). Surveys indicated that population densities in oiled areas were reduced, and abundance declined in oiled areas but not in unoiled areas. In 1998, assays of the cytochrome P4501A biomarker showed elevated levels in ducks from oiled areas, indicating continued exposure to hydrocarbons (Trust et al. 2000). Further, survival of adult females over the winter was depressed in oiled, relative to unoiled areas, from 1995 to 1998 (Esler et al. 2000). However, from 2000 to 2003, winter survival rates of female harlequin ducks were similar between oiled and unoiled areas (Esler and Iverson 2010), suggesting that negative effects of lingering oil on survival had subsided after about a decade post-spill.

The significance of the effects at the population level were best demonstrated by modeling, where data from studies of survival were combined with other demographic data on harlequin ducks, including movements (Iverson and Esler 2006) and fecundity, to develop a population model estimating the timeline and process of population recovery (Iverson and Esler 2010). Results showed that female survival likely was poorest during the first year after the spill, but that effects of the spill on survival persisted for more than a decade. Most importantly, mortality associated with the chronic phase of the spill was estimated at 772 females, almost double the 400 females estimated to have died during the first post-spill year. The population model also provided an estimated timeline of 24 years (to 2013) for full recovery of the population, with a range of 16 to 32 years (2005 to 2021) under best-case and worst-case scenarios, respectively (Iverson and Esler 2010).

The harlequin duck research has generated a unique and long-term data set evaluating PAH exposure with the cytochrome P4501A biomarker, with liver samples collected from 1998 through 2009, over an 11 year span. In 1998, 9 years after the spill, harlequin ducks from oiled areas of PWS had indicators of the CYP1A biomarker that averaged nearly three times higher than those from unoiled areas (Trust et al. 2000), and similar patterns were observed through 2009 (Esler et al. 2010), suggesting some individuals were exposed to lingering *Exxon Valdez* oil up to 20 years post-spill. Studies are ongoing (2014) to evaluate whether or not exposure to lingering oil continues for harlequin ducks in PWS.

In conclusion, it was discovered that harlequin duck recovery after the oil spill was influenced by persistent, chronic exposure to oil, with demographic consequences for at least a decade after the initial event. The importance of documenting chronic exposure and associated long-term effects is significant, as these studies provide insight into the various mechanisms of population-level effects from oil spills. In addition, these results heighten concerns over the potential for similar effects that could result from chronic exposure to non-point source pollution.

Sea Otters

Sea otters, the smallest marine mammal, reside in shallow nearshore marine habitats of the north Pacific. They occupy relatively small home ranges (from 10 to 100 km^2), with adult males defending exclusive territories from other males, and female home ranges typically extending across several male territories. Females usually give birth to a single pup annually, and both sexes may live to about 20 years of age. They prey almost exclusively on large benthic marine invertebrates, including bivalves, gastropods, echinoderms, and crustaceans, foraging mainly in waters less than 40 m in depth (Bodkin et al. 2004). In mixed sediment habitats, such as PWS, clams can account for more than 75% of their diet (Dean et al. 2002). Unlike other marine mammals, sea otters have no blubber but rely on their dense fur and a high metabolic rate to stay warm in the cold northern waters. This reliance on their fur for warmth renders them particularly vulnerable to spilled oil, as the insulating capacity of the fur is severely impaired by oiling (Costa and Kooyman 1982; Siniff et al. 1982; Williams et al. 1995). Their high metabolic rate is maintained by intensive foraging and consumption of prey equal to about 25% of their body mass each day, and thus sea otters have little scope for increasing energy intake in response to environmental or physiological stress.

Acute effects of the EVOS on sea otters were significant, demonstrated by recovery of about 1000 carcasses throughout the spill area, and capture of several hundred sea otters from oiled areas for placement in rescue and rehabilitation facilities within a few months of the spill (Ballachey et al. 1994). Estimates of the total number of sea otters within PWS dying from acute exposure ranged from 750 to 2,650 animals (Garrott et al. 1993; Garshelis 1997), with the disparity among acute mortality estimates generally reflecting the lack of accurate estimates of the abundance of sea otters prior to the spill. In 1993, aerial survey methods were established (Bodkin and Udevitz 1999) and thereafter, abundance was monitored annually in PWS. In the most heavily oiled areas, numbers remained depressed until 2011 (Ballachey et al. 2014), whereas in the larger area of western PWS where many shorelines were less severely affected by the spill, a return to prespill abundance (based on best available prespill estimates) occurred by 2009 (Bodkin et al. 2002, 2011).

Several lines of research demonstrated linkages between lingering oil and elevated sea otter mortality. Ballachey et al. (2003) and Rotterman and Monnett (1991) found low survival of juvenile sea otters in oiled areas of PWS in the early 1990s, relative to their counterparts in unoiled reference areas. Carcasses were collected annually from shorelines to obtain ages-at-death data (based on cementum annuli in teeth; Bodkin et al. 1997). Similar collections had been made in years prior to the spill, thus providing one of the few prespill data sets. Monson et al. (2000, 2011) developed models incorporating both pre- and post-spill ages-at-death data, and found elevated mortality post-spill, not only for animals that were alive at the time of the spill but also among those born after, implicating lingering oil as a contributing factor. Monson et al. (2011) estimated that the mortality associated with chronic exposure or long-term effects of acute exposure was about 900 animals, a number similar in magnitude to the known immediate mortality based on the number of carcasses recovered in 1989 following the spill.

Causes of sea otter mortality from acute exposure, based on necropsies of carcasses recovered in the weeks and months immediately after the spill, were documented by Lipscomb et al. (1994), and included hypothermia from fouling of the pelage and direct toxicity from the oil itself. Over the following years, as it became evident that recovery was not progressing as anticipated, chronic exposure to lingering oil was implicated as a limiting factor (Bodkin et al. 2002). However, the mechanisms involved with exposure to relatively small amounts of oil lingering in intertidal habitats in the decades following the spill were not readily understood. Sea otters excavate sediments when foraging for clams, and frequently forage in intertidal areas (Bodkin et al. 2012). This excavation can routinely disturb sediments at the depths at which lingering oil occurs (Short et al. 2004). Thus, if sea otters forage in areas where oil persists, they may be exposed through consumption of prey that have assimilated hydrocarbons or by disturbing oiled sediments and releasing lingering oil, which could then adhere to their fur and affect its ability to repel water, and perhaps be ingested upon grooming (Ballachey et al. 2013).

In the mid-2000s, Bodkin et al. (2012) implemented a study to determine if spatial and temporal overlap of foraging sea otters and lingering oil would elucidate pathways of exposure. To assess foraging patterns, sea otters were implanted with time-depth recorders, providing data on over a million foraging dives from 19 sea otters over a several year span (Bodkin et al. 2012). Sea otters foraged in intertidal areas to varying extents, with individuals averaging 8 to 91 intertidal foraging dives per day. Further, there was a strong seasonal component to intertidal foraging, with a pronounced peak from late spring to early summer when most adult females have small pups. This places females and pups at greater risk of encountering oil, presumably incurring added metabolic costs and perhaps greater chronic problems for pups from early exposure to toxic compounds (Bodkin et al. 2012; Ballachey et al. 2013). Utilizing data on foraging dive depths in conjunction with data on the distribution of lingering oil during 2001 and 2003 (Short et al. 2004, 2006), Bodkin et al. (2012) estimated that sea otters would encounter subsurface lingering oil an average of 10 times each year, ranging from 2 to 24 times, depending on individual foraging routines.

During the same time period, Bodkin et al. (2012) surveyed mixed sediment beaches in western PWS, including beaches that were known to contain patches of lingering oil, and found widespread evidence of sea otter foraging activity. Sediment samples were collected from within or adjacent to sea otter foraging pits on a subset of beaches, and in a few instances, sediments were subsequently confirmed to have elevated levels of PAHs (Bodkin et al. 2012). Collectively, these findings confirm that sea otter exposure to oil in sediments during foraging likely was a pathway for chronic exposure. Bodkin et al. (2012) concluded that the population of sea otters in western PWS has been an important factor in disruption and depletion of patches of lingering oil, through their excavation of intertidal foraging pits, estimated to number more than a million annually since 1989.

For sea otters, gene transcription studies were used to evaluate potential molecular responses to hydrocarbon exposure, based on laboratory studies with mink (Bowen et al.

2007, 2011). Gene transcript profiles indicated that sea otters sampled from PWS in 2008 had differential transcription relative to captive otters in aquaria, or to wild otters from a reference area on the Alaska Peninsula (Miles et al. 2012), implicating exposure to organic compounds for sea otters in oiled regions. Some differential gene expression noted in 2008 could be the result of historical, not necessarily current, exposure in these long-lived animals. Resampling of PWS sea otters in 2012 suggested that lingering effects of exposure had abated (Ballachey et al. 2014).

In summary, sea otter populations residing in the path of the oil spill suffered high rates of mortality from acute exposure, resulting in large-scale declines in abundance. For about two decades after the spill, the abundance remained depressed, relative to the best estimates of prespill numbers. Studies showed that foraging in intertidal sediments to obtain clams, a preferred prey item, provided a direct pathway for exposure to lingering oil. Chronic exposure to oil and possibly latent effects of acute exposure appear to have decreased survival and constrained recovery of the population for more than two decades. The processes of population injury and recovery were similar between sea otters and harlequin ducks, suggesting that life history traits and ecological attributes may be of greater importance in predicting susceptibility to oil spill effects than taxonomic relationships.

Lessons Learned

Over the two and a half decades since the EVOS, there has been a tremendous effort to determine factors contributing to the accident, develop technologies to assess and remediate shorelines, and understand and evaluate the significance and long-term consequences for populations and ecosystems. The extent of this body of work is evident in the many hundreds of reports and publications now available, including several recent review articles (Peterson et al. 2003; Rice 2009; Bohn 2011; Shigenaka 2014). An enduring legacy of the EVOS is the change in how we attempt to prevent oil spills (e.g., the Oil Spill Pollution Act of 1990 requires double hull tankers), implementation of additional safeguards (e.g., escort of tankers through PWS by two tugboats), and our approach to studies of spill effects—recognizing the potential for long-term consequences—as new spill events occur. Long-term EVOS studies of oil persistence on shorelines and effects on biota, some still ongoing in PWS, have significantly influenced the time line of spill events and increased the emphasis on prevention. Despite differences between the EVOS and the *Deepwater Horizon* spill, in terms of the environment, the material spilled, the location and depths of the spill, and the species affected, lessons from the body of research following the EVOS are relevant for understanding biological consequences of the *Deepwater Horizon* spill. A prime example of the application of results from EVOS to Deepwater Horizon research is presented by Incardona et al. (2014), which builds on the studies of vulnerability of pink salmon embryos to oil. Incardona et al. (2014) demonstrate that environmentally relevant levels of oil exposure cause cardiac defects in embryonic stages of several fish species that inhabit the Gulf of Mexico.

A common impediment in the assessment of environmental damages is a deficiency of baseline data, and certainly this deficit was a limitation for studies of injured biological resources following the EVOS. Further, there was a delay in implementing post-spill studies, which exacerbated the deficit of data characterizing conditions prior to contamination by oil. Although pre-spill data sets were limited, when available (including some information for chemical baselines, killer whales, pink salmon, and sea otters), they were invaluable in assessing damage. For EVOS research, limited pre-spill data required greater emphasis on comparisons of individuals and conditions in oiled versus unoiled areas. In general, biological systems tend to be dynamic, increasing the effort required to obtain reliable and relevant baseline information. Nevertheless, evaluation of damages and recovery in the aftermath of future spills and other environmental perturbations will be vastly facilitated by the availability of baseline data sets.

One of the least expected but most significant findings from the body of EVOS research has been the long-term persistence of oil in intertidal sediments, which in turn has consequences for the chronic exposure and delayed recovery of some species. Long-term effects of the spill are related to the life history and natural history characteristics of individual species, and where mechanisms and pathways of exposure have been identified, the study process has been extended and costly. In some cases, the methods available for study have improved over the course of several decades. The EVOS Trustee Council's support for long-term studies, employing multiple approaches at both the individual and population level, has been essential for clarifying the extent of chronic effects and status of recovery, and for better understanding cascading effects. Further, and of great importance, it now is apparent that the magnitude of chronic effects may be similar to or greater than the more obvious acute effects. Based on the EVOS experience and long-term studies of other spills, we recognize that spilled oil may persist in nearshore areas for decades, delaying recovery and potentially influencing the structure of coastal marine communities (Bodkin et al. 2014). This has led to a paradigm shift in how we study damage assessment following spills: our concerns now encompass not only the immediate effects caused by fouling of fauna at the surface (marine mammals, birds, and shorelines) but also the significant effects to fauna over a much longer period of time, resulting from chronic persistence and toxicity of oil.

Over the past decades, our understanding of the subtlety of effects of chronic exposure, and complexity of interactions among species, has greatly increased. Assessing spill impacts over a lengthy time frame necessitates knowledge of the presence and persistence of oil in the environment, and the direct and indirect effects on individual organisms and populations at a range of trophic levels. Research programs to identify long-term effects, including pathways and mechanisms of exposure and injury for affected species, will benefit from an ecosystem perspective. Additionally, studies must consider the potential complications of combined multiple environmental stressors that may compound acute and chronic effects of exposure to oil (Hylland 2006; Ballachey et al. 2013; Whitehead 2013).

Finally, as emphasized previously by Peterson et al. (2003), given observations of delayed recovery of multiple species and the minute amounts of oil that were linked to ecological effects following the EVOS, and the continuing release of hydrocarbon contaminants (point and non-point sources of pollution) into marine environments, development of ecosystem-based toxicology is required to understand and ultimately predict chronic, delayed, and indirect long-term risks and effects of hydrocarbons and other contaminants. Studies of impacts of the *Exxon Valdez* spill are applicable to the development of long-term research efforts ensuing from the *Deepwater Horizon* and other spill events, and more broadly, are relevant for understanding chronic toxicity and impacts of pollution in coastal marine areas in general.

References

Atlas, R. M. and T. C. Hazen. 2011. Oil biodegradation and bioremediation: A tale of the two worst spills in US history. *Environmental Science & Technology* 45:6709–6715.

Ballachey, B. E., J. L. Bodkin, and A. R. DeGange. 1994. An overview of sea otter studies. In *Marine Mammals and the Exxon Valdez*, ed. T. R. Loughlin, 47–59. San Diego: Academic Press.

Ballachey, B. E., J. L. Bodkin, S. Howlin, A. M. Doroff, and A. H. Rebar. 2003. Correlates to survival of juvenile sea otters in Prince William Sound, Alaska. *Canadian Journal of Zoology* 81:1494–1510.

Ballachey, B. E., J. L. Bodkin, and D. H. Monson. 2013. Quantifying long-term risks to sea otters from the 1989 *Exxon Valdez* oil spill: Reply to Harwell & Gentile. *Marine Ecology Progress Series* 488:297–301.

Ballachey, B. E., D. H. Monson, G. G. Esslinger, et al. 2014. 2013 update on sea otter studies to assess recovery from the 1989 *Exxon Valdez* oil spill, Prince William Sound, Alaska. U.S. Geological Survey Open-File Report 2014-1030, 40 p. doi: 10.3133/ofr20141030.

Bixler, K. S., D. D. Roby, D. B. Irons, et al. 2010. Pigeon guillemot restoration research in Prince William Sound, Alaska. *Exxon Valdez* Oil Spill Restoration Project Final Report (Restoration Project 070853), Oregon State University, Corvallis, Oregon. Available at: http://www.evostc.state.ak.us/Store/FinalReports/2007-070853-Final.pdf

Bodkin, J. L., B. E. Ballachey, T. A. Dean, et al. 2002. Sea otter population status and the process of recovery from the *Exxon Valdez* oil spill. *Marine Ecology Progress Series* 241:237–253.

Bodkin, J. L., G. G. Esslinger, and D. H. Monson. 2004. Foraging depths of sea otters and implications to coastal marine communities. *Marine Mammal Science* 20:305–321.

Bodkin, J. L., B. E. Ballachey, H. A. Coletti, et al. 2012. Long-term effects of the *Exxon Valdez* oil spill: Sea otter foraging in the intertidal as a pathway of exposure to lingering oil. *Marine Ecology Progress Series* 447:273–287.

Bodkin, J. L., B. E. Ballachey, and G. G. Esslinger. 2011. Trends in sea otter population abundance in western Prince William Sound, Alaska: Progress toward recovery following the 1989 *Exxon Valdez* oil spill. U.S. Geological Survey Scientific Investigations Report 2011–5213 14 p.

Bodkin, J. L., D. Esler, S. D. Rice, et al. 2014. The effects of spilled oil on coastal ecosystems: Lessons from the *Exxon Valdez* spill. In *Coastal Conservation*, eds. B. Maslo and J. L. Lockwood, 311–346. Cambridge: Cambridge Univ Press.

Bodkin, J. L., J. A. Ames, R. J. Jameson, et al. 1997. Estimating age of sea otters with cementum layers in the first premolar. *Journal of Wildlife Management* 61:967–973.

Bodkin, J. L. and M. S. Udevitz. 1999. An aerial survey method to estimate sea otter abundance. In *Marine Mammal Survey and Assessment Methods*, eds. G. W. Garner, S. C. Amstrup, J. L. Laake, B. J. F. Manly, L. L. McDonald, and D. G. Robertson, 13–26. Rotterdam: A. A. Balkema.

Bohn, D. 2011. The *Exxon Valdez* oil spill experience: Lessons learned from a cold-water spill in subarctic waters. In *An Evaluation of the Science Needs to Inform Decisions on Outer Continental Shelf Energy Development in the Chukchi and Beaufort Seas,Alaska*, eds. L. Holland-Bartels and B. Pierce, Appendix D, 253–262. U.S. Geological Survey Circular 1370.

Boufadel, M. C. and A. M. Bobo. 2011. Feasibility of high pressure injection of chemicals into the subsurface for the bioremediation of the *Exxon Valdez* oil. *Ground Water Monitoring & Remediation* 31:59–67.

Boufadel, M. C., A. M. Bobo, and Y. Q. Xia. 2011. Feasibility of deep nutrients delivery into a Prince William Sound beach for the bioremediation of the *Exxon Valdez* oil spill. *Ground Water Monitoring & Remediation* 31:80–91.

Bowen, L., A. K. Miles, M. Murray, et al. 2011. Gene transcription in sea otters (*Enhydra lutris*)—Development of a diagnostic tool for sea otter and ecosystem health. *Molecular Ecology Resources* 12: 67–74.

Bowen, L., F. Riva, C. Mohr, et al. 2007. Differential gene expression induced by exposure of captive mink to fuel oil: A model for the sea otter. *EcoHealth Jounal Consortium* 4:298–309.

Bowyer, R. T., G. M. Blundell, M. Ben-David, et al. 2003. Effects of the *Exxon Valdez* oil spill on river otters: Injury and recovery of a sentinel species. *Wildlife Monographs* 153:1–53.

Bragg, J. R., R. C. Prince, E. J. Harner, et al. 1994. Effectiveness of bioremediation for the *Exxon Valdez* oil spill. *Nature* 368:413–418.

Bue, B. G., S. Sharr, S. D. Moffitt, et al. 1994. Assessment of injury to pink salmon embryos and fry. In *Proceedings of the 16th Northeast Pacific Pink and Chum Salmon Workshop*, 173–176. AK-SG-94–02.

Bue, B. G., S. Sharr, S. D. Moffitt, et al. 1996. Effects of the *Exxon Valdez* oil spill on pink salmon embryos and pre-emergent fry. In *Proceedings of the Exxon Valdez Oil Spill Symposium*, eds. S. D. Rice, R. B. Spies, D. A. Wolfe, and B. A. Wright, 619–627. Bethesda: American Fisheries Society.

Bue, B. G., S. Sharr, and J. E. Seeb. 1998. Evidence of damage to pink salmon populations inhabiting Prince William Sound, Alaska, two generations after the *Exxon Valdez* oil spill. *Transactions of the American Fisheries Society* 127:35–43.

Carls, M. G., M. M. Babcock, P. M. Harris, et al. 2001. Persistence of oiling in mussel beds after the *Exxon Valdez* oil spill. *Marine Environmental Research* 51:167–190.

Carls, M. G., R. A. Heintz, G. D. Marty, et al. 2005. Cytochrome P4501A induction in oil-exposed pink salmon *Oncorhynchus gorbuscha* embryos predicts reduced survival potential. *Marine Ecology Progress Series* 301:253–265.

Carls, M. G., G. D. Marty, and J. E. Hose. 2002. Synthesis of the toxicological impacts of the *Exxon Valdez* oil spill on Pacific herring (Clupea pallasi) in Prince William Sound, Alaska, USA. *Canadian Journal of Fisheries and Aquatic Science* 59:153–172.

Carls, Mark G., G. D. Marty, and S. D. Rice. 2003. Is pink salmon spawning habitat recovering from the *Exxon Valdez* oil spill? In *Proceedings of the Twenty-sixth Arctic and Marine Oil Spill Program Technical Seminar*. Victoria: Environment Canada.

Carls, M. G., A. C. Wertheimer, J. W. Short, et al. 1996. Contamination of juvenile pink and chum salmon by hydrocarbons in Prince William Sound after the *Exxon Valdez* oil spill. In *Proceedings of the Exxon Valdez Oil Spill Symposium*, eds. S. D. Rice, R. B. Spies, D. A. Wolfe, and B. A. Wright, 593–607. Bethesda: American Fisheries Society.

Costa, D. P. and G. L. Kooyman. 1982. Oxygen consumption, thermoregulation, and the effect of fur oiling and washing on the sea otter, *Enhydra lutris*. Canadian Journal of Zoology 60:2761–2767.

Dean, T. A., J. L. Bodkin, A. K. Fukuyama, et al. 2002. Food limitation and the recovery of sea otters following the *Exxon Valdez* oil spill. *Marine Ecology Progress Series* 241:255–270.

DeGange, A. R., A. M. Doroff, and D. H. Monson. 1994. Experimental recovery of sea otter carcasses at Kodiak Island, Alaska, following the *Exxon Valdez* oil spill. *Marine Mammal Science* 10:492–496.

Esler, D., B. E. Ballachey, K. A. Trust, et al. 2011. Cytochrome P4501A biomarker indication of the timeline of chronic exposure of Barrow's goldeneyes to residual *Exxon Valdez* oil. *Marine Pollution Bulletin* 62:609–614.

Esler, D., T. D. Bowman, K. A. Trust, et al. 2002. Harlequin duck population recovery following the *Exxon Valdez* oil spill: Progress, process and constraints. *Marine Ecology Progress Series* 241:271–286.

Esler D. and S. A. Iverson. 2010. Female harlequin duck winter survival 11 to 14 years after the *Exxon Valdez* oil spill. *Journal of Wildlife Management* 74:471–478.

Esler D., J. A. Schmutz, R. L. Jarvis, et al. 2000. Winter survival of adult female harlequin ducks in relation to history of contamination by the *Exxon Valdez* oil spill. *Journal of Wildlife Management* 64:839–847.

Esler, D., K. A. Trust, B. E. Ballachey, et al. 2010. Cytochrome P4501A biomarker indication of oil exposure in harlequin ducks up to 20 years after the *Exxon Valdez* oil spill. *Environmental Toxicology and Chemistry* 29:1138–1145.

Frost, K. J., L. F. Lowry, E. H. Sinclair, et al. 1994. Impacts on distribution, abundance and productivity of harbor seals. In *Marine Mammals and the Exxon Valdez*, eds. T. R. Loughlin, 97–118. San Diego: Academic Press.

Fukuyama, A. K., G. Shigenaka, and R. Z. Hoff. 2000. Effects of residual *Exxon Valdez* oil on intertidal Protothaca staminea: Mortality, growth, and bioaccumulation of hydrocarbons in transplanted clams. *Marine Pollution Bulletin* 40:1042–1050.

Garrott, R. A., L. L. Eberhardt, and D. M. Burn. 1993. Mortality of sea otters in Prince William Sound following the *Exxon Valdez* oil spill. *Marine Mammal Science* 9:343–359.

Garshelis, D. L. 1997. Sea otter mortality estimated from carcasses collected after the *Exxon Valdez* oil spill. *Conservation Biology* 11:905–916.

Golet, G. H, P. E. Seiser, A. D. McGuire, et al. 2002. Long-term direct and indirect effects of the *Exxon Valdez* oil spill on pigeon guillemots in Prince William Sound, Alaska. *Marine Ecology Progress Series* 241:287–304.

Hayes, M. O. and J. Michel. 1999. Factors determining the long-term persistence of *Exxon Valdez* oil in gravel beaches. *Marine Pollution Bulletin* 38:92–101.

Heintz, R. A., S. D. Rice, A. C. Wertheimer, et al. 2000. Delayed effects on growth and marine survival of pink salmon *Oncorhynchus gorbuscha* after exposure to crude oil during embryonic development. *Marine Ecology Progress Series* 208:205–216.

Heintz, R. A., J. W. Short, and S. D. Rice. 1999. Sensitivity of fish embryos to weathered crude oil: Part II. Incubating downstream from weathered *Exxon Valdez* crude oil caused increased mortality of pink salmon (*Oncorhynchus gorbuscha*) embryos. *Environmental Toxicolology and Chemistry* 18:494–503.

Hoff, R. Z. and G. Shigenaka. 1999. Lessons from 10 years of post-*Exxon Valdez* monitoring on intertidal shorelines. 1999. *International Oil Spill Conference Proceedings* 111–117.

Holmes, W. N., J. Gorsline, and J. Cronshaw. 1979. Effects of mild cold stress on the survival of seawater-adapted mallard ducks (*Anas platyrhynchos*) maintained on food contaminated with petroleum. *Environmental Research* 20:425–444.

Hylland, K. 2006. Polycyclic aromatic hydrocarbon (PAH) ecotoxicology in marine ecosystems. *Journal of Toxicology and Environmental Health. Part A* 69:109–123.

Incardona, J. P., L. D. Gardner, T. L. Linbo, et al. 2014. *Deepwater Horizon* crude oil impacts the developing hearts of large predatory pelagic fish. *Proceedings of the National Academy of Science of the United States of America* 111:E1510-E1518.

Incardona, J. P., C. A. Vines, B. F. Anulacion, et al. 2012. Unexpectedly high mortality in Pacific herring embryos exposed to the 2007 *Cosco Busan* oil spill in San Francisco Bay. *Proceedings of the National Academy of Science of the United States of America* 109:E51-E58.

Irvine, G. V., D. H. Mann, M. Carls, C. Reddy, and R. K. Nelson. 2014. *Exxon Valdez* oil after 23 years on rocky shores in the Gulf of Alaska: Boulder armor stability and persistence of slightly weathered oil. Abstract, presented at the American Geophysical Union meeting, 23-28 February 2014, Honolulu, Hawaii.

Irvine, G. V., D. H. Mann, and J. W. Short. 2006. Persistence of ten-year old *Exxon Valdez* oil on Gulf of Alaska beaches: The importance of boulder armoring. *Marine Pollution Bulletin* 52:1011–1022.

Iverson, S. A. and D. Esler. 2010. Harlequin duck population injury and recovery dynamics following the 1989 *Exxon Valdez* oil spill. *Ecological Applications* 20:1993–2006.

Iverson, S. A., D. Esler, and D. J. Rizzolo. 2004. Winter philopatry of harlequin ducks in Prince William Sound, Alaska. *The Condor* 106:711–715.

Iverson, S. A. and D. Esler. 2006. Site fidelity and the demographic implications of winter movements by a migratory bird, the harlequin duck *Histrionicus histrionicus*. *Journal of Avian Biology* 37:219–228.

Jenssen, B. M. 1994. Review article: Effects of oil pollution, chemically treated oil, and cleaning on the thermal balance of birds. *Environmental Pollution* 86:207–215

Jewett S. C., T. A. Dean, B. R. Woodin, et al. 2002. Exposure to hydrocarbons ten years after the *Exxon Valdez* oil spill: Evidence from cytochrome P4501A expression and biliary FACs in nearshore demersal fishes. *Marine Environmental Research* 54:21–48.

Lees, D. C., J. P Houghton, and W. B. Drickell. 1996. Short-term effects of several types of shoreline treatment on rocky intertidal biota in Prince William Sound. In *Proceedings of the Exxon Valdez Oil Spill Symposium*, eds. S. D. Rice, R. B. Spies, D. A. Wolfe, and B. A. Wright, 329–348. Bethesda: American Fisheries Society.

Li, H. L. and M. C. Boufadel. 2010. Long-term persistence of oil from the *Exxon Valdez* spill in two-layer beaches. *Nature Geoscience* 32:96–99.

Lindeberg, M., J. N. Harney, J. R. Harper, et al. 2009. ShoreZone Coastal Habitat Mapping Data Summary Report – Prince William Sound. *Exxon Valdez* Oil Spill Restoration Project Final Report (Restoration Project 070805). NOAA National Marine Fisheries Service, Juneau, Alaska, 135p. Available at: http://alaskafisheries.noaa.gov/shorezone/logs/PWS_2004-07_SummaryRpt.pdf

Lipscomb, T. K., R. K. Harris, A. H. Rebar, et al. 1994. Pathology of sea otters. In *Marine Mammals and the Exxon Valdez*, eds. T. R. Loughlin, 265–280. San Diego: Academic Press.

Loughlin, T. R., B. E. Ballachey, and B. A. Wright. 1996. Overview of studies to determine injury caused by the *Exxon Valdez* oil spill to marine mammals. In *Proceedings of the Exxon Valdez Oil Spill Symposium*, eds. S. D. Rice, R. B. Spies, D. A. Wolfe, and B. A. Wright, 798–808. Bethesda: American Fisheries Society.

Marty, G. D., D. E. Hinton, J. W. Short, et al. 1997. Ascites, premature emergence, increased gonadal cell apoptosis, and cytochrome P4501A induction in pink salmon larvae continuously exposed to oil-contaminated gravel during development. Canadian Journal of Zoology 75:989–1007.

Matkin, C. O., J. W. Durban, E. L. Saulitis, et al. 2012. Contrasting abundance and residency patterns of two sympatric populations of transient killer whales (*Orcinus orca*) in the northern Gulf of Alaska. *Fishery Bulletin* 110:143–155.

Matkin, C. O., G. M. Ellis, M. E. Dahlheim, et al. 1994. Status of killer whales in Prince William Sound, 1985–1992. In *Marine Mammals and the Exxon Valdez*, ed. T. R. Loughlin, 141–162. San Diego: Academic Press.

Matkin, C. O., E. L. Saulitis, G. M. Ellis, et al. 2008. Ongoing population-level impacts on killer whales Orcinus orca following the *Exxon Valdez* oil spill in Prince William Sound, Alaska. Marine Ecology Progress Series 356:269–281.

Matkin, C. O., J. W. Testa, G. M. Ellis, et al. 2014. Life history and population dynamics of southern Alaska resident killer whales (Orcinus orca). Mar Mamm Sci 30:460–479.

Mearns, A. J. 1996. *Exxon Valdez* shoreline treatment and operations: Implicaitons for response, assessment, monitoring, and research. In *Proceedings of the Exxon Valdez Oil Spill Symposium*, eds. S. D. Rice, R. B. Spies, D. A. Wolfe, and B. A. Wright, 309–328. Bethesda: American Fisheries Society.

Miles, A. K., L. Bowen, B. E. Ballachey, et al. 2012. Variation in transcript profiles in sea otters (*Enhydra lutris*) from Prince William Sound, Alaska and clinically normal reference otters. *Marine Ecology Progress Series* 451:201–212.

Moles, A. and S. D. Rice. 1983. Effects of crude oil and naphthalene on growth, caloric content and fat content of pink salmon juveniles in seawater. *Transactions of the American Fisheries Society* 112:205–211.

Monson, D. H., D. F. Doak, B. E. Ballachey, et al. 2000. Long-term impacts of the *Exxon Valdez* oil spill on sea otters, assessed through age-dependent mortality patterns. *Proceedings of the National Academy of Science of the United States of America* 97:6562–6567.

Monson, D. H., D. F. Doak, B. E. Ballachey, et al. 2011. Could residual oil from the *Exxon Valdez* spill create a long-term population "sink" for sea otters in Alaska? *Ecological Applications* 21: 2917–2932.

Morris, B. F. and T. R. Loughlin. 1994. Overview of the *Exxon Valdez* oil spill 1989–1992. In *Marine Mammals and the Exxon Valdez*, ed. T. R. Loughlin, 1–22. San Diego: Academic Press.

Mundy, P. R. (ed). 2005. *The Gulf of Alaska: Biology and Oceanography*. Fairbanks: Alaska Sea Grant College Program, Univ. of Alaska.

Murphy, M. L., R. A. Heintz, J. W. Short, et al. 1999. Recovery of pink salmon spawning after the *Exxon Valdez* oil spill. *Transactions of the American Fisheries Socities* 128:909–918.

Neff, J. M., E. H. Owens, S. W. Stoker, et al. 1995. Shoreline oiling conditions in Prince William Sound following the *Exxon Valdez* oil spill. In *Exxon Valdez Oil Spill: Fate and Effects in Alaskan Waters*, eds. P. G. Wells, J. N. Butler, and J. S. Hughes, 312–346. Philadelphia: American Society for Testing and Materials.

Nixon, Z., Michel, J. Hayes, et al. 2013. Geomorphic Factors Related to the Persistence of Subsurface Oil from the *Exxon Valdez* Oil Spill. *Journal of Coastal Research* 69:115–127.

Paine, R. T., J. L. Ruesink, A. Sun, et al. 1996. Trouble on oiled waters: Lessons from the *Exxon Valdez* oil spill. *Annual Review of Ecology and Systematics* 27:197–235.

Peterson, C. H., S. D. Rice, J. W. Short, et al. 2003. Long-term ecosystem response to the *Exxon Valdez* oil spill. *Science* 302:2082–2086.

Piatt J. F. and R. G. Ford. 1996. How many seabirds were killed by the *Exxon Valdez* oil spill? In *Proceedings of the Exxon Valdez Oil Spill Symposium*, eds. S. D. Rice, R. B. Spies, D. A. Wolfe, and B. A. Wright, 712–719. Bethesda: American Fisheries Society.

Piatt, J. F., C. J. Lensink, W. Butler, et al. 1990. Immediate impact of the *Exxon Valdez* oil spill on marine birds. *The Auk* 107:387–397.

Plater, Z. J. 2010. Learning From Disasters: Twenty-One Years After the *Exxon Valdez* Oil Spill, Will Reactions to the Deepwater Horizon Blowout Finally Address the Systemic Flaws Revealed in Alaska? *Environmental Law Reporter* 40:11041–11047.

Rice, S. D. 2009. Persistence, toxicity, and long-term environmental impact of the *Exxon Valdez* oil spill. *University of St Thomas Law Journal* 7:55–67.

Rice, S. D., R. B. Spies, D. A. Wolfe, et al. 1996. Proceedings of the *Exxon Valdez* Oil Spill Symposium. American Fisheries Society Symposium 18.

Rice, S. D., R. E. Thomas, M. G. Carls, et al. 2001. Impacts to pink salmon following the *Exxon Valdez* oil spill: Persistence, toxicity, sensitivity, and controversy. *Reviews in Fisheries Societies* 9:165–211.

Rotterman, L. M. and C. Monnett. 1991. Mortality of sea otter weanlings in eastern and western Prince William Sound, Alaska, during the winter of 1990–91, *Exxon Valdez* Oil Spill State/Federal Natural Resource Damage Assessment Final Report (Marine Mammal Study 6-18). Anchorage: U.S. Fish and Wildlife Service.

Saether, B. E. and O. Bakke. 2000. Avian life history variation and contribution of demographic traits to the population growth rate. *Ecology* 81:642–653.

Shigenaka, G. 2014. *Twenty-five years after the Exxon Valdez: NOAA's Scientific Support, Monitoring, and Research*. Seattle: NOAA Office of Response and Restoration. 78 pp.

Short, J. W. and M. M. Babcock. 1996. Pre-spill and post-spill concentrations of hydrocarbons in mussels and sediments in Prince William Sound. In *Proceedings of the Exxon Valdez Oil Spill Symposium*, eds. S. D. Rice, R. B. Spies, D. A. Wolfe, and B. A. Wright, 149–166. Bethesda: American Fisheries Society.

Short, J. W., M. R. Lindeberg, P. A. Harris, et al. 2004. Estimate of oil persisting on beaches of Prince William Sound, 12 years after the *Exxon Valdez* oil spill. *Environmental Science & Technology* 38:19–25.

Short, J. W., J. M. Maselko, M. R. Lindeberg, et al. 2006. Vertical distribution and probability of encountering intertidal *Exxon Valdez* oil on shorelines of three embayments within Prince William Sound, Alaska. *Environmental Science & Technology* 40:3723–3729.

Short, J. W., G. V. Irvine, D. H. Mann, et al. 2007. Slightly weathered *Exxon Valdez* oil persists in Gulf of Alaska beach sediments after 16 years. *Environmental Science & Technology* 41:1245–1250.

Siniff, D. B., T. D. Williams, A. N. Johnson, et al. 1982. Experiments on the response of sea otters *Enhydra lutris* to oil contamination. *Biological Conservation* 23:261–272.

Spies, R. B., S. D. Rice, D. A. Wolfe, et al. 1996. The effects of the *Exxon Valdez* oil spill on the Alaskan coastal environment. In *Proceedings of the Exxon Valdez Oil Spill Symposium*, eds. S. D. Rice, R. B. Spies, D. A. Wolfe, and B. A. Wright, 1–16. Bethesda: American Fisheries Society.

Stekoll, M. S., L. Deysher, R. C. Highsmith, et al. 1996. Coastal habitat community assessment: Intertidal communities and the *Exxon Valdez* oil spill. In *Proceedings of the Exxon Valdez Oil Spill Symposium*, eds. S. D. Rice, R. B. Spies, D. A. Wolfe, and B. A. Wright, 177–192. Bethesda: American Fisheries Society.

Sturdevant, M. V., A. C. Wertheimer, and J. L. Lum. 1996. Diets of juvenile pink and chum salmon in oiled and non-oiled nearshore habitats in Prince William Sound, 1989 and 1990. In *Proceedings of the Exxon Valdez Oil Spill Symposium*, eds. S. D. Rice, R. B. Spies, D. A. Wolfe, and B. A. Wright, 578–592. Bethesda: American Fisheries Society.

Sylves, R. T. and L. K. Comfort. 2012. The *Exxon Valdez* and BP *Deepwater Horizon* oil spills: Reducing risk in socio-technical systems. *American Behavioural Science* 56:76–103.

Thorne, R. E. and G. L. Thomas. 2008. Herring and the *Exxon Valdez* oil spill: An investigation into historical data conflicts. *ICES Journal of Marine Science: Journal du Conseil* 65:44–50.

Trust, K. A., D. Esler, B. R. Woodin, et al. 2000. Cytochrome P4501A induction in sea ducks inhabiting nearshore areas of Prince William Sound, Alaska. *Marine Pollution Bulletin* 40:397–403.

Venosa, A. D., P. Campo, and M. T. Suidan. 2010. Biodegradability of lingering crude oil 19 years after the *Exxon Valdez* oil spill. *Environmental Science & Technology* 44:7613–7621.

Wertheimer, A. C. and A. G. Celewycz. 1996. Abundance and growth of juvenile pink salmon in oiled and non-oiled locations of western Prince William Sound after the *Exxon Valdez* oil spill. In *Proceedings of the Exxon Valdez Oil Spill Symposium*, eds. S. D. Rice, R. B. Spies, D. A. Wolfe, and B. A. Wright, 518–532. Bethesda: American Fisheries Society.

Wiedmer, M., J. J. Fink, J. J. Stegeman, et al. 1996. Cytochrome P-450 induction and histopathology in pre-emergent pink salmon from oiled sites in Prince William Sound. In *Proceedings of the Exxon Valdez Oil Spill Symposium*, eds. S. D. Rice, R. B. Spies, D. A. Wolfe, and B. A. Wright, 509–517. Bethesda: American Fisheries Society.

Willette, M. 1996. Impacts of the *Exxon Valdez* oil spill on migration, growth, and survival of juvenile pink salmon in Prince William Sound. In *Proceedings of the Exxon Valdez Oil Spill Symposium*, eds. S. D. Rice, R. B. Spies, D. A. Wolfe, and B. A. Wright, 533–550. Bethesda: American Fisheries Society.

Williams, T. M., D. J. O'Connor, and S. W. Nielsen. 1995. The effects of oil on sea otters: Histopathology, toxicology and clinical history. In *Emergency Care and Rehabilitation of Sea Otters: A Guide for Oil Spills Involving Fur-bearing Marine Mammals*, eds. T. M. Williams, and R. W., 3–22. Fairbanks: Univ. of Alaska Press.

Whitehead, A. 2013. Interactions between oil-spill pollutants and natural stressors can compound ecotoxicological effects. *Integrative and Comparative Biology* 53:635–647.

Wolfe, D. A, M. J. Hameedi, J. A. Galt, et al. 1994. The fate of the oil spilled from the *Exxon Valdez*. *Environmental Science & Technology* 28:561A–568A.

10

The Exxon Valdez *Oil Spill and the Collapse of the Prince William Sound Herring Stock: A Reexamination of Critical Biomass Estimates*

Richard E. Thorne and Gary L. Thomas

CONTENTS

Introduction

On March 23, 1989, the oil tanker *Exxon Valdez* hit Bligh Reef and spilled 11 million gallons (38,800 t) of crude oil into Prince William Sound (PWS). Over the subsequent 3 years, commercial herring fishers harvested about 65,000 t of Pacific herring (*Clupea pallasi*). It appeared at that time that the *Exxon Valdez* oil spill (EVOS) had no impact on herring populations. In 1993, fishery managers forecasted an adult Pacific herring biomass of 133,852 t based on an age-structured assessment (ASA) model (Funk 1993; Quinn and Deriso 1999; Brown 2007; Hulson et al. 2008). However, commercial fishers were unable to locate fishable concentrations that year. An acoustic survey in fall 1993, the first of 20 annual acoustic surveys conducted by the senior author, estimated the adult population to be only 18,812 (95% CI ± 3140). Subsequently, the ASA model estimate was revised downward and the plunge from the 1992 estimate to the new 1993 estimate was referred to as the 1993 herring collapse (Brown 2007; Hulson et al. 2008). Because this collapse occurred 4 years after EVOS, it was not attributed to the oil spill. The 1989 year class was a recruitment failure in 1993 that was clearly associated with EVOS (Peterson et al. 2003; Brown 2003, 2007). However, this recruitment failure was too small to explain the collapse. Instead, a disease outbreak was hypothesized as the cause, even though there were no observations of surface mats of dead Pacific herring as seen in many other, much smaller, disease outbreaks and disease monitoring programs did not start until 1994 (Quinn et al. 2001).

In early 2002, after a decade of acoustic surveys, we initiated a search and subsequent analyses to find and verify other sources of information to compare with our results. We found our best correlation with a relatively obscure index referred to as the "mile-days of spawn" (Brown 2007) (see the description of the method in the following section). Subsequent examination of this index revealed a multiyear decline that began in 1989. With this information, we first challenged the concept of a 1993 collapse and the lack of impact by EVOS (Thomas and Thorne 2003). In this chapter, first we argue that the collapse of the PWS Pacific herring stock was temporally associated with EVOS. Then we present evidence for a causal relationship. Finally, we argue that indirect fishery assessment techniques, such as ASA, are limited in their ability to detect acute mortalities in juvenile and adult fish. This analysis includes comparisons of the estimates from the acoustic surveys, mile-days of spawn, and ASA models, as well as the original management forecasts.

Methods

The general use of acoustic methods for fisheries assessment is commonplace and is described by Simmonds and MacLennan (2005) and Zale et al. (2012). Applications to Pacific herring have been well documented (Thorne 1977a,b; Thorne et al. 1983; Trumble et al. 1983) and details of the methods we used in PWS can be found in Thomas et al. (1997), Thomas and Thorne (2003), and Thorne and Thomas (2008, 2012). Our methods are similar to standard acoustic methods except that we use a three-stage adaptive sampling methodology rather than systematic transects. Adult Pacific herring during the extended winter in PWS are typically located in a few select bays and inlets and are distributed primarily in large, mid-water schools or dense layers at night. Thus, the acoustic surveys were conducted during the late winter/early spring prespawning distribution when the Pacific herring are most concentrated. The initial survey stage focused on the location of these adult Pacific herring aggregations within PWS and included aerial surveys of foraging marine mammals, especially Steller sea lions and humpback whales, sonar surveys, and observations from fishers, hunters, and others transiting PWS, coupled with a database of historic locations. After the Pacific herring were located, the second stage consisted of multiple echo integration surveys over the areas occupied by the herring schools. These surveys were generally conducted at night with a dark vessel since Pacific herring are up in the water column at night, but are very light sensitive. The multiple surveys were used to determine the precision of the biomass estimates. The focus on prespawning Pacific herring simplifies species composition problems as the concentrations were virtually all adult Pacific herring. After the echo integration surveys, the Pacific herring schools were subsampled primarily with a commercial purse seine for biological information. The size composition of the Pacific herring in the net catches was used to estimate target strengths for converting backscatter to biomass. The target strength that was used to scale echo integration had been derived from many years of research that culminated in the *ex situ* target strength measurements (described by Thomas et al. [2002]).

The mile-days of spawn index is an aerial survey of the cumulative linear extent of herring spawn (milt) along beaches that has been collected by the Alaska Department of Fish and Game (ADFG) since 1973 (Becker and Biggs 1992). Multiple aerial surveys (>20) were conducted during the approximate 1-month duration of spawning. The database contains over 6000 independent observations of herring spawning events in PWS. The current methodology, used since 2001 when geographic information systems (GIS) tracking was incorporated, records spawning events as arcs (linestrings) using GIS technology. Recently, the older data were reanalyzed by reconstructing flight paths and recalculating lengths using ArcMap. The mile-days of spawn was never viewed as a primary measure of stock abundance (Brown 2007) and was given low weight as an auxiliary input in the development of the ASA model (Quinn et al. 2001).

Age data from herring in PWS have been collected by ADFG since 1973 (Funk and Sandone 1990). The ASA model has been run to forecast the PWS adult herring biomass most years since 1993 and has included several versions (Quinn et al. 2001; Hulson et al. 2008). Standard practice for model runs is to reconstruct the population history of the herring since 1980. Recent versions of the ASA model incorporate the mile-days of spawn index, the acoustic estimates, and disease indices (Quinn et al. 2001; Hulson et al. 2008). The current model used by ADFG is described by Hulson et al. (2008), where it is referred to as the M4 model (also called the Moffett 2005 version). All mile-days and ASA values used in this chapter were provided courtesy of Steve Moffitt (lead herring biologist for ADFG, Cordova).

Management of the PWS herring stock is based on a harvest policy established by the Alaska Board of Fisheries that specifies a maximum 20% exploitation rate (Brown 2007). The allowable harvest is based on biomass estimates established the previous year modified by the expected growth and survival over the year. Aerial surveys (visual estimates from spotter pilots) were used to estimate the biomass of herring schools from 1973 to 1987 (Biggs and Baker 1993; Brown 2003, 2007). However, egg deposition surveys were a primary estimation technique from 1988 to 1992 (Biggs and Baker 1993; Brown 2003, 2007). During the same period (1988–1992), virtual population analysis was used to develop an ASA model to track survival of adult herring by age (Doubleday 1976; Deriso et al. 1985; Funk and Sandone 1990; Funk and Harris 1992). Population estimates from egg deposition and aerial surveys along with regional age structure information were inputs for the ASA models. The ASA modeling estimates became increasing important as a primary tool for setting fishing quotas and estimating population size and became the primary basis for management forecasts in 1992 when an age-structured model was first used to forecast the 1993 stock biomass (Funk 1993, 1994; Brown 2007). Various age-structured models have been used subsequently (Quinn et al. 2001; Hulson et al. 2008). Subsequent to 1992, all the fishery management forecast data used in this chapter come from real-time ADFG publications of annual forecasts (www.cf.adfg.state.ak.us).

Thomas and Thorne (2003) showed that the cumulative mile-days of milt index from the aerial surveys correlated well with the acoustic estimates from 1993 to 2002. We subsequently converted the mile-days index to an absolute measure through its regression with the acoustic estimates, often referred to as the mile-days biomass model (Hulson et al. 2008). This procedure allows us to compare the index in absolute terms with the other databases, which are also expressed in absolute values of herring biomass. In this chapter, we use the regression with the best correlation ($r = 0.68$, $p \leq 0.05$) based on the 1993–2006 update of the model, including the recalculated index values. It is important to note that the mile-days index and associated model estimates are postfishery measures; comparisons with prefishery estimates need to account for harvests. The acoustic surveys have been prefishery since 1995, but there has not been a fishery since 1998.

Results

The initial acoustic survey estimate in 1993 verified a population decline (Figure 10.1). The population began a recovery in 1996, and then declined again after 1997. The mile-days biomass model estimates agree with the acoustic measurements and/or the direction of change in almost all cases (Figure 10.2).

Timing of the Collapse

Both the mile-days index and associated mile-days biomass model estimates declined every year after its 1988 peak until 1995 with a single exception, 1991, when it experienced a slight increase associated with a large recruitment from the 1988 (pre-EVOS) year class (Figure 10.3). The fishery management forecast data substantially lag the post-EVOS pattern shown by the mile-days estimates. The mile-days biomass estimates drop precipitously after the 1989 oil spill, while the management forecasts portray a continuing high population through 1993 (although the 1993 forecast was revised downward after the 1993 run failure). Fishery harvests during this time were the highest during the days of the reduction fishery (Brown 2007) with 65,000 t removed. Back-calculated or hindcast estimates from the current ASA model actually agree well with the mile-days biomass model estimates when adjusted to the same postharvest basis (Figure 10.3). In fact, the

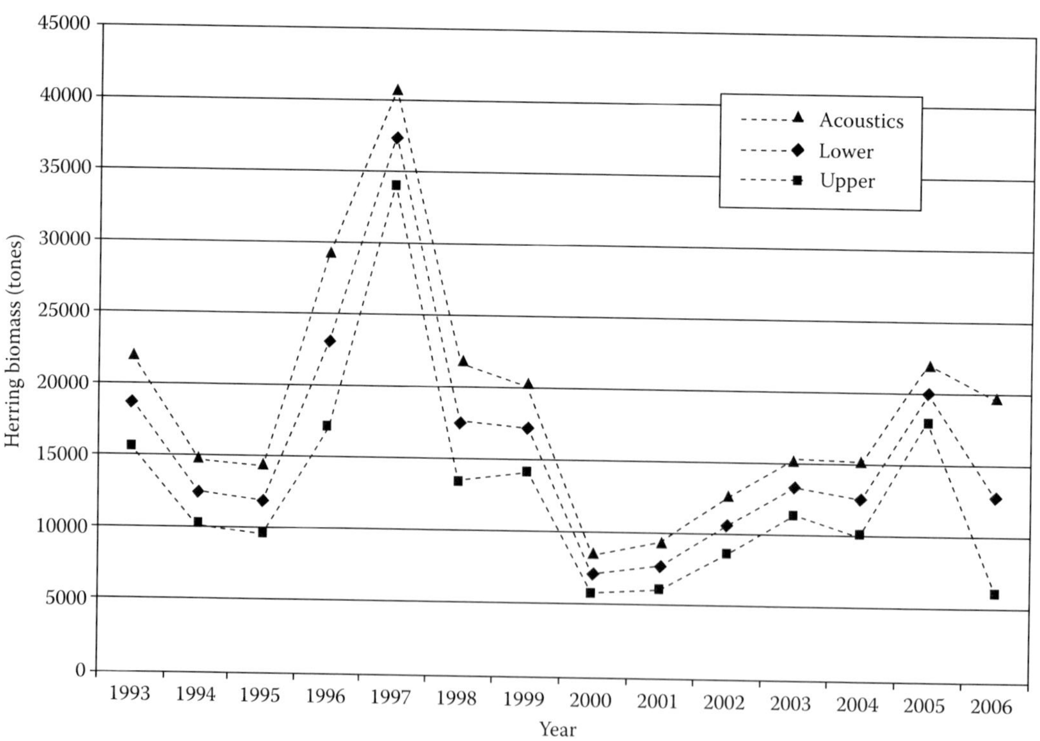

FIGURE 10.1
Prince William Sound Pacific herring biomass estimates with upper and lower 95% confidence intervals from the 1993–2006 acoustic surveys. Fishery harvests in 1997 and 1998 have been subtracted to standardize to a postharvest measure.

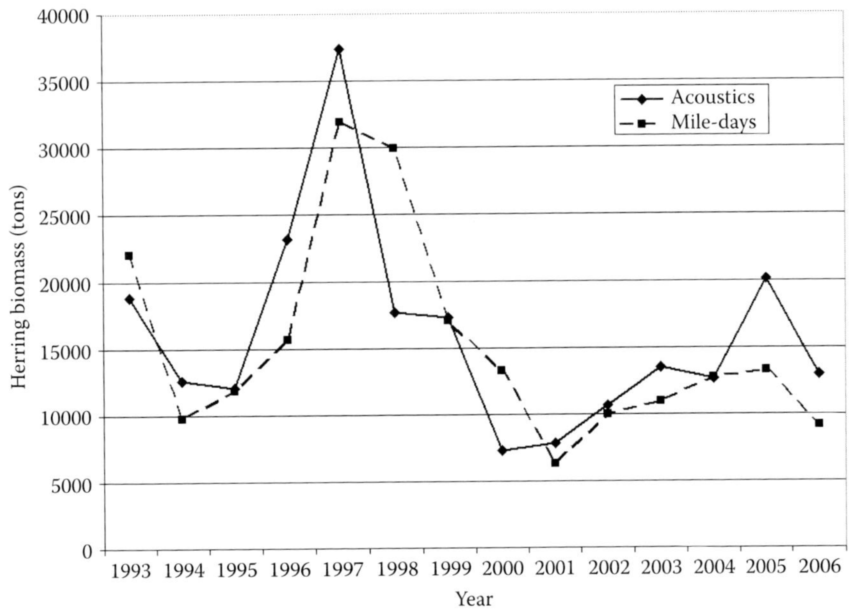

FIGURE 10.2
Prince William Sound Pacific herring biomass estimates from acoustic surveys and the mile-days model, 1993–2006, both standardized to postharvest.

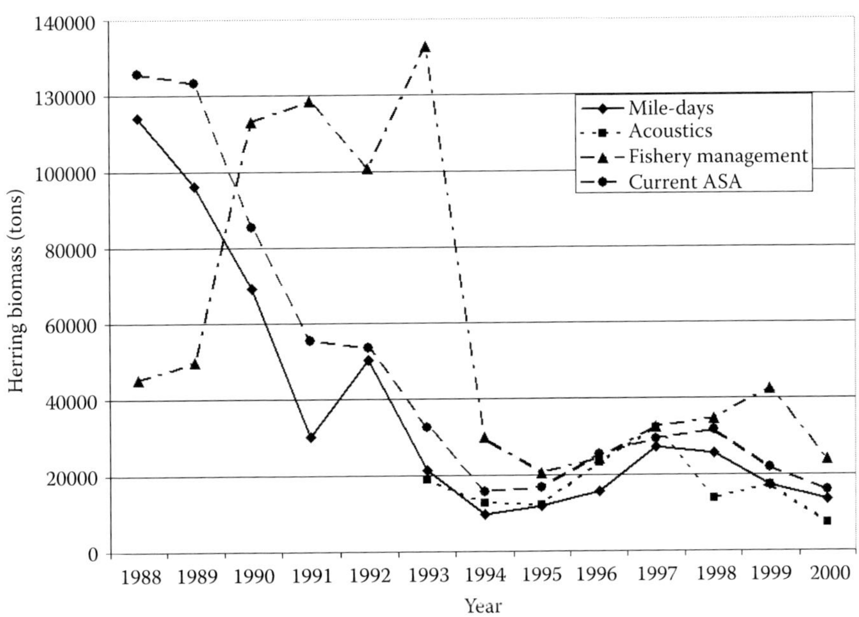

FIGURE 10.3
Prince William Sound Pacific herring biomass estimates from the miles-days biomass model, acoustic surveys, original management forecasts, and the 2005 ASA model, 1988–2000, all standardized to postfishery.

decline calculated from the current ASA model between 1989 and 1991 is 64% greater than the magnitude of the decline during the so-called year of collapse (1992–1993). Thus, both the mile-days data and the updated ASA estimates are in agreement that the collapse was multiyear beginning shortly after the EVOS.

Possible Mechanism for Damage by *Exxon Valdez* Oil Spill

In 1990, we described an interesting phenomenon concerning gas bubble release by physostomous Pacific herring (Thorne and Thomas 1990). We hypothesized that Pacific herring would have to surface to replace their reduced swim bladder gas levels as they did not have gas production capability. There have been many subsequent verifications of this phenomenon in both Pacific and Atlantic herring (Nottestad 1998; Wilson et al. 2004; Thorne and Thomas 2008).

It is well documented that the EVOS oil spill covered the primary areas of herring concentration in PWS (Peterson et al. 2003; Thomas and Thorne 2003). Newspaper accounts at the time (*Cordova Times*) reported observations of Pacific herring in distress underneath the oil. There were only two options for herring: either surface and find oil rather than air, or continue to lose gas and sink to the bottom. Although there have been no studies to directly address the impact of such circumstances on adult Pacific herring, it is highly likely that the impact would be deleterious. There were measurements of high polycyclic aromatic hydrocarbon levels in adult Pacific herring after the spill (Brown 2003).

Biomass Estimates and Lag Times

The fishery management forecast data clearly lagged the pattern of decline shown by the mile-days estimates (Figure 10.3), and a similar but reduced delayed response was repeated in the 1998–1999 record. The Pacific herring population began to recover after a low in 1994. By 1997 estimates from the acoustics, the mile-days biomass model, and the ASA model forecast all agreed that the stock had recovered to a biomass of more than 30,000 t, and a commercial fishery was reopened. However, after the resumption of the commercial fishery in 1997, both the acoustic and mile-days estimates declined in 1998. In the case of the acoustics, the decline was substantial. By 1999, both the acoustics and mile-days estimates showed a substantial decline, but the ASA model-based forecasts did not decline until 2000 (Figure 10.3). In 1989, the management forecast model moved in the wrong direction (upward and increasing direction) and did not detect the decline until 5 years later. For the 1998 decline, the lag time was 2 years.

The ASA model has been revised several times (Hulson et al. 2008). Hindcasts using the ASA model in current use come closer to the historical biomass values of the acoustics and the mile-days model, but still lags and overestimates slightly relative to the other two estimators (Figure 10.3).

Discussion

The original concept of a 1993 collapse was the result of an apparent drop from the 1993 forecast of 133,852 t to the 18,812 t estimate from the first acoustic survey in fall 1993. However, there is universal agreement that the original forecast was in error. The revised ASA in 1994 suggested that the decline was from 91,792 t to 32,049 t (Moffitt 2005). However, when the 27,700 t commercial harvest of spring 1992 is subtracted, the actual natural decline between April 1992 and March 1993 was only about 30,000 t, comparable to the previous year's fishery harvest. As the recruitment from the 1989 year class was a total failure due to EVOS, this decline does not meet the standards of a collapse and does not require the intervention of a major disease catastrophe.

Nottesad (1998) reported only gas bubble release by Atlantic herring. He did not observe gulping of air at the surface. However, our original observation of gas bubble release by Pacific herring was followed by literally hundreds of observations of surface air gulping by both adult and juvenile herring, including night observations using infrared cameras (Thomas and Thorne 2001). The research by Ben Wilson (University of British Columbia) and Lawrence Dill (Simon Fraser University) focused on sound production associated with gas bubble release, but did include verification of surface gulping. It is unfortunate that no experiments have been conducted, to our knowledge, on the direct impact of surface oil on herring in general. It would seem a logical avenue of experimentation given the widespread suspicions of damage to clupeids from oil spills associated with locations such as Cherry Point in Washington State and Sakhalin Island in Russia (Thorne and Thomas 2008). The lack of such experimental efforts associated with PWS is likely a product of the delayed recognition of injury to Pacific herring stocks.

Pearson et al. (2012) discuss several hypotheses regarding the decline of the PWS Pacific herring population. They reference our earlier suggestion of surface gulping by herring as a mechanism, but discount any impacts of the EVOS oil spill on herring. They argue that no mechanism for damage by gulping oil is shown. However, as we noted previously, the impact has not been studied, so it cannot be discounted. Pearson et al. (2012) also argue that any impact would be transient. Our data show that any explanation of the herring decline must address its multiyear nature rather than a 1-year collapse. Again, the possibility that sublethal damage to herring from gulping oil could result in extended mortality cannot be discounted. The analyses by Pearson et al. (2012) continue an emphasis on a 1-year (1992–1993) mortality event and do not explain the population trend documented in this chapter.

We believe the lag times we observed for detecting changes in stock biomass may be a serious structural flaw in the preseason fishery forecast methods that are dominant in fisheries management today (Quinn and Deriso 1999). The possibility of a 1-year lag time when using preseason forecasting was identified briefly in the 1998 National Research Council review of fish stock assessment, but it was not pursued as an inherent flaw of the method (Anonymous 1998). The possibility of longer lag times was not discussed. The ASA forecasts do have an inherent 1-year lag time as they are based on the previous year's data, and actual measures of subsequent mortality and recruitment are not included. In contrast, the acoustic surveys provide a fishery-independent, real-time estimate of fishery abundance in the same season as the fishery so lag time is not an issue. The mile-days of spawn index provides a postfishery measure of population size. The inherent 1-year lag likely played a role in the initial erroneous ASA estimate, as the EVOS oil spill added an undetected mortality to both juveniles and adults. The original 133,852 t estimate for 1993 was driven by the observation of a high ratio of 1988 year class herring compared to older adults. In retrospect, this high ratio did not reflect a large 1988 year class as much as it reflected an unexpectedly low older adult abundance due to EVOS oil spill mortality. As the ASA model makes an assumption about natural mortality, it is especially sensitive to undetected higher mortality of adults. Our data suggest the lag time is more than 1 year, which warrants further investigation. It is possible that inherent averaging functions in the model structure add inertia that leads to greater lag. The current ASA model used for Pacific herring management in PWS includes inputs from the acoustics and mile-days of spawn index, and appears to do a better job. However, many fishery management models are not this sophisticated, and although they generally accomplish their management function, they have very limited capability to detect in a timely fashion the acute mortalities that may be associated with a catastrophe such as an oil spill.

Acknowledgments

Support for this program over the years has come from Cordova District Fisherman United, the Alaska Department of Fish and Game, the Oil Spill Recovery Institute, the National Marine Fisheries Service, and the *Exxon Valdez* Oil Spill Trustee Council.

References

Anonymous. 1998. *Improving Stock Assessment*. National Research Council. Washington, DC: National Academy Press.

Becker, K. E. and E. D. Biggs. 1992. Prince William Sound herring spawn deposition survey manual. Regional Information Report 2A92-05, 2C92-02, Alaska Department of Fish and Game, Anchorage, AK.

Biggs, E. D. and T. T. Baker. 1993. Studies on Pacific herring *Clupea pallasi* spawning in Prince William Sound following the 1989 Exxon Valdez Oil Spill, 1989–1992. Report to Exxon Valdex Oil Spill Trustee Council, Anchorage, AK.

Brown, E. D. 2003. Stock structure and environmental effects on year class formation and population trends of Pacific herring, *Clupea pallasi*, in Prince William Sound, Alaska. PhD Dissertation, University of Alaska, Fairbanks, AK.

Brown, E. D. 2007. Pacific herring. In *Long-term Ecological Change in the Northern Gulf of Alaska*, ed. R. Spies, 290–299, Amsterdam: Elsevier.

Deriso, R. B., T. J. Quinn II, and P. R. Neal. 1985. Catch-age analysis with auxiliary information. *Canadian Journal of Fisheries and Aquatic Sciences* 42:815–824.

Doubleday, W. G. 1976. A least squares approach to analyzing catch at age data. *Research Bulletin of the International Commission on Northwest Atlantic Fisheries* 12:68–81.

Funk, F. 1993. Preliminary forecasts of catch and stock abundance for 1993 Alaska herring fisheries. Regional Information Report 5J93-06, Alaska Department of Fish and Game, Juneau, AK.

Funk, F. 1994. Forecast of the Pacific herring biomass in Prince William Sound, Alaska, 1993. Regional Information Report No. 5J94-04, Alaska Department of Fish and Game, Juneau, AK.

Funk, F. and M. Harris. 1992. Preliminary forecasts of catch and stock abundance for 1992 Alaska herring fisheries. Regional Information Report 5J92-04, Alaska Department of Fish and Game, Juneau, AK.

Funk, F. and G. Sandone. 1990. Catch at age analysis of Prince William Sound, Alaska Herring 1973–1988, Fisheries Information Report 90-1, Alaska Department of Fish and Game, Juneau, AK.

Hulson, P-J. F., S. E. Miller, T. J. Quinn II, G. D. Marty, S. D. Moffitt, and F. Funk. 2008. Data conflicts in fishery models: Incorporating hydroacoustic data into the Prince William Sound Pacific herring assessment model. *ICES Journal of Marine Science* 65:25–43.

Moffitt, S. 2005. Historical prefishery run biomass from catch-age models used by ADF&G 1993–2005. Unpublished data, Alaska Department of Fish and Game, Cordova, AK.

Nottestad, L. 1998. Extensive gas bubble release in Norwegian spring-spawning herring (*Clupea harengus*) during predator avoidance. *ICES Journal of Marine Science* 55:1133–1140.

Pearson, W. H., R. B.Deriso, R. A. Elston, S. E. Hook, K. R. Parker, and J. W Anderson. 2012. Hypotheses concerning the decline and poor recovery of Pacific herring in Prince William Sound, Alaska. *Reviews in Fish Biology and Fisheries* 22:95–135.

Peterson, C. H., S. D. Rice, J. W Short, D. Esler, J. L Bodkin, B. E.Belliachey, and D. B Irons. 2003. Long-term ecosystem response to the Exxon Valdez Oil Spill. *Science* 302:2082–2086.

Quinn, T. J. II and R. B. Deriso. 1999. *Quantitative Fish Dynamics*, New York: Oxford University Press.

Quinn, T. J. II, G. D. Marty, J. Wilcock, and M. Willette. 2001. Disease and population assessment of Pacific herring in Prince William Sound, Alaska. In *Herring Expectations for a New Millennium*, eds. F. Funk, J. Blackburn, D. Hay, A. J. Paul, R Stephanson, R. Toresen, and D. Witherell, 363–379, Fairbanks: Alaska Sea Grant College Program.

Simmonds, J. and D. MacLennon. 2005. *Fisheries Acoustics: Theory and Practice*, 2nd edition, London: Blackwell Science.

Thomas, G. L, J. Kirsch, and R. E. Thorne. 2002. Ex situ target strength measurements of Pacific herring and Pacific sand lance. *North American Journal of Fisheries Management* 22:1136–1145.

Thomas, G. L., E. V. Patrick, J. Kirsch, and J. R. Allen. 1997. Development of a multi-species model for managing the fisheries resources of Prince William Sound. In *Developing and Sustaining World Fisheries Resources: The State of Science and Management*, 2nd World Fisheries Congress, eds. D. A. Hancock, D. C. Smith, A. Grant, and J. P. Beumer, 606–613, Collingwood, Australia: CSIRO Publishing.

Thomas, G. L. and R. E. Thorne. 2001. Night-time predation by Steller Sea Lions. *Nature* 411:1013.

Thomas, G. L. and R. E Thorne. 2003. Acoustical-optical assessment of Pacific herring and their predator assemblage in Prince William Sound, Alaska. *Aquatic Living Resources* 16:247–253.

Thorne, R. E. 1977a. Acoustic assessment of hake and herring stocks in Puget Sound, Washington and southeastern Alaska. In *Hydroacoustics in Fisheries Research*, ed. A. R. Margets, 265–278, ICES Rapp. Et P.-v., vol. 170.

Thorne, R. E. 1977b. A new digital hydroacoustic data processor and some observations on herring in Alaska. *Journal of Fisheries Research Board Canada* 34:2288–2294.

Thorne, R. E. and G. L. Thomas. 1990. Acoustic observation of gas bubble release by Pacific herring. *Canadian Journal of Fisheries and Aquatic Sciences* 47(10):1920–1928.

Thorne, R. E. and G. L Thomas. 2008. Herring and the "Exxon Valdez" oil spill: An investigation into historical data conflicts. *ICES Journal of Marine Science* 65:44–50.

Thorne, R. E. and G. L Thomas. 2012. The role of fishery independent data. In *Fisheries Management*, ed. J. S. Intilli, 121–138, Hauppauge: Nova Science Publishers.

Thorne, R. E., R. Trumble, N. Lemberg, and D. Blankenbeckler. 1983. Hydroacoustic assessment and management of herring fisheries in Washington and southeastern Alaska. FAO Fisheries Report 300, 217–222, Rome, Italy.

Trumble, R., R. E. Thorne, and N. Lemberg. 1983. The strait of Georgia herring fishery: A case history of timely management aided by hydroacoustic surveys. *Fishery Bulletin* 80(2):381–388.

Wilson, B., R. S.Batty, and L. M Dill 2004. Pacific and Atlantic herring produce burst pulse sounds. *Proceedings of the Royal Society B*, 271(3):S95–S97.

Zale, A. V., D. L. Parrish, and T. M. Sutton, eds. 2012. *Fisheries Techniques*, 3rd edition, Bethesda: American Fisheries Society.

11

Effects of the Ixtoc I *Oil Spill on Fish Assemblages in the Southern Gulf of Mexico*

Felipe Amezcua-Linares, Felipe Amezcua, and Brigitte Gil-Manrique

CONTENTS

Introduction

The main nonrenewable natural resource of Mexico is oil. Most of its production comes from the oil wells located in the Gulf of Mexico (GOM), especially in the region known as Sonda de Campeche. This area contains the majority of the oil rigs installed by PEMEX (Petroleos Mexicanos), the state-owned Mexican Petroleum Company. In December 1978, the Sedco 135F oil rig started drilling the *Ixtoc I* exploratory well for PEMEX. The oil well was located in the Bay of Campeche of the GOM, about 100 km (60 mi.) northwest of Ciudad del Carmen, Campeche, and 600 mi. south of Texas (92° 13′W–19° 24′N) in waters 50-m (160-ft) deep (Milne 2008) (Figure 11.1).

On June 3, 1979, the oil well suffered a blowout at a depth of approximately 3600 m below the seafloor. Drilling fluid, commonly known as drilling mud, was being pumped from surface mud tanks down the inside of the drill pipe, which continued down the drill collars. The mud then flowed up and returned to the surface mud tanks (Garmon 1980). Mud circulation was lost when the drill hit a region of soft strata. Without the hydrostatic pressure of the mud column, oil and gas were able to flow unrestricted to the surface. The oil and gas fumes exploded on contact with the operating pump motors, a fire broke out, and the drilling tower collapsed and sank onto the wellhead area on the seabed (Garmon 1980). Over the next approximately 10 months, 140 million gallons of oil flowed from *Ixtoc I* into the GOM until March 24, 1980, when it was controlled (Garmon 1980).

The *Ixtoc I* spill caused a massive contamination from June 3, 1979 to March 23, 1980 as the oil, composed mainly of tar, remained on the sea surface. Prevailing winds and

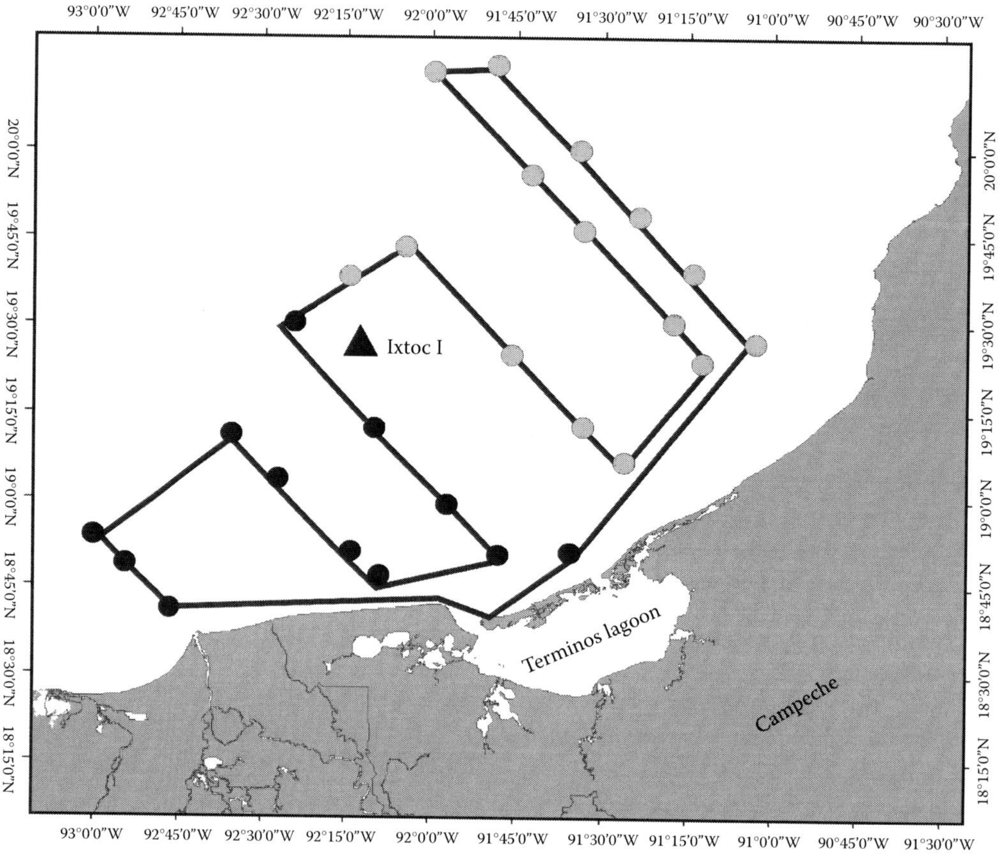

FIGURE 11.1
Location of the *Ixtoc I* oil rig and the 27 sampling stations (black circles indicate Zone A and gray circles indicate Zone B).

currents dispersed the oil and caused extensive damage to the flora and fauna along the Texas coast in the United States, and the coasts of Tamaulipas, Veracruz, Tabasco, and Campeche in Mexico. By June 12, the oil slick measured 180 km × 80 km and sand beaches and their intertidal fauna were particularly affected (Office of Response and Restoration 2012).

This large oil spill had impacted the marine environment because of the toxicity of the complex fractions of hydrocarbons on marine organisms (Blumer and Sass 1972a,b; Mironov 1972; Anderson et al. 1978, 1979). Previous oil studies demonstrated that this spill adversely affected zooplankton (Teal and Howarth 1984; Guzmán del Próo et al. 1986), benthos and infauna (Teal and Howarth 1984), shrimps and crabs (Jernelöv and Linden 1981), and turtles and birds (Garmon 1980; Teal and Howarth 1984). Oil spills also affect fish larvae and eggs (Teal and Howarth 1984), which can affect or disrupt recruitment, and therefore have an impact on the fisheries and the ecosystem in general (Teal and Howarth 1984; Hjermann et al. 2007). Accordingly, to assess possible effects of the *Ixtoc I* spill on the fish stocks, PEMEX, together with the Mexican Department of Fisheries (PESCA) and Universidad Nacional Autónoma de México (UNAM), began a series of ecological studies on demersal fishes, undertaken in the area after the spill was controlled. One of the main concerns of the government and academia was the potential negative effects on the fishing industry including the trawling boats that landed 4000 metric tons per year of shrimp plus a large quantity of finfish, and the artisanal finfish fishery that captured around 100 species of fish.

However, results from these studies were inconclusive (Jernelöv and Linden 1981) and these were only limited to describe the species found before and after the spill (Yáñez-Arancibia et al. 1982). In general, the information generated about biological effects on fish of the *Ixtoc I* spill was scarce, particularly for Mexican waters (Jernelöv and Linden 1981). Thus, the goal of this study was to analyze available data from the UNAM fish surveys undertaken prior and after the spill, as well as data from the commercial landings from 1977 to 1989. The data were compared through measures of fish diversity and multivariate analyses to determine if the fish assemblage composition and the landings composition changed as a consequence of the *Ixtoc I* oil spill.

Materials and Methods

Prior to the oil spill, UNAM conducted a demersal fish survey in the area around the *Ixtoc I* oil rig. The survey lasted from May to June 1978. During this survey, 27 stations, designated by a depth-stratified design, were sampled with a commercial shrimp trawler fitted with two trawl nets each 18.3 m long and 9.15 m wide, and with a mesh size of 3.4 cm (Figure 11.1). These stations were trawled monthly for 30 minutes each at three knots and at depths up to 70 m. Two zones were designated: Zone A was to the west of the oil rig and Zone B was to the east. Following the spill, UNAM, with funding from PEMEX, completed five more surveys at the same 27 stations in August 1980, November 1980, July 1981, October 1981, and March 1982, using the exact prespill procedures. In the laboratory, all fish were identified to species, weighted (g), and measured (total length, cm). All common fish names and their presentation follow Page et al. (2013).

To standardize all UNAM/PEMEX samples, the catch was converted into a relative abundance estimate (catch per unit effort) that was estimated by dividing the total number (n) or biomass of each species (kg) by the sampled area. The sampled area was estimated with the following equation:

$$D = 60\sqrt{[(\mathrm{Lat}_s - \mathrm{Lat}_e)^2 + (\mathrm{Lon}_s - \mathrm{Lon}_e)^2 \cos^2(0.5\{\mathrm{Lat}_s + \mathrm{Lat}\})]}$$

where D = distance, Lat_s = initial latitude, Lat_e = final latitude, Lon_s = initial longitude, and Lon_e = final longitude (Sparre and Venema 1998). Using the standardized data, a matrix containing the biomass of each fish species captured in the different surveys was constructed.

In addition, fish landings data from the commercial fisheries were available in the states of Tamaulipas, Veracruz, Campeche, and Tabasco from 1977 to 1989 through the fisheries yearbooks edited by PESCA. The information used was total landings (biomass, kg) per year per state, fish species landed (g) per year per state, and the number of fishing boats per year per state to standardize the data. With this information, a matrix containing landings (kg) of marine fish species per fishing boat was constructed per state per year. For the diversity and the multivariate analyses, the collections were divided into four sample groups: before the spill (1977–1978), during the spill (1979–1980), up to 3 years after the spill

(1981–1983), and >3 years after the spill (1984–1989). These groups were formed considering that the cohort of most fish species reaches a fishable age at about 3 years (Hjermann et al. 2007).

The catch per year per state was plotted to observe trends in the landings and to determine if the trend changed across the years following the spill as a possible consequence of this event. Diversity of each fish survey was estimated as well as commercial landing data from the different years available with the Shannon index (H'); the form of the index is

$$H' = \sum p_i \ln p_i$$

where p_i refers to the proportion of individuals found in the ith species. Because the true value of p_i is unknown for the sample, a maximum likelihood estimator was used, where $p_i = n_i/N$, n_i is the abundance of the ith species and N is the total abundance of fish in this case (Magurran 2004).

To test the hypothesis that the diversity was different before and after the spill, for both sets of data, two tests were used (UNAM surveys and commercial landings). For the UNAM survey data, the information from every survey was pooled and a randomization test for pairs of data proposed by Solow (1993) was used to compare the diversity around the *Ixtoc I* area before the spill (June 1978) with the diversity after the spill (August and November 1980, July and October 1981, and March 1982). The combined data set was randomly partitioned into two subsets, the diversity index was calculated for each subset, and the difference was recorded. The procedure was repeated 10,000 times and the estimated p-value was calculated with the following equation:

$$p = \left(\frac{m_1 + m_2}{m_1} \right)^{-1}$$

where m_1 (before) and m_2 (after) represent the fish assemblage for both time periods (dates). The test was performed using the version 2.3 of the Diversity software (PISCES Conservation Ltd.).

For the commercial landings data, because information was available from the years previous, during, and after the spill, a two-way analysis of variance (ANOVA) was performed using state (n = 4, Tamaulipas, Veracruz, Campeche, and Tabasco), and date(s) (n = 4, before, during, 0–3 years after the spill, and >3 years after the spill) as factors, with years as replicates and the Shannon index as the independent variable. The commercial landings data set was standardized using the number of boats and transformed into catch per boat. ANOVA was performed using Statistica 10 software (StatSoft) and homoscedasticity was tested using Cochran's C test. According to Taylor (1978), when the Shannon index is calculated for a number of sites or times, the indices will be normally distributed and thus it is possible to use ANOVA to compare differences in diversity.

The fish assemblage data were compared among states (n = 4, Tamaulipas, Veracruz, Campeche, and Tabasco) and dates of the spill (n = 4, before, during, 0–3 years after, and >3 years after) for the effort-adjusted commercial landings data, and among dates (before or after the spill) for the UNAM survey data using nonmetric multidimensional scaling analysis (MDS) based on Bray–Curtis similarity coefficients calculated from 4th-root-transformed standardized abundance data. To test for differences in the fish standardized abundance between factors, an analysis of similarity (ANOSIM) was employed using the *R*-statistic values for pairwise comparisons to determine the degree of dissimilarity among groups (Clarke 1993). Similarity of percentages (SIMPER) was used to determine which species account for most of the dissimilarities between group compositions (Collins and Williams 1982; Clarke and Warwick 1994). All multivariate analyses were performed in the PRIMER 6 suite of programs (Clarke and Warwick 1994).

Results

Survey Descriptions

From the UNAM survey data, a total of 99 fish species were captured in the area surrounding the *Ixtoc I* in May–June 1978 (before the spill, see Appendix 11.1). The rough scad (*Trachurus lathami*, 24.7%), the hardhead catfish (*Ariopsis felis*, 23%), the inshore lizardfish (*Synodus foetens*, 13.1%), the Atlantic bumper (*Chloroscombrus chrysurus*, 11.7%), the silver seatrout (*Cynoscion nothus*, 6.1%), and the silver jenny (*Eucinostomus gula*, 4.9%) accounted for approximately 80% of the total biomass captured prior to the oil spill.

The number of fish species captured in August 1980 (first survey after the spill) decreased to 61 with the rough scad (38.9%), the Atlantic bumper (16.6%), the shoal flounder (*Syacium gunteri*, 11.7%), the inshore lizardfish (7.8%) and the Atlantic bigeye (*Priacanthus arenatus*, 6.8%) accounting for approximately 80% of the total biomass captured. In contrast, collections in November 1980 produced 82 species. Six species accounted for approximately 80% of the total biomass captured and included the hardhead catfish (28.8%), the shoal flounder (17.6%), the Atlantic threadfin (*Polydactylus octonemus*, 14%), the dwarf goatfish (*Upeneus parvus*, 9.7%), the Bean's searobin (*Prionotus beanii*, 6.5%) and the silver seatrout (4.4%). In July 1981, the total number of fish species captured was 74 with the hardhead catfish (24.4%), the inshore lizardfish (17.2%), the Atlantic threadfin (14.5%), the silver seatrout (14.13%), and the Bean's searobin (7.22%) accounting for approximately 80% of the captured biomass. In October 1981, the total number of fish species captured during the survey was 70 with six species comprising approximately 80% of the captured biomass. These were the silver jenny (24.1%), the hardhead catfish (19.9%), the Atlantic bumper (17.2%), the silver seatrout (6.6%), the dusky anchovy (*Anchoa lyolepis*, 6.4%), and the hardhead halfbeak (*Chriodorus atherinoides*, 5.2%). Finally, in March 1982, 108 fish species were captured, the largest number found in all surveys, and nine species accounted for approximately 80% of the captured biomass. These were the silver jenny (20.36%), the hardhead halfbeak (18.8%), the dusky anchovy (10.4%), the silver seatrout (5.7%), the Atlantic bumper (5.7%), the largescale fat snook (*Centropomus mexicanus*, 4.9%), the red snapper (*Lutjanus campechanus*, 4.7%), the hardhead catfish (4.4%), and the blue runner (*Caranx crysos*, 4.3%).

Commercial Fisheries Landings

The commercial fisheries data of the states of Tamaulipas, Veracruz, Campeche, and Tabasco for the years 1976–1989 included 66 marine and estuarine fish species (Appendix 11.1). The catch per boat differed between states with Campeche having the higher landings but with more marked fluctuations (Figure 11.2). The landings in Campeche at the time of the spill decreased, when in all the other states landings remained relatively constant, although it can be observed that a decreasing landings trend was already occurring in

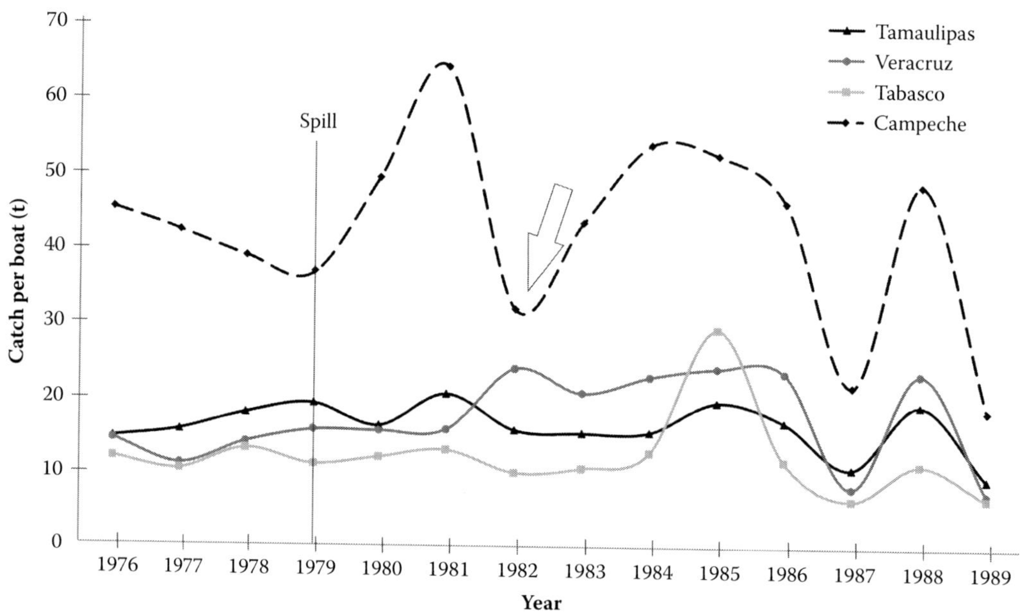

FIGURE 11.2
Effort-adjusted landings trends (tons per boat) of the states of Tamaulipas, Veracruz, Tabasco, and Campeche in different years. Arrow indicates a decrease in landings 3 years after the spill.

Campeche. After the spill, the landings increased in Campeche and Tamaulipas, but remained similar in Veracruz and Tabasco. However, in 1982 a decrease in the landings occurred in Campeche, Tamaulipas, and Tabasco, being more marked in Campeche, where the landings decreased from approximately 65 to approximately 30 t/boat, a decrease >50%. In Tamaulipas and Tabasco, the decrease in landings was not as great as in Campeche, however, it was still considerable (~20 to ~15 t/boat in Tamaulipas and ~12 to ~10 t/boat in Tabasco). On the contrary, in Veracruz the landings increased during the same year. In all states, the landings reached a high value by 1984–1985, decreased by 1987, increased again in 1988, but decreased in 1989.

Diversity, Assemblage Structure, and Composition

The Shannon diversity index from the UNAM survey data showed that the diversity was higher before the spill (4.59) than after the spill (4.11). According to the pairwise comparisons, these values were statistically different ($\delta = 0.484$, $p < .01$), indicating that the diversity before the spill was higher than the diversity after the spill. The Shannon diversity index of the effort-adjusted commercial landings was significantly different between states ($F_{3,36} = 6.83$, $p < .01$), but not by date relative to the spill ($F_{3,36} = 1.07$, $p > .01$), although there was a significant interaction ($F_{9,36} = 3.09$, $p < .01$). In general, the landings from Tabasco were less diverse than those from the other states, and based on the interaction term, the diversity in Tabasco before the spill was only different from the diversity in Veracruz found in the years (0–3 years) following the spill (Figure 11.3).

The fish assemblage based on UNAM collections around the *Ixtoc I* well was different before and after the spill. The MDS plot shows clear groups by date of the survey (Figure 11.4) and ANOSIM indicated the assemblage was significantly different before and

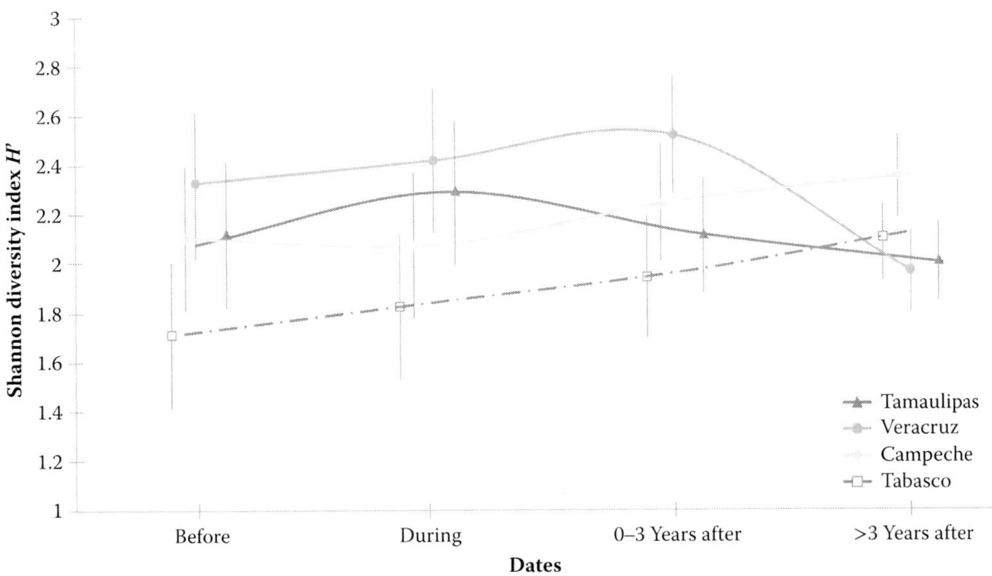

FIGURE 11.3
Variation in the Shannon index of diversity of the commercial landings trends (mean ± SD) in the states of Tamaulipas, Veracruz, Tabasco, and Campeche by date(s) relative to the *Ixtoc I* oil spill: before (years 1977–1978), during (1979–1980), immediately after (1981–1983), and after (1984–1989).

after the spill (global $R = 0.86$, $p < .05$). SIMPER indicated that the fish more responsible for discrimination of the assemblage/date groups were the rough scad, the Atlantic thread herring (*Opisthonema oglinum*), the blackcheek tonguefish (*Symphurus plagiusa*), the Atlantic bigeye, the Atlantic bumper, and the inshore lizardfish, which were more abundant before the spill. In fact, the first three species were never caught again after the spill. After the spill, the discriminating species were the Atlantic threadfin, the Bean's searobin, and the stardrum (*Stellifer colonensis*), which appeared from 1980 onward (Table 11.1, Figure 11.4).

In regard to the effort-adjusted commercial landings, the MDS plot shows clear groups according to the catch composition of the fish landed per state, and within these groups, smaller groups were formed based on the catch composition in relation to the date of the spill (Figure 11.5). The groups were corroborated through ANOSIM, and significant differences were found in the catch composition among states (global $R = 0.941$, $p < .01$, Table 11.2) and in relation to the date of the spill (global $R = 0.671$, $p < .01$, Table 11.3).

The fish more responsible for the differences in commercial landings between states varied depending on the location (SIMPER, Table 11.4), however every state showed characteristic species. For Tamaulipas, the striped mullet (*Mugil cephalus*), the sand seatrout (*Cynoscion arenarius*), and the hardhead catfish were characteristic. For Veracruz, the wahoo (*Acanthocybium solandri*) and the white mullet (*Mugil curema*) were characteristic. In Tabasco, the characteristic species were the ground croaker (*Bairdiella ronchus*) and the gafftopsail catfish (*Bagre marinus*). Finally, in Campeche, the characteristic species were sharks, such as silky shark (*Carcharhinus falciformis*), tiger shark (*Galeocerdo cuvier*), smooth dogfish (*Mustelus canis*), and Atlantic angel shark (*Squatina dumeril*) among others, along with the mutton snapper (*Lutjanus analis*), the crevalle jack (*Caranx hippos*), and the Atlantic Spanish mackerel (*Scomberomorus maculatus*).

The commercial landings composition at different dates varied according to the state, but in general, there was no difference between the composition before and during the

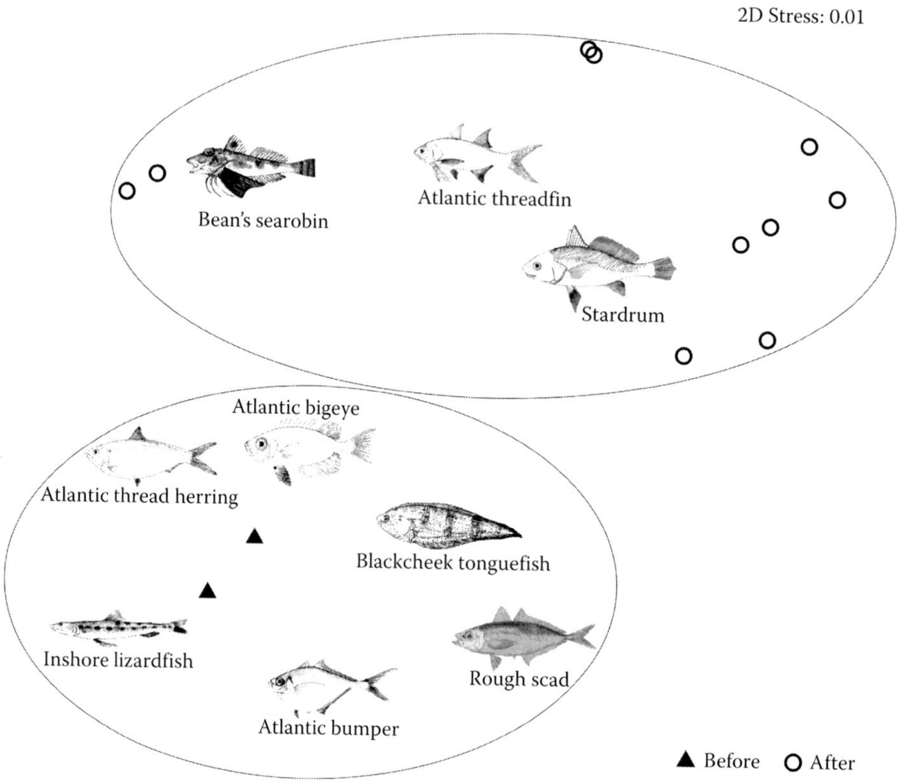

FIGURE 11.4
Multidimensional scaling analysis plot of the Universidad Nacional Autónoma de México fish assemblages found around the *Ixtoc I* oil well before and after the spill. (Fish illustrations obtained from Fischer, W., *FAO Species Identification Sheets for Fishery Purposes. Western Central Atlantic (Fishing Area 31) Vols. I–VII*, FAO, Rome, Italy, 1978.)

TABLE 11.1

Mean Pairwise Abundance of Important Fish Species before and after the ***Ixtoc I*** Oil Spill from the Surveys Undertaken around the Oil Rig Based on Similarity of Percentages Analysis

Species	Mean Abundance Before	Mean Abundance After	Mean Dissimilarity	Mean Dissimilarity/SD	Contribution (%)
Stardrum	0	1.23	2.76	23.47	5.09
Bean's searobin	0	1.48	3.36	12.1	6.2
Rough scad	2.59	0	5.88	6.69	10.85
Blackcheek tonguefish	0.84	0	1.9	5.81	3.51
Atlantic bigeye	1.55	0	3.51	4.74	6.48
Inshore lizardfish	1.92	1.3	1.41	3.74	2.6
Atlantic threadfin	0	1.53	3.52	3.62	6.49
Atlantic thread Herring	1.35	0	3.07	3.5	5.66
Atlantic bumper	0.83	1.65	2.22	1.3	4.1

Note: Mean dissimilarity was 49.18. Species are listed in order of their contribution to the mean dissimilarity divided by the standard deviation between pairs of fish species.

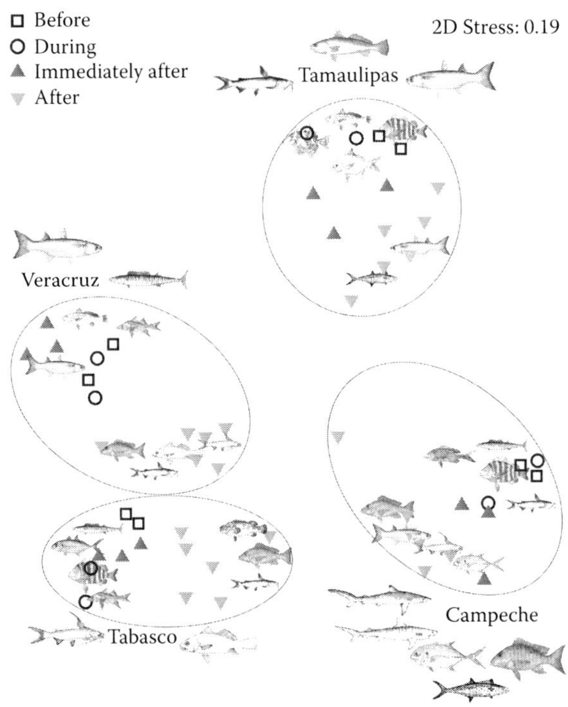

FIGURE 11.5
Multidimensional scaling plot of the commercial landings composition of the states of Tamaulipas, Veracruz, Tabasco, and Campeche in different times in relation to the *Ixtoc I* oil spill: before (1977–1978), during (1979–1980), immediately after (1981–1983), and after (1984–1989). The fish outside the circles are the characteristic species for every state, whereas the fish inside the circles are the ones responsible for the differences before and after the spill within each state. (Fish illustrations obtained from Fischer W., *FAO Species Identification Sheets for Fishery Purposes. Western Central Atlantic (Fishing Area 31) Vols. I–VII*, FAO, Rome, Italy, 1978.)

TABLE 11.2
Results of Analysis of Similarity Testing for Differences in the Commercial Landings Composition in Relation to Different States

Groups	*R* Statistic	*p*
Tam, Ver	0.925	<.01
Tam, Tab	0.977	<.01
Tam, Camp	0.979	<.01
Ver, Tab	0.677	<.01
Ver, Camp	0.93	<.01
Tab, Camp	0.952	<.01

Tam, Tamaulipas; Ver, Veracruz; Tab, Tabasco; Camp, Campeche.

spill, although the composition after the spill was different (Table 11.5). In Tamaulipas before the spill, the characteristic species were the grey triggerfish (*Balistes capriscus*), the Atlantic croaker (*Micropogonias undulatus*), the sheepshead (*Archosargus probatocephalus*), and the Irish pompano (*Diapterus auratus*), and after the spill the species responsible for the differences were the mullet and the Atlantic Spanish mackerel. In Veracruz, the

TABLE 11.3

Results of Analysis of Similarity Testing for Differences in the Commercial Landings Composition in Relation to the Date of the Spill for the Different States

Groups	*R* Tamaulipas	*p*	*R* Veracruz	*p*	*R* Campeche	*p*	*R* Tabasco	*p*
B, D	0.032	>.05	0.489	>.05	0.024	>.05	0.014	>.05
B, IA	0.333	>.05	0.297	>.05	0.459	>.05	0.111	>.05
B, A	0.956	<.05	0.867	<.05	0.911	<.05	0.922	<.05
D, IA	0.167	>.05	0.357	>.05	0.333	>.05	0.167	>.05
D, A	1.0	<.05	0.963	<.05	1.0	<.05	0.917	<.05
IA, A	0.519	<.05	0.901	<.05	0.741	<.05	0.778	<.05

Before (B, 1977–1978); during (D, 1979–1980); immediately after (IA, 1981–1983); and after (A, 1984–1989).

species more responsible for differences in the landings composition before the spill were the largescale fat snook, the Atlantic croaker, and the mullets, while after the spill the characteristic species were the gafftopsail catfish, the hardhead catfish, the ground croaker, and the mutton snapper. In Campeche, the wahoo, the sheepshead, the yellowtail snapper (*Ocyurus chrysurus*), and the hardhead catfish were important before the spill. After the spill, the crevalle jack, the gafftopsail catfish, the red snapper, and the mullet were important. Finally, in Tabasco the wahoo, the sheepshead, the blue runner, and the largescale fat snook were important before the spill, but after the spill, the red snapper, the hardhead catfish, the Atlantic goliath grouper (*Epinephelus itajara*), and the common snook (*Centropomus undecimalis*) among others became important (Table 11.5).

Discussion

The *Ixtoc I* oil spill is considered the third worst in the world, only after the oil spill in the Persian Gulf that occurred during the Gulf War of 1991, and the *Deepwater Horizon* spill of 2010 that occurred in the GOM near the Mississippi River Delta, United States. Immediate effects of the spill were detected in the area around of the *Ixtoc I* oil well through the sampling surveys. The number of fish species and the diversity decreased drastically after the spill and the fish assemblage changed. With respect to the effort-adjusted commercial landings, the trend for all the states showed that the spill did not seem to affect the landings immediately; on the contrary, they increased. However, a significant decrease in the landing biomass occurred in 1982 in Campeche, Tamaulipas, and Tabasco, although in Veracruz, the landings per boat increased that year. The observed decreases were not trivial, as the combined sum of these accounted for about 40 t/boat less than those landed the previous year, and from these 30 t/boat less were captured in Campeche, the place where the spill occurred. Translated into currency values, surely these diminished returns had impacts on the economy of these states, mainly Campeche. The diversity of the commercial landings did not appear to be affected by the spill; on the contrary, it increased in the

TABLE 11.4

Mean Pairwise Biomass of Commercially Important Fish Species from the Mexican States Affected by the *Ixtoc I* Oil Spill Based on SIMPER Analysis

Species	Pair	Mean Abundance	Mean Abundance	Mean Dissimilarity	Mean Dissimilarity/SD	Contribution (%)
	Tamaulipas versus Veracruz	**Tamaulipas**	**Veracruz**	**41.21**		
Sand seatrout		0.58	0.04	1.49	4.49	3.62
Wahoo		0.07	0.56	1.35	3.92	3.26
Striped mullet		0.02	0.71	1.84	2.97	4.47
Ground croaker		1.02	0.61	1.15	2.97	2.79
Crevalle jack		1.04	1.41	1.02	2.13	2.48
White mullet		0.09	0.55	1.33	2.11	3.22
Silky shark, tiger shark, and so forth		0	0.44	1.17	2.04	2.84
Irish pompano		0	0.46	1.24	1.97	3
Atlantic goliath grouper		0.17	0.63	1.26	1.95	3.06
	Tamaulipas versus Tabasco	**Tamaulipas**	**Tabasco**	**42.89**		
Gafftopsail catfish		0	0.89	2.9	10.1	6.77
Sand seatrout		0.58	0	1.9	8.6	4.42
Striped mullet		1.02	0.52	1.67	3.82	3.89
Hardhead catfish		0.56	0.04	1.67	3.56	3.9
White mullet		0.49	0.01	1.57	2.94	3.67
Silky Shark, tiger shark, so forth		1.3	0.75	1.79	2.92	4.17
Common snook		1.04	1.43	0.63	2.37	1.47
Ground croaker		0.54	0.77	0.73	2.2	1.7
Crevalle jack		0.66	0.93	0.89	2	2.08

(*Continued*)

TABLE 11.4

Mean Pairwise Biomass of Commercially Important Fish Species from the Mexican States Affected by the *Ixtoc I* Oil Spill Based on SIMPER Analysis (*Continued*)

Species	Pair	Mean Abundance	Mean Abundance	Mean Dissimilarity	Mean Dissimilarity/SD	Contribution (%)
	Tamaulipas versus Campeche	**Tamaulipas**	**Campeche**	**45.14**		
Sand seatrout		0	1.01	2.52	7.17	5.59
Common snook		0.07	0.9	2.06	4.33	4.57
Silky shark, tiger shark, so forth		0.58	0.92	0.84	3.18	1.86
Mutton snapper		0.54	0.84	0.73	2.74	1.61
Crevalle jack		0.62	0.95	0.84	2.66	1.86
Atlantic Spanish mackerel		0.23	0.76	1.32	2.37	2.92
Hardhead catfish		0.75	0.17	1.51	2.32	3.34
Gafftopsail catfish		0.48	1.1	1.54	2.28	3.41
Striped mullet		1.02	0.44	0.96	2.12	2.13
	Veracruz versus Tabasco	**Veracruz**	**Tabasco**	**35.68**		
White mullet		0.84	0.01	2.34	5.66	6.57
Gafftopsail catfish		0.24	0.89	1.83	2.64	5.13
Atlantic Spanish mackerel		0.85	0.66	0.64	2.07	1.78
Wahoo		0.44	0.02	1.15	1.98	3.24
Crevalle jack		0.79	0.73	0.37	1.82	1.05
Silky shark, tiger shark, and so forth		0.71	0.3	1.24	1.69	3.49
Irish pompano		0.63	0.46	0.49	1.68	1.38
Hardhead catfish		0.79	0.75	0.49	1.68	1.38
Striped mullet		0.76	0.6	0.51	1.59	1.42

	Veracruz versus Campeche	**Veracruz**	**Campeche**	**42.62**		
Sand seatrout		0.04	0.92	1.95	5.97	4.57
Wahoo		0.71	0	1.53	3.07	3.6
Ground croaker		0.55	0.03	1.16	2.6	2.71
White mullet		0.84	0.16	1.52	2.51	3.57
Mutton snapper		0.31	0.76	1	1.75	2.34
Gafftopsail catfish		0.24	0.6	0.96	1.75	2.26
Silky shark, tiger shark, and so forth		0.66	0.9	0.53	1.73	1.24
Guachinango		0.57	0.6	0.38	1.73	0.9
Irish pompano		0.94	1.11	0.59	1.64	1.37
	Tabasco versus Campeche	**Tabasco**	**Campeche**	**43.7**		
Sand seatrout		0	0.92	2.33	9.19	5.34
Crevalle jack		0.46	0.75	0.93	3.12	2.13
Silky shark, tiger shark, and so forth		0.6	0.95	0.9	2.42	2.05
Atlantic Spanish mackerel		0.73	1.1	0.95	2.19	2.17
Mutton snapper		0.33	0.76	1.11	1.79	2.53
Ground croaker		0.31	0.03	0.75	1.61	1.71
Irish pompano		0.91	1.11	0.58	1.58	1.33
Blue runner		0.07	0.48	1.09	1.55	2.5
Striped mullet		0.52	0.44	0.79	1.55	1.81

Note: Species are listed in order of their contribution to the mean dissimilarity divided by the standard deviation between pairs of fish species.

TABLE 11.5

Mean Pairwise Biomass of Commercially Important Fish Species before and after the ***Ixtoc*** Oil Spill from the Four States Affected Based on SIMPER Analysis

Species	State	Mean Abundance Before	Mean Abundance After	Mean Dissimilarity	Mean Dissimilarity/ SD	Contribution (%)
	Tamaulipas			**27.9**		
Atlantic croaker		0.81	0	2.73	139.61	9.78
Irish pompano		0.67	0	2.28	139.61	8.19
Sheepshead		0.58	0	1.95	139.61	7
Largescale fat snook		0.54	0	1.84	139.61	6.58
Grey triggerfish		0.54	0	1.83	139.61	6.56
Atlantic Spanish mackerel		0	0.64	2.15	19.33	7.72
White mullet		0	0.55	1.85	13.72	6.65
Striped mullet		0.92	1.04	0.41	4.19	1.48
Little tunny (*Euthynnus alletteratus*)		0.33	0.33	0.73	1.84	2.61
	Veracruz			**33.03**		
Hardhead catfish		0	0.41	1.19	12.66	3.61
Largescale fat snook		0.55	0	1.61	11.87	4.88
Atlantic croaker		0.18	0	0.53	11.87	1.6
Ground croaker		0	0.56	1.61	8.37	4.88
Mutton snapper		0	0.44	1.28	7.26	3.88
Gafftopsail catfish		0	0.41	1.2	6.4	3.64
White mullet		0.92	0.68	0.71	1.98	2.16
Striped mullet		0.68	0.5	0.53	1.96	1.61
Atlantic Spanish mackerel		0.84	0.63	0.61	1.91	1.84

Campeche			**34.48**		
Crevalle jack	0	0.75	1.78	111.61	5.16
Gafftopsail catfish	0	0.69	1.65	18.09	4.78
Wahoo	0.73	0	1.42	13.66	5.41
Sheepshead	0.55	0	1.08	6.87	4.12
Yellowtail snapper	0.56	0	1.08	4.75	4.12
Hardhead catfish	0.83	0.12	1.37	4.3	5.22
Red snapper	0.45	0.76	0.6	1.92	2.27
Striped mullet	0.3	0.75	0.86	1.87	3.29
Irish pompano	0.9	1.21	0.26	1.67	1.01
Tabasco			**34.48**		
Red snapper	0.3	0.53	0.77	10.98	2.59
Wahoo	0.55	0	1.81	9.5	6.08
Sheepshead	0.36	0	1.18	9.5	3.95
Blue runner	0.35	0	1.14	9.5	3.82
Largescale fat snook	0.34	0	1.13	9.5	3.78
Mutton snapper	0.32	0.48	0.52	4.68	1.73
Hardhead catfish	0.35	0.06	0.96	3.61	3.22
Atlantic goliath grouper	0.39	0.43	0.99	2.83	3.33
Common snook	0.76	0.65	0.38	1.61	1.27

Note: Species are listed in order of their contribution to the mean dissimilarity divided by the standard deviation between pairs of fish species.

years following the spill in all the states with the exception of Tamaulipas. This reduction does not seem to be linked to the oil spill, as it happened in the state which was furthest to the wellhead and was similar to other trends in the data at that time. However, the landings composition was different in all states before, during, and after the spill.

The changes on the fish assemblages and landings composition that occurred immediately after the spill seemed to be related not to a direct mortality of the fish, but rather to a migration of the fish out of the affected area. Laboratory studies have shown that adult fish are able to detect petroleum at very low concentrations (Hellstrøm and Døving 1983; Dauble et al. 1985; Beitinger 1990; Farr et al. 1995), and large numbers of dead fish were seldom reported after oil spills (IPIECA 1997). Thus, juvenile and adult fish appear to be capable of avoiding water with high hydrocarbon concentrations. In general, it is recognized that finfish, due to their mobility, can avoid adverse zones, thus reducing the impact of oil spills on their abundance. On the other hand, Stickle (2002) indicated that marine fauna located near oil rigs have developed the ability to degrade hydrocarbons so minor spills and leaks (e.g., flows upward through faults and cracks in sedimentary rocks) do not affect them. These observations could explain why the commercial landings and their diversity did not decrease during and immediately after the spill. However, after large oil spills heavy petroleum and hydrocarbons also deposit on the bottom (Stickle 2002) and may be a problem for demersal species. Also, the fleets from the affected states could have traveled to distant fishing grounds where they do not normally fish thus avoiding the spill, and this is why the landings remained similar or increased. This is speculative as there are no logbooks from those dates available that might indicate the fishing behavior and location of the fleets during the months following the spill.

The alterations in commercial landings that occurred 2–3 years after the spill may be related to high mortality of the larval stages that occurred at the time of the accident. It is known that the main impact of spills on fish populations is related to the impact on larval stages because they are unable to avoid the affected area and die (Hjermann et al. 2007). Natural mortality of larval stages of many fish is usually enormous under normal natural conditions and has high annual and spatial variation. Thus, if the oil spill killed larvae that would have been transported to ecologically unfavorable areas, and thus have suffered 100% mortality, then the real effect of the oil spill would be close to zero when the cohort reaches fishable age (e.g., age 3 years). On the other hand, if the larvae killed were those that would have been transported to favorable areas and thus have had the best chance of survival, then the effect would be high, so an oil spill does not affect all eggs or larvae equally (Hjermann et al. 2007). In the case of our study, it seems that the *Ixtoc I* oil spill did affect the larvae that would have survived under normal conditions and that is why there was an evident reduction in the landings and diversity, as well as shifts in fish assemblage structure and composition between 1982 and 1983, mainly in the states of Veracruz and Campeche. Accordingly, it appears that the spill had an adverse effect on the fish assemblages, and therefore, in the economy of the region 2–3 years later.

Changes that occurred in the landings after 1984 (no more surveys were undertaken in the area after 1982) seem to be related to phenomena other than the spill. During the oil spill, the main impact on the fishing industry was the inability to fish in the area, although the landings did not decrease, probably because the fishermen used other fishing grounds as noted previously. However, results from the postspill sampling, as well as for the landings after 1984, indicated that the apparent severe effects to the marine fauna were negligible. A permanent decrease in landings occurred in the southern GOM around 1985–1986 although this decrease was likely not related to the *Ixtoc I* oil spill, but to other causes (García-Cuellar et al. 2004). Such variations are often due to abiotic factors such as seasonal changes (dry and rainy seasons, hurricane seasons, etc.) (Yáñez-Arancibia et al. 1982).

Although a spill of this dimension is considered to cause effects on the fish and consequently on the fisheries in the affected area (Teal and Howarth 1984; Hjermann et al. 2007), there are few reported cases in which oil spills have conclusively had a significant impact on fish stocks. Our study shows that the fish assemblages and the composition of the landings changed immediately after the spill in all the affected states with another major impact occurring 2–3 years later in the commercial landings, likely due to impacts to the larval stages. However, the fish populations appeared to recuperate rapidly as the number of species from the UNAM fish surveys increased to similar values found previous to the spill within 3 years after the event.

An explanation for this is that tropical environments have high biological productivity and high oil degradation rates (Reijnhart and Rose 1982) due to high temperature (29°C), ecosystems are relatively complex, and fish stocks often spawn over a longer period of the year or even year-round. Although it is necessary to note that the composition of the commercial fish landings was different before and after the spill, the results of the SIMPER analysis suggested some species that were being captured before the spill were replaced by others along contiguous regions. For example, in Campeche and Tabasco, the wahoo and the sheepshead were common before the spill and after the event these two species were not captured again. Instead, the red snapper appeared as an important species and in Veracruz, Campeche, and Tabasco, catfishes seem to become important after the spill. The available fisheries data precluded us from knowing if these changes are related to a decrease of some of the species as a consequence of the spill or other oceanographic or environmental factors in the region, or if they are more related to market phenomena (such as prices, supply, and demand). However, the fact is that the commercial landings composition was different before and after the spill, so this event had an impact on the fishery. Something similar occurred with the data from the UNAM surveys, with the number of species decreasing after the spill, but some species that were previously captured were not recorded again. The problem with the UNAM surveys is that these lasted only 3 years after the spill, so it is not possible to account for more permanent changes in the fish assemblage composition.

We cannot ascertain if the changes are related to biological or ecological factors, economic factors, or a combination of both. The changes on the commercial landings composition and UNAM fish assemblages could be related to abiotic or ecologic factors that could not be detected with the available information. The effect on a fish population of an oil spill in an area depends to a great extent on oceanographic and ecological conditions. The extent of the spill, the weather conditions at the moment of impact, the time of year, and several ecological aspects all influence the extent of the impact on the year-class affected (Hjermann et al. 2007). Unfortunately, these factors were not accounted for at the time of the surveys or by the commercial fisherman.

However, it is clear that the *Ixtoc I* oil spill changed the landings composition in the region. In the last few decades, the fisheries in the area of the *Ixtoc I* oil rig continue to be adversely affected by the oil industry. Activities such as drilling, shipping activities, and oil storage pollute the environment, and small oil-spill accidents occur frequently (García-Cuellar et al. 2004). Since 2001, PEMEX has a certificate of accident prevention that is aimed to reduce the environmental impacts of the petroleum industry, however it is known that accidents and minor spills continue to occur in the area. Presently, the fishing industry and the quantity of seafood landings has decreased in the area, and the general consensus is that the combined effect of the oil industry together with an overexploitation of the fishing resources are responsible for this. In conclusion, the *Ixtoc I* oil spill had an immediate impact on adults and their larvae. It seems that the impact on the larvae caused reduced abundances 2–3 years later, but the impacts from the spill after that appear negligible in terms of commercial landing biomass, although it seems that a change in the captured species composition occurred.

Acknowledgments

We thank B. Bellgraph for the English editing and his helpful comments, and G. Ramirez and C. Suarez for editing the figures.

Appendix

Phylogenetic Listing of All Fish Species Collected by Survey Type

Order	Family	Species	Common Name	UNAM Survey Fishes	Commercial Data Fishes
Carcharhiniformes	Carcharhinidae	*Carcharhinus brachyurus* (Günther, 1870)	Narrowtooth shark	x	
		Carcharhinus spp.	Requiem sharks	x	
	Sphyrnidae	*Sphyrna lewini* (Griffith and Smith, 1834)	Scalloped hammerhead		x
		Sphyrna tiburo (Linnaeus, 1758)	Bonnethead	x	
Torpediniformes	Narcinidae	*Narcine brasiliensis* (Olfers, 1831)	Brazilian electric ray	x	
Rajiformes	Rhinobatidae	*Rhinobatos lentiginosus* (Garman, 1880)	Atlantic guitarfish	x	
	Rajidae	*Raja texana* (Chandler, 1921)	Roundel skate	x	

Order	Family	Species	Common Name	UNAM Survey Fishes	Commercial Data Fishes
Myliobatiformes	Urotrygonidae	*Urobatis jamaicensis* (Cuvier, 1816)	Yellow stingray	x	
	Dasyatidae	*Dasyatis americana* (Hildebrand and Schroeder, 1928)	Southern stingray		x
		Dasyatis sabina (Lesueur, 1824)	Atlantic stingray	x	
	Myliobatidae	*Aetobatus narinari* (Euphrasen, 1790)	Spotted eagle ray	x	
Elopiformes	Megalopidae	*Megalops atlanticus* (Valenciennes, 1847)	Tarpon		x
Anguilliformes	Ophichthidae	*Ophichthus puncticeps* (Kaup, 1859)	Palespotted eel	x	
	Congridae	*Rhynchoconger flavus* (Goode and Bean, 1896)	Yellow conger	x	
	Nettastomatidae	*Hoplunnis diomediana* (Goode and Bean, 1896)	Blacktail pike-conger	x	
Clupeiformes	Engraulidae	*Anchoa hepsetus* (Linnaeus, 1758)	Broad-striped Anchovy	x	
		Anchoa lamprotaenia (Hildebrand, 1943)	Bigeye anchovy	x	
		Anchoa mitchilli (Valenciennes, 1848)	Bay anchovy	x	
		Anchoa pectoralis (Hildebrand, 1943)	Bigfin anchovy	x	
		Cetengraulis edentulus (Cuvier, 1829)	Atlantic anchoveta	x	
	Clupeidae	*Harengula clupeola* (Cuvier, 1829)	False pilchard		x
		Harengula jaguana (Poey, 1865)	Scaled sardine	x	x
		Jenkinsia lamprotaenia (Gosse, 1851)	Dwarf herring		x
		Opisthonema oglinum (Lesueur, 1818)	Atlantic thread herring	x	
		Sardinella aurita (Valenciennes, 1847)	Spanish sardine		x
		Sardinella brasiliensis (Steindachner, 1879)	Brazilian sardinella		x

(Continued)

Order	Family	Species	Common Name	UNAM Survey Fishes	Commercial Data Fishes
Siluriformes	Ariidae	*Ariopsis felis* (Linnaeus, 1766)	Hardhead catfish	x	x
		Bagre marinus (Mitchill, 1815)	Gafftopsail catfish	x	x
Aulopiformes	Synodontidae	*Saurida brasiliensis* (Norman, 1935)	Largescale lizardfish	x	
		Synodus foetens (Linnaeus, 1766)	Inshore lizardfish	x	
Gadiformes	Bregmacerotidae	*Bregmaceros atlanticus* (Goode and Bean, 1886)	Antenna codlet	x	
Ophidiiformes	Ophidiidae	*Brotula barbata* (Bloch and Schneider, 1801)	Atlantic bearded brotula	x	
		Lepophidium brevibarbe (Cuvier, 1829)	Blackedge cusk-eel	x	
		Lepophidium marmoratum (Goode and Bean, 1885)	Marbled cusk-eel	x	
Batrachoidiformes	Batrachoididae	*Nautopaedium porosissimum* (Valenciennes, 1837)	Toadfish	x	
Lophiiformes	Antennariidae	*Antennarius striatus* (Shaw, 1794)	Striated frogfish	x	
		Fowlerichthys ocellatus (Bloch and Schneider, 1801)	Ocellated frogfish	x	
	Ogcocephalidae	*Halieutichthys aculeatus* (Mitchill, 1818)	Pancake batfish	x	
		Ogcocephalus radiatus (Mitchill, 1818)	Polka-dot batfish	x	
		Ogcocephalus vespertilio (Linnaeus, 1758)	Seadevil	x	
Mugiliformes	Mugilidae	*Joturus pichardi* (Poey, 1860)	Bobo mullet		x
		Mugil cephalus (Linnaeus, 1758)	Striped mullet		x
		Mugil curema (Valenciennes, 1836)	White mullet		x

Order	Family	Species	Common Name	UNAM Survey Fishes	Commercial Data Fishes
Gasterosteiformes	Syngnathidae	*Hippocampus erectus* (Perry, 1810)	Lined seahorse	x	
	Fistulariidae	*Fistularia petimba* (Lacepède, 1803)	Red cornetfish	x	
Dactylopteriformes	Dactylopteridae	*Dactylopterus volitans* (Linnaeus, 1758)	Flying gurnard	x	
Scorpaeniformes	Scorpaenidae	*Scorpaena brasiliensis* (Cuvier, 1829)	Barbfish	x	
		Scorpaena calcarata (Goode and Bean, 1882)	Smoothhead scorpionfish	x	
		Scorpaena dispar (Longley and Hildebrand, 1940)	Hunchback scorpionfish	x	
		Scorpaena plumieri (Bloch, 1789)	Spotted scorpionfish	x	
Scorpaeniformes	Triglidae	*Bellator militaris* (Goode and Bean, 1896)	Horned searobin	x	
		Prionotus beanii (Goode, 1896)	Bean's searobin	x	
		Prionotus carolinus (Linnaeus, 1771)	Northern searobin	x	
		Prionotus ophryas (Jordan and Swain, 1885)	Bandtail searobin	x	
		Prionotus punctatus (Bloch, 1793)	Bluewing searobin	x	
		Prionotus roseus (Jordan and Evermann, 1887)	Bluespotted searobin	x	
		Prionotus scitulus (Jordan and Gilbert, 1882)	Leopard searobin	x	
		Prionotus stearnsi (Jordan and Swain, 1885)	Shortwing searobin	x	
		Prionotus spp.	Searobins	x	
	Peristediidae	*Peristedion gracile* (Goode and Bean, 1896)	Slender searobin	x	

(Continued)

Order	Family	Species	Common Name	UNAM Survey Fishes	Commercial Data Fishes
Perciformes	Centropomidae	*Centropomus ensiferus* (Poey, 1860)	Swordspine snook		x
		Centropomus mexicanus (Bocourt, 1868)	Largescale fat snook		x
		Centropomus parallelus (Poey, 1860)	Smallscale fat snook		x
		Centropomus poeyi (Chávez, 1961)	Mexican snook		x
		Centropomus undecimalis (Bloch, 1792)	Common snook	x	x
	Epinephelidae	*Epinephelus guttatus* (Linnaeus, 1758)	Red hind	x	
		Epinephelus itajara (Lichtenstein, 1822)	Atlantic goliath grouper		x
		Epinephelus morio (Valenciennes, 1828)	Red grouper		x
		Hyporthodus niveatus (Valenciennes, 1828)	Snowy grouper	x	
		Mycteroperca bonaci (Poey, 1860)	Black grouper		x
	Serranidae	*Centropristis ocyurus* (Jordan and Evermann, 1887)	Bank sea bass	x	
		Diplectrum formosum (Linnaeus, 1766)	Sand perch	x	
		Diplectrum radiale (Quoy and Gaimard, 1824)	Pond perch	x	x
		Serranus atrobranchus (Cuvier, 1829)	Blackear bass	x	
	Priacanthidae	*Priacanthus arenatus* (Cuvier, 1829)	Atlantic bigeye	x	
		Pristigenys alta (Gill, 1862)	Short bigeye	x	
	Malacanthidae	*Caulolatilus guppyi* (Beebe and Tee-Van, 1937)	Reticulated tilefish	x	
	Rachycentridae	*Rachycentron canadum* (Linnaeus, 1766)	Cobia		x
	Echeneidae	*Echeneis naucrates* (Linnaeus, 1758)	Sharksucker	x	

Order	Family	Species	Common Name	UNAM Survey Fishes	Commercial Data Fishes
	Carangidae	*Caranx crysos* (Mitchill, 1815)	Blue runner	x	x
		Caranx hippos (Linnaeus, 1766)	Crevalle jack		x
		Caranx latus (Agassiz, 1831)	Horse-eye jack	x	
		Chloroscombrus chrysurus (Linnaeus, 1766)	Atlantic bumper	x	
		Decapterus punctatus (Cuvier, 1829)	Round scad	x	
		Selar crumenophthalmus (Bloch, 1793)	Bigeye scad	x	
		Selene setapinnis (Mitchill, 1815)	Atlantic moonfish	x	
		Selene vomer (Linnaeus, 1758)	Lookdown	x	
		Seriola dumerili (Risso, 1810)	Greater amber jack		x
		Trachinotus carolinus (Linnaeus, 1766)	Florida pompano		x
		Trachurus lathami (Nichols, 1920)	Rough scad	x	
		Trichiurus lepturus (Linnaeus, 1758)	Largehead hairtail	x	x
	Lutjanidae	*Etelis oculatus* (Valenciennes, 1828)	Queen snapper		x
		Lutjanus analis (Cuvier, 1828)	Mutton snapper		x
		Lutjanus campechanus (Poey, 1860)	Red snapper	x	x
		Lutjanus cyanopterus (Cuvier, 1828)	Cubera snapper		x
		Lutjanus synagris (Linnaeus, 1758)	Lane snapper	x	x
		Ocyurus chrysurus (Bloch, 1791)	Yellowtail snapper		x
		Pristipomoides aquilonaris (Goode and Bean, 1896)	Wenchman		x
		Pristipomoides macrophthalmus (Müller and Troschel, 1848)	Cardinal snapper	x	
		Rhomboplites aurorubens (Cuvier, 1829)	Vermilion snapper	x	x

(Continued)

Order	Family	Species	Common Name	UNAM Survey Fishes	Commercial Data Fishes
	Gerreidae	*Diapterus auratus* (Ranzani, 1842)	Irish pompano	x	x
		Diapterus rhombeus (Cuvier, 1829)	Rhombic mojarra	x	
		Eucinostomus argenteus (Baird and Girard, 1855)	Spotfin mojarra	x	
		Eucinostomus gula (Quoy and Gaimard, 1824)	Silver jenny	x	
		Eucinostomus melanopterus (Bleeker, 1863)	Flagfin mojarra	x	
	Haemulidae	*Conodon nobilis* (Linnaeus, 1758)	Barred grunt	x	
		Haemulon aurolineatum (Cuvier, 1830)	Tomtate	x	
		Haemulon plumierii (Lacepède, 1801)	White grunt	x	
		Haemulon sciurus (Shaw, 1803)	Bluestriped grunt		x
		Orthopristis chrysoptera (Linnaeus, 1766)	Pigfish	x	x
	Sparidae	*Archosargus probatocephalus* (Walbaum, 1792)	Sheepshead		x
		Archosargus rhomboidalis (Linnaeus, 1758)	Sea bream	x	
		Calamus calamus (Valenciennes, 1830)	Saucereye porgy	x	
		Calamus penna (Valenciennes, 1830)	Sheepshead porgy	x	
		Stenotomus caprinus (Jordan and Gilbert, 1882)	Long spined porgy	x	
	Polynemidae	*Polydactylus octonemus* (Girard, 1858)	Atlantic threadfin	x	
	Sciaenidae	*Bairdiella ronchus* (Cuvier, 1830)	Ground croaker		x
		Cynoscion arenarius (Ginsburg, 1930)	Sand Seatrout	x	x
		Cynoscion nebulosus (Cuvier, 1830)	Spotted seatrout		x
		Cynoscion nothus (Holbrook, 1848)	Silver seatrout	x	x

Order	Family	Species	Common Name	UNAM Survey Fishes	Commercial Data Fishes
		Equetus lanceolatus (Linnaeus, 1758)	Jack-knife fish	x	
		Larimus fasciatus (Holbrook, 1855)	Banded drum	x	x
		Menticirrhus americanus (Linnaeus, 1758)	Southern kingfish	x	
		Menticirrhus saxatilis (Bloch and Schneider, 1801)	Northern kingfish	x	
		Micropogonias undulatus (Linnaeus, 1766)	Atlantic croaker	x	x
		Micropogonias spp.	Croaker	x	x
		Pareques acuminatus (Bloch and Schneider, 1801)	High-hat	x	
		Pogonias cromis (Linnaeus, 1766)	Black drum		x
		Sciaenops ocellatus (Linnaeus, 1766)	Red drum		x
		Stellifer colonensis (Meek and Hildebrand, 1925)	Stardrum	x	
		Umbrina broussonnetii (Cuvier, 1830)	Striped drum	x	
	Mullidae	*Upeneus parvus* (Poey, 1852)	Dwarf goatfish	x	
	Labridae	*Lachnolaimus maximus* (Walbaum, 1792)	Hogfish		x
		Nicholsina usta usta (Valenciennes, 1840)	Emerald parrotfish	x	
	Gobiidae	*Bollmannia boqueronensis* (Evermann and Marsh, 1899)	White-eye goby	x	
	Ephippidae	*Chaetodipterus faber* (Broussonet, 1782)	Atlantic spadefish	x	
	Sphyraenidae	*Sphyraena barracuda* (Edwards, 1771)	Great barracuda		x
		Sphyraena guachancho (Cuvier, 1829)	Guachanche	x	x

(Continued)

Order	Family	Species	Common Name	UNAM Survey Fishes	Commercial Data Fishes
	Scombridae	*Acanthocybium solandri* (Cuvier, 1832)	Wahoo		x
		Euthynnus alletteratus (Rafinesque, 1810)	Little tunny		x
		Scomber colias (Gmelin, 1789)	Atlantic chub mackerel	x	
		Scomberomorus maculatus (Mitchill, 1815)	Spanish mackerel		x
		Thunnus albacares (Bonnaterre, 1788)	Yellowfin tuna		x
	Stromateidae	*Peprilus paru* (Linnaeus, 1758)	Harvestfish	x	
		Peprilus triacanthus (Peck, 1804)	Butterfish	x	
Pleuronectiformes	Paralichthyidae	*Ancylopsetta dilecta* (Goode and Bean, 1883)	Three-eye flounder	x	
		Ancylopsetta ommata (Jordan and Gilbert, 1883)	Ocellated flounder	x	x
		Citharichthys macrops (Dresel, 1885)	Spotted whiff	x	x
		Citharichthys spilopterus (Günther, 1862)	Bay whiff	x	x
		Cyclopsetta chittendeni (Bean, 1895)	Mexican flounder	x	x
		Cyclopsetta fimbriata (Goode and Bean, 1885)	Spotfin flounder	x	
		Etropus crossotus (Jordan and Gilbert, 1882)	Fringed flounder	x	x
		Syacium gunteri (Ginsburg, 1933)	Shoal flounder	x	
		Syacium micrurum (Ranzani, 1842)	Channel flounder	x	
		Syacium papillosum (Linnaeus, 1758)	Dusky flounder	x	x
	Bothidae	*Bothus ocellatus* (Agassiz, 1831)	Eyed flounder	x	x
		Engyophrys senta (Ginsburg, 1933)	Spiny flounder	x	
		Trichopsetta ventralis (Goode and Bean, 1885)	Sash flounder	x	

Order	Family	Species	Common Name	UNAM Survey Fishes	Commercial Data Fishes
	Achiridae	*Achirus lineatus* (Linnaeus, 1758)	Lined sole	x	
		Gymnachirus spp.	American soles	x	
		Trinectes maculatus (Bloch and Schneider, 1801)	Hogchoker	x	
	Cynoglossidae	*Symphurus plagiusa* (Linnaeus, 1766)	Blackcheek tonguefish	x	
Tetraodontiformes	Balistidae	*Balistes capriscus* (Gmelin, 1789)	Grey triggerfish	x	
	Monacanthidae	*Aluterus schoepfii* (Walbaum, 1792)	Orange filefish	x	
		Stephanolepis hispidus (Linnaeus, 1766)	Planehead filefish	x	
	Tetraodontidae	*Lagocephalus laevigatus* (Linnaeus, 1766)	Smooth puffer	x	x
		Sphoeroides dorsalis (Longley, 1934)	Marbled puffer	x	
		Sphoeroides greeleyi (Gilbert, 1900)	Green puffer	x	
		Sphoeroides nephelus (Goode and Bean, 1882)	Southern puffer	x	
	Diodontidae	*Chilomycterus schoepfii* (Walbaum, 1792)	Striped burrfish	x	

References

Anderson, J. W., S. L. Kiesser, and J. W. Blaylock. 1979. Comparative uptake of naphthalenes from water and oiled sediment in benthic amphipods. In *Proceedings of the 1979 Oil Spill Conference*, ed. American Petroleum Institute, 579–584. Publ. No. 4308. Washington, DC: American Petroleum Institute Publication.

Anderson, J. W., G. Roesijadi, and E. A. Crecelius. 1978. Bioavailability of hydrocarbons and heavy metals to marine detritivores from oil-impacted sediments. In *Marine Biological Effects of OCS Petroleum Development*, ed. D. A. Wolfe, 130–148. Washington, DC: National Oceanic and Atmospheric Administration.

Beitinger, T. L. 1990. Behavioral reactions for the assessment of stress in fishes. *Journal of Great Lakes Research* 16: 495–528.

Blumer, M. and J. Sass. 1972a. Indigenous and petroleum-derived hydrocarbons in a polluted sediment. *Marine Pollution Bulletin* 3:92–94.

Blumer, M. and J. Sass. 1972b. Oil pollution: Persistence and degradation of spilled fuel oil. *Science* 176:1120–1122.

Clarke, K. R. 1993. Non-parametric multivariate analyses of changes in community structure. *Australian Journal of Ecology* 18:117–143.

Clarke, K. R. and R. M. Warwick. 1994. *Change in Marine Communities: An Approach to Statistical Analysis and Interpretation*. Plymouth, United Kingdom: PRIMER-E Ltd.

Collins, N. R. and R. Williams. 1982. Zooplankton communities in the Bristol Channel and Severn Estuary. *Marine Ecology Progress Series* 9:1–11.

Dauble, D. D., R. H. Gray, J. R. Skalski, E. W. Lusty, and M. A. Simmons. 1985. Avoidance of a water-soluble fraction of coal liquid by fathead minnows. *Transactions of the American Fisheries Society* 114:754–760.

Farr, A. J., C. C. Chabot, and D. H. Taylor. 1995. Behavioral avoidance of fluoranthene by fathead minnows (*Pimephales promelas*). *Neurotoxicology and Teratology* 17:265–271.

Fischer, W. 1978. *FAO Species Identification Sheets for Fishery Purposes. Western Central Atlantic (Fishing Area 31) Vols. I–VII*. Rome, Italy: FAO.

García-Cuellar, J. A., F. Arreguín–Sánchez, S. Hernández-Vázquez, and D. Lluch-Cota. 2004. Impacto ecológico de la industria petrolera en la sonda de Campeche, México, tras tres décadas de actividades: Una revisión. *Interciencia* 29(6):311–319.

Garmon, L. 1980. Autopsy of an oil spill. *Science News* 118(17):267–270.

Guzmán del Próo, S. A., E. A. Chávez, F. M. Alatriste, S. de la Campa, G. De la Cruz, L. Gómez et al. 1986. The impact of the Ixtoc-1 oil spill on zooplankton. *Journal of Plankton Research* 8:557–581.

Hellstrøm, T. and K. B. Døving. 1983. Perception of diesel oil by cod (*Gadus morhua* L.). *Aquatic Toxicology* 4:303–315.

Hjermann, D. O., A. Melsom, G. E. Dingsør, J. M. Durant, A. M. Eikeset, L. P. Røed et al. 2007. Fish and oil in the Lofoten–Barents Sea system: Synoptic review of the effect of oil spills on fish populations. *Marine Ecology Progress Series* 339:283–299.

IPIECA. 1997. *Biological Impacts of Oil Pollution: Fisheries*. IPIECA Rep Ser 8, International Petroleum Industry Environmental Conversation Association, London, United Kingdom.

Jernelöv, A and O. Linden. 1981. Ixtoc I: A case study of the world's largest oil spill. *Ambio* 10(6):299–306.

Magurran, A. E. 2004. *Measuring Biological Diversity*. Oxford, United Kingdom: Blackwell Publishing.

Milne, J. C. 2008. Sedco 135 Series. OilCity. Available at: http://www.oilcity.co.uk/home/article.asp?pageid=645 (last accessed September 11, 2012).

Mironov, O. G. 1972. Effect of oil pollution on flora and fauna of the Black Sea. In *Marine Pollution and Sea Life. Fish*, ed. M. Ruivo, 222–224. London, United Kingdom: Fishing News Books Ltd.

Office of Response and Restoration, National Ocean Service. "Ixtoc I." *IncidentNews*. National Oceanic and Atmospheric Administration, US Department of Commerce. Available at: http://incidentnews.noaa.gov/incident/6250/response (last accessed on June 24, 2014).

Page, L. M., H. Espinosa-Pérez, L. T. Findley, C. R. Gilbert, R. N. Lea, N. E. Mandrak et al. 2013. *Common and Scientific Names of Fishes from the United States, Canada, and Mexico, 7th ed*. Bethesda, MD: American Fisheries Society, Special Publication 34.

Reijnhart, R. and R. Rose. 1982. Evaporation of crude oil at sea. *Water Research* 16:1319–1325.

Solow, A. R. 1993. A simple test for change in community structure. *Journal of Animal Ecology* 62:191–193.

Sparre, P. and S. C. Venema. 1998. *Introduction to Tropical Fish Stock Assessment, Part 1-Manual*. FAO Fisheries Technical Paper 306-1. Rome, Italy: FAO.

Stickle, W. B. 2002. *The Effects of Simultaneous Exposure to Petroleum Hydrocarbons, Hypoxia and Prior Exposure on the Tolerance and Sub Lethal Responses of Marine Animals, Blue Crabs and Killifish: Final Report U.S. Dept. of the Interior*. Minerals Management Service, Gulf of Mexico OCS Region, New Orleans, Louisiana OCS Study MMS 2002-009.

Taylor, L. R. 1978. Bates, Williams, Hutchinson - a variety of diversities. In *Diversity of Insect Faunas: 9th Symposium of the Royal Entomological Society*, eds. L. A. Mound and N. Waloff, 1–18. Oxford, United Kingdom: Blackwell Publishing.

Teal, J. M. and R. W. Howarth. 1984. Oil spill studies: A review of ecological effects. *Environmental Management* 8:27–44.

Yánez Arancibia, A., F. Amezcua Linares, A. Lara Domínguez, P. Sánchez-Gil., M Tapia, H. Álvarez, et al. 1982. *Análisis de la dinámica ambiental y estructura de las comunidades de peces en variaciones estaciónales, resultados de los cruceros en la Sonda de Campeche*. Proyecto de investigación: Análisis comparativo de las poblaciones de peces de la Sonda de Campeche y la Laguna de Términos, antes y después del derrame petrolero del pozo IXTOC I. Quinto informe. PCEESC/UNAM/CCML.

12

Impacts of the Deepwater Horizon *Oil Spill on the Reproductive Biology of Spotted Seatrout* (Cynoscion nebulosus)

Nancy J. Brown-Peterson, Rachel A. Brewton, Robert J. Griffitt, and Richard S. Fulford

CONTENTS

Introduction

The April 2010 *Deepwater Horizon* (DWH) oil spill released an estimated 4.1 million barrels of crude oil as well as 2.1 million gallons of the dispersant Corexit 9500 into the northern Gulf of Mexico (GOM). The marshes and coastal shorelines along the north central GOM were impacted as crude and dispersed oil from the offshore blowout moved onshore. Although Louisiana waters received the heaviest impacts of bioavailable polycyclic aromatic compounds (PAHs) immediately after the spill, Mississippi coastal areas were also significantly impacted (Allan et al. 2012). Although the amount of measured PAHs in coastal waters in Mississippi decreased to prespill levels by August 2010, PAH levels did not return to prespill levels in Louisiana waters until March 2011 (Allan et al. 2012). Despite these impacts, juvenile fish communities in Louisiana and Mississippi seagrass meadows did not show a dramatic decline in 2010 collections compared with historical (2006–2009) collections (Fodrie and Heck 2011), suggesting fish that were spawned during the oil spill were more resilient than expected. However, sublethal effects of oil exposure on adult fishes may alter reproductive capacity, resulting in future impacts on fish populations. For instance, PAHs are known to have a deleterious effect on vitellogenesis (the production of egg-yolk proteins) in fishes, which can substantially reduce ovarian development and reproductive output (Nicholas 1999). Laboratory exposure of Atlantic croaker (*Micropogonias undulatus*) to diesel fuel oil impaired ovarian growth and oocyte development (Thomas and Budiantara 1995), while wild-caught yellowfin sole (*Limanada aspera*) and Dolly Varden (*Salevelinus malma*) exhibited reproductive impairment for several years after the *Exxon Valdez* oil spill (Sol et al. 2000). Examining the reproductive biology of GOM fishes affected by the DWH oil spill is a first step in helping to determine potential long-term population impacts of the oil spill.

Spotted seatrout (*Cynoscion nebulosus*) is an abundant commercially and recreationally important species along the northern GOM. As a resident estuarine species, spotted seatrout has been suggested as an excellent indicator of estuarine conditions (Bortone 2003). Spotted seatrout generally reach sexual maturity within their first year of life (Brown-Peterson 2003) and live 5–9 years along the northern GOM (Bedee et al. 2003), thus allowing a multiple-year examination of potential effects of the DWH oil spill on their reproduction. The reproductive biology of spotted seatrout along the northern GOM in DWH-impacted areas has been well described (Brown-Peterson and Warren 2001; Brown-Peterson et al. 2002; Nieland et al. 2002; Brown-Peterson 2003). These data provide a historical database for comparison to post-DWH reproductive parameters from the same estuaries. Spotted seatrout in this region are batch spawners and spawn from April through September (Brown-Peterson et al. 2002), and thus 2010 young-of-the-year were probably exposed to oiled waters as larvae and juveniles. This exposure potentially impacted their subsequent sexual maturation and reproductive ability.

The objective of this study was to compare reproductive parameters (gonadosomatic index [GSI], ovarian development, and spawning frequency) of spotted seatrout captured 1-year post-DWH oil spill from Mississippi and Louisiana estuaries to previous reproductive data from the same estuaries. We hypothesized that spotted seatrout captured in 2011, 1-year post-DWH oil spill, would show significantly altered reproductive parameters compared to fish captured from the same locations in the 1990s. We also hypothesized that these differences would be more pronounced in age-1 fish that had been spawned during the time of the DWH oil spill.

Materials and Methods

Sample Collections

Spotted seatrout were collected pre- and post-DWH spill from both Louisiana and Mississippi waters (Figure 12.1). Historical (prespill) data for Louisiana were from collections made in 1995 in Barataria Bay, Louisiana, and published in Nieland et al. (2002) and Brown-Peterson et al. (2002). Historical data from Mississippi were from collections made in 1999 along the Mississippi Gulf Coast and published in Brown-Peterson and Warren (2001) and Brown-Peterson et al. (2002). The post-DWH oil spill collection sites were matched as closely to prespill collection sites as possible. Post-DWH spill Louisiana fish were captured from five sites within Barataria Bay in April and June 2011. Post-DWH spill Mississippi fish were collected from St. Louis Bay (two sites), Biloxi Bay (three sites), Grand Bay (one site), and the north side of the barrier islands (two sites) from March through July 2011. All Mississippi post-DWH spill collection sites exhibited light oiling during

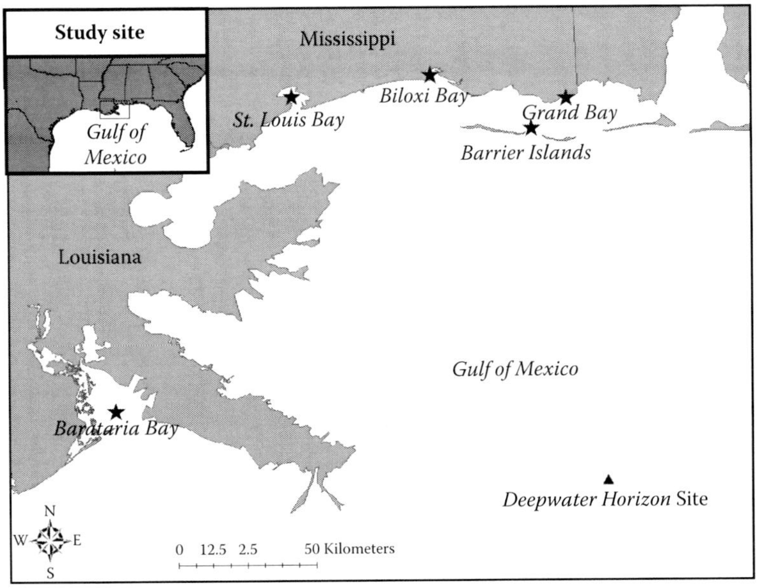

FIGURE 12.1
Map of collection areas for spotted seatrout samples in Mississippi and Louisiana. Fish were collected from these areas prior to the *Deepwater Horizon* oil spill (1995, Louisiana; 1999, Mississippi) and after the oil spill (2011, both states).

2010, while the Louisiana sites exhibited moderate oiling in 2010, as determined from the web-based Environmental Response Management Application (www.restorethegulf.gov/release/2011/10/28/erma-gulf-response).

Sampling gear was similar between pre- and post-DWH spill collections. Fish from Louisiana were captured with gill nets or by hook and line in 1995 and 2011. Fish from Mississippi were captured with gill nets in 1999 and with gill nets or hook and line in 2011. In all cases, fish were placed on ice immediately after capture and transported to the laboratory for dissection. Surface water temperature (°C) and salinity were recorded at each sampling event. Water temperature and salinity were collected from the same sites in months prior to collections as part of routine monitoring and sampling by personnel from Gulf Coast Research Laboratory (GCRL; Mississippi sites) and Louisiana Department of Wildlife and Fisheries (LDWF; Louisiana sites). Water samples were collected in amber bottles and stored on ice during March, April, and June 2011 collections for subsequent water chemistry analysis.

In the laboratory, total length (TL, mm), standard length (SL, mm), and wet weight (W, g) were recorded for each fish, and gonads were excised and weighed (GW, 0.1 g). The GSI was calculated for each fish as $GSI = [GW/(W-GW)] \times 100$. Fish <300 mm TL were considered to be 1 year old, based on length-at-age data from Mississippi (Brown-Peterson and Warren 2001) and Louisiana (Nieland et al. 2002).

Water Chemistry Analysis

Duplicate water samples from five collections at three sites (Barataria Bay, Louisiana; St. Louis Bay, Mississippi; and Grand Bay, Mississippi) were analyzed within 48 hours of collection by Micro-Methods Laboratory, Inc. (Ocean Springs, Mississippi) for 17 semivolatile organics using GC/MS following U.S. Environmental Protection Agency protocol 8270C. Minimum detection limit for all organics was 1.03 μg L^{-1}.

Histological Analysis

A portion of the midsection of each ovary collected in 2011 was placed into an individually labeled histological cassette and preserved in 10% neutral buffered formalin for a minimum of 7 days. Tissues were rinsed in running tap water overnight, dehydrated in a series of graded ethanols, cleared, embedded in paraffin, sectioned at 4 μm and stained with hematoxylin and eosin following standard histological techniques. Ovarian tissue was examined for oocyte stages (including postovulatory follicle complex [POF] and atresia) and classified into reproductive phases following Brown-Peterson et al. (2011). Histological data from 1995 (Louisiana) and 1999 (Mississippi) were reexamined for the occurrence of atresia and POF and classified into reproductive phases. Females were considered to be sexually mature when cortical alveolar oocytes were observed as fish were entering the developing ovarian phase.

Spawning frequency estimates were made on females in the spawning capable reproductive phase based on the percentage of ovaries containing POF <24 hours. The POFs were staged and spawning frequency calculations made following Hunter and Macewicz (1985).

Statistical Analyses

All GSI values were arcsine square-root transformed prior to analysis (Sokal and Rohlf 1995). All data were tested for normality (Kolmogorov–Smirnov one-sample test) and homogeneity of variance (Levene's test). Differences in monthly GSI values, water temperatures, and salinity between pre- and post-DWH spill collections were tested with Student's t-test. The difference in the monthly distribution of reproductive phases was tested for pre- and post-DWH spill collections using a Kolmogorov–Smirnov independent samples test. Differences in the distribution of reproductive phases of age-1 fish (<300 mm TL) from pre- and post-DWH spill collections was tested using a Kolmogorov–Smirnov independent samples test; months were combined for these analyses due to small sample sizes. Differences in spawning frequency and the percent occurrence of atresia were tested between pre- and post-DWH spill collections using Pearson χ^2. All statistics were computed with SPSS version 18 (SPSS Inc., Chicago, Illinois) and differences were considered significant if $p \leq .05$.

Results

Water Chemistry

Concentrations of semivolatile organics in water samples at all sites were below the detection limit of 1.03 μg L^{-1} in almost all cases (Table 12.1). Monthly temperature and salinity did not differ between pre- and post-DWH spill collections in Mississippi or Louisiana during the reproductive season (Table 12.2), with two exceptions. In Mississippi, temperature was 5.6°C higher in April 2011 compared to April 1999, a significant difference

TABLE 12.1

Mean (± SE) Semivolatile Organic Compound Concentrations of Water Samples Collected in 2011

Site	Date	Concentration (μg·L^{-1})
St. Louis Bay, Mississippi	March 22, 2011	1.080 ± 0.020
Grand Bay, Mississippi	March 24, 2011	BDL
Barataria Bay, Louisiana	April 8, 2011	BDL
Grand Bay, Mississippi	June 2, 2011	BDL
Barataria Bay, Louisiana	June 13, 2011	1.045 ± 0.035

Note: Minimum detection limit is 1.03 μg L^{-1}.
BDL, below detection limit.

TABLE 12.2

Monthly Temperature and Salinity (Mean ± SE) for Collection Areas in Mississippi and Louisiana Pre- and Post-DWH Oil Spill

		Temperature (°C)			Salinity		
Location	**Month**	**Pre**	**Post**	***p***	**Pre**	**Post**	***p***
Mississippi	January	14.1 ± 1.3	11.8 ± 0.6	.070	9.6 ± 2.2	19.7 ± 1.9	**.002**
	February	15.3 ± 0.8	13.1 ± 0.6	**.031**	8.9 ± 1.8	15.6 ± 2.2	**.029**
	March	20.6 ± 2.29	19.4 ± 2.18	.558	14.5 ± 4.5	7.9 ± 2.6	.074
	April	20.9 ± 0.5	25.3 ± 0.6	**.006**	8.0 ± 4.0	8.1 ± 0.7	.172
	May	25.6 ± 0.6	25.1 ± 0.7	.686	10.0 ± 2.9	10.5 ± 1.1	.248
	June	27.9 ± 0.9	29.4 ± 0.5	.163	16.0 ± 6.0	7.2 ± 3.0	.355
	July	29.0 ± 0.2	31.4 ± 1.0	.243	17.5 ± 3.9	17.4 ± 7.3	.994
Louisiana	January	14.4 ± 0.3	12.9 ± 0.1	**<.001**	15.6 ± 1.1	16.8 ± 1.0	.474
	February	16.1 ± 0.4	14.8 ± 0.4	**.025**	18.9 ± 1.6	14.5 ± 0.9	**.019**
	March	19.3 ± 0.4	20.6 ± 0.2	**.017**	17.6 ± 1.0	12.0 ± 0.8	<.00
	April	22.6 ± 0.4	24.3 ± 0.7	.231	19.6 ± 0.4	8.4 ± 1.9	**<.001**
	June	29.0 ± 0.3	29.8 ± 0.3	.519	12.5 ± 0.5	13.1 ± 5.4	.922

Note: Prespill—1999, Mississippi; 1995, Louisiana. Postspill—2011 in Mississippi and Louisiana. Bold font indicates significant difference between pre- and postvalues (Student's t-test).

($t_{13} = -3.239$, $p = .006$). In Louisiana, salinity was 8.4 in April 2011 compared to 19.6 in April 1995, a significant difference ($t_{31} = 8.058$, $p < .001$) probably related to the opening of the Bonnet Carré spillway due to high March and April 2011 rainfall upstream. However, there are some significant differences in water quality from both locations prior to the reproductive season. In Mississippi, temperature was significantly lower in February in 2011 than in 1999, and salinity was significantly lower in both January and February in 1999 compared with 2011 (Table 12.2). In Louisiana, temperature was significantly lower in January and February but significantly higher in March in 1995 compared with 2011; salinity was significantly lower in February and March in 1995 compared to 2011 (Table 12.2).

Spawning Season and Histological Observations

The mean monthly GSI values of female spotted seatrout captured from Mississippi after the oil spill were significantly lower than 1999 GSI values in March ($t_{39} = 3.15$, $p = .007$), April ($t_{120} = 2.34$, $p = .021$), May ($t_{97} = 6.319$, $p < .001$), and June ($t_{58} = 3.52$, $p = .001$); by July there was no difference in female GSI between pre- and postspill fish (Figure 12.2a). There was no significant difference in mean monthly GSI values of Mississippi males captured pre- and post-DWH spill, although values from 1999 were slightly higher in March, May, and June (Figure 12.2b).

Spotted seatrout were only collected in 2011 in April and June from Louisiana, and thus GSI was only compared for those 2 months. In April, GSI of females collected in 2011 was significantly lower than those collected in 1995 ($t_{43} = 2.74$, $p = .01$), but there was no significant difference in GSI in June between years (Figure 12.3a). However, GSI values of males from Louisiana collected in 2011 were significantly lower in both April ($t_{31} = 3.69$, $p = .001$) and June ($t_{86} = 4.22$ $p < .001$; Figure 12.3b). Overall, the GSI data from both Mississippi and Louisiana showed lower values in post-DWH spill fish than in fish collected prior to the oil spill, particularly at the beginning of the reproductive season.

The significant difference in female GSI values in Mississippi fish was associated with differences in the percentage of females in each reproductive phase between pre- and post-DWH spill collections. The distribution of reproductive phases in March is significantly different between 1999 and 2011 ($p = .007$), with no spawning capable fish observed in 2011, and a much higher percentage of fish in 2011 in the early developing subphase, indicating these individuals are just beginning to enter the reproductive cycle (Figure 12.4). Although there was no significant difference in reproductive phase distribution in April ($p = .064$),

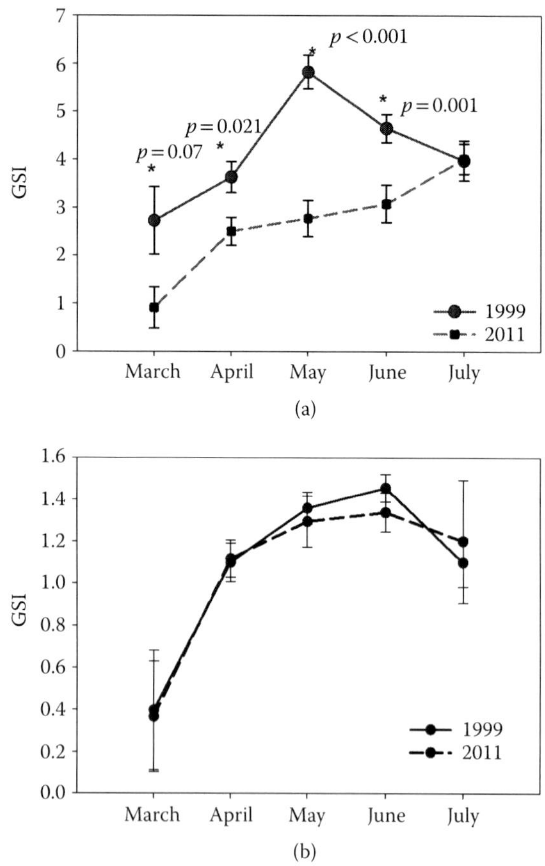

FIGURE 12.2
Mean (± SE) gonadosomatic index (GSI) of female (a) and male (b) spotted seatrout from the Mississippi Gulf coast prior to (1999) and 1 year after (2011) the *Deepwater Horizon* oil spill.

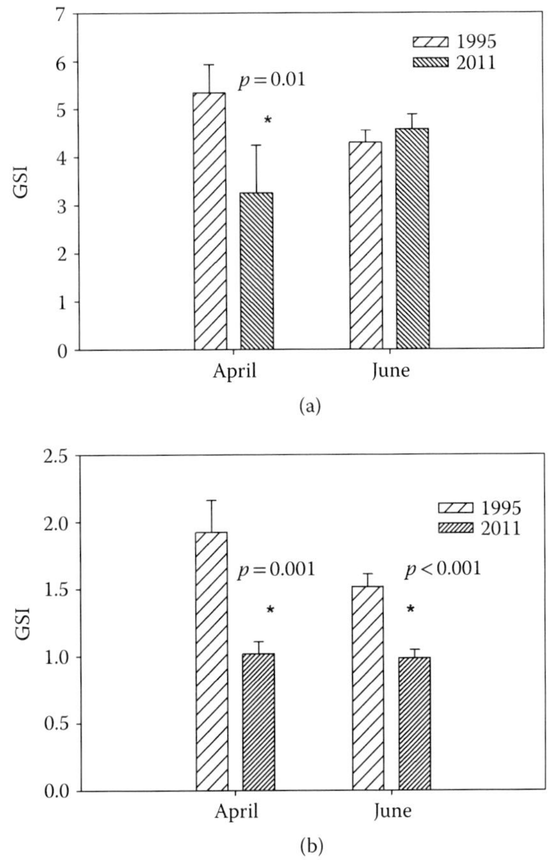

FIGURE 12.3
Mean (± SE) gonadosomatic index (GSI) of female (a) and male (b) spotted seatrout from Barataria Bay, Louisiana prior to (1995) and 1 year after (2011) the *Deepwater Horizon* oil spill.

there was a lower percentage of fish in the actively spawning phase and a much greater percentage of females in the developing phase in 2011 compared to 1999 (Figure 12.4). There was a difference in the distribution of reproductive phases in May ($p = .008$), with a higher percentage of fish in the immature and early developing phases in 2011 than 1999. In addition, no fish were found in the regressing phase, which indicates cessation of spawning for the year, in May 1999 (Figure 12.4). By the middle of the spawning season (June and July), there was no significant difference in the distribution of reproductive phases between 1999 and 2011, although some fish in the regressing and regenerating phases were found in 2011 while none were present in either month in 1999 (Figure 12.4). Overall, the reproductive phase data suggest female spotted seatrout in Mississippi exhibited developmental delay at the beginning of the reproductive season 1 year after the DWH oil spill compared with historical, prespill data.

Surprisingly, there was no significant difference in the distribution of female reproductive phases between pre- and post-DWH spill collections from Louisiana in April or June (Figure 12.5). However, there was a much higher percentage of females in the regenerating phase in April 2011 than in April 1995, suggesting these fish had not yet begun gonadal recrudescence for the current spawning season. Similarly, some females in June 2011 were in the regressing phase, indicating these fish were no longer reproductively active,

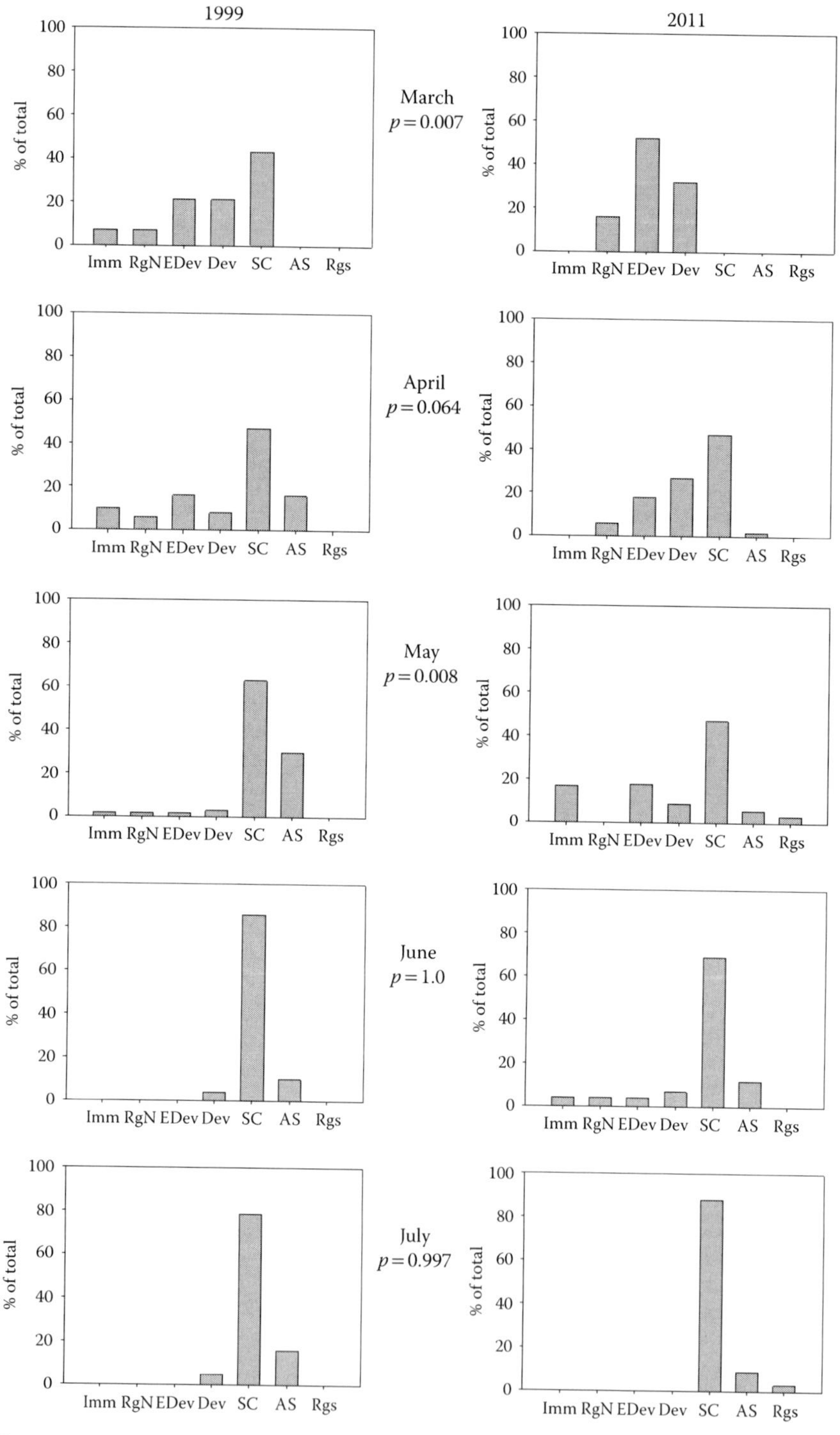

FIGURE 12.4
Monthly distribution of ovarian reproductive phases of spotted seatrout from the Mississippi Gulf coast prior to (1999) and 1 year after (2011) the *Deepwater Horizon* oil spill. Differences in distributions between years were tested with a Kolmogorov–Smirnov independent samples test. Imm, immature; RgN, regenerating; EDev, early developing; Dev, developing; SC, spawning capable; AS, actively spawning; Rgs, regressing.

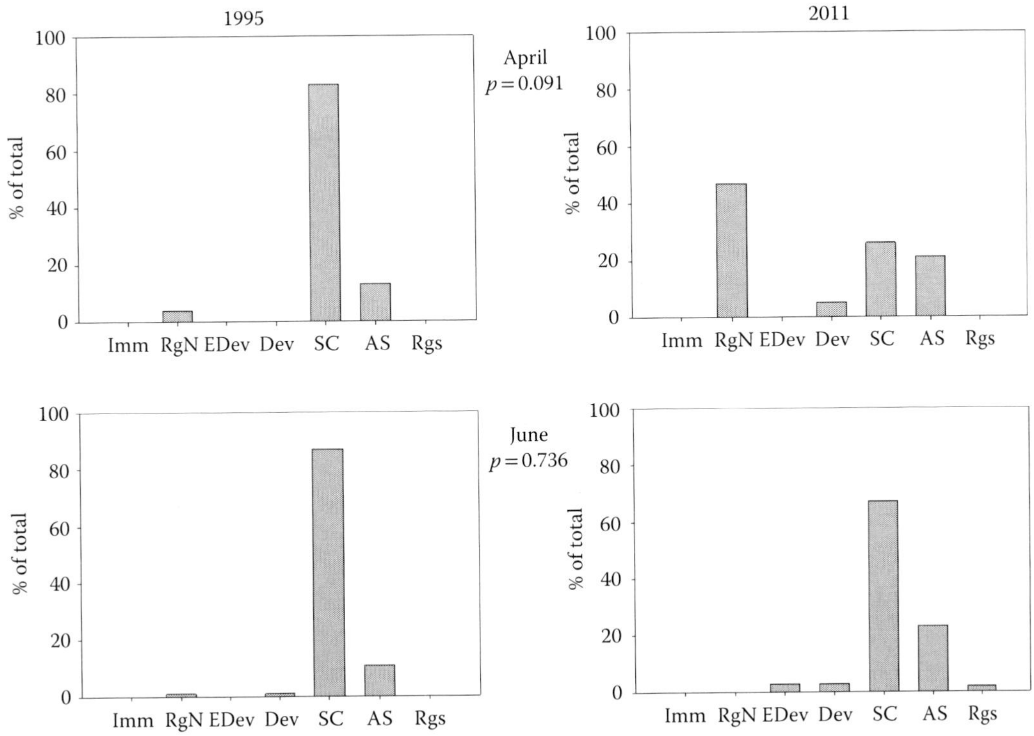

FIGURE 12.5
Monthly distribution of ovarian reproductive phases of spotted seatrout from Barataria Bay, Louisiana prior to (1995) and 1 year after (2011) the *Deepwater Horizon* oil spill. Differences in distributions between years were tested with a Kolmogorov–Smirnov independent samples test. Imm, immature; RgN, regenerating; EDev, early developing; Dev, developing; SC, spawning capable; AS, actively spawning; Rgs, regressing.

whereas no females were found in the regressing phase in June 1995. These data suggest that, for at least some females, the spawning season was shorter in 2011 in Louisiana than it was in 1995.

Spotted seatrout captured the year after the oil spill exhibited a higher percentage of atretic oocytes than those captured in the 1990s (Table 12.3). The percentage of atretic oocytes was significantly higher in Mississippi fish in March ($\chi^2 = 10.77$, $p = .001$), May ($\chi^2 = 4.92$, $p = .026$) and June ($\chi^2 = 5.01$, $p = .025$) 2011 compared to the same months in 1999. By July 2011, however, atresia was similar to that seen in July 1999. The percentage of atretic oocytes in Louisiana fish in 1995 was very low; there was a significantly higher percentage of atresia in both April ($\chi^2 = 11.14$, $p = .001$) and June ($\chi^2 = 58.49$, $p < .001$) 2011 compared to the same months in 1995. These results suggest that a greater percentage of vitellogenic oocytes in spotted seatrout from both Mississippi and Louisiana were unable to develop normally and thus underwent atresia in 2011 compared to pre-DWH spill oocytes.

Spawning Frequency

There is a dramatic reduction in the spawning frequency of spotted seatrout captured after the DWH oil spill compared to pre-DWH spill spawning frequencies as calculated by the percentage of POF observed in the ovaries (Table 12.4). Spawning frequency in Mississippi at the beginning of the spawning season (April–May) decreased significantly from every

TABLE 12.3

Prevalence (%) of Atretic Oocytes in Female Spotted Seatrout Ovaries Pre- and Post-DWH Oil Spill

Location	Month	Prespill (1999—Mississippi, 1995—Louisiana)	Postspill (2011)	p
Mississippi	March	0	58.8	**.001**
	April	42.6	60.9	.065
	May	21.6	46.9	**.026**
	June	13.6	42.8	**.025**
	July	18.8	24.2	.483
Louisiana	April	4.0	47.0	**.001**
	June	6.7	70.1	**<.001**

Note: Bold font indicates significant difference between pre- and postvalues (Pearson χ^2).

TABLE 12.4

Pre-and Post-DWH Oil Spill Comparison of Spawning Frequency in ***Cynoscion nebulosus***

		Prespill (1999—Mississippi, 1995—Louisiana)			Postspill (2011)			
Location	**Month**	**N**	**% POF**	**Days**	**N**	**% POF**	**Days**	p
Mississippi	April–May	164	17.7	5.6	39	2.6	38.5	**.017**
	June–July	112	22.3	4.5	54	7.4	13.5	**.018**
Louisiana	April	47	31.9	3.1	9	0	—	**.048**
	June	125	20.0	5.0	52	3.8	26.3	**.006**

Note: Bold font indicates significant difference between pre- and postvalues (Pearson χ^2).
POF, postovulatory follicle complex.

5.6 days to every 38.5 days ($\chi^2 = 5.72$, $p = .017$). Spawning frequency increased in the middle of the spawning season (June–July), but the 2011 frequency of every 13.5 days was still significantly lower than the 1999 frequency of every 4.5 days ($\chi^2 = -5.62$, $p = .018$). The results for Louisiana are even more dramatic; few spawning capable fish were captured in April 2011, and none had POFs, resulting in a significant difference in spawning frequency between 1995 and 2011 ($\chi^2 = 3.92$, $p = .048$). Spawning frequency decreased significantly from every 5 days in June 1995 in Louisiana to every 26.3 days in June 2011 ($\chi^2 = 7.41$, $p = .006$). These results, in combination with the GSI, reproductive phases, and percentage of atretic oocytes, indicate that spotted seatrout in both Louisiana and Mississippi showed significant negative impacts on their reproductive potential in 2011 compared to previous years.

Age-1 Fish

The mean GSI of Mississippi age-1 female fish was significantly lower in 2011 than in 1999 ($t_{46} = 2.41$, $p = .02$; Table 12.5). Although mean GSI of age-1 Louisiana female fish was also lower in 2011 than 1995, this difference was not significant ($p = .256$). In contrast to the GSI results, there was no significant difference in the distribution of reproductive phases in age-1 females from Mississippi (Figure 12.6), despite the fact that all age-1 fish captured in 1999 were sexually mature, while 15% of age-1 fish were immature in 2011. Furthermore,

TABLE 12.5

Gonadosomatic Index (Mean ± SE) of Age-1 Spotted Seatrout (<300 mm TL) Pre- and Post-DWH Oil Spill

Location	Prespill (1999—Mississippi, 1995—Louisiana)	Postspill (2011)	p
Mississippi	2.07 ± 0.45	1.14 ± 0.19	**.020**
Louisiana	2.67 ± 0.47	2.10 ± 0.46	.256

Note: Bold font indicates significant difference between pre- and postvalues (Student's t-test).

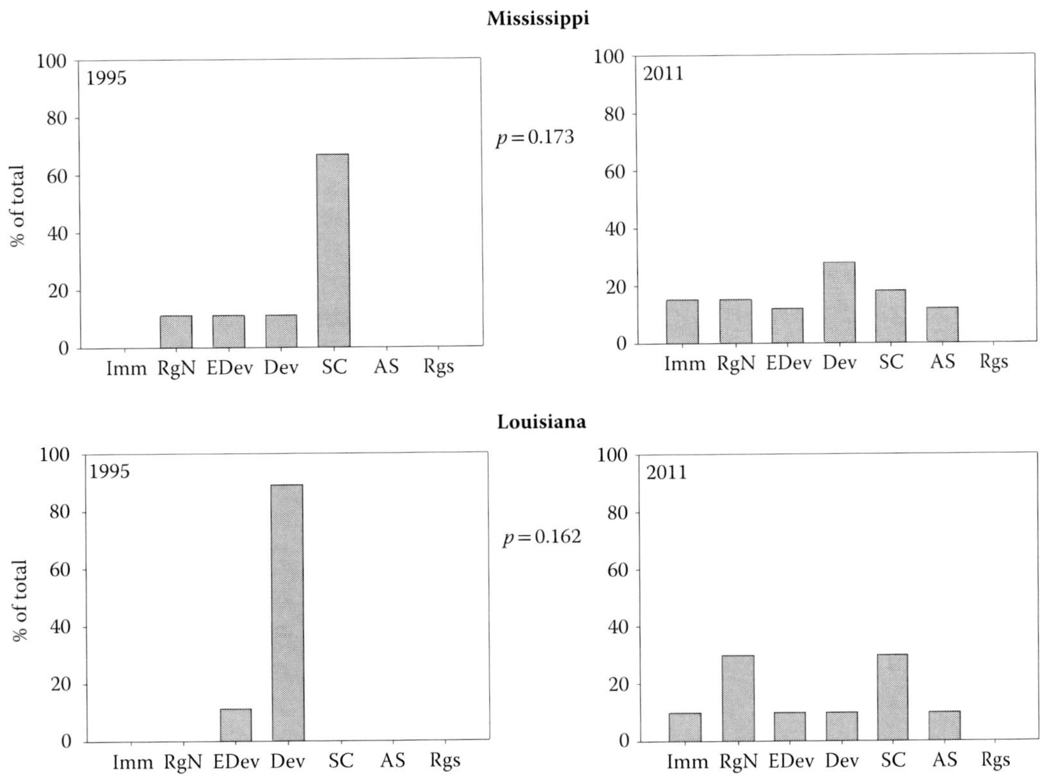

FIGURE 12.6
Distribution of ovarian reproductive phases of age-1 spotted seatrout from the Mississippi Gulf coast (top) and Barataria Bay, Louisiana (bottom) prior to and 1 year after the *Deepwater Horizon* oil spill. Differences in distributions between years were tested with a Kolmogorov–Smirnov independent samples test. Imm, immature; RgN, regenerating; EDev, early developing; Dev, developing; SC, spawning capable; AS, actively spawning; Rgs, regressing.

a much higher percentage of age-1 fish were in the spawning capable phase in 1999 than in 2011 in Mississippi (87% vs. 18%). There was also not a significant difference in the distribution of reproductive phases of age-1 females pre- and post-DWH spill in Louisiana, although almost all (89%) age-1 fish were in the spawning capable phase in 1995, whereas 50% of the age-1 fish in 2011 were in the immature, regenerating, or early developing phases. These results suggest that spotted seatrout spawned during 2010 showed significant negative impacts on their reproductive biology in both Mississippi and Louisiana. However, these impacts were not more dramatic than the impacts seen on spotted seatrout of all ages.

Discussion

There appears to be a substantial impact on reproductive parameters of spotted seatrout 1 year after the DWH oil spill when compared with historical, prespill data. The availability of prespill reproductive data allowed direct comparisons of specific reproductive parameters from the same sites before and after the DWH oil spill. Since spotted seatrout are estuarine residents (Helser et al. 1993), this resulted in sampling the same populations at two distinct time points. Although using baseline data to make postevent comparisons can be problematic, consistent sampling methodology at the same sites alleviates much of this concern (Wiens and Parker 1995). Furthermore, environmental variables (temperature and salinity) that can significantly impact reproductive development were not different at the sites pre- and post-DWH spill during most months of the reproductive season. However, warmer water temperatures during late winter and early spring (January–March) can accelerate ovarian recrudescence, which could result in higher GSI values and an earlier spawning season. Temperatures were slightly warmer in both Mississippi and Louisiana in February during prespill years compared to 2011, which may have contributed to the lower GSI values and delayed development seen in females from 2011; however, by March there was no significant difference in temperature between years in Mississippi, and in Louisiana March 2011 temperatures were significantly warmer than March 1995 temperatures. Thus, any delay in gonadal development as a result of cooler February 2011 temperatures should not have persisted through April in the form of lower GSI values. The two significant differences noted in environmental variables during the reproductive season are not biologically important. The fact that lower GSI values observed in April 2011 compared to April 1999 in Mississippi, despite significantly warmer 2011 water temperatures, suggest temperature differences between years cannot explain the observed differences in GSI values at this point in the reproductive season. The significant reduction in salinity during April in Louisiana in 2011 was most likely related to the opening of the Bonnet Carré spillway during that time. However, the April 2011 salinity value from Louisiana was similar to values from Mississippi in April of both 1999 and 2011, and clearly within the range to allow spotted seatrout spawning. Finally, differences in reproductive parameters of spotted seatrout have been previously shown to vary little within a 7–8-year period for fish from the same estuarine system (Brown-Peterson et al. 2002). Thus, differences in the observed reproductive parameters solely as a function of natural variability between sampling events is unlikely. Other impacts on the spotted seatrout populations in the years between the two sampling events, such as heavy fishing pressure, habitat loss, and changes in salinity due to drought and flood conditions, may also contribute to differences in reproductive parameters, although impacts from DWH seems to be the most parsimonious explanation for the observed differences.

The PAH concentration of water samples taken from sampling sites in 2011 were generally below the 1.03 $\mu g\ L^{-1}$ detection range of our analyses, indicating very low levels of PAHs in the water. This was confirmed by published reports from May 2011 that showed PAH concentrations were not significantly different from preoiled (April 2010) values (3.8 ± 0.64 ng L^{-1} in Barataria Bay, Louisiana and 7.3 ± 0.41 ng L^{-1} in the Mississippi Sound) (Allan et al. 2012). Thus, although there was still some evidence of oiling in the marshes and sediments in 2011, particularly in Louisiana, no collections occurred in PAH-contaminated waters during 2011, although concentrations of PAH were significantly elevated during the summer of 2010 at both sites (Allen et al. 2012). However, it should be noted that larval spotted seatrout may have been exposed to DWH oil during the summer of 2010, leading to altered reproductive dynamics in 2011 as the fish reached sexual maturity.

There are surprisingly few published reports of the effects of oil on the reproductive parameters of fish; most information is concentrated on egg and larval effects of oil exposure. Young Atlantic croaker exposed to the water-soluble fraction (WSF) of crude oil in the laboratory during the time of initial gonadal recrudescence showed significantly impaired ovarian development compared to control fish (Thomas and Budientara 1995). This is similar to our results for age-1 spotted seatrout in both Mississippi and Louisiana, although the oil exposure for spotted seatrout occurred when they were juveniles, well before they began gonadal recrudescence. In addition, WSF-exposed Atlantic croaker showed a significantly higher percentage of atresia in the ovaries compared to controls (Thomas and Budientara 1995), results that were mirrored in spotted seatrout from this study. Mummichog (*Fundulus heteroclitus*) collected from sites contaminated by PAHs in New Jersey showed both significantly lower GSI values in males and females as well as developmental delay in the form of a significantly lower percentage of vitellogenic oocytes when compared to control fish (Bugel et al. 2010). These results are similar to the lower GSI values and developmental delay exhibited in spotted seatrout 1 year after the DWH oil spill. Finally, Whitehead et al. (2011) found significant down-regulation of a suite of genes associated with reproduction in Gulf killifish (*F. grandis*) collected from oiled marshes within 4 months of the DWH spill, suggesting reproductive impairment.

Several studies have examined reproductive effects on fish several months to years after a major oil spill. However, these studies have been conducted in northern locations, where fish have a shorter reproductive season than in the GOM and are generally not batch spawners. Collections of plaice (*Pleuronectes platessa*) from France after the *Amoco Cadiz* oil spill in March 1978 showed a decreased percentage of mature (i.e., tertiary vitellogenic) oocytes during the spawning season both 7 and 25 months postspill compared to a nearby reference site (Stott et al. 1983). Furthermore, plaice from the oiled sites exhibited developmental delay when compared with fish from the reference site, similar to our observations for post-DWH spill spotted seatrout, particularly during the first half of the spawning season. There was a negative relationship between female GSI values and oil exposure in Dolly Varden and yellowfin sole collected more than a year after the *Exxon Valdez* oil spill (Sol et al. 2000), also suggesting long-term effects on reproductive parameters. This is similar to GSI patterns seen in both Mississippi and Louisiana spotted seatrout during the first half of the reproductive season a year after the DWH oil spill. Thus, the data presented here support earlier speculations that sublethal exposure to oil spills has population-level impacts that can be manifested in effects on fish reproduction a year or more after the spill (Peterson et al. 2003; Whitehead et al. 2011).

The results presented here show reproductive impairment in spotted seatrout measured on several levels. Unfortunately, no fecundity measurements were obtained during the 2011 sampling, so we are unable to evaluate the effect of the DWH oil spill on this important metric. Most significantly, the dramatic reduction in spawning frequency in post-DWH spill spotted seatrout in both Louisiana and Mississippi suggests reduced year class strength for the 2011 cohort. A spawning frequency of once or twice per month, as calculated here for 2011, compared with the normally expected GOM spawning frequency of six to seven times per month (Brown-Peterson et al. 2002) suggests a potential impact on spotted seatrout recruitment into Mississippi and Louisiana estuaries. The ultimate impact of reduced reproductive activity on recruitment will be mediated by larval and juvenile survival, but the impact of PAH exposure on these variables has not been evaluated. Although it appears that spotted seatrout reproductive parameters were returning to prespill levels by July in Mississippi, the effective reproductive season was reduced from the normal 6 months to 3 months after the DWH oil spill. It is important to note that these effects were observed 1 year after the spill. Young Pacific herring exposed to the *Exxon Valdez* oil spill exhibited significant reproductive impairment 3 years postspill

(Kocan et al. 1996), suggesting that oil-exposed fish may continue to show reproductive impairment for several years after a major oil spill.

Contribution by age-1 spotted seatrout to the reproductive population in 2011 was lower than observed prior to the DWH spill, which may be correlated to their exposure to oil as juveniles. Although there was no difference in community composition of juvenile fishes in seagrass meadows immediately after the DWH oil spill (Fodrie and Heck 2011), there are few seagrass meadow nursery areas along the Mississippi and Louisiana coast. Marshes serve as the primary nursery areas along this section of the northern GOM, and some areas remained impacted by the DWH oil spill 1.5 years postspill (Misha et al. 2012; Silliman et al. 2012). Laboratory studies of larval and juvenile spotted seatrout exposed to dispersed oil showed no differences in mortality but did demonstrate significantly decreased growth after 96 hours of exposure compared to controls (Brewton 2012). While these results demonstrate the resiliency of young spotted seatrout, mortality during the larval and juvenile stage is known to be size-dependent in fishes (Gislasen et al. 2010) and even a slight impact on growth as juveniles may influence cohort size and the timing of sexual maturation with concomitant effects on population structure (Pecquerie et al. 2009).

Our data suggest that oil released from the DWH platform may have produced severe effects on the reproductive capacity and potentially population structure of spotted sea-trout. The amount of bioavailable PAHs along the Louisiana coastline was eight times greater than that received along the Mississippi coastline during June and July 2010 (Allan et al. 2012). Furthermore, PAH concentrations in Mississippi waters were at pre-DWH spill levels by August 2010, whereas PAH concentrations did not reach prespill levels in Louisiana waters until March 2011 (Allan et al. 2012). However, despite these differences in PAH concentrations and duration, the reproductive impairment observed in spotted seatrout was similar between Mississippi and Louisiana. Therefore, even a relatively short-term, low-level exposure to oiled waters may have significant impacts on the reproductive capability of fishes a year later. A large diversity of fishes, including many commercially and recreationally important species in families such as Lutjanidae, Sciaenidae, and Scombridae use the northern GOM as their spawning grounds. Importantly, the northern GOM is the only known spawning ground for bluefin tuna (*Thunnus thynnus*); impacts of the DWH spill on eggs and larvae of this species have been investigated (Muhling et al. 2012), but there is no research on potential impacts to the adult spawning stocks. Although all species may not exhibit the reproductive impairment demonstrated in spotted seatrout, the implications for reduced spawning, and thus decreased stock sizes, are enormous. Continued monitoring and assessment of reproductive impacts and stock assessments will be necessary for a more thorough understanding of the long-term impacts of the DWH oil spill. Fourteen years after the *Exxon Valdez* oil spill the full impact of delayed, chronic, and indirect effects resulted in a new paradigm for understanding the consequences of large spills (Peterson et al. 2003); research following the DWH oil spill should add substantially to that knowledge.

Acknowledgments

We thank D. L. Nieland, Louisiana State University, for sharing data from the 1995 Louisiana spotted seatrout collections. Numerous people helped with sampling in 2011, including personnel from the LDWF (R. Boothe, C. Edds, S. Dartez, E. Newman,

B. Hardcastle, C. Davis, M. Tumlin, M. Fischer, and S. Maillian) and the University of Southern Mississippi GCRL (W. Dempster, Z. Olsen, J. Dieterich, J. Green, C. Somerset, T. Albaret, M. Andres, C. Matten, P. Gillam, B. Ennis, and L. Antoni). We thank L. Bustamante, Texas A&M Integrated Veterinary Pathology Laboratory, for histological processing. Water quality data for January–March was provided by C. Butler, E. Anderson, and R. Hendon (GCRL) for Mississippi sampling sites and by G. Thomas and N. Smith (LDWF) for Louisiana sampling sites. This project was supported by a grant from the Northern Gulf Institute.

References

Allan, S. E, B. W. Smith, and K. A. Anderson. 2012. Impact of the *Deepwater Horizon* Oil Spill on bioavailable polycyclic aromatic hydrocarbons in Gulf of Mexico coastal waters. *Environmental Science and Technology* 46:2033–2039.

Bedee, C. D., D. A. DeVries, S. A. Bortone, and C. L. Palmer. 2003. Estuary-specific age and growth of spotted seatrout in the northern Gulf of Mexico. In *Biology of the Spotted Seatrout*, ed. S. A. Bortone, 57–77. Boca Raton, FL: CRC Press.

Bortone, S. A. 2003. Spotted seatrout as a potential indicator of estuarine conditions. In *Biology of the Spotted Seatrout*, ed. S. A. Bortone, 297–300. Boca Raton, FL: CRC Press.

Brewton, R. A. 2012. Gene expression and growth as indicators of effects of the BP *Deepwater Horizon* oil spill on *Cynoscion nebulosus*. Master's Thesis, University of Southern Mississippi, Hattiesburg.

Brown-Peterson, N. J. 2003. The reproductive biology of spotted seatrout. In *Biology of the Spotted Seatrout*, ed. S. A. Bortone, 99–133. Boca Raton, FL: CRC Press.

Brown-Peterson, N. J., M. S. Peterson, D. L. Nieland, M. D. Murphy, R. G. Taylor, and J. R. Warren. 2002. Reproductive biology of female spotted seatrout, *Cynoscion nebulosus*, in the Gulf of Mexico: Differences among estuaries? *Environmental Biology of Fishes* 63:405–415.

Brown-Peterson, N. J. and J. W. Warren. 2001. The reproductive biology of spotted seatrout, *Cynoscion nebulosus*, along the Mississippi Gulf coast. *Gulf of Mexico Science* 2001:61–73.

Brown-Peterson, N. J., D. M. Wyanski, F. Saborido-Rey, B. J. Macewicz, and S. K. Lowerre-Barbieri. 2011. A standardized terminology for describing reproductive development in fishes. *Marine and Coastal Fisheries* 3:52–70.

Bugel, S. M., L. A. White, and K. R. Cooper. 2010. Impaired reproductive health of killifish (*Fundulus heteroclitus*) inhabiting Newark Bay, NJ, a chronically contaminated estuary. *Aquatic Toxicology* 96:182–193.

Fodrie, F. J. and K. L. Heck, Jr. 2011. Response of coastal fishes to the Gulf of Mexico oil disaster. *PLoS One* 6(7):e21609. doi:10.1371/journal.pone.0021609.

Gislason, H., N. Daan, J. C. Rice, and J. G. Pope. 2010. Size, growth, temperature and the natural mortality of marine fish. *Fish and Fisheries* 11:149–158.

Helser, T. E., R. E. Condrey, and J. P. Geaghan. 1993. Spotted seatrout distribution in four coastal Louisiana estuaries. *Transactions of the American Fisheries Society* 122:99–111.

Hunter, J. R. and B. J. Macewicz. 1985. Rates of atresia in the ovary of captive and wild northern anchovy, *Engraulis mordax. U.S. Fishery Bulletin* 83:119–136.

Kocan, R. M., G. D. Marty, M. S. Okihiro, E. D. Brown, and T. T. Baker. 1996. Reproductive success and histopathology of individual Prince William Sound Pacific herring 3 years after the *Exxon Valdez* oil spill. *Canadian Journal of Fisheries and Aquatic Sciences* 53:2388–2393.

Misha, D. R., H. J. Cho, S. Ghosh, A. Fox, C. Downs, P. B. T. Merani, et al. 2012. Post-spill state of the marsh: Remote estimation of the ecological impact of the Gulf of Mexico oil spill on Louisiana salt marshes. *Remote Sensing of the Environment* 118:176–185.

Muhling, B. A., M. A. Roffer, J. T. Lamkin, G. W. Ingram Jr., M. A. Upton, G. Gawlikowski, et al. 2012. Overlap between Atlantic bluefin tuna spawning grounds and observed *Deepwater Horizon* surface oil in the northern Gulf of Mexico. *Marine Pollution Bulletin*. 64(4):679–687. doi:10.1016/j.marpolbul.2012.01.034.

Nicolas, J.-M. 1999. Vitellogenesis in fish and the effects of polycyclic aromatic hydrocarbon contaminants. *Aquatic Toxicology* 44:77–90.

Nieland, D. L., R. G. Thomas, and C. A. Wilson. 2002. Age, growth, and reproduction of spotted seatrout in Barataria Bay, Louisiana. *Transactions of the American Fisheries Society* 131:245–259.

Pecquerie, L., P. Petitgas, and S. A. L. M. Kooijman. 2009. Modeling fish growth and reproduction in the context of the Dynamic Energy Budget theory to predict environmental impact on anchovy spawning duration. *Journal of Sea Research* 62:93–105.

Peterson, C. H., S. D. Rice, J. W. Short, D. Esler, J. L. Bodkin, B. E. Ballachey, et al. 2003. Long term ecosystem response to the *Exxon Valdez* oil spill. *Science* 302:2082–2086.

Silliman, B. R., J.van de Koppel, M. W. McCoy, J. Diller, G. N. Kasozi, K. Earl, et al. 2012. Degradation and resilience in Louisiana salt marshes after the BP–*Deepwater Horizon* oil spill. *Proceedings of the National Academy of Sciences of the United States of America*. doi:10.1073/pnas.1204922109.

Sokal, R. R. and F. J. Rohlf. 1995. *Biometry: The Principles and Practice of Statistics in Biological Research*, 3rd edition. New York: W. H. Freeman.

Sol, S. Y., L. L. Johnson, B. H. Horness, and T. K. Collier. 2000. Relationship between oil exposure and reproductive parameters in fish collected following the *Exxon Valdez* oil spill. *Marine Pollution Bulletin* 40:1139–1147.

Stott, G. G., W. E. Haensly, J. M. Neff, and J. R. Sharp. 1983. Histopathologic survey of ovaries of plaice, *Pleuronectes platessa* L., from Aber Wrac'h and Aber Benoit, Brittany, France: Long-term effects of the *Amoco Cadiz* crude oil spill. *Journal of Fish Diseases* 6:429–437.

Thomas, P. and L. Budiantara. 1995. Reproductive life history stages sensitive to oil and naphthalene in Atlantic croaker. *Marine Environmental Research* 39:147–150.

Whitehead, A., B. Dubanski, C. Bodinier, T. I Garcia, S. Miles, C. Pilley, et al. 2011. Genomic and physiological footprint of the *Deepwater Horizon* oil spill on resident marsh fishes. *Proceedings of the National Academy of Sciences of the United States of America*. 109(50):20298–20302. doi:10.1073/pnas.1109545108.

Wiens, J. A. and K. R. Parker. 1995. Analyzing the effects of accidental environmental impacts: Approaches and assumptions. *Ecological Applications* 5:1069–1083.

13

Impacts of the Deepwater Horizon *Oil Spill on Blue Crab,* Callinectes sapidus, *Larval Settlement in Mississippi*

Richard S. Fulford, Robert J. Griffit, Nancy J. Brown-Peterson, Harriet Perry, and Guillermo Sanchez-Rubio

CONTENTS

Introduction

Accidental oil spills resulting from the extraction and transportation of petroleum products unfortunately are common events, and can have strong and long-lasting effects on coastal marine ecosystems (Carls et al. 2000; Neuparth et al. 2012). These include direct mortality of ecosystem component species (McGurk and Brown 1996) as well as indirect effects such as habitat loss (Roth and Baltz 2009; McCall and Pennings 2012), interference with reproduction (Kocan et al. 1996; Brown-Peterson et al., Chapter 12, this volume), and reduced individual and population growth (Carls et al. 2002; Pearson et al. 2012). The evaluation of ecosystem response and recovery from an oil spill is complex because spill impacts are only one of a myriad of stressors affecting an ecosystem, and because the vulnerability of individual ecosystem components to oil will vary based on biology, diet, mobility, habitat, and timing (Harwell et al. 2010). Understanding ecosystem-level effects therefore requires an understanding of population-level effects for important ecosystem species and this requires a comprehensive approach involving field and laboratory analyses utilizing both rapid response and long-term monitoring frameworks (Peterson et al. 1996; Carls et al. 2002). In this chapter, we examine potential effects of the *Deepwater Horizon (DWH)* oil spill based on rapid response field data collection and laboratory assays with the intent of measuring population-level responses of an economically important marine invertebrate, *Callinectes sapidus*. We also make maximum use of available historical data in a "before–after" analytical framework.

The largest accidental release of crude oil in history occurred in the north-central Gulf of Mexico (GOM) between April 20 and July 15, 2010 (Alford et al., this volume). The DWH spill was unprecedented due to both its magnitude (>600,000 metric tons released) and its occurrence in deeper offshore waters (>1000 m), where the fate of oil is not well understood. Regardless of the debate on the fate of subsurface oil (OSAT 2010), the observed surface slick measured approximately 10,000 km^2 at its maximum extent and was highly likely to have had an impact on near surface planktonic larvae, such as *C. sapidus*, that are typically present in offshore surface waters of the GOM between March and October (Figure 13.1) (Sulkin and VanHeukelem 1986). The *C. sapidus* fishery in the GOM was valued at more than $45 million in 2009 and is the fourth largest

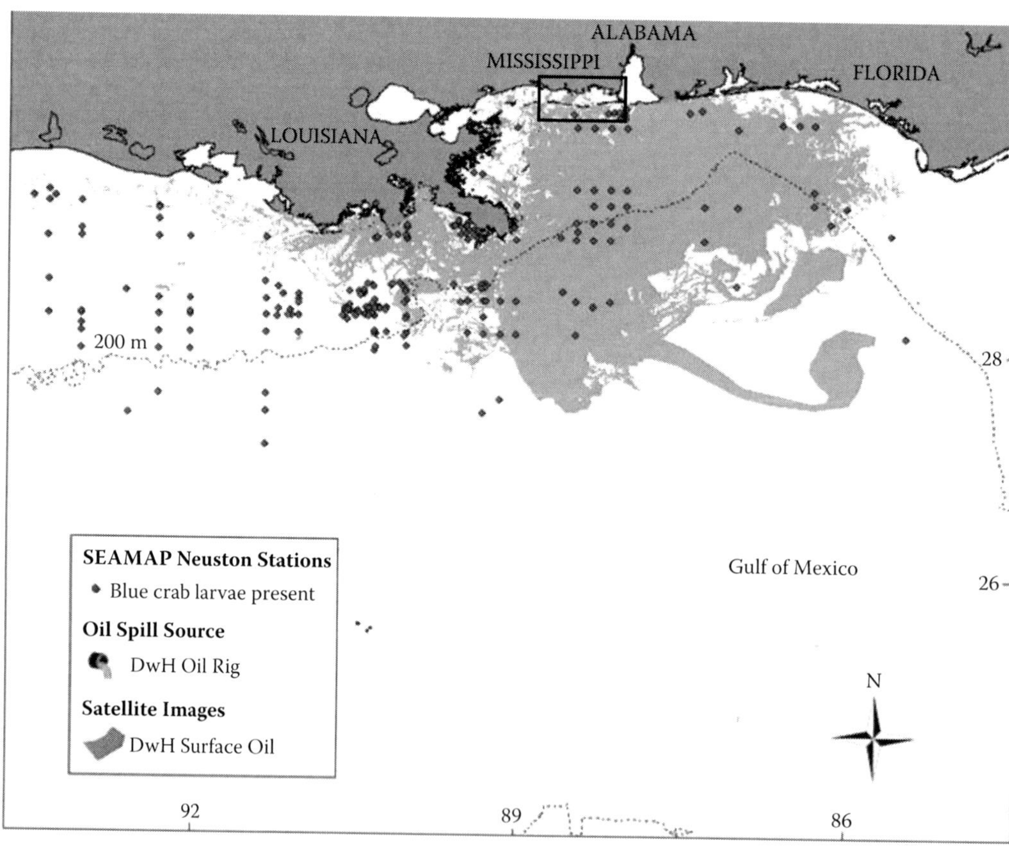

FIGURE 13.1
(See color insert.) Map showing largest extent of the *Deepwater Horizon* oil spill April 20-July 15, 2010 (from National Oceanic and Atmospheric Administration [NOAA]) overlaid with distribution of *Callinectes sapidus* larvae in the summer based on SEAMAP sampling (from NOAA). The area used in this study for measurement of daily settlement is indicated by the black rectangle.

fishery in the GOM in terms of value (National Marine Fisheries Service landings data; st.nmfs.noaa.gov/st1/commercial/index.html). An understanding of oil spill effects on *C. sapidus* recruitment may have economic implications. More generally, the effects of the DWH spill on *C. sapidus* larval survival are a potentially important indicator of both impacts of surface oil on invertebrate larvae and effects of the DWH spill on future fishery production.

Measuring or predicting population-level effects of an oil spill is highly complex due to the myriad of potential ecological drivers that may impact population trajectories. For example, *C. sapidus* population size is primarily driven by three major factors: fishery harvest, annual larval recruitment, and postsettlement biotic factors (predation) associated with habitat structure and suitability (e.g., salt marsh) (Guillory et al. 2001; Bunnell and Miller 2005). All three were potentially affected by the DWH spill, particularly harvest, which was stopped for over 6 months in response to the spill. Realized effects will be a combination of specific effects on all three factors and only by partitioning them out can we draw inferences about which factors are most important.

Analysis of factors affecting recruitment dynamics in marine planktonic larvae is complicated by the naturally high interannual variability in survival (Gaines and Roughgarden 1985; Heck et al. 2001). Larval *C. sapidus* are released as eggs in offshore waters and drift for 20–40 days as they metamorphose through seven zoeal stages (Epifanio et al. 1984). Later-stage larvae begin to migrate inshore before the final transition to megalopae, which settle into nearshore benthic habitat. The settlement rate of megalopae is a useful indicator of the importance of disturbance events during the marine pelagic stage because larval survival is correlated with the megalopal settlement rate (Perry et al. 1995; Rabalais et al. 1995). Megalopae are relatively easy to sample using passive collectors, and a valuable historical dataset exists for settlement patterns in previous years (Perry et al. 1995; H. Perry, unpublished data). Megalopal settlement patterns are a potentially valuable tool for examining the immediate effects of the DWH spill, after accounting for background settlement variation in the analysis.

Pelagic larvae, such as *C. sapidus*, are exposed to a wide variety of environmental conditions and have a mortality rate typically in excess of 99% before inshore settlement (Heck and Coen 1995; Bunnell and Miller 2005). During life stages with such a high mortality rate, relatively small changes in mortality (<1%) can yield large differences in the number of megalopae that recruit to nearshore habitat. Megalopal settlement rates in the GOM are highly episodic with 3- to 5-day peaks separated by indeterminate periods with little or no settlement, but settlement does follow a predictable seasonal pattern with a small peak in May and a larger peak in August–September (Perry et al. 1995). Interannual variability in settlement rate is high with order of magnitude differences in settlement rate among years (H. Perry, unpublished data). Settlement is also highly spatially variable with large differences in both timing and magnitude of peaks at large (>100 km) (Rabalais et al. 1995) and small (<10 km) (Perry et al. 1995) spatial scales, although spatial patterns are consistent between years. Our study takes advantage of an existing time series of daily megalopal settlement (1991–1997) to compare settlement rates of larvae in 2010 to years before the DWH spill at the same sites. Field data analysis was combined with laboratory assays of the effect of crude oil and dispersant on *C. sapidus* larval mortality to examine association of settlement patterns with the oil spill in time and estimate the probable effects of oil/dispersant exposure on larval survival. The objective was a rapid assessment of short-term survival of the larval stage of *C. sapidus* that should be useful for estimating population-level effects following the DWH spill.

Methods

Laboratory Preparation and Analysis of Oil/Dispersant Solutions

Crude source oil obtained from the DWH riser on May 22, 2010 (CAS # 8002-05-9) was received in the laboratory on August 19, 2010 and stored in amber bottles at room temperature in the dark. Dispersant (NalcoVX9831, lot 071210-1BB), hereafter referred to as Corexit, was received from Nalco Holding Company on August 24, 2010 and stored in an amber bottle at room temperature in the dark. A chemically enhanced water-accommodated fraction (CEWAF) solution was prepared by stirring 2.4 mL crude oil plus 200 μL of Corexit in 2 L of artificial sea water (ASW) in an aspirator bottle in the dark at room temperature for 18–24 hours; the stir vortex was about 25% of the depth of the aspirator bottle. The solution was allowed to settle for 3–6 hours in the dark, then the lower layer was drained, stored in glass containers in the dark at room temperature, and used within 48 hours. The nominal concentration of the CEWAF solution was 1 g/L. A dispersant-only stock solution was prepared by mixing 100 μL of Corexit with 1 L of ASW for 5 minutes at room temperature, which produced a dispersant-only solution of 100 mg/L and was used to test the effects of dispersant on crab larvae.

Polycyclic aromatic hydrocarbon (PAH) analyses were conducted on water samples from the CEWAF stock preparation by using modifications to previously described methods (Spier et al. 2011). Duplicate water samples (500 mL) were transferred to precleaned separatory funnels, spiked with surrogate standards (1,4-dichlorobenzene-d_4, naphthalene-d_8, acenaphthene-d_{10}, phenanthrene-d_{10}, chrysene-d_{12}, perylene-d_{12}, PCB 30, PCB 65, PCB 204, 1,1′-binaphthyl, perinaphthenone, and 1-chlorooctadecane; Ultra Scientific, Kingston, Rhode Island and AccuStandard, New Haven, Connecticut), and extracted three times with 40 mL of dichloromethane (Burdick & Jackson, Muskegon, Michigan). Each set of extractions also contained a laboratory blank of deionized water spiked with surrogate standards. The combined extract volume was reduced under a gentle stream of nitrogen using a TurboVap® evaporator (Zymark Corp., Hopkinton, Massachusetts) and the internal standard p-terphenyl (Chem Service, West Chester, Pennsylvania) was added. The extracts were analyzed for selected PAHs on a Varian 3800 Gas Chromatograph (GC) using a Varian CP-8400 Autosampler coupled to a Saturn 2000 GC/MS/MS ion trap mass spectrometer (Varian Inc., Walnut Creek, California) operated in electron ionization mode (70 eV). It was equipped with a split/splitless injector maintained at 320°C. The carrier gas was helium and injections were made in splitless mode on a DB5, 60 m × 0.32 mm × 0.25 μm film thickness capillary column from J&W Scientific (Folsom, California). The GC temperature program was 75°C –320°C at 4°C/min with an initial hold of 1 minute. The MS trap, manifold, and the transfer line temperatures were 220°C, 80°C, and 320°C, respectively. Scans were 100–500 m/z for 8–49 minutes and then 100–650 m/z from 49 to 62.25 minutes; selected ions were used to quantify the targeted analytes. Seven-point calibration curves were prepared for the individual analytes (0.1–20 μg/mL, Aldrich, Milwaukee, Wisconsin) and the internal standard, p-terphenyl (Chem Service, West Chester, Pennsylvania). These calibration samples and all experimental sample analyses of individual analytes were performed using the Varian MS Workstation software package, version 6.8 (Varian Inc., Walnut Creek, California).

Crab Culture and Exposures

Crab zoea and megalopae for experiments were obtained from ovigorous *C. sapidus* females captured in oil-free areas of the Mississippi Sound and spawned under controlled laboratory conditions (25°C, salinity 28, 14L:10D photoperiod). In culture, larval crabs were fed a mixture of rotifers (*Branchionus rotundiformis* ss) and instant algae (*Nanochloropsis*, *Isochrysis*) twice daily, regardless of crab stage. Once zoea reached stage Z4, newly hatched brine shrimp (*Artemia*) nauplii were added to the feeding schedule once daily. At stage Z5, both newly hatched and lipid-enriched *Artemia* were fed once daily and frozen copepods (CYCLOP-EEZE) were added to the diet with twice daily feedings.

Zoea in stages Z3–Z4 and newly transformed megalopae were tested in static exposures with four log-order concentrations of CEWAF for 96 hours (10, 100, 1,000, and 10,000 µg/L nominal concentration). Tests included an ASW-only control, as well as a dispersant-only control (megalopae only), and were performed in a dark incubator at 25°C, salinity 28 ppt. No water changes were done during the exposures, and megalopae were not fed. Zoea were fed *Artemia* nauplii once 48 hours following the onset of the experiment.

Zoea-stage trials were performed in covered 150 mL glass crystallizing dishes containing 100 mL of CEWAF solution. Three replicates of 10 zoea each were tested for each concentration. Megalopae-stage trials were conducted in 20 mL capped scintillation vials containing 10 mL of CEWAF solution. Ten replicates of one megalopae each were tested for each concentration. All zoea and megalopae were visually assessed daily under a dissecting microscope to document mortality; dead crabs were removed from the test containers. The resulting data consisted of proportional survival at each stage and CEWAF concentration.

Field Data Collection

Daily megalopal settlement rate was measured at four sites along the Mississippi Gulf Coast (Figure 13.2) in a manner identical to collection of historical settlement data (Perry et al. 1995). Sample sites were all piers within 30 m of the shoreline at a maximum total depth of 3 m. Settlement collectors comprised of synthetic filter material (0.25 m^2) wrapped around a PVC pipe were suspended in the water at a depth of 0.5 m. The collectors were deployed in triplicate at each site and were retrieved and replaced every 24 hours. Retrieved collectors were returned to the laboratory in individual sealed containers and thoroughly rinsed through a 0.5 mm sieve to retain all megalopae present. The sieved material was preserved in 10% buffered formalin then examined under 4X magnification. *Callinectes* megalopae were separated into *C. sapidus* and *C. similis* and counted. Daily settlement was calculated as the mean settlement count across triplicate collectors for each site-day.

Data Analysis

Differences in survival at 96 hours between CEWAF concentrations and the control treatments (ASW, dispersant only) were examined with a χ^2 goodness of fit test employing the control outcome as the expected frequency of survivors. If a significant result was obtained, further analysis of specific differences was conducted with a partition of the χ^2 (Zar 2010). These tests were conducted separately for zoea and megalopal stages and all statistical tests were conducted at a Type-I error rate of 0.05. A 96-hour LC_{50} (lethal concentration resulting in 50% mortality after 96 hours) was also calculated with the trimmed Spearman–Karber test (Hamilton et al. 1977).

Daily settlement data collected in 2010 were combined for analysis with complementary data from the same six sites collected from 1991 to 1997. Mean daily settlement was compared by collection month, year, and sample site. Differences in mean daily settlement rate among months (within year) and years were examined with an analysis of variance (ANOVA). Differences among sites were tested in an ANOVA but only for years

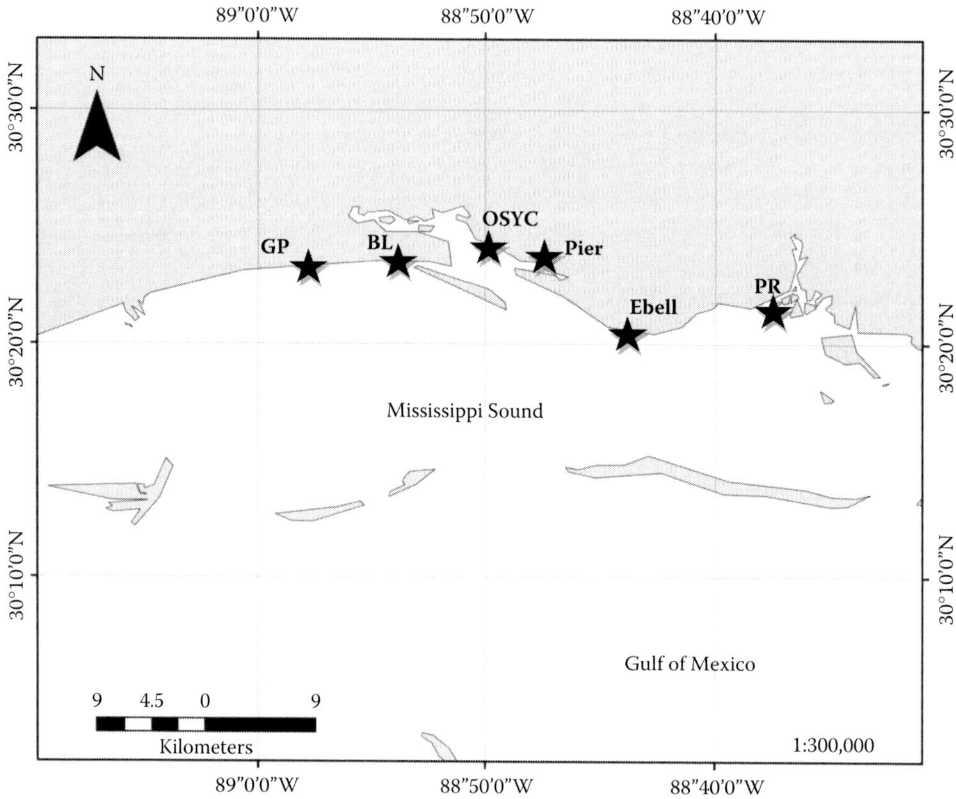

FIGURE 13.2
Map showing six sample sites along the Mississippi coast used for measurement of daily settlement of *Callinectes sapidus* megalopae. Site names are from left to right: Gulfport (GP), Biloxi (BL), Ocean Springs Yacht Club (OSYC), GCRL pier (Pier), East Bellfountaine (Ebell), and Pascagoula River (PR).

with sufficient data across all sites. Data were tested for normality and homogeneity of variance and if assumptions could not be verified analyses were conducted with a Kruskal–Wallis test. All statistical analyses of field data were conducted at a Type-I error rate of 0.05. Years in the data series were also identified with respect to Loop Current incursion (yes/no; Johnson and Perry 1999; Perry et al. 2003) and multidecadal climatic cycles (wet/dry) identified by Sanchez-Rubio et al. (2011a). Years were assigned to a category either in the cited references or based on consultation with the authors. Differences in mean daily settlement rate among years were interpreted with respect to differences in these two factors.

Daily settlement data were compared to previous years with spectral time series analysis. Time series analysis was conducted with an emphasis on comparison between definable year types (climate and oil). Daily settlement pattern was characterized based on dominant spectral frequencies from a discrete Fourier transform of daily settlement data and a spectral model selection based on Bayesian information criterion (BIC) (Shumway and Stoffer 2011). The outcome of model selection was summarized with a partial autocorrelation function (PACF) for each year to measure differences in the relative contribution of the dominant frequencies among years. Time series analysis was conducted in the R programming language (http://www.r-project.org/) with the package "tsa3" (Shumway and Stoffer 2011).

Results

Chemically Enhanced Water-Accommodated Fraction Characterization

Analysis of the prepared oil stock solution showed the presence of 124 different PAH peaks. PAH analytes with the highest concentration included C3-benzene, 2-methyl napthalene, and 1,3-dimethyl naphthalene (Table 13.1). No PAHs were detected in ASW control water or a 10% Corexit-only solution. Total PAH concentration of all analytes in the nominal 1 g/L stock solution was 9.977 ± 0.267 mg/L for CEWAF.

Laboratory Assays of Oil Dispersant Exposure

Sensitivity of *C. sapidus* to oil and dispersant varied by lifestage, but both displayed increased mortality as oil concentration increased (Figure 13.3). The zoeal stage (Z3–Z4) displayed a significant increase in 96-hour mortality at 10% and 100% of the stock CEWAF (Figure 13.3a, $\chi^2_2 = 13.9$, $p = .007$). On the basis of our quantified PAH data, this is estimated to be equivalent to 1.0 and 10 mg/L ΣPAH, respectively. Mortality at lower concentrations was not significantly different than the control ($\chi^2_2 = 1.3$, $p = .52$). The difference in mortality was evident as early as 24 hours into the trials, but variability was quite high at earlier time points. Megalopal mortality at 96 hours was only significantly different from the ASW control at the highest concentration tested (10% CEWAF stock, ~1 mg/L [Figure 13.3b]; $\chi^2_1 = 5.4$, $p = .02$). The dispersant-only treatment was not significantly different than the ASW control ($\chi^2_1 = 0.1$, $p = .74$). The 96-hour LC_{50} was 315 μg/L (95% CL not calculable) and 269 μg/L (95% CL—186–389) for zoea and megalopae, respectively.

Natural Settlement Rate and Pattern

Daily settlement of *C. sapidus* megalopae displayed seasonal, as well as spatial and interannual, patterns (Figure 13.4). Mean daily settlement was significantly higher in August and September across all study years (Figure 13.4a) ($F_{3,803} = 48.4$, $p < .0001$) indicating a late summer settlement peak. Sites located near the mouth of Biloxi Bay (Gulfport and Biloxi, Figure 13.4b) also displayed significantly higher mean daily settlement than sites either east of Biloxi Bay and sites deeper within the Bay (1996: $F_{3,207} = 4.8$, $p < .0016$; 2010: $F_{3,410} = 44.8$, $p < .0001$). Daily settlement rate was different among years (Figure 13.4a) ($F_{7,803} = 41.9$, $p < .0001$), and 1991 had significantly higher mean daily settlement than all other years examined (Tukey; $p < .0001$). Mean daily settlement from 1993 to 1995 was lower than in 1991, but also significantly different (Tukey; $p < .0001$) from daily settlement in 1996, 1997, and 2010; 1996 had the lowest mean daily settlement of all years examined.

Three year-group comparisons were examined as a part of the before–after analysis of mean daily settlement rate. The comparison of 2010 to all other years resulted in no significant difference (*t*-test, $p = .09$). Mean daily settlement in 2010 was most similar to more recent years in the historical dataset (1996, 1997). Second, a comparison of years before and after 1995 based on Sanchez-Rubio et al. (2011b) did yield a significant difference (*t*-test, $p < .0001$) indicating a threshold in 1995 (Figure 13.4a). Finally, years reported as strong Loop Current incursion years (1992, 1993, 1996, and 1997) by Perry et al. (2003) were compared to all other years to test for an association of annual settlement with this more episodic climatic variable. They reported a potential negative association of annual settlement with the timing of Loop Current incursion, particularly in August. No significant

TABLE 13.1

Concentration (μg/L) of PAH Analytes in Two Replicate Samples of Corexit and CEWAF Stock Solution Used in Laboratory Experiments

Peak Name	Corexit #1	Corexit #2	CEWAF #1	CEWAF #2
C3-Benzene	0	0	1,076.3	1,073.4
C3-Benzene	0	0	327.4	321.5
C4-Benzene	0	0	263.6	260.2
C4-Benzene	0	0	126.7	126.6
C4-Benzene	0.1	0.1	183.1	184.6
C4-Benzene	0	0	127.7	125.8
C4-Benzene	0.2	0	104.3	112.1
C4-Benzene	0.1	0.1	142.6	134.8
C4-Benzene	0.2	0.2	172.4	171.7
1,2,4,5-Tetramethyl benzene	0	0	159.1	115.0
C4-Benzene	0	0	92.7	158.1
C5-Benzene	0	0	109.8	102.8
1,2,3,4-Tetramethyl benzene	0	0	126.8	126.4
Naphthalene	0	0	260.5	250.4
2-Methyl napthalene	0	0	474.2	508.2
1-Methyl napthalene	0	0	335.2	346.4
2-Ethylnaphthalene	0	0	83.7	97.9
2,6-Dimethylnapthalene	0	0	318.2	336.8
1,3-Dimethylnaphthalene	0	0	344.9	377.9
C2-Naphthalene	0	0	228.2	245.6
2,3- and 1,4-Dimethylnaphthalene	0	0	129.2	139.1
3-Methyl biphenyl	0	0	159.3	143.0
C3-Naphthalene	0	0	176.0	150.7
C3-Naphthalene	0	0	240.1	215.9
C3-Naphthalene	0	0	281.8	248.5
C3-Naphthalene	0	0	229.0	208.4
C3-Naphthalene	0	0	198.2	174.4
2,3,5,-Trimethylnapthalene	0	0	290.0	252.5
Phenanthrene	0	0	134.4	106.6
2-Methyl phenanthrene	0	0	122.4	97.1
4-Methyl phenanthrene	0	0	141.9	115.0
C2-178	0	0	209.7	159.4
Sum of analytes (μg/L)	0.6	0.4	10,166	9,788

Note: Only analytes with concentration >1% of total PAH concentration are listed.

differences were detected for this variable in our dataset (ANOVA, $p = .1$) independent of the pre–post 1995 settlement differences.

Analysis of smoothed periodigrams and partial autocorrelation functions by year indicate two dominant frequencies in pre-2010 years (Figure 13.5a–g) based on BIC scores. The data are summarized in Figure 13.5 in units of period (approximate days between settlement pulses) but were analyzed with respect to frequency patterns (proportion of total

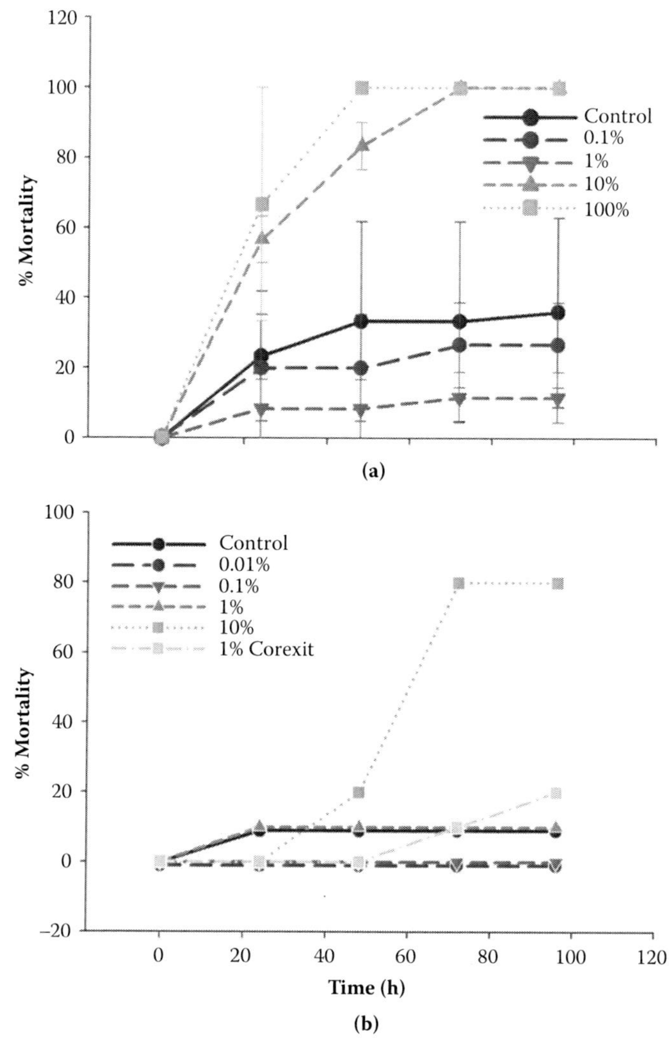

FIGURE 13.3
Ninety-six-hour mortality (±SE, zoea only) trajectories for (a) *Callinectes sapidus* zoea and (b) megalopae exposed to four concentrations of the chemically enhanced water-accommodated fraction (CEWAF) of oil and dispersant, as well as dispersant only (megalopae only) and a negative control. Stock CEWAF concentration was measured at 9.977 ± 0.267 mg/L and treatments are described as a percent dilution from stock. Control and lowest concentration (0.01%) offset by 1 in lower panel for clarity.

time series), which is the inverse of period. The first was near a frequency of 0.01 (100 days) and indicates the monotonic seasonal trend that peaks in August–September. The seasonal frequency is not shown in Figure 13.5, but can be seen in the monthly means shown in Figure 13.4. The second dominant frequency was near a mean of 0.13, which is consistent with a pulsed settlement pattern with a period of approximately 8 days. The observed pattern in 2010 also had two dominant frequencies although position and strength varied from previous years. The seasonal frequency was present but not dominant as in previous years. In 2010, the dominant frequency was 0.06 or a period of 17 days between settlement

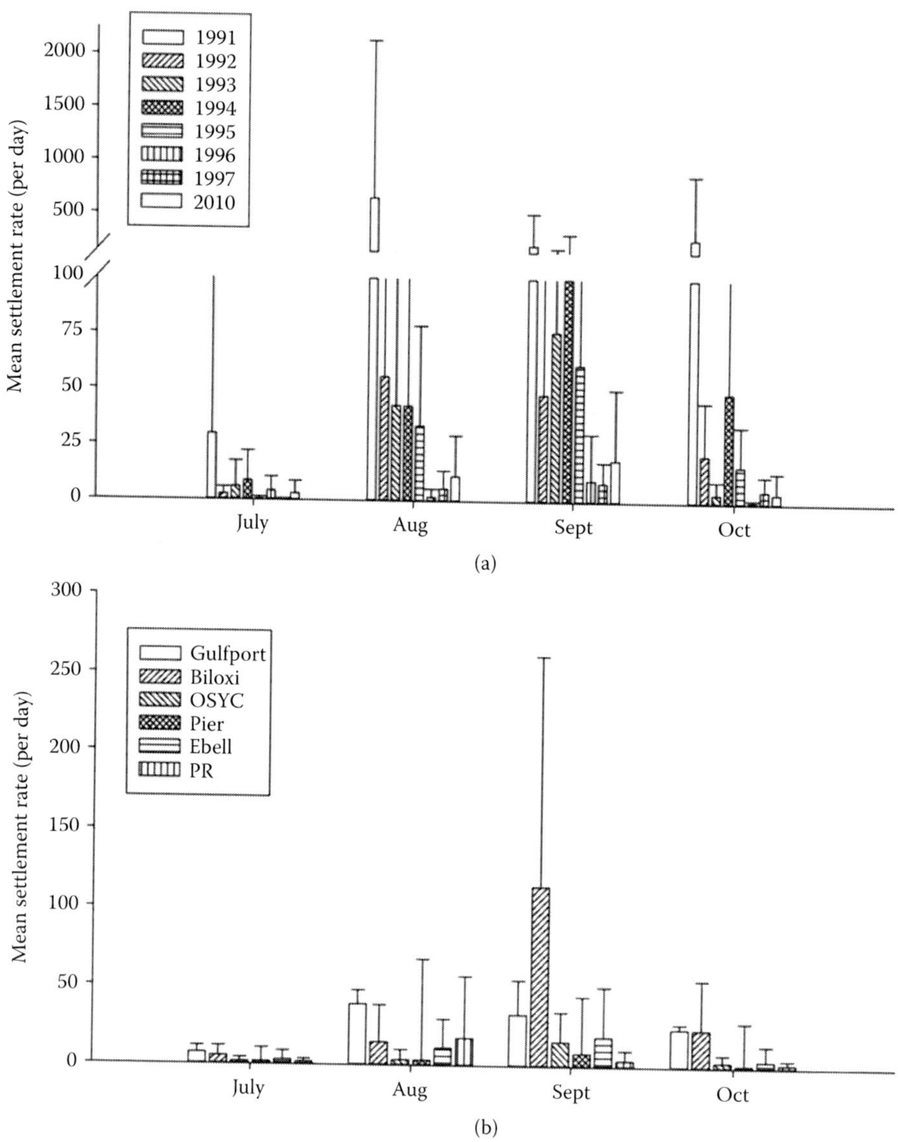

FIGURE 13.4
Mean settlement (per day ± SE) by (a) year and (b) site for *Callinectes sapidus* megalopae and Z7 zoea in coastal Mississippi. Site names are Pascagoula River (PR), East Bellfountaine (Ebell), Biloxi, Gulfport, GCRL pier (Pier), and Ocean Springs Yacht Club (OSYC).

pulses. Only 1997 had a similar settlement pattern to 2010 (Figure 13.5g), with a longer period between megalopal settlement pulses. In addition to the dominant frequencies in each year, analysis of the PACF shows the short-term autocorrelation indicative of the length of a single settlement pulse. In most previous years, this short-term autocorrelation was significant for up to 3 days suggesting a settlement pulse lasting this long (Figure 13.5a–g). In contrast, the pulse peak in 2010 was only significant for 1 day (Figure 13.5h) suggesting a shorter mean pulse duration in 2010. A similarly short pulse duration was also observed in 1991 and 1996.

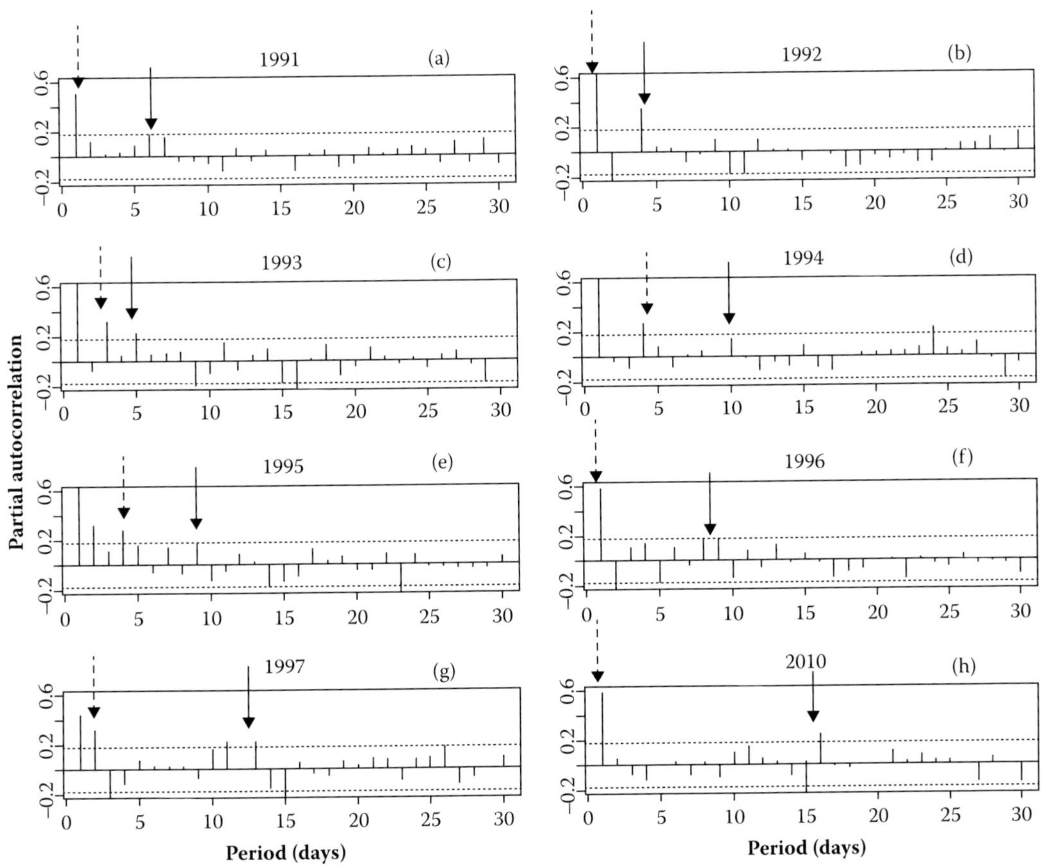

FIGURE 13.5
Partial autocorrelation function for *Callinectes sapidus* daily settlement time series for all sample years (a-1991, b-1992, c-1993, d-1994, e-1995, f-1996, g-1997, and h-2010). Reference lines (dash) indicate 95% confidence interval for partial autocorrelation. The solid arrow indicates the period between megalopal settlement pulses associated with the dominant pulse frequency and the dashed arrow indicates the period associated with the estimated megalopal settlement pulse width. See text for details.

Discussion

A comparison of megalopal settlement in 2010 to previous years suggested a complex set of differences. Total settlement intensity in 2010 was lower than years before 1995 but similar to more recent years, suggesting that known differences in climatic conditions before and after 1995 (Sanchez-Rubio et al. 2011b) are a more likely cause for the reduction in mean daily settlement. Sanchez-Rubio et al. (2011a) observed significant reductions in numbers of juvenile blue crabs in the years following 1995. The pattern of settlement in 2010 was different than all previous years with the exception of 1997 and suggested a weaker seasonal peak, a 5- to 7-day wider gap between settlement pulses, and a 1- to 2-day shorter pulse duration than in previous years. Thus, daily settlement was in line with historical data, but on a seasonal basis less settlement was observed in the study area.

An association between fishery recruitment and climatic cycles has been established for multiple species including *C. sapidus* in Louisiana (Sanchez-Rubio et al. 2011a;

Eriksen et al. 2012; Smart et al. 2012). The primary driver seems to be variation in freshwater flow, but this hypothesis has not been directly tested. Nonetheless, cyclic uncertainty in recruitment of decades or longer is important, and as our analysis suggests, should be accounted for in interpreting data time series. Mean daily settlement in 2010 was low by historical standards but was actually the highest of the three post-1995 years in the dataset, suggesting a slightly higher survival and delivery rate for *C. sapidus* larvae in 2010. The dynamics of Loop Current incursion have also been associated with variation in larval delivery (Johnson and Perry 1999), but we could find no interannual pattern in the current dataset with regard to this variable. Johnson and Perry (1999) reported Loop Current incursion strong enough to negatively impact megalopal settlement in 1992, 1993, 1996, and 1997, but the observed differences in settlement between these years were more strongly related to longer term climatic cycling before and after 1995. Detected differences in larval survival potentially associated with the oil spill would always need to be interpreted in the context of such interacting natural factors. It is important to note that the historical dataset ended in 1997 or 13 years before the DWH spill in 2010. Changes between the historical period and 2010 may be due to other periodic factors or changes in the *C. sapidus* population over time. However, a continuation of the analysis of Sanchez-Rubio et al. (2011a) indicates that climatic cycles have remained relatively stable in the GOM region since the 1995 shift and annual monitoring data for juvenile *C. sapidus* collected in the same region of Mississippi Sound indicates annual recruitment has been low but stable over the same period (H. Perry, unpublished data). Although it is not possible to rule out all potential differences to the system over this 13-year period, this is evidence that the comparison we make here is a reasonable one.

A low level of larval mortality due to PAH exposure in 2010 in the wild is supported by the laboratory assays that show a mortality effect only at PAH concentrations of 1 mg/L or higher. Exposure of both zoea and megalopae to oil concentrations below this level for 96 hours showed no mortality effect, and reported concentrations of surface oil from coastal waters in 2010 were several orders of magnitude lower (maximum reported 0.17 μg/L) (Allen et al. 2012). Allen et al. (2012) collected their data very close to shore rather than in the open water where *C. sapidus* larvae are found during the spring and summer, so some uncertainty exists regarding realized exposure. Concentrations in deeper waters (>1000 m) close to the spill site were high enough to yield an increase in mortality (OSAT 2010), but larvae are not found at that depth (Goodrich et al. 1989). It is hard to state with any certainty the concentrations of oil to which larvae were exposed in 2010, but the results of the laboratory study suggest that acute concentrations above 1 mg/L would be needed to yield a mortality effect. However, laboratory assays only tested exposure in static conditions with no renewal for 96 hours; longer, chronic exposure to lower levels of PAH has been shown to impact embryonic and larval fish (Barron et al. 2004; Hendon et al. 2008). Thus, results from the our exposures suggest direct mortality due to acute exposure to oil was low in 2010, but do not provide a comprehensive estimate of toxicity, particularly for crab larvae exposed to chronic, lower levels of oil. The current research also does not provide data on the potential for exposure to crude or dispersed oil to produce developmental, reproductive, immunological, or behavioral changes in blue crab populations.

Although significant differences in larval survival at realized PAH concentrations were not observed in the laboratory or indicated by observed settlement intensity, differences in settlement pattern between 2010 and previous years are suggestive of an indirect association with larval recruitment. An observed difference in settlement pattern, but not daily settlement rate, suggests that any population-level reduction in total megalopal delivery to the nearshore habitat in 2010 was more strongly related to the larval source than to survival during the larval stage. Perry et al. (1995) compared the periodicity of megalopal settlement to a variety of natural cycles including moon phase, tidal cycles, wind direction, and current direction and magnitude and found only partial association with settlement. Wind, tide, and current conditions thought optimal for larval delivery did not always result in a concomitant settlement pulse. Such discordance between settlement and nearshore physical drivers suggests an underlying periodicity related to egg development and synchrony in the offshore waters before hatching. An increase in the gap between settlement pulses and a decrease in the length of the pulse were both indicated as differences between 2010 and all previous years except 1997. Although no data were collected on exposure of ovigerous females to oil or dispersant, reduced synchrony in egg release or survival is consistent with a change in the pattern but not the intensity of settlement and should be explored.

In general, multiple drivers will impact larval survival and inshore delivery including potential exposure to pollutants such as those released by the DWH spill. Partitioning out the specific impact of a single stressor can be difficult particularly if that stressor acts in synergy with natural variability. The judicious use of historical data, combined with rapid assessments and laboratory experiments, allow us to make an assessment of the association of the DWH oil with patterns in *C. sapidus* megalopal settlement in Mississippi. The data suggest that the vector of effect most consistent with settlement patterns in 2010 was at the larval source, which would have impacted periodicity of settlement without necessarily altering pulse intensity. Such indirect effects were noted as a part of the analysis of the collapse of the Pacific herring fishery following the *Exxon Valdez* Spill in Alaska (Hose et al. 1996; Carls et al. 2002). In that case, direct adult mortality was present but low and it was reduced egg and larval viability that may have interacted with other factors to generate a population-level effect. Most significantly, an interaction with harvest was considered to be an exacerbating factor, and should be a note of caution for management of *C. sapidus* fishery in the next few years. Rapid assessments such as this do not provide clear evidence of an impact from the spill, and determination of a population-level effect will require longer term monitoring. Yet, this study provides important insight in that direct larval exposure to oil in the pelagic zone may not be the most important point of vulnerability for this population. Such insights are critical as the basis for structuring future work intended to investigate the population-level effects of an oil spill on this important marine invertebrate.

Acknowledgments

This work would not have been possible without the support of many people including Dyan Gibson and Bobby Trigg who coordinated the sampling effort and the army of student volunteers from the USM Center for Fisheries Research and Development who assisted them. This work was funded by an NSF RAPID grant. We thank Michael. A. Unger (Virginia Institute of Marine Science) for PAH analysis of the CEWAF stock solution and Idrissa Boube for assistance with the laboratory assays. The views expressed in this chapter are those of the authors and do not necessarily reflect the views or policies of the U.S. Environmental Protection Agency.

References

Allen, S. E., B. W. Smith, and K. A. Anderson. 2012. Impact of the *Deepwater Horizon* oil spill on bioavailable polycyclic aromatic hydrocarbons in Gulf of Mexico coastal waters. *Environmental Science and Technology* 46:2033–2039.

Barron, M. G., M. G. Carls, R. Heintz, and S. D. Rice. 2004. Evaluation of fish early life-stage toxicity models of chronic embryonic exposures to complex polycyclic aromatic hydrocarbon mixtures. *Toxicological Sciences* 78:60–67.

Bunnell, D. B. and T. J. Miller. 2005. An individual-based modeling approach to spawning-potential per recruit models: An application to blue crab (*Callinectes sapidus*) in Chesapeake Bay. *Canadian Journal of Fisheries and Aquatic Sciences* 62:2560–2572.

Carls, M. G., J. E. Hose, R. E. Thomas, and S. D. Rice. 2000. Exposure of Pacific herring to weathered crude oil: Assessing effects on ova. *Environmental Toxicology and Chemistry* 19:1649–1659.

Carls, M. G., G. D. Marty, and J. E. Hose. 2002. Synthesis of the toxicological impacts of the *Exxon Valdez* oil spill on Pacific herring (*Clupea pallasi*) in Prince William Sound, Alaska, U.S.A. *Canadian Journal of Fisheries and Aquatic Sciences* 59:153–172.

Epifanio, C. E., C. C. Valenti, and A. E. Pembroke. 1984. Dispersal and recruitment of the blue crab larvae in the Delaware Bay, USA. *Estuarine Coastal and Shelf Science* 18:1–12.

Eriksen, E., R. Ingvaldsen, J. E. Stiansen, and G. O. Johansen. 2012. Thermal habitat for 0-group fish in the Barents Sea; how climate variability impacts their density, length, and geographic distribution. *ICES Journal of Marine Science* 69:870–879.

Gaines, S. and J. Roughgarden. 1985. Larval settlement rate: A leading determinant of structure in an ecological community of the marine intertidal zone. *Proceedings of the National Academy of Sciences of the United States of America* 82:3707–3711.

Goodrich, D. M., J. Van Montfrans, and R. J. Orth. 1989. Blue crab megalopal influx to Chesapeake Bay: Evidence for a wind driven mechanism. *Estuarine Coastal and Shelf Science* 29:247–260.

Guillory, V., H. Perry, P. Steele, T. Wagner, W. Keithly, B. Pellegrin et al. 2001. *The Blue Crab Fishery of the Gulf of Mexico, United States: A Regional Management Plan*. Ocean Springs: Gulf States Marine Fishery Commission.

Hamilton, M. A., R. C. Russo, and R. V. Thurston. 1977. Trimmed Spearman–Karber method for estimating median lethal concentrations in toxicity bioassays. *Environmental Science and Technology* 11:714–719.

Harwell, M. A., J. H. Gentile, K. W. Cummins, R. C. Highsmith, R. Hilborn, C. P. McRoy et al. 2010. A conceptual model of natural and anthropogenic drivers and their influence on the Prince William Sound, Alaska. *Ecosystem. Human and Ecological Risk Assessment* 16:672–726.

Heck, K. L. and L. D. Coen. 1995. Predation and the abundance of juvenile blue crabs: A comparison of selected east and Gulf coast (USA) studies. *Bulletin of Marine Science* 57:877–883.

Heck, K. L., L. D. Coen, and S. G. Morgan. 2001. Pre- and post-settlement factors as determinants of juvenile blue crab, *Callinectes sapidus*, abundance: Results from the north-central Gulf of Mexico. *Marine Ecology Progress Series* 222:163–176.

Hendon, L. A., E. A. Carlson, S. Manning, and M. Brouwer. 2008. Molecular and developmental effects of exposure to pyrene in the early live stages of *Cyprinodon variegatus*. *Comparative Biochemistry and Physiology Part C* 147:205–215.

Hose, J. E., M. D. McGurk, G. D. Marty, D. E. Hinton, E. D. Brown, and T. T. Baker. 1996. Sublethal effects of the *Exxon Valdez* oil spill on herring embryos and larvae: morphological, cytogenetic, and histopathological assessments, 1989–1991. *Canadian Journal of Fisheries and Aquatic Sciences* 53:2355–2365.

Johnson, D. R. and J. M. Perry. 1999. Blue crab larval dispersion and retention in the Mississippi Bight. *Bulletin of Marine Science* 65:129–149.

Kocan, R. M., G. D. Marty, M. S. Okihiro, E. D. Brown, and T. T. Baker. 1996. Reproductive success and histopathology of individual Prince William Sound Pacific herring 3 years after the *Exxon Valdez* oil spill. *Canadian Journal of Fisheries and Aquatic Sciences* 53:2388–2393.

McCall, B. D. and S. C. Pennings. 2012. Disturbance and recovery of salt marsh arthropod communities following BP *Deepwater Horizon* oil spill. *Plos One* 7:e32735. doi:32710.31371/journal.pone.0032735.

McGurk, M. D. and E. D. Brown. 1996. Egg–larval mortality of Pacific herring in Prince William Sound, Alaska, after the *Exxon Valdez* oil spill. *Canadian Journal of Fisheries and Aquatic Sciences* 53:2343–2354.

Neuparth, T., S. M. Moreira, M. M. Santos, and M. A. Reis-Henriques. 2012. Review of oil and HNS accidental spills in Europe: Identifying major environmental monitoring gaps and drawing priorities. *Marine Pollution Bulletin* 64:1085–1095.

OSAT, O. S. A. T. 2010. Summary report for sub-sea and sub-surface oil and dispersant detection: Sampling and monitoring. Unified Area Command, New Orleans: United States Coast Guard.

Pearson, W. H., R. B. Deriso, R. A. Elston, S. E. Hook, K. R. Parker, and J. W. Anderson. 2012. Hypotheses concerning the decline and poor recovery of Pacific herring in Prince William Sound, Alaska. *Reviews in Fish Biology and Fisheries* 22:95–135.

Perry, H. M., C. K. Eleuterius, C. B. Trigg, and J. R. Warren. 1995. Settlement patterns of *Callinectes sapidus* megalopae in Mississippi Sound: 1991, 1992. *Bulletin of Marine Science* 57:821–833.

Perry, H. M., D. R. Johnson, K. Larsen, C. Trigg, and F. Vukovich. 2003. Blue crab larval dispersion and retention in the Mississippi Bight: Testing the hypothesis. *Bulletin of Marine Science* 72:331–346.

Peterson, C. H., M. C. Kennicutt, R. H. Green, P. Montagna, D. E. Harper, E. N. Powell et al. 1996. Ecological consequences of environmental perturbations associated with offshore hydrocarbon production: A perspective on long-term exposures in the Gulf of Mexico. *Canadian Journal of Fisheries and Aquatic Sciences* 53:2637–2654.

Rabalais, N. N., F. R. Burditt, L. D. Coen, B. E. Cole, C. Eleuterius, K. L. Heck et al. 1995. Settlement of *Callinectes sapidus* megalopae on artificial collectors in four Gulf of Mexico estuaries. *Bulletin of Marine Science* 57:855–876.

Roth, M. F. and D. M. Baltz. 2009. Short-term effects of an oil spill on marsh edge fishes and decapod crustaceans. *Estuaries and Coasts* 32:565–572.

Sanchez-Rubio, G., H. M. Perry, P. M. Biesiot, D. R. Johnson, and R. N. Lipcius. 2011a. Climate-related hydrological regimes and their effects on abundance of juvenile blue crabs (*Callinectes sapidus*) in the northcentral Gulf of Mexico. *Fishery Bulletin* 109:139–146.

Sanchez-Rubio, G., H. M. Perry, P. M. Biesiot, D. R. Johnson, and R. N. Lipcius. 2011b. Oceanic-atmospheric modes of variability and their influence on riverine input to coastal Louisiana and Mississippi. *Journal of Hydrology* 396:72–81.

Shumway, R. H. and D. S. Stoffer. 2011. *Time Series Analysis and Its Applications with R Examples.* New York: Springer Science.

Smart, T. I., J. T. Duffy-Anderson, J. K. Horne, E. V. Farley, C. D. Wilson, and J. M. Napp. 2012. Influence of environment on walleye pollock eggs, larvae, and juveniles in the southeastern Bering Sea. *Deep-Sea Research Part II—Topical Studies in Oceanography* 65–70:196–207.

Spier, C. S., G. G. Vadas, S. L. Kaattari, and M. A. Unger. 2011. Near-real-time, on-site, quantitative analysis of PAHs in the aqueous environment using an antibody-based biosensor. *Environmental Toxicology and Chemistry* 30:1557–1563.

Sulkin, S. D. and W. F. Van Heukelem. 1986. Variability in the length of the megalopal stage and its consequence for dispersal and rectruitment in the portunid crab, *Callinectes sapidus*. *Bulletin of Marine Science* 39:269–278.

Zar, J. H. 2010. *Biostatistical Analysis*. 5th edition. Upper Saddle River, NJ: Prentice Hall.

14

Oyster Responses to the Deepwater Horizon *Oil Spill across Coastal Louisiana: Examining Oyster Health and Hydrocarbon Bioaccumulation*

Jerome La Peyre, Sandra Casas, and Scott Miles

CONTENTS

Introduction

Eastern oysters (*Crassostrea virginica*) support an extensive commercial fishery in coastal Louisiana with a dockside value typically in excess of $35 million (LDWF 2012). Oyster beds are also vital constituents of estuarine environments (Dame 1996). This keystone species creates essential habitats that provide critical spawning, nursery, and foraging grounds for many species including economically important macroinvertebrates and fish species (Coen et al. 1999; Plunket and La Peyre 2005), and contributes to water quality and protection of shorelines (Piazza et al. 2005; Grizzle et al. 2008). Maintenance of oyster beds ultimately depends on the ability of individual oysters to survive, grow, and reproduce.

In summer 2010, oil from the *Deepwater Horizon* (DWH) oil spill reached the Louisiana coastline and entered Barataria Bay, a productive oyster-growing estuary resulting in the closure of harvest areas (Louisiana Department of Health and Hospitals Oil Spill Shellfish Harvest Area Maps, http://www.dhh.louisiana.gov/index.cfm/resource/category/5/n/90). Polynuclear aromatic hydrocarbons (PAHs) that make up 0.2%–7% of oil (Albert 1995; Allan et al. 2012) are believed to pose the greatest ecological threat because of their toxic, mutagenic, and/or carcinogenic activities (Arfsten et al. 1996, but see Connell and Miller 1981; Barron et al. 1999; Pickering 1999; ATSDR 2005). Although low-molecular-weight PAHs preferentially dissolved in seawater can be taken up through epithelial surfaces, the majority of PAHs, which are heavier, lipophilic, and more toxic, preferentially bind to particulate organic matter, sediment, and microalgae, and thus enter the food web as these particles are ingested (Baumard et al. 1999). Sedentary benthic filter-feeding bivalves such as oysters, given their substantial filtration capacity and low detoxification capability, are therefore particularly effective in taking up and accumulating these persistent contaminants via filtration and ingestion, and are susceptible to their negative effects (Stegeman and Teal 1973; Boehm and Quinn 1977; Livingston 1985). For this reason, oysters and mussels are used worldwide as sentinel organisms to assess spatial and temporal trends of contaminant concentrations in coastal and estuarine environments by national programs such as the U.S. National Status and Trends program administered by the National Oceanic and Atmospheric Administration (NOAA) or France's Réseau National d'Observation (Claisse and Beliaeff 2000; O'Connor and Lauenstein 2006).

Past work has shown that PAHs can have a wide range of effects on oysters from sublethal effects, including reduced growth, decreased immunocompetence, increased susceptibility to disease, and impaired reproduction, to death (Mahoney and Noyes 1982; Moore et al. 1987; Winstead and Couch 1988; Chu and Hale 1994; Fisher et al. 1999). Biomarkers have been used to examine contaminant stress including the effects of PAHs in bivalves (Ringwood et al. 2004; Bocchetti et al. 2008; Yeats et al. 2008; Baussant et al. 2009). Alterations of hemocyte counts, lysosomal stability, and phagocytosis, for example, have been reported to be among the most reliable indicators of polluted sites (Ringwood et al. 1998; Auffret et al. 2006; Moore et al. 2006; Bocchetti et al. 2008; Donaghy et al. 2009). Hemocytes of oysters and other bivalves not only participate in host defenses but are also involved in a wide range of physiological processes (Cheng 1996; Mount et al. 2004). Immunosuppression upon exposure to chemical contaminants even at low concentrations is of prime concern as it has implications for susceptibility to oyster pathogens and ultimately mortality of the host (Chu and Hale 1994; Fisher et al. 1999; Baier-Anderson and Anderson 2000). Infections can in turn drastically affect oyster physiology including host responses to contaminants (Desclaux-Marchand et al. 2007; Minguez et al. 2012).

We assessed the responses of Louisiana oysters to the DWH oil spill using a suite of biological effects measurements or biomarkers. The responses of both wild-collected oysters and oysters produced by the Louisiana Sea Grant hatchery and deployed in three Louisiana estuaries were determined. Specifically, we measured whole organism responses (survival of caged oysters, condition, *Perkinsus marinus* infection intensities of caged and wild oysters), tissue responses (histopathology alterations), cellular responses (hemocyte density, viability, phagocytosis capacity), and subcellular responses (lysosomal stability, heat shock protein 70 [HSP70] proteins) in oysters from nonoiled sites and an oiled site post-spill in relation to PAH concentrations. It is important to note that the responses measured indicate general stress and can be affected by nonchemical stressors in addition to oil or other pollutants. Many of the biomarkers were selected based on 2010 recommendations by the International Council for the Exploration of the Sea working group on the biological effects of contaminants for use in monitoring, and none of the biomarkers recommended for bivalves respond specifically to oil (Martinez-Gomez et al. 2010). Measurements were done at several sites over two consecutive years (i.e., 2010, 2011) to evaluate potential confounding factors (e.g., salinity) affecting the biomarkers at the different sites. Although some discrete lab and field data exist for general health, disease, and reproductive state of oysters (Supan and Wilson 2001; La Peyre et al. 2009; Volety et al. 2009), very little seasonally and spatially relevant data exist for oysters along the coast of Louisiana. Data on the bioaccumulation and effects on oil on oysters in subtropical regions such as ours are also limited (Del Castillo et al. 1992; Norena-Barroso et al. 1999).

Materials and Methods

Experimental Design

Study 1: Caged Oysters Sampling Pre-Coastal Oiling in May 2010

Oysters have been maintained in cages in Sister (or Caillou) Lake since April 2009, off the north shore of Grand Isle since November 2009, and in Bay Gardene since April 2010 for projects unrelated to the DWH oil spill. These oysters were sampled for this study in May 2010 before the oil reached the Louisiana coast (Table 14.1). Sites in Sister Lake (29°14′11.09″N, 90°55′16.48″W) in Terrebonne Parish, and Grand Isle (29°12′45.93″N, 90°56′15.34″W) in Barataria Bay are located west of the Mississippi River whereas Bay Gardene (29°35′28.80″N, 89°37′31.62″W) in Breton Sound is located east of the Mississippi River (Figure 14.1). All oysters deployed in cages originated from the Louisiana Sea Grant oyster hatchery in Grand Isle and were the progeny of oysters spawned in summer 2008.

In mid-May 2010, 10 oysters from each site were collected and brought to the Department of Veterinary Science at Louisiana State University Agricultural center where they were placed in recirculating artificial seawater (ASW) systems (Crystal Sea Marinemix, Marine Enterprises International, Baltimore, Maryland), adjusted to the temperature and salinity of each site sampled (20°C and salinities of 10 ppt for Bay Gardene and Sister Lake and

TABLE 14.1

Months Oysters Deployed in Cages for Sampling Pre-Coastal Oiling in May 2010 (i.e., Study 1) at Three Sites and Post-Coastal Oiling (i.e., Study 2) in October 2010 and 2011 at Four Sites in Louisiana Estuaries

Sites	Deployment Times		
	Study 1: May 2010 sampling	Study 2: October 2010 sampling	Study 2: October 2011 sampling
Bay Gardene	April 2010	May 2010	August 2011
Sister Lake	April 2009	May 2010	August 2011
Grand Isle	November 2009	May 2010	August 2011
Grande Terre	—	August 2010	August 2011

Note: Oysters deployed were obtained for the Louisiana Sea Grant hatchery and were the progeny of Louisiana oysters spawned in summer 2008.

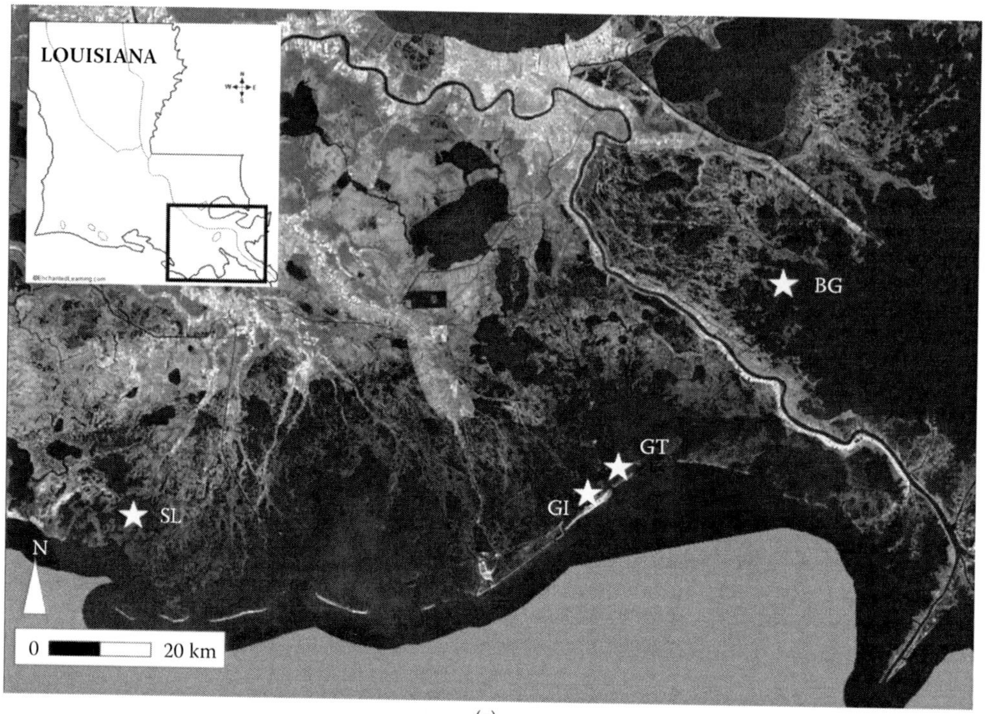

(a)

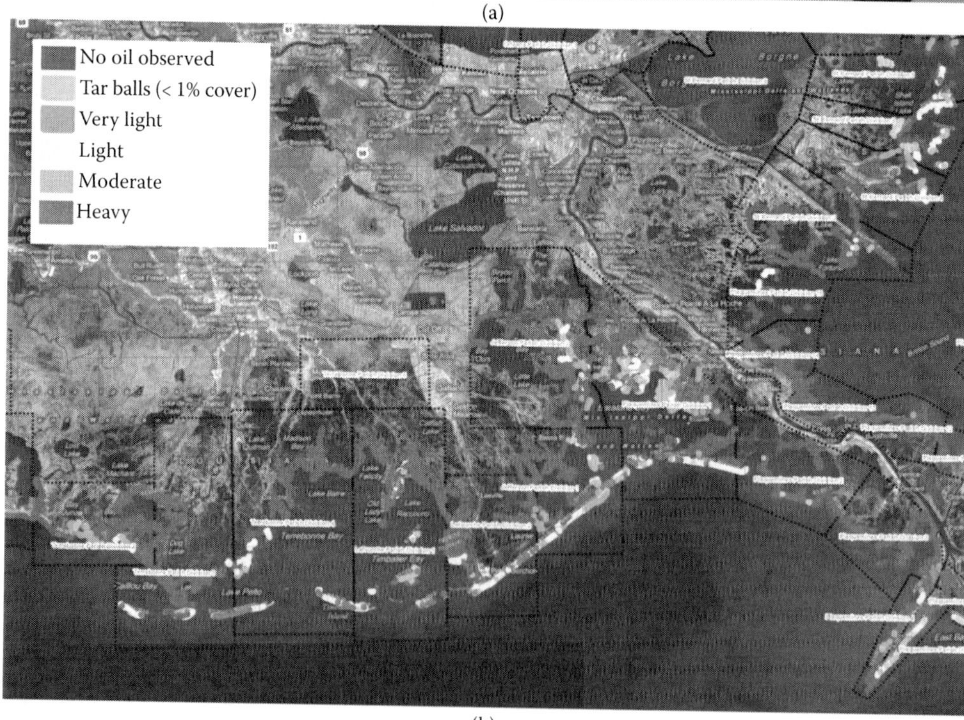

(b)

FIGURE 14.1
(See color insert.) Maps indicating study sites where caged oysters were deployed and wild oysters were collected in Louisiana estuaries (a) and *Deepwater Horizon* shoreline oiling on September 20, 2010 obtained from the Shoreline Cleanup Assessment Teams (SCAT) (b). (From http://www.noaa.gov/deepwaterhorizon/maps/pdf/scat_la/Reduced_MC252_ShorelineCurrentOilingSituation_092010.pdf.)

20 ppt for Grand Isle). After about a week of acclimation, the oyster weights and shell heights were measured. Their hemolymph were withdrawn from the adductor muscle sinus through a notch on their dorsal sides with a syringe equipped with a 25-gauge 3.8-cm long needle. These samples were immediately placed on ice, and used to determine hemocyte density (cells/mL), percentage of granulocyte, viability, phagocytic index, and lysosomal stability as described in the Hemocyte Measurements section. Oysters were then carefully opened and their meat weighed (g). Five-mm wide transverse cross sections of their gills and digestive gland were cut and fixed for standard histological processing. The slides obtained were used to assess histopathology and gonadal condition (Howard and Smith 1983; Kim et al. 2006). Sections of gills were cut and stored at −80°C until used to determine HSP70 concentrations as described in the HSP70 Measurements section. The rest of the body was homogenized in 20 mL of 0.2-μm filtered ASW, and tissue homogenates were used to determine *Perkinsus marinus* infection intensity, oyster condition index, and PAH concentrations as described in the *Perkinsus marinus* Infection Intensity and Condition index, Histopathology Analyses, and PAH Concentrations sections. All measurements were done on individual oysters except for PAHs that were determined for three composite samples per site with each composite sample made up of three oysters.

Study 2: Caged Oysters Sampling Post-Coastal Oiling in October 2010 and 2011

Caged oysters at the three historic pre-spill sites and a new site just off Grande Terre (29 °16′42.52″N, 89°56′36.08″W) in Barataria Bay that received oil from the DWH oil spill (Shoreline Cleanup Assessment Teams ground oil survey on August 18, 2010, United States Geological Survey Geographic Information System team) were sampled at the end of October 2010 (Figure 14.1). The oysters off Grande Terre were deployed from the Louisiana Sea Grant oyster hatchery in three cages containing 25 oysters each in August 2010. Cages with oysters from the Louisiana Sea Grant oyster hatchery were deployed at the four sites (three cages per site) in August 2011 and sampled in October 2011.

At the end of October 2010 and 2011, mortality of the caged oysters was determined at each site, and 30 oysters were collected and transferred to recirculating water systems (20°C and salinities of 10 ppt for Bay Gardene and Sister Lake and 20 ppt for Grand Isle and Grande Terre) for acclimation until they were evaluated as described in study 1. For study 2, hemocyte measurements were done on 30 oysters, *P. marinus* infection intensity and condition index were determined for 15 oysters, histopathology was analyzed in 10 oysters, and HSP70 protein concentration was measured in 6 oysters collected at each site. PAH concentrations of three composite samples made up of five oysters each were measured at each site.

Study 3: Wild Oysters Sampling Post-Coastal Oiling in October 2010 and August 2011

In mid-October 2010 and mid-August 2011, 30 wild subtidal oysters were dredged from Sister Lake and Bay Gardene and hand collected along the northwest shore of Grande Terre. The oysters were then transferred to recirculating water systems for acclimation until they were evaluated as described in study 2.

Environmental Parameters

Continuous water salinity and temperature were obtained from the USGS recorders located in Sister Lake (USGS 07381349), Bay Gardene (USGS 87374527), and Barataria Pass (USGS 073802516) as a proxy for Grand Isle and the Grande Terre sites. Temperature and salinity data from the Barataria Pass USGS recorder and discrete values measured with aYSI 556 (YSI Incorporated, Yellow Springs, Ohio) in Grand Isle and Grande Terre at times of deployment and sampling showed no significant differences.

Hemocyte Measurements

Hemolymph (1–3 mL) was withdrawn from the adductor muscle sinus of each oyster using a 3-mL syringe with a 25-gauge 3.8-mm needle through a notch on the dorsal side of the shell and immediately placed into vials on ice to reduce hemocyte clumping. Hemocyte density was determined with an improved Neubauer hemocytometer (Reichert, Buffalo, New York) as described by La Peyre et al. (1995). Hemocyte viability was determined using the trypan blue exclusion test (Volety et al. 1999; Akaishi et al. 2007). Briefly, 5 μL of trypan blue (3%) was added to a slide containing 5 μL of hemolymph and incubated for 10 min in a humid chamber, and the percentage of live cells (unstained) was determined with a microscope at a magnification of 400×. At least 100 hemocytes were counted per sample.

The ability of oyster hemocytes to phagocytose zymosan particles was determined for individual oysters using hemocyte monolayers as described by La Peyre et al. (1995). Briefly, two samples of 50 μL of hemolymph containing 1×10^5 hemocytes each were added to a slide and hemocytes were allowed to adhere for 30 min in a humidified chamber at room temperature. Plasma was then carefully discarded and 50 μL containing 1×10^6 zymosan particles in ASW was added to each hemocyte monolayer for a final 1:10 hemocyte to zymosan ratio. After a 60-minute incubation in a humidified chamber, the hemocyte monolayers were rinsed with ASW, fixed with Hemacolor solution 1 fixative (Harleco, EMD Chemicals, Gibbstown, New Jersey) for 1 minute, washed with distilled water, and stained with Hemacolor solution 2 and 3 for 2 minutes each. Slides were dried and mounted with Permont (Fisher Scientific, Hampton, New Hampshire). The percentage of hemocytes ingesting one or more zymosan particles and the number of zymosan particles per phagocytic hemocyte were determined microscopically at a magnification of 400×. At least five fields of view or a minimum of 200 hemocytes were counted per monolayer.

Hemocyte lysosomal destabilization was measured using the neutral red retention assay of Lowe et al. (1995). Neutral red is a cationic lipophilic dye that accumulates in cell lysosomes. Briefly, hemolymph from each oyster containing 1×10^5 hemocytes was added to 50 μL of ASW in duplicate wells of a 96-well plate. The plates were centrifuged at 100*g* for 2 minutes and hemocytes were allowed to adhere for 15 minutes. The supernatant was carefully removed and 100 μL of neutral red solution (50 μg/mL ASW) was added to each well. The wells were then inspected with an inverted microscope at 400× at 30-minute intervals for a maximum of 180 minutes and the time recorded, at which point at least 50% of granulocytes became spherical and had neutral red loss from the lysosomes to the cytosol. At least 100 cells per well were counted to verify lysosome destabilization in at least 50% of the cells. Lysosomal stability was scored as "neutral red retention time," defined as the time of the inspection previous to the inspection when granulocyte lysosome destabilization exceeded 50%.

HSP70 Measurements

Gill HSP70 concentrations were measured with an enzyme-linked immunosorbent assay (ELISA) and characterized by western blot using the mouse monoclonal antibody MA3-006 from Pierce Biotechnology (Rockford, Illinois) produced against human HSP70 that recognizes two oyster HSP70 proteins, hsc72 and hsp69 (Encomio and Chu 2007). Gill tissues were thawed and homogenized with a glass tissue grinder held on ice using 0.5 g/mL of tissue bicarbonate homogenization buffer (pH 9.5). The bicarbonate homogenization buffer contained 0.1 M $NaHCO_3$, 1% Nonidet-P40, and 1% protease inhibitor cocktail set I (i.e., 500 μM AEBSF hydrochloride, 150 nM aprotinin, 1 μM E-64, 0.5 mM EDTA disodium, and 1 μM Leupeptin hemisulfate in bicarbonate homogenization buffer) from CalBiochem (San Diego, California). The homogenates were centrifuged at 3000*g* for 10 minutes at 4°C and the supernatants collected and centrifuged at 20,000*g* for 30 minutes at 4°C. The resulting gill lysates were saved and their protein concentrations were measured with the Micro BCA Protein Assay Reagent kit (Pierce Biotechnology) using bovine serum albumin (BSA) as a standard.

For the ELISA, protein concentrations were adjusted to 2 mg/mL with bicarbonate buffer (0.1 M $NaHCO_3$, pH 9.5) and the gill lysates were diluted 1:1 (v/v) with a reducing buffer consisting of 5% β-mercaptoethanol and 4 mM SDS in Tris-buffered saline (TBS; 10 mM NaCl, 50 mM Tris-HCl, pH 7.5) and boiled for 3 minutes. The reduced gill lysates diluted 1/200 in bicarbonate buffer and recombinant human heat shock protein 70 (rhHSP70) (Sigma-Aldrich, St. Louis, Missouri) used as a standard were added to triplicate wells, which consisted of a 96-well plate (50 μL per well). After an overnight incubation at 4°C, the wells were rinsed three times with TTBS (i.e., TBS with 0.05% Tween-20) and incubated for 1 hour at room temperature with 150 μL of 1% BSA in TBS. The blocking buffer was removed, wells were rinsed three times with TTBS, and 50 μL of mouse monoclonal antibody MA3-006 diluted 1/5000 in TBS containing 0.5% BSA was added to each well. After a 90-minute incubation, each well was rinsed three times and received 50 μL of rabbit anti-Mouse IgG1 alkaline phosphatase conjugate (BioFX Laboratories, Owings Mills, Maryland) diluted 1/5000 in TBS with 0.5% BSA. After another 90-minute incubation, the wells were rinsed three times with TTBS and once with TBS. Finally, 100 μL of P-nitrophenyl phosphate (Zymed Laboratories, San Francisco, California) was added to each well as a substrate and absorbance at 405 nm was read after 30 minutes of incubation. Gill lysate HSP70 concentrations (ng/mL) were calculated using the standard curve and data are presented as ng/mg of total HSP70 protein.

For the western blot, gill lysates or rhHSP70 used as a reference, were mixed 1:1 with 2× Laemmli sample buffer (65.8 mM Tris-HCl, pH 6.8, 2.1% SDS, 26.3% [w/v] glycerol, 0.01% bromophenol blue, Bio-Rad Laboratories, Hercules, California) containing 5% mercaptho-ethanol. After boiling for 5 minutes, 10 μg of gill lysates or 0.35 μg of rhHSP70 were loaded onto 8% acrylamide resolving gels, and proteins were separated by SDS-PAGE in running buffer (3.03 g Tris Base, 14.4 g glycine, 1.0 g SDS per liter) at 35 mA for about 55 minutes. Precision plus protein standards (Bio-Rad) were included in all gels as molecular weight markers. After equilibrating the gels in transfer buffer (100 mL of 10× running buffer, 200 mL methanol, 700 mL dH_2O) for 30 minutes, the separated proteins were transferred onto nitrocellulose membranes at 100 V for 1 hour. The resulting membranes were blocked with 5% BSA in TBS overnight at 4°C and rinsed three times for 5 minutes each in TTBS. Membranes were probed with the same concentrations of primary and secondary antibodies used in the ELISA. The bound antibodies were then detected by the addition of 5-bromo-4-chloro-3′-indolyphosphate p-toluidine salt/nitro-blue tetrazolium chloride (BCIP/NBT) substrate (Zymed Laboratories) and incubation at room temperature for 2–3 minutes. The color reaction was stopped by washing the membrane with distilled water. Digital photographs of the membrane were taken with a ChemiDoc XRS instrument (Bio-Rad) and the area, volume, and intensity of each visible band were analyzed with "Quantity One" software (Bio-Rad). Levels of hsc72 and hsp69 isoforms were standardized relative to the level of the reference rhHSP70 band and calculated by dividing the volume (OD × area) of each sample band by rhHSP70 volume. Data are expressed as relative levels of hsc72 or hsp69 isoforms.

Perkinsus marinus Infection Intensity and Condition Index

The number of *Perkinsus marinus* parasites per gram of oyster tissue was determined using the whole-oyster procedure as described by Fisher and Oliver (1996) and modified by La Peyre et al. (2003). All chemicals were from Sigma unless otherwise indicated. Each oyster was homogenized in ASW and 1 mL of tissue homogenate was resuspended in alternate fluid thioglycollate medium (ARFTM) supplemented with 15 g/L marine salts, 5 mL/L of commercial lipid concentrate 1000×, and 50 mg/L chloramphenicol at a final ratio of about 0.2 g oyster tissue per 10 mL of ARFTM. After 1 week of incubation in ARFTM, samples were centrifuged at 1500*g* for 10 minutes and the ARFTM was discarded. The resulting pellets were incubated in 2 N NaOH at 60°C for 6 hours to digest oyster tissues, leaving the parasites intact. The samples were rinsed twice with 0.1 M phosphate buffer saline containing 0.5 mg/mL of BSA to prevent parasite clumping. Samples were then serially diluted in 96-well plates and stained with Lugol's solution. The number of parasites was counted from wells containing 100–300 parasites (i.e., hypnospores) with an inverted microscope at a magnification of 200×. Infection intensity of individual oysters is reported as number of parasites/g of oyster tissue wet weight.

Condition index (CI) was calculated as the ratio of dry tissue weight to dry shell weight multiplied by 100 (Mann 1978; Lucas and Beninger 1985). For each oyster, a 10 mL aliquot of oyster tissue homogenate in ASW was dried at 65°C for 48 hours and dry weight determined by subtracting the dry weight of ASW only. The dry weight for the whole oyster was calculated based on the total volume of homogenized tissue in ASW.

Histopathology Analyses

Histological slides were examined by light microscopy for common parasites and pathologies as described by Kim et al. (2006). Gametic stages were assigned based on the gametogenic classification of Kennedy and Krantz (1982) and were classified as early development, late development, spawning, advanced spawning, and regressing. The density of brown cells in the digestive gland and connective tissues surrounding the intestinal tract of individual oyster was quantified in 10 random microscopic fields at 400×. Brown cells contain translucent brown lysosomes used in detoxification and are potential indicators of pollution and stress (Zaroogian and Yevich 1993).

PAH Concentrations

PAH concentrations of oyster tissue were determined using standard tissue preparation, chromatographic cleanup, and gas chromatograph/mass spectrometry (GC/MS) techniques as described in AOAC Official 2007.01 Method (2007). Multiple validation studies have been performed, using the QuEChERS-based extraction method, for the determination of PAHs in edible seafood (Ramalhosa et al. 2009; Anderson et al. 2011; Johnson 2012). Individual oyster tissue homogenate was placed in aluminum pans and lyophilized for 72 hours (condenser temperature–30°C, 300 mTorr). The lyophilized tissues of individual oysters were combined to prepare composite samples as described earlier and weighed, PAH concentrations of three composite samples of three oysters each (study 1) or five oysters each (studies 2 and 3) were measured at each site. Analytical analyses of the extract were done on an Agilent 7890A GC system configured with a 5% diphenyl/95% dimethyl polysiloxane high resolution capillary column directly interfaced to an Agilent 5975 inert XL MS detector system. The data are expressed as total PAHs (tPAHs) after adding all

PAHs measured. The ranges of individual PAH measured in oyster and sediment sampled at the end of the project were also reported (Tables 14.2 and 14.3). Data obtained were compared with reported long-term PAH concentrations of Louisiana oysters sampled by NOAA's National Status and Trends (NST) Mussel Watch program (http://ccma.nos.noaa.gov/about/coast/nsandt/download.aspx) (Kimbrough et al. 2006; Johnson et al. 2009). Most of the PAHs reported by the NST program and those used to calculate tPAHs were also measured in our study (Tables 14.2 and 14.3). Two sediment samples per site were also collected in October 2011, at the end of the study, and analyzed for tPAHs and alkanes. tPAHs levels were described as low (<100 ng/g·dw), moderate (100–1000 ng/g·dw), high (1000–5000 ng/g·dw), and very high (>5000 ng/g·dw) according to Baumard et al. (1998) classification.

Statistical Analyses

Temperature, salinity, hemocyte density, granulocyte (%), phagocytic index, neutral red retention (NRR) time, HSP70 concentration, hsc72 and hsp69 isoform concentrations, number of brown cells per microscopic field, condition index, and PAH concentrations were examined for normality and homogeneity of variance, transformed as required, and analyzed with a one-factor (site) analysis of variance (ANOVA). Each sampling date was analyzed separately. When data did not fulfill the ANOVA requirements, they were analyzed with Kruskal–Wallis one-way analysis of variance on ranks. *P. marinus* body burden concentrations were compared with Kruskal–Wallis one-factor (site) ANOVA on ranks. Tukey's test was used for pairwise multiple comparison. Mortality and gregarine prevalence were compared with χ^2 analysis. All data are presented as mean ± standard error.

Results

Study 1: Caged Oysters Sampling Pre-Coastal Oiling in May 2010

Salinity during the 30 days preceding May 2010 sampling differed significantly between sites with Bay Gardene (BG, 6.6 ± 0.1 ppt) being the lowest, Sister Lake (SL, 11.6 ± 0.5 ppt) being intermediate, and Grand Isle (GI, i.e., Barataria Pass) being the highest (16.0 ± 0.2 ppt) (Figure 14.2). Water temperature in Sister Lake (24.8°C ± 0.1°C) was slightly but significantly higher than Grand Isle (24.2°C ± 0.1°C) or Bay Gardene (24.2°C ± 0.1°C) (Figure 14.2).

No differences between sites could be found in any of the hemocyte parameters measured including hemocyte density (cells/mL, BG: 4.7 ± 1.2 × 10^6, SL: 4.8 ± 0.8 × 10^6, GI: 5.4 ± 1.4 × $10^{6)}$, viability (% live, BG: 99.6 ± 0.2, SL: 99.3 ± 0.3, GI: 99.6 ± 0.2), lysosomal stability (minutes, BG: 96 ± 10, SL: 99 ± 13, GI: 123 ± 9), percentage of phagocytic hemocytes (BG: 51 ± 5, SL: 52 ± 4, GI: 55 ± 3), and number of zymosan per phagocytic hemocytes (BG: 5.0 ± 0.5, SL: 4.8 ± 0.4, GI: 5.2 ± 0.3).

There were significant differences in the relative levels of HSP70 isoforms detected by western blot between sites and in total gill HSP70 concentrations measured by ELISA (ng HSP70 per mg total protein, BG: 464 ± 18, SL: 483 ± 35, GI: 335 ± 31). The relative level of hsc72 isoform in Bay Gardene (0.370 ± 0.033) was significantly greater than in Grand Isle (0.186 ± 0.027), and the level of hsp69 isoform in Bay Gardene (0.243 ± 0.026) was significantly greater than in Grand Isle (0.064 ± 0.0130) and Sister Lake (0.145 ± 0.020).

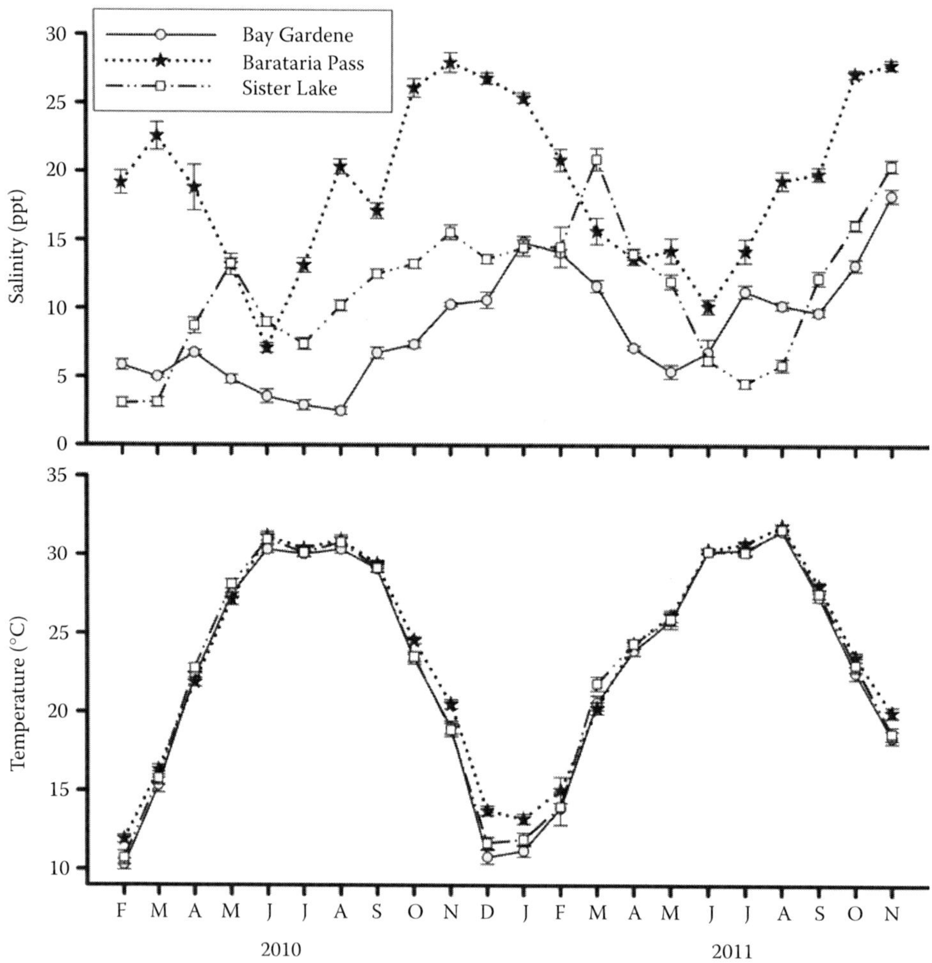

FIGURE 14.2
Mean monthly water salinity and temperature (°C) for Bay Gardene (BG), Sister (or Caillou) Lake (SL), and Barataria Bay Pass (BB) as a proxy for Grand Isle and Grande Terre. All data are from USGS continuous water quality recorders: BG-USGS 87374527, SL-USGS 07381349, BB-USGS 073802516.

No obvious differences in tissue parasites or pathologies were observed between sites, nor could any differences be found in the density of brown cells (cells/field, BG: 4.3 ± 0.8, SL: 4.2 ± 1.4, GI: 3.7 ± 0.6) in digestive glands and connective tissues surrounding intestinal tracts. Gregarines prevalence was about 10% at all sites. Both early and late stages of gonad development were observed at each site, which is not unusual for this time of the year in our subtropical region.

P. marinus infection prevalence (i.e., percentage of oysters infected) was greater in Grand Isle oysters (90%) than Sister Lake (50%) or Bay Gardene oysters (60%). The percentage of oysters with light *P. marinus* infection intensities (i.e., <10,000 parasites per gram wet oyster tissue) were about the same (50%–60%) at all three sites. Only Grand Isle oysters (20%) had heavy infection intensities of *P. marinus* (i.e., >500,000 parasites per gram wet oyster tissue).

Condition index in Sister Lake (2.7 ± 0.1) was significantly greater than in Bay Gardene (1.5 ± 0.1) and Grand Isle (1.8 ± 0.2). Although there were no significant differences in shell heights between sites (mm, BG: 99 ± 6, SL: 96 ± 5, GI: 98 ± 94), oysters from Sister Lake had significantly lighter dried shells than Grand Isle oysters (g, BG: 138 ± 20, SL: 87 ± 7, GI: 173 ± 22), and oysters from Grand Isle had significantly heavier dried meats than Bay Gardene oysters (g, BG: 1.97 ± 0.27, SL: 2.33 ± 0.19, GI: 2.96 ± 0.33).

tPAH concentrations in all oyster composite samples were low (<100 ng/g·dry oyster weight), but Bay Gardene tPAH concentrations (42.5 ± 3.8 ng/g·dry oyster weight) were higher than Grand Isle (18.4 ± 1.3 ng/g·dry oyster weight) or Sister Lake (16.1 ± 1.4 ng/g·dry oyster weight).

Study 2: Caged Oysters Sampling Post-Coastal Oiling in October 2010 and 2011

October 2010

Salinity during the 30 days preceding October 2010 sampling differed significantly between sites with Bay Gardene (6.2 ± 0.3) being the lowest, Sister Lake (12.9 ± 0.3) being intermediate, and Grand Isle (i.e., Barataria Pass) being the highest (22.6 ± 0.8), whereas mean water temperature in Grand Isle (24.5°C ± 0.2°C) was slightly, but significantly, higher than in Bay Gardene (23.3°C ± 0.3°C) and Sister Lake (23.5°C ± 0.3°C). Between May and October 2010 sampling, salinity showed a similar trend with Bay Gardene mean salinity (5.0 ± 0.3 ppt) being significantly lower than Sister Lake (11.1 ± 0.2 ppt) and Barataria Pass (16.2 ± 0.6 ppt), whereas differences in temperatures between sites were not observed (Figure 14.2).

Oyster mortality was significantly greater in Bay Gardene (64%) than in Sister Lake (8%) and Grand Isle (8%) during the period between May and October 2010. Grande Terre mortality was 12% from August 2010 to October 2010.

Hemocyte density of caged oysters in Grande Terre (8.1 ± 0.7 × 10^6 cells/mL) was significantly higher than in Sister Lake (4.7 ± 0.5 × 10^6 cells/mL) or Bay Gardene (5.2 ± 0.4 × 10^6 cells/mL) and Grand Isle hemocyte density (7.0 ± 0.9 × 10^6 cells/mL) was intermediate (Figure 14.3). The percentage of phagocytic hemocytes in Grande Terre oysters (44 ± 2%) was also greater than in Bay Gardene (34 ± 2) and Grand Isle (34 ± 2), and percentage of phagocytic hemocytes in Sister Lake (38 ± 2) was intermediate (Figure 14.3). Hemocyte lysosomal stability measured as the neutral red retention time was significantly lower in Bay Gardene caged oysters (122 ± 8 minutes) than in Grand Isle (153 ± 6) and Sister Lake caged oysters (158 ± 5) (Figure 14.3). No differences between sites could be found in hemocyte viability (% live, BG: 96.0 ± 0.6, SL: 95.6 ± 0.7, GI: 97.1 ± 0.4, GT: 96.6 ± 0.6) and number of zymosan particles per phagocytic hemocytes (BG: 5.2 ± 0.2, SL: 5.5 ± 0.2, GI: 5.8 ± 0.2, GT: 5.4 ± 0.2).

The relative level of hsc72 isoform in Bay Gardene (0.653 ± 0.033) and Sister Lake (0.619 ± 0.052) oysters were significantly higher than in Barataria Bay oysters (GI: 0.329 ± 0.030, GT: 0.360 ± 0.050) (Figure 14.3). No significant differences in HSP70 concentrations however could be found in caged oysters between sites (ng HSP70/mg·total protein, BG: 565 ± 35, SL: 544 ± 43, GI: 480 ± 38, GT: 504 ± 84), nor could differences be found in hsp69 isoform levels (BG: 0.161 ± 0.050, SL: 0.282 ± 0.041, GI: 0.232 ± 0.056, GT: 0.143 ± 0.036).

The density of brown cells in Grande Terre oysters (6.7 ± 0.6 cells per field) was significantly greater than in Grand Isle oysters (3.6 ± 0.4 cells per field) (Figure 14.3). The density of brown cells from Bay Gardene oysters (5.0 ± 1.2) and Sister Lake (5.3 ± 0.6) were intermediate (Figure 14.3). The prevalence of gregarines in Grande Terre (80%) was significantly higher than in Grand Isle (50%) and both sites had significantly higher prevalence than Sister Lake (10%) and Breton Sound (0%); gregarines were located mainly in the gills. No other differences in tissue parasites or pathologies were observed between sites. Oyster gonads were all in the regressing stage at each site.

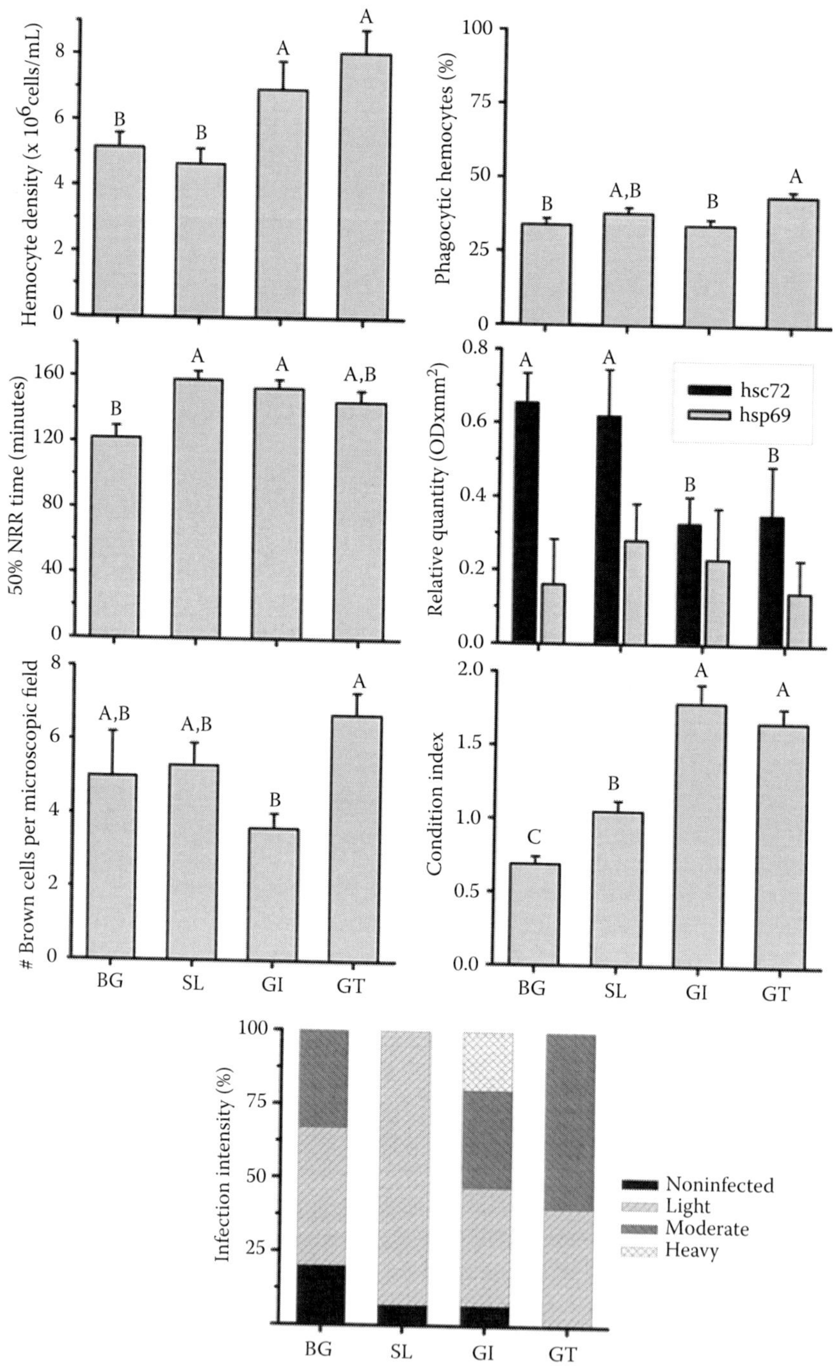

FIGURE 14.3
Mean (± SE) hemocyte density and lysosomal stability expressed as neutral red retention time, percentage of phagocytic hemocytes, relative quantity of hsc72 and hsp69 in gill tissues, density of brown cells per microscopic field of view, condition index, and *Perkinsus marinus* infection intensity categorized as light (<10,000 parasite per gram wet tissue), moderate (10,000–500,000 parasites per gram wet tissue), and heavy (>500,000 parasites per gram wet tissue) in caged oysters sampled in October 2010 from Bay Gardene (BG), Sister Lake (SL), Grand Isle (GI), and Grande Terre (GT).

The intensity of *P. marinus* infection was greater in Grande Terre oysters ($1.17 \pm 0.3 \times 10^5$ parasites per gram wet oyster tissue, median: 4.1×10^4) than in Sister Lake oysters ($8.92 \pm 4.72 \times 10^2$ parasites per gram wet oyster tissue, median: 1.0×10^2). All Grande Terre oysters were infected and over half of them had moderate infection intensities (i.e., 10,000–500,000 parasites per gram wet oyster tissue) (Figure 14.3). Although most (93%) Sister Lake oysters were infected, all had light infection intensities (Figure 14.3).

Condition indices of oysters in Barataria Bay (GI: 1.8 ± 0.1, GT: 1.7 ± 0.1) were significantly higher than in the oysters at the lower salinity sites with Bay Gardene (BG: 0.7 ± 0.1) being significantly lower than Sister Lake (1.1 ± 0.1) (Figure 14.2). Dried shells of Sister Lake oysters (115 ± 7 g) were lighter than that of Grande Terre (170 ± 10 g), Grand Isle (199 ± 13 g), and Bay Gardene (212 ± 16 g) oysters. Dried meats of GI oysters (3.38 ± 0.20 g) were significantly heavier than that of GT oysters (2.75 ± 0.17 g), and both groups of oysters had significantly heavier dried meats than Bay Gardene (1.40 ± 0.09 g) and Sister Lake (1.21 ± 0.10 g) oysters.

tPAH concentrations in all oyster composite samples were low with Bay Gardene's (29.6 ± 2.7 ng/g dry oyster weight) and Grande Terre's (30.1 ± 1.4 ng/g dry oyster weight) tPAH concentrations being significantly higher than Grand Isle's (13.1 ± 1.8 ng/g dry oyster weight) and Sister Lake's (13.4 ± 1.8 ng/g dry oyster weight).

October 2011

Salinity during the 30 days preceding October 2011 sampling was significantly greater in Barataria Pass (i.e., Grand Isle/Grande Terre area, 26.1 ± 0.5 ppt) than in Sister Lake (15.3 ± 0.6 ppt) and Bay Gardene (12.1 ± 0.6 ppt). Overall mean salinity from the time the cages were deployed in August 2011 to the time the oysters were sampled in October 2011 followed the same trend with significantly greater mean salinity at Barataria Pass (21.3 ± 0.5 ppt) than Sister Lake (10.9 ± 0.5 ppt) and Bay Gardene (10.6 ± 0.2 ppt) (Figure 14.2). Temperatures at all sites were comparable (Figure 14.2).

No significant differences in oyster mortalities could be found between sites (BG: 11%, SL: 4%, GI: 4%, GT: 7%).

Hemocyte density of oysters deployed at the Barataria Bay sites (cells/mL, GI: $10.5 \pm 0.6 \times 10^6$, GT: $9.5 \pm 0.8 \times 10^6$) were significantly higher than in Bay Gardene ($6.3 \pm 0.6 \times 10^6$ cells/mL) and Sister Lake ($4.8 \pm 0.7 \times 10^6$ cells/mL) oysters. No significant differences in hemocyte viability (% live, BG: 99.3 ± 0.1, SL: 99.3 ± 0.3, GI: 99.6 ± 0.1, GT: 99.6 ± 0.1), lysosomal stability (min, BG: 176 ± 3, SL: 173 ± 5, GI: 179 ± 1, GT: 175 ± 4), percentage of phagocytic hemocytes (BG: 39 ± 1, SL: 35 ± 2, GI: 36 ± 2, GT: 37 ± 2), and number of zymosan particles per phagocytic hemocytes (BG: 7.3 ± 0.3, SL: 8.1 ± 0.3, GI: 7.2 ± 0.2, GT: 7.5 ± 0.3) could be found between sites.

No significant differences in oyster gill HSP70 concentrations (ng HSP70 per mg total protein, BG: 612 ± 26, SL: 660 ± 25, GI: 658 ± 27, GT: 691 ± 21), relative level of hsc72 isoform (BG: 0.500 ± 0.027, SL: 0.468 ± 0.027, GI: 0.401 ± 0.031, GT: 0.509 ± 0.033) and hsp69 isoform (BG: 0.206 ± 0.042, SL: 0.347 ± 0.084, GI: 0.362 ± 0.065, GT: 0.428 ± 0.010) could be found between sites.

The density of brown cells in oysters deployed in Sister Lake (6.3 ± 0.7 cells per field) oysters was significantly greater than at the other sites (cells per field, BG: 3.5 ± 0.4, GI: 2.7 ± 0.5, GT: 3.5 ± 0.5). The prevalence of gregarines in Grande Terre oysters (60%) was significantly higher than at the three other sites, which all had a prevalence of 10%. No other differences in tissue parasites or pathologies were observed between sites. Oyster gonads were all in the regressing stage at each site.

The intensity of *P. marinus* infection was significantly greater in oysters deployed in Grande Terre (2.76 ± 0.13 × 10^5 parasites per gram wet oyster tissue, median: 10.5 × 10^3) than in oysters deployed in Sister Lake oysters (8 ± 2, median: 7) and Grand Isle (1.34 ± 0.86 × 10^3, median: 1.6 × 10^1). Most (93%) oysters deployed in Grande Terre were infected and over half of them had either moderate infection intensities (40%) or heavy infection intensities (13%). In contrast, 67% of oysters deployed in Sister Lake oysters were infected and all had light infection intensities. Oysters deployed in Grand Isle and Bay Gardene were mostly lightly infected (BG: 67%, GI: 73%) with a low percentage of oysters with moderate (BG: 13%, GI: 7%) and heavy infection (BG: 7%).

Condition index of the oysters deployed in the Barataria Bay sites (GI: 1.9 ± 0.1, GT: 2.0 ± 0.1) were significantly higher than that of Bay Gardene (1.2 ± 0.1) and Sister Lake (1.2 ± 0.1) deployed oysters. Dried meats of Grand Isle (2.99 ± 0.22 g) and Grande Terre (2.33 ± 0.10 g) oysters were significantly heavier than that of Bay Gardene (2.10 ± 0.14 g) and Sister Lake (1.65 ± 0.07 g) oysters. Dried shells in Bay Gardene oysters (192 ± 12 g) were significantly heavier than in Grande Terre (122 ± 6 g) and Sister Lake (141 ± 7 g) oysters, and Grand Isle (167 ± 13 g) oysters had heavier dried shells than Grande Terre oysters.

tPAH concentrations of oyster composite samples were low (<100 μg/g dry oyster weight). Grande Terre oyster (55.3 ± 1.7 ng/g dry oyster weight) tPAHs concentration was significantly higher and Sister Lake (23.0 ± 1.0 ng/g dry oyster weight) significantly lower than that of Bay Gardene oysters (44.2 ± 1.3 ng/g dry oyster weight) or Grand Isle oysters (34.9 ± 1.7 ng/g dry oyster weight). The ranges of individual PAH measured in oyster samples are reported in Table 14.2.

tPAHs concentration of sediment samples from Grande Terre (394.5 ± 22.5 ng/g) was higher than from Grand Isle (137.0 ± 8.0 ng/g), Bay Gardene (271.5 ± 14.5 ng/g), or Sister Lake (116.5 ± 6.5 ng/g). A similar trend was found when sediment total alkanes were measured (GT: 7.8 ± 0.7 μg/g, GI: 2.6 ± 0.2 μg/g, BG: 4.4 ± 0.3 μg/g, SL: 3.5 ± 0.3 μg/g).

Moisture of sediment samples ranged from 54% to 64%. The ranges of individual PAHs measured in sediment samples are reported in Table 14.3.

Study 3: Wild Oysters Sampling Post-Coastal Oiling in October 2010 and August 2011

October 2010

Salinity during the 30 days before October 2010 sampling differed significantly between sites with Bay Gardene (6.1 ± 0.2) being the lowest, Sister Lake being intermediate (12.5 ± 0.3), and Grande Terre being the highest (i.e., Barataria Pass, 20.2 ± 0.8). The same results were obtained when salinities for the 3 months preceding the sampling were analyzed (BG: 5.2 ± 0.3, SL: 11.1 ± 0.3, BP: 18.3 ± 0.6, Figure 14.2). Temperatures at all sites were comparable (Figure 14.2).

Hemocyte viability in Sister Lake (98.8 ± 0.3%) was significantly higher than in Bay Gardene (97.0 ± 0.4%) with Grande Terre being intermediate (97.4 ± 0.5%). No differences between sites could be found in the other hemocyte parameters measured, including hemocyte density (cells/mL, BG: 7.5 ± 0.6 × 10^6, SL: 6.7 ± 0.6 × 10^6, GT: 6.9 ± 0.8 × 10^6), lysosomal stability (minutes, BG: 140 ± 6, SL: 128 ± 9, GT: 133 ± 9), percentage of phagocytic hemocytes (BG: 42 ± 2, SL: 36 ± 2, GT: 37 ± 2), and number of zymosan per phagocytic hemocytes (BG: 4.4 ± 0.1, SL: 4.8 ± 0.1, GT: 4.6 ± 0.1).

No significant differences in oyster gill HSP70 concentrations (ng HSP70 per mg total protein, BG: 647 ± 21, SL: 531 ± 22, GT: 657.6 ± 26.6, GT: 632 ± 48) and hsc72 isoform (BG: 0.412 ± 0.048, SL: 0.485 ± 0.084, GT: 0.535 ± 0.079) and hsp69 isoform relative levels (BG: 0.201 ± 0.032, SL: 0.091 ± 0.030, GT: 0.120 ± 0.036) could be found between sites.

TABLE 14.2

Range (ng/g·dw) of Polycyclic Aromatic Hydrocarbons (PAHs) Measured in Caged Oysters Sampled in October 2011 from Bay Gardene (BG), Sister Lake (SL), Grand Isle (GI), and Grande Terre (GT)

Compound	BG	SL	GI	GT
Naphthalene	5.1–5.7	1.0–1.2	1.9–2.3	2.6–2.7
C1-Naphtalenes	8.2–10.6	nd	nd	13.9–14.5
C2-Naphtalenes	nd	nd	nd	nd
C3-Naphtalenes	nd	0.1–0.2	nd	0.1
C4-Naphtalenes	nd	0.2	0.1–0.2	0.2–0.3
Fluorene	nd	nd	1.5–1.9	nd
C1-Fluorenes	nd	nd	0.6–0.8	nd
C2-Fluorenes	nd	nd	1.7–2.2	nd
C3-Fluorenes	3.4–4.2	1.4–1.8	nd	0.5
Dibenzothiophene	nd	nd	0.6–0.7	nd
C1-Dibenzothiophenes	nd	nd	0.2–0.3	nd
C2- or C3-Dibenzothiophenes	nd	nd	nd	nd
Phenanthrene	7.0–8.1	3.1–4.0	4.9–5.5	8.1–9.4
C1-Phenanthrenes	nd	1.9–2.2	6.1–7.7	nd
C2-Phenanthrenes	0.5–0.7	2.2–2.8	nd	nd
C3-Phenanthrenes	0.8–1.1	1.2–3.6	nd	1.0
C4-Phenanthrenes	nd	nd	0.2	nd
Anthracene	1.0–1.3	0.9–1.1	2.0–2.4	2.2–2.3
Fluoranthene	2.9–3.7	1.5–1.8	2.7–4.1	2.4–3.0
Pyrene	2.3–2.8	1.3–1.7	2.3–2.5	4.2–4.7
C1- or C2-Pyrenes	nd	nd	nd	nd
C3-Pyrenes	0.1–0.2	nd	nd	nd
C4-Pyrenes	nd	nd	nd	nd
Naphthobenzothiophene	nd	nd	nd	nd
C1-, C2-, or C3-Naphthobenzothiophenes	nd	nd	nd	nd
Benzo(a)Anthracene	nd	1.6–1.9	2.2–2.3	5.5–7.0
Chrysene	1.1–1.4	0.2	nd	4.0–5.4
C1-, C2-, C3-, or C4-Chrysenes	nd	nd	nd	nd
Benzo(b)Fluoranthene	1.9–2.1	0.2	nd	nd
Benzo(k)Fluoranthene	1.7–2.0	1.3–1.4	nd	nd
Benzo(e)Pyrene	1.4–1.5	nd	1.3–1.6	nd
Benzo(a)Pyrene	nd	0.2–0.3	1.2–1.7	2.6–3.3
Perylene	2.7–2.8	1.4–1.5	1.5–2.2	4.0–5.0
Indeno(1,2,3-cd)Pyrene	nd	nd	nd	nd
Dibenzo(a,h)Anthracene	nd	nd	nd	nd
Benzo(g,h,i)Perylene	nd	nd	nd	nd
Total PAHs	42.1–46.7	21.4–24.8	31.7–37.3	52.5–58.4

nd, not detected.

Gregarines prevalence was significantly higher in Grande Terre (88%) than in Breton Sound (20%) and Sister Lake (11%). No other differences in tissue parasites or pathologies were observed between sites, nor could any differences be found in the density of

TABLE 14.3

Range (ng/g·dw) of Polycyclic Aromatic Hydrocarbons (PAHs) Measured in Sediment Collected in October 2011 from Bay Gardene (BG), Sister Lake (SL), Grand Isle (GI), and Grande Terre (GT) Nearby Caged Oysters

Compound	BG	SL	GI	GT
Naphthalene	nd	nd	nd	nd
C1-, C2-, C3-, or C4-Naphthalenes	nd	nd	nd	nd
Fluorene	0.3–0.4	0.3	1.4–1.6	0.3–0.4
C1-Fluorenes	5.2–5.9	2.5–2.7	0.2–0.3	6.5–7.4
C2-Fluorenes	2.9–3.3	1.5–1.6	1.0–1.1	3.5–3.9
C3-Fluorenes	1.9–2.2	1.2–1.3	0.9–1.0	3.2–3.5
Dibenzothiophene	16.6–18.6	1.5–1.7	1.8–2.1	7.5–8.2
C1-Dibenzothiophenes	10.9–12.2	5.1–5.7	2.7–3.0	11.5–12.9
C2-Dibenzothiophenes	9.0–10.3	5.4–6.1	3.2–3.6	13.2–15.2
C3-Dibenzothiophenes	9.4–10.5	7.1–7.9	4.8–5.5	15.8–18.0
Phenanthrene	12.6–14.6	11.5–13.1	11.8–13.3	25.9–29.6
C1-Phenanthrenes	6.3–7.3	1.0–1.1	1.1–1.2	4.7–5.4
C2-Phenanthrenes	3.2–3.8	1.2–1.3	1.1–1.2	3.9–4.2
C3-Phenanthrenes	1.2–1.3	0.7	0.4–0.5	1.8–1.9
C4-Phenanthrenes	nd	0–1.0	nd	nd
Anthracene	2.3–2.5	2.0–2.2	1.5–1.7	5.9–6.5
Fluoranthene	1.2–1.4	1.6–1.7	0.9–1.0	2.1–2.4
Pyrene	21.2–23.5	7.1–7.8	10.0–11.3	16.4–18.2
C1-Pyrenes	7.1–7.8	0.9–1.0	1.4–1.6	5.2–5.8
C2-Pyrenes	4.3–4.7	1.9–2.1	1.6–1.7	3.2–3.7
C3-Pyrenes	1.4–1.7	0.6–0.7	1.5–1.7	2.2–2.6
C4-Pyrenes	0–2.7	nd	nd	0–2.3
Naphthobenzothiophene	26.6–29.6	2.0–2.3	4.0–4.6	15.5–17.5
C1-Naphthobenzothiophenes	2.0–2.3	1.3–1.5	8.5–9.7	18.0–19.8
C2-Naphthobenzothiophenes	1.1–1.3	2.4–2.7	8.7–9.7	19.4–22.3
C3-Naphthobenzothiophenes	1.1–1.2	2.7–3.0	14.8–16.5	29.2–33.0
Benzo(a)anthracene	20.6–23.5	11.1–12.5	9.5–10.6	43.4–49.1
Chrysene	7.9–8.7	4.6–5.1	8.6–9.6	32.5–36.7
C1-Chrysenes	5.9–6.8	7.2–8.2	1.7–1.9	5.2–5.8
C2-Chrysenes	25.9–28.5	6.0–6.9	2.2–2.4	3.1–3.6
C3-Chrysenes	19.4–22.3	4.1–4.7	2.3–2.6	3.2–3.6
C4-Chrysenes	nd	nd	2.0–2.2	3.4–3.8
Benzo(b)fluoranthene	4.5–5.0	2.3–2.6	3.8–4.1	20.1–22.3
Benzo(k)fluoranthene	4.5–5.0	2.8–3.1	2.4–2.7	7.8–8.9
Benzo(e)pyrene	6.4–7.3	3.6–4.0	5.0–5.7	12.8–14.2
Benzo(a)pyrene	3.0–3.5	2.2–2.4	3.8–4.3	7.6–8.4
Perylene	7.6–8.3	4.2–4.6	4.0–4.6	14.3–16.1
Indeno(1,2,3-cd)pyrene	0.4–0.5	0–0.2	0.5–0.6	1.3–1.5
Dibenzo(a,h)anthracene	0.1	0–0.1	0–0.1	nd
Benzo(g,h,i)perylene	nd	0–0.1	0–0.1	0.4
Total PAHs	256.6–285.9	110.5–122.8	129.4–145.4	372.1–416.8

nd, not detected.

brown cells (cells/field, BG: 8.8 ± 1.0, SL: 11.0 ± 2.9, GT: 6.7 ± 0.6) in digestive glands and connective tissues surrounding intestinal tracts. Oyster gonads were all in the regressing stage at each site.

The intensity of *P. marinus* infection was significantly greater in wild oysters from Grande Terre (8.04 ± 3.14 × 10^5 parasites per gram wet oyster tissue, median: 2.69 × 10^5) than in Sister Lake oysters (5.89 ± 5.72 × 10^5 parasites per gram wet oyster tissue, median: 2.71 × 10^2) with Bay Gardene being intermediate (2.56 ± 1.71 × 10^5 parasites per gram wet oyster tissue, median: 1.11 × 10^3). All Grande Terre oysters were infected and over half of them had either moderate (40%) or heavy infection intensities (33%). In contrast, 66% of Sister Lake oysters were either not infected or had light infection intensities, whereas 27% had moderate infection intensities and 7% had heavy infection intensities.

Condition index of Grande Terre oysters (1.2 ± 0.1) was significantly higher than that of Bay Gardene (0.6 ± 0.1) and Sister Lake (0.6 ± 0.1) oysters. Shell heights of the wild Grande Terre oysters (89 ± 2 mm) were smaller than Bay Gardene (98 ± 2 mm) and Sister Lake (100 ± 3 mm) wild oysters. Dried shell weights followed the same pattern, with Grande Terre (123 ± 7 g) being the lightest followed by Bay Gardene (200 ± 9 g) and Sister Lake (210 ± 13 g). However, dried meat weights in Grande Terre (1.49 ± 0.08 g) oysters were significantly heavier than in Bay Gardene (1.16 ± 0.06 g) and Sister Lake (1.18 ± 0.08 g) oysters.

tPAH concentrations in all oyster composite samples were very low (<100 ng/g dry oyster weight), but Grande Terre (47.1 ± 4.6 ng/g dry oyster weight) and Bay Gardene (36.8 ± 1.4 ng/g dry oyster weight) tPAH concentrations were significantly higher than Sister Lake (16.6 ± 1.1 ng/g dry oyster weight).

August 2011

Salinity during the 30 days preceding sampling in August 2011 differed significantly between sites with Sister Lake (3.4 ± 0.2) being the lowest, Bay Gardene being intermediate (9.0 ± 0.3 ppt), and Grande Terre being the highest (i.e., Barataria Pass, 17.0 ± 1.0). The same results were obtained when salinity was analyzed for the 3 months prior to sampling (SL: 6.0 ± 0.5, BG: 10.3 ± 0.4, GI: 15.6 ± 0.8) (Figure 14.2). Temperatures at all sites for either the month or the 3 months before sampling were comparable (Figure 14.2).

The percentage of phagocytic hemocytes in Grande Terre oysters (20 ± 1) was significantly lower than in Bay Gardene (27 ± 1) and Sister Lake (28 ± 2) oysters. Hemocyte lysosomal stability of Bay Gardene oysters (155 ± 8 minutes) was significantly greater than that of Sister Lake oysters (128 ± 9 minutes), whereas it was intermediate in Grande Terre oysters (131 ± 10 minutes). No significant differences in hemocyte density (cells/mL, BG: 3.6 ± 0.4 × 10^6, SL: 2.1± 0.2 × 10^6, GT: 3.3± 0.4 × 10^6), hemocyte viability (% live, BG: 99.4 ± 0.4, SL: 99.4 ± 0.3, GT: 99.6 ± 0.3), and number of zymosan particles per phagocytic hemocytes (BG: 6.1 ± 0.2, SL: 6.3 ± 0.4, GT: 6.2 ± 0.4) could be found between sites.

Gill HSP70 concentrations in Grande Terre oysters (949 ± 90 ng/mg) were significantly greater than in Bay Gardene (726 ± 32 ng/mg), whereas Sister Lake HSP70 concentrations (735 ± 51 ng/mg) were intermediate. However, no significant differences in the relative level of hsc72 (BG: 0.449 ± 0.039, SL: 0.487 ± 0.046, GT: 0.470 ± 0.033) or hsp69 isoforms (BG: 0.218 ± 0.008, SL: 0.278 ± 0.037, GT: 0.316 ± 0.047) were found between sites.

No differences were found in the density of brown cells (cells/field, BG: 5.6 ± 0.8, SL: 8.7 ± 1.1, GT: 7.2 ± 0.8) in oyster digestive glands and connective tissues surrounding intestinal tracts between sites. The prevalence of gregarines in Grande Terre oysters was 90% whereas no gregarines were found in Bay Gardene and Sister Lake oysters. Oyster gonads were in the spawning stage.

The intensity of *P. marinus* infection was significantly higher in wild oysters from Grande Terre ($1.66 \pm 0.70 \times 10^6$ parasites per gram wet oyster tissue, median: 2.80×10^4) than in Bay Gardene (86 ± 59 parasites per gram wet oyster tissue, median: 15) and Sister Lake oysters ($2.41 \pm 2.37 \times 10^4$ parasites per gram wet oyster tissue, median: 23). Most (93%) oysters deployed in Grande Terre were infected and over half of them had either moderate infection intensities (27%) or heavy infection intensities (40%). In contrast, 60% of Bay Gardene oysters were infected and all had light infection intensities. Infection also had a prevalence of 60% in Sister Lake oysters, and the infected oysters were mostly light cases of infection (53%), with a small percentage of heavy infections (7%).

Condition index of Grande Terre oysters (1.4 ± 0.1) was significantly greater than that of Bay Gardene (0.9 ± 0.1) and Sister Lake (0.9 ± 0.1) oysters. No differences in shell heights were observed between sites (mm, BG: 89 ± 1, SL: 84 ± 8, GT: 86 ± 2); however, Bay Gardene dry shell weights (135 ± 6 g) were heavier than that of Sister Lake (83 ± 3 g) and Grande Terre (73 ± 4 g). Oyster dried meat weights differed significantly between sites, with Bay Gardene oysters having the highest (1.19 ± 0.07 g), Grande Terre the intermediate (0.99 ± 0.04 g), and Sister Lake the lowest (0.69 ± 0.04 g).

tPAH concentrations in all oyster composite samples were low (<100 ng/g dry oyster weight), but they differed significantly between sites, with Grande Terre tPAH concentrations (38.7 ± 3.1 ng/g dry oyster weight) being highest, Bay Gardene intermediate (27.6 ± 1.2 ng/g dry oyster weight), and Sister Lake lowest (17.7 ± 1.2 ng/g dry oyster weight).

Discussion

tPAH concentrations of caged and wild oysters were greater at the oiled site, just off Grande Terre in Barataria Bay in 2010 and 2011, but the concentrations were low (<100 ng/g dw) and at the lower end of the range reported by NOAA National Status and Trends "Mussel Watch" Program for Barataria Bay oysters before the DWH oil spill (Jackson et al 1994; Kimbrough et al. 2006; Johnson et al. 2009). This program, which was initiated in 1986, reported oyster tPAH concentrations in Barataria Bay (i.e., Middle Bank, Bayou Saint Denis sites) that varied considerably from year to year, but with high tPAH concentrations (>1000 ng/g dw) in the late 1980s followed by a trend of decreasing concentrations (O'Connor and Lauenstein 2006). Occasional high oyster tPAH concentrations are not surprising in Louisiana estuaries where the oil and gas industry is well established. Indeed, there have been about 60 cases of localized oyster contamination (oily taste) since the 1940s (Soniat 1988).

It is noteworthy that oysters in our study were sampled starting in October 2010 several months after the oil reached Louisiana estuaries. tPAH concentrations would have likely been higher had oysters been sampled during or shortly after the oiling event but limited funding and inaccessibility to oiled sites during the extended response and cleanup phases of the oil spill precluded more intensive sampling. Xia et al. (2012) did report higher PAH concentrations in Mississippi oysters sampled in July and August 2010 (average: 34 ng/g dw) than in September 2010 and following months (average: 15 ng/g dw). Oysters can reduce accumulated hydrocarbons from their tissues by over 90% within weeks (Stegeman and Teal 1973; Neff et al. 1976; Sericano et al. 1996; Michel and Henry 1997). In terms of public health, concentrations of individual PAH in oysters at all of our sites were far below levels of concern (LOC) for oyster consumption established by NOAA, FDA, and state agencies, which agrees with the findings of Xia et al. (2012) for Mississippi oysters. Extensive sampling of oysters along Gulf of Mexico estuaries through the NOAA Natural Resource Damage Assessment (NRDA) process should provide additional tPAH concentrations for comparison once the data become usable (http://www.gulfspillrestoration.noaa.gov/oil-spill/gulf-spill-data/).

Oyster tPAH concentrations in our studies also appear to be well below those reported to impact oyster biomarker responses (Chu et al. 2002; Hwang et al. 2008; Hannam et al. 2009; Ramdine et al. 2012). For example, catalase was inhibited and lipid peroxidation was increased in field-collected mangrove oysters with tPAH concentrations of 961 ng/g·dw but not in those with PAHs in the 67–241 ng/g dw range (Ramdine et al. 2012). Critical hemocyte lysosomal destabilization (i.e., 50% destabilized after 1 hour) was reported in eastern oysters with 2100 ng/g·dw or greater tPAH concentrations following laboratory exposure (Hwang et al. 2008). Hemocyte neutral red uptake also increased in hemocytes of eastern oysters with tPAH concentrations of about 2500 ng/g wet tissues but not in oysters with below 1500 ng/g wet tissues following laboratory exposure to field-contaminated sediments (Chu et al. 2002). Hemocyte phagocytic activity was reduced in artic scallops with tPAH body burden of 5700 ng/g dw but not 3365 ng/g dw following exposure to low (0.06 mg/L) and high (0.25 mg/L) dispersed oil for 15 days (Hannam et al. 2009). Interestingly, when contaminated arctic scallops were placed in clean water for 7 days hemocyte phagocytic activity returned to control levels (Hannam et al. 2009), and destabilized oyster hemocytes were shown to recover rapidly within weeks of being placed in clean seawater concomitant with decreasing body tPAH concentrations (Hwang et al 2004). Similarly, field studies on the impact of oil spills indicate hemocyte parameters and other biomarkers can recover within a few months (Moore et al. 1987; Dyrynda et al. 2000; Donaghy et al. 2010).

Overall, most of our selected biomarkers showed no significant differences in values between sites at the different sampling times, but there were exceptions. Hemocyte density of caged oysters was significantly lower at the lowest salinity sites (Bay Gardene, Sister Lake) in October 2010 and 2011 and also tended to be lower in May 2010 at the lowest salinities. Hemocyte density of wild oysters sampled in August 2011 also tended to be lower at the lowest salinity site. Finding of lower hemocyte density in eastern oysters maintained at low salinity has been reported previously (Chu et al. 1993). Environmental factors such as salinity and temperature, as well as physiological conditions (e.g., spawning), are known to affect the hemocyte densities and activities of oysters and other bivalves (Chu and La Peyre 1993; Chu 2000; Matozzo and Marin 2011). Hemocyte densities and activities are most often reduced at the extreme ends of a species environmental range. There were also trends of lowered hemocyte lysosomal stabilities and condition index, and increased stress proteins at the lowest salinities (<7) in caged oysters but not consistently in wild oysters. Additional sampling of wild oysters may help determine whether those oysters are better adapted to their local environmental conditions and less affected by shift to more extreme conditions (e.g., extreme low salinity <5 for oysters growing in low salinity water <15). The use of hatchery-produced oysters have generally been recommended because of common genetic background, age, and reproductive status, which allow for more sensitive and accurate measurements of hydrocarbon bioaccumulation and biomarker responses between sites, thus leading to a better assessment of ecological impacts (Andral et al. 2004; Viarengo et al. 2007; Tsangaris et al. 2010). Differences in oyster responses to environmental and anthropogenic conditions because of their genetic background, however, will need to be addressed more thoroughly in future studies considering the diverse and varying environmental conditions encountered in Louisiana estuaries.

In conclusion, adult oysters sampled at one oiled site and non-oiled sites a few months after the DWH oil spill had tPAH concentrations that were low and within ranges not expected to have negative impact on oyster health. The concentrations of the individual PAH measured in sediments were also below the concentrations associated with adverse biological effects in oysters whether adult or embryo, which is the most sensitive stage (Long et al. 1995; Geffard et al. 2002). However, additional oiled sites need to be sampled to validate our conclusions more broadly to Louisiana estuaries. High oyster mortality was recorded only in Bay Gardene, Breton Sound, a non-oiled site, between May and October 2010 and was attributed to an extended period of low salinity (<5) during the summer months concomitant with high water temperature. The low salinity was the result of the unprecedented opening of the Caernarvon freshwater diversion structure to near capacity discharge of 225 $m^3 \cdot s^{-1}$ for several summer months in response to the DWH oil spill (La Peyre et al. 2013). Extensive sampling of oysters on public grounds and privately leased areas by the Louisiana Department of Wildlife and Fisheries (LDWF) in August 2010 also revealed elevated mortalities in part of Breton Sound and Barataria Bay most affected by freshwater infuxes (Banks 2011). In Bay Gardene, LDWF reported a higher mortality (100%) than we measured for caged oysters (64%) suggesting that factors such as sedimentation and predation, in addition to depressed salinities, may have contributed to the elevated mortality. Increased particulate matter leading to increased sedimentation likely increased as a result of increased freshwater influxes from diversion structures (i.e., Caernarvon in Breton Sound, Davis Pond in Barataria Bay) and of increased boat traffic and dredging at cleanup sites. Low salinity appeared to have the greatest stressor impact on the biomarkers used in our studies. Those biomarker responses indicate general stress and can be affected by nonchemical stressors. More intensive sampling throughout the year of Gulf of Mexico estuaries will be needed to characterize the effects of salinity and other potential confounding factors such as temperature, *P. marinus* infection intensities, and suspended particulate matter on biomarker responses. This information will be essential to interpret biological effects data collected after oil spills or other pollution events that may occur in the future in our region characterized by high temporal and spatial environmental variability.

Acknowledgments

We thank two anonymous reviewers for their helpful comments. This research was funded by the Louisiana Sea Grant College Developmental Program in 2010 and by Gulf of Mexico Research Initiative (GoMRI) funds administered by the Louisiana State University Office of Research and Economic Development in 2011. Sediment PAHs collected in 2011 were analyzed through funding to the Coastal Waters Consortium from GoMRI.

References

Akaishi, F. M., S. D. St-Jean, F. Bishay, J. Clarke, I. S. Rabitto, and C. A. Oliveira Ribeiro. 2007. Immunological responses, histopathological finding and disease resistance of blue mussel (*Mytilus edulis*) exposed to treated and untreated municipal wastewater. *Aquatic Toxicology* 82:1–14.

Albert, P. H. 1995. Petroleum and individual polycyclic aromatic hydrocarbons. In *Handbook of Ecotoxicology*, 2nd edition. D. T. Haffman, B. A. Rattner, G. A. Burton, and J. Cairns (Eds.), 330–55. London: Lewis Publishers.

Allan, S. E., B. W. Smith, and K. A. Anderson. 2012. Impact of the *Deepwater Horizon* oil spill on bioavailable polycyclic aromatic hydrocarbons in Gulf of Mexico coastal waters. *Environmental Science and Technology* 46:2033–9.

Anderson, K. E., N. D. Forsberg, and G. R. Wilson. 2011. Determination of parent and substituted polycyclic aromatic hydrocarbons in high-fat salmon using a modified QuEChERS extraction, dispersive SPE and GC-MS. *Journal of Agricultural and Food Chemistry* 59:8108–16.

Andral, B., J. Y. Stanisiere, D. Sauzade, E. Damier, H. Thebault, F. Galgani et al. 2004. Monitoring chemical contamination levels in the Mediterranean based on the use of mussel caging. *Marine Pollution Bulletin* 49:704–12.

AOAC Official Method 2007.01, Pesticide residues in foods by acetonitrile extraction and partitioning with magnesium sulfate. Gas chromatography/mass spectrometry and liquid chromatography/tandem mass spectrometry first action 2007.

Arfsten, D. P., D. J. Schaeffer, and D. C. Mulveny. 1996. The effects of near ultraviolet radiation on the toxic effects of polycyclic aromatic hydrocarbons in animals and plants: A review. *Ecotoxicology and Environmental Safety* 33:1–24.

ATSDR (Agency for toxic substances and disease registry). 2005. *CERCLA Priority List of Hazardous Substances*. Washington, DC: U.S. Department of Health and Human Services.

Auffret, M., S. Rousseau, I. Boutet, A. Tanguy, J. Baron, D. Moraga et al. 2006. A multiparametric approach for monitoring immunotoxic responses in mussels from contaminated sites in Western Mediterranean. *Ecotoxicology and Environmental Safety* 63:393–405.

Baier-Anderson, C. and R. S. Anderson. 2000. Immunotoxicity of environmental pollutants in marine invertebrates. In *Recent Advances in Marine Biotechnology, Immunobiology and Pathology*, volume 5. M. Fingerman and R. Nagabhushanam (Eds.), 189–225. Enfield, NH/Plymouth, UK: Science Publishers.

Banks, P. 2011. Comprehensive report of the 2010 oyster mortality study in Breton Sound and Barataria Basins—May 2011. Louisiana Department of Wildlife and Fisheries. Available at http://www.wlf.louisiana.gov/sites/default/files/pdf/document/34262-oyster-mortality-study/ldwf_-_oyster_mortality_study_-_exec_summary__full_report_combined.pdf (accessed September 2013).

Barron, M. G., T. Podrabsky, S. Ogle, and R. W. Ricker. 1999. Are aromatic hydrocarbons the primary determinant of petroleum toxicity to aquatic organisms? *Aquatic Toxicology* 46:253–68.

Baumard, P., H. Budzinski, and P. Garrigues. 1998. Polycyclic aromatic hydrocarbons (PAHs) in sediments and mussels of the western Mediterranean Sea. *Environmental Toxicology and Chemistry* 17:765–76.

Baumard, P., H. Budzinski, P. Garrigues, J. F. Narbonne, T. Burgeot, X. Michel et al. 1999. Polycyclic aromatic hydrocarbon (PAH) burden of mussels (*Mytilus* sp.) in different marine environments in relation with sediment PAH contamination and bioavailability. *Marine Environmental Research* 47:415–39.

Baussant, T., R. K. Bechmann, I. C. Taban, B. K. Larsen, A. H. Tandberg, A. Bjornstad et al. 2009. Enzymatic and cellular responses in relation to body burden of PAHs in bivalve mollusks: a case study with chronic levels of North Sea and Barents Sea dispersed oil. *Marine Pollution Bulletin* 58:1796–807.

Bocchetti, R., D. Fattorini, B. Pisanelli, S. Macchia, L. Oliviero, F. Pilato et al. 2008. Contaminant accumulation and biomarker responses in cages mussels, *Mytilus galloprovincialis*, to evaluate bioavailability and toxicological effects of remobilized chemicals during dredging and disposal operations in harbor areas. *Aquatic Toxicology* 89:257–66.

Boehm, P. D. and J. G. Quinn. 1977. The persistence of chronically accumulated hydrocarbons in the hard shells clam *Mercenaria mercenaria*. *Marine Biology* 44:227–33.

Cheng, T. C. 1996. Hemocytes: forms and functions. In *The Eastern Oyster Crassostrea virginica*. V. S. Kennedy, R. I. E. Newell, and A. F. Eble (Eds.), 299–333. College Park, MD: Maryland Sea Grant.

Chu, F.-L. E. 2000. Defense mechanism of marine bivalves. In *Recent Advances in Marine Biotechnology, Immunology and Pathology*, volume 5. M. Fingerman and R. Nagabhushman (Eds.), 1–42. Enfield, NH/Plymouth, UK: Science Publishers.

Chu, F.-L. E. and R. C. Hale. 1994. Relationship between pollution and susceptibility to infectious disease in eastern oyster, *Crassostrea virginica*. *Marine Environmental Research* 38:243–56.

Chu, F.-L. E. and J. F. La Peyre. 1993. *Perkinsus marinus* susceptibility and defense-related activities in eastern oysters (*Crassostrea virginica*): temperature effects. *Diseases of Aquatic Organisms* 16:223–34.

Chu, F.-L. E., J. F. La Peyre, and C. S. Burreson. 1993. *Perkinsus marinus* infection and potential defense-related activities in eastern oysters, *Crassostrea virginica*: salinity effects. *Journal of Invertebrate Pathology* 62:226–32.

Chu, F.-L. E., A. K. Volety, R. C. Hale, and Y. Huang. 2002. Cellular responses and disease expression in oysters (*Crassostrea virginica*) exposed to suspended field-contaminated sediments. *Marine Environmental Research* 53:17–35.

Claisse, D. and B. Beliaeff. 2000. Tendances temporelles des teneurs en contaminants dans les mollusques du littoral français. In *Surveillance du Milieu Marin. Travaux du RNO.* édition 2000, 9–32. Ifremer et Ministère de l'Aménagement du territoire et de l'Environnement. ISSN 1620-1124.

Coen, L. D., M. W. Luckenbach, and D. L. Breitburg. 1999. The role of oyster reefs as essential fish habitats: a review of current knowledge and some new perspectives. In *Fish Habitat: Essential Fish Habitat and Rehabilitation.* L. R. Bernaka (Ed.), 438–54. Bethesda, MD: American Fisheries Society, Symposium 22.

Connell, D. W. and G. J. Miller. 1981. Petroleum hydrocarbons in aquatic ecosystems—behavior and effects of sublethal concentrations Part 2. *Critical Reviews in Environmental Control* 11:105–62.

Dame, R. 1996. *Ecology of Marine Bivalves: An Ecosystem Approach.* Boca Raton, FL: CRC Marine Science Series Press.

Del Castillo, C., J. Corredor, and J. Morell. 1992. Accumulation and depuration of hydrocarbons in the mangrove oyster *Crassostrea rhizophorae*. 2. Conference Internationale sur la Purification des Coquillages, Rennes, France. Available at http://archimer.ifremer.fr/doc/00000/1602/ (accessed September 2013).

Desclaux-Marchand, C., I. Paul-Pont, P. Gonzalez, M. Baudrimont, J. P. Bourdineaud, and X. De Montaudouin. 2007. Metallothionein gene identification and expression in the cockle (*Cerastoderma edule*) under parasitism (trematodes) and cadmium contaminations. *Aquatic Living Resources* 20:43–9.

Donaghy, L., H.-K. Hong, H.-J. Lee, J.-C. Jun, Y.-J. Park, and K.-S., Choi. 2010. Hemocyte parameters of the Pacific oyster *Crassostrea gigas* a year after the Hebei Spirit oil spill off the west coast of Korea. *Helgoland Marine Research* 64:349–55.

Donaghy, L., B.-K. Kim, H.-K. Hong, H.-S. Park, and K.-S. Choi. 2009. Flow cytometry studies on the populations and immune parameters of the hemocytes of the Suminoe oyster, *Crassostrea ariakensis*. *Fish and Shellfish Immunology* 27:296–301.

Dyrynda, E. A., R. J. Law, P. E. J. Dyrynda, C. A. Kelly, R. K. Pipe, and N. A. Ratcliffe. 2000. Changes in immune parameters of natural mussel *Mytilus edulis* populations following a major oil spill ('Sea Empress', Wales, UK). *Marine Ecology-Progress Series* 206:155–70.

Encomio, V. G. and F.-L. E. Chu. 2007. Heat shock (HSP70) expression and thermal tolerance in sublethally heat-shocked eastern oysters *Crassostrea virginica* infected with the parasite *Perkinsus marinus*. *Diseases of Aquatic Organisms* 76:251–60.

Fisher, W. S. and L. M. Oliver. 1996. A whole-oyster procedure for diagnosis of *Perkinsus marinus* disease using Ray's fluid thioglycollate culture medium. *Journal of Shellfish Research* 15:109–18.

Fisher, W. S., L. M. Oliver, W. W. Walker, S. Manning, and T. Lytle. 1999. Decreased resistance of eastern oysters (*Crassostrea virginica*) to a protozoan pathogen (*Perkinsus marinus*) after sub-lethal exposure to tributyltin oxide. *Marine Environmental Research* 47:185–201.

Geffard, O., H. Budzinski, E. His, M. N. L. Seaman, and P. Garrigues. 2002. Relationships between contaminant levels in marine sediments and their biological effects on embryos of oysters, *Crassostrea gigas*. *Environmental Toxicology and Chemistry* 21:2310–8.

Grizzle, R. E. J., J. K. Greene, and L. D. Coen. 2008. Seston removal by natural and constructed intertidal Eastern oyster (*Crassostrea virginica*) reefs: a comparison with previous laboratory studies and the value of in situ methods. *Estuaries and Coasts* 31:1208–20.

Hannam, M. L., S. D. Bamber, J. A. Moody, T. S. Galloway, and M. B. Jones. 2009. Immune function in the arctic scallop, *Chlamys islandica*, following dispersed oil exposure. *Aquatic Toxicology* 92:187–94.

Howard, D. W. and C. S. Smith. 1983. Histological techniques for marine bivalve mollusks. NOAA Technical Memorandum NMFS-F/NEC-25. U.S. Department of Commerce, Woods Hole, MA.

Hwang, H.-M., T. Wade, and J. L. Sericano. 2004. Destabilized lysosomes and elimination of polycyclic aromatic hydrocarbons and polychlorinated biphenyls in eastern oysters (*Crassostrea virginica*). *Environmental Toxicology and Chemistry* 23:1991–5.

Hwang, H.-M., T. Wade, and J. L. Sericano. 2008. Residue-response relationship between PAH body burdens and lysosomal membrane destabilization in eastern oysters (*Crassostrea virginica*) and toxicokinetics of PAHs. *Journal of Environmental Science and Health part A-Toxic/Hazardous Substances and Environmental Engineering* 43:1373–80.

Jackson, T. J., T. L. Wade, T. J. McDonald, D. L. Wilkinson, and J. M. Brooks. 1994. Polynuclear aromatic hydrocarbon contaminants in oysters from the Gulf of Mexico (1986–1990). *Environmental Pollution* 83:291–8.

Johnson, W. E., K. L. Kimbrough, G. G. Lauenstein, and J. Christensen. 2009. Chemical contamination assessment of Gulf of Mexico oysters in response to hurricanes Katrina and Rita. *Environmental Monitoring and Assessment* 150:211–25.

Johnson, Y. S. 2012. Determination of polycyclic aromatic hydrocarbons in edible seafood by QuEChERS-based extraction and gas chromatography-tandem mass spectrometry. *Journal of Food Science* 77:T131–7.

Kennedy, V. S. and L. B. Krantz. 1982. Comparative gametogenic and spawning patterns of the oyster *Crassostrea virginica* (Gmelin) in central Chesapeake Bay. *Journal Shellfish Research* 2:133–40.

Kim, Y., E. N. Powell, and K. A. Ashton-Alcox. 2006. Histopathology analysis. In *Histological Techniques for Marine Bivalve Molluscs: Update*. 19–52, Silver Spring, MD: NOAA Technical Memorandum NOS NCCOS 27.

Kimbrough, K. L., G. G. Lauenstein, and W. E. Johnson (Eds.). 2006. Organic Contaminant Analytical Methods of the National Status and Trends Program: Update 2000–2006. Silver Spring, MD: NOAA Technical Memorandum NOS NCCOS 30.

La Peyre, J. F., F.-L. E. Chu, and W. K. Vogelbein 1995. In vitro interaction of *Perkinsus marinus* merozoites with Eastern and Pacific oyster hemocytes. *Developmental and Comparative Immunology* 19:291–304.

La Peyre, M. K., B. S. Eberline, T. M. Soniat, and J. F. La Peyre. 2013. Differences in extreme low salinity timing and duration differentially affect eastern oyster (*Crassostrea virginica*) size class growth and mortality in Breton Sound, LA. *Estuarine Coastal and Shelf Science* 135:146–57.

La Peyre, M. K., B. Gossman, and J. F. La Peyre. 2009. Defining optimal freshwater flow for oyster production: effects of freshet rate and magnitude of change and duration on eastern oysters and *Perkinsus marinus* infection. *Estuaries and Coasts* 32:522–34.

La Peyre, M. K., A. D. Nickens, A. K. Volety, G. S. Tolley, and J. F. La Peyre. 2003. Environmental significance of freshets in reducing *Perkinsus marinus* infection in eastern oysters *Crassostrea virginica*: potential management applications. *Marine Ecological Progress Series* 248:165–76.

Livingston, D. R. 1985. Responses of the detoxification/toxication enzyme systems of molluscs to organic pollutants and xenobiotics. *Marine Pollution Bulletin* 16:158–64.

Long, E. R., D. D. MacDonald, S. L. Smith, and F. D. Calder. 1995. Incidence of adverse biological effects within ranges of chemical concentrations in marine and estuarine sediments. *Environmental Management* 19:81–97.

Lowe, D. M., C. Soverchia, and M. N. Moore. 1995. Lysosomal membrane responses in the blood and digestive cells of mussels experimentally exposed to fluoranthene. *Aquatic Toxicology* 33:105–12.

Lucas, A. and P. G. Beninger. 1985. The use of physiological condition indices in marine bivalve aquaculture. *Aquaculture* 44:187–200.

Mahoney, B. M. S. and G. S. Noyes. 1982. Effects of petroleum on feeding and mortality of the American oyster. *Archives of Environmental Contamination and Toxicology* 11:527–31.

Mann, R. 1978. A comparison of morphometric, biochemical, and physiological indices of condition in marine bivalve mollusks. In *Energy and Environmental Stress in Aquatic Systems*. J. H. Thorp and J. W. Gibbons (Eds.), 484–97. Springfield, VA: National Technical Information Service, D. O. E. Symposium series (Conf-771114).

Martinez-Gomez, C., A. D. Vethaak, K. Hylland, T. Burgeot, A. Köhler, B. P. Lyons et al. 2010. A guide to toxicity assessment and monitoring effects at lower levels of biological organization following marine oil spills in European waters. *ICES Journal of Marine Science* 67:1105–18.

Matozzo, V. and M. G. Marin. 2011. Bivalve immune responses and climate changes: Is there a relationship? *Invertebrate Survival Journal* 8:70–77.

Michel, J. and C. B. Henry. 1997. Oil uptake and depuration in oysters after use of dispersants in shallow water in El Salvador. *Spill Science and Technology Bulletin* 4:57–70.

Minguez, L., T. Buronfosse, J.-N. Beisel, and L. Giamberini. 2012. Parasitism can be a confounding factor in assessing the responses of zebra mussels to water contamination. *Environmental Pollution* 162:234–40.

Moore, M. N., J. I. Allen, and A. McVeigh. 2006. Environmental prognostics: an integrated model supporting lysosomal stress responses as predictive biomarkers of animal health status. *Marine Environmental Research* 61:278–304.

Moore, M. N., D. R. Livingstone, J. Widdows, D. M. Lowe, and R. K. Pipe. 1987. Molecular, cellular and physiological effects of oil-derived hydrocarbons on molluscs and their use in impact assessment. *Philosophical Transactions of the Royal Society London B-Biologic* 316:603–23.

Mount, A. S., A. P. Wheeler, R. P. Paradkar, and D. Snider. 2004. Hemocyte-mediated shell mineralization in the Eastern oyster. *Science* 304(5668):297–300

Neff, J. M., B. A. Cox, and J. W. Anderson. 1976. Accumulation and release of petroleum-derived aromatic hydrocarbons by four species of marine animals. *Marine Biology* 38:279–89.

Norena-Barroso, E., G. Gold-Bouchot, O. Zapata-Perez, and J. L. Sericano. 1999. Polynuclear aromatic hydrocarbons in American oysters *Crassostrea virginica* from the Terminos Lagoon, Campeche, Mexico. *Marine Pollution Bulletin* 38:637–45.

O'Connor, T. and G. G. Lauenstein. 2006. Trends in chemical concentrations in mussels and oysters collected along the US coast: update to 2003. *Marine Environmental Research* 62:261–85.

Piazza, B. P., P. D. Banks, and M. K. La Peyre. 2005. The potential for created oyster shell reefs as a sustainable shoreline protection strategy in Louisiana. *Restoration Ecology* 13:499–506.

Pickering, R. W. 1999. A toxicological review of polycyclic aromatic hydrocarbons. *Journal of Toxicology-Cutaneous and Ocular Toxicology* 18:101–35.

Plunket, J. and M. K. La Peyre. 2005. Oyster beds as fish and macroinvertebrate habitat in Barataria Bay, Louisiana. *Bulletin of Marine Science* 77:155–64.

Ramalhosa, M. J., P. Paiga, S. Morais, and C. D. Matos. 2009. Analysis of polycyclic aromatic hydrocarbons in fish: evaluation of a quick, easy, cheap, effective, rugged, and safe extraction method. *Journal of Separation Science* 32:3529–38.

Ramdine, G., D. Fichet, M. Louis, and S. Lemoine. 2012. Polycyclic aromatic hydrocarbons (PAHs) in surface sediment and oysters (*Crassostrea rhizophorae*) from mangrove of Guadeloupe: levels, bioavailability, and effects. *Ecotoxicology and Environmental Safety* 79:80–9.

Ringwood, A. H., D. E. Conners, and J. Hoguet. 1998. Effects of natural and anthropogenic stressors on lysosomal destabilization in oysters *Crassostrea virginica*. *Marine Ecology Progress Series* 166:163–71.

Ringwood, A. H., J. Hoguet, C. Keppler, and M. Gielazyn. 2004. Linkages between cellular biomarker responses and reproductive success in oysters—*Crassostrea virginica*. *Marine Environmental Research* 58:151–6.

Sericano, J. L., T. L. Wade, and J. M. Brooks. 1996. Accumulation and depuration of organic contaminants by the American oyster (*Crassostrea virginica*). *Science of the Total Environment* 179:149–60.

Soniat, T. M. 1988. Oil and oyster industry conflicts in coastal Louisiana. *Journal of Shellfish Research* 7:511–4.

Stegeman, J. J. and J. M. Teal. 1973. Accumulation, release and retention of petroleum hydrocarbons by the oyster *Crassostrea virginica*. *Marine Biology* 22:37–44.

Supan, J. E. and C. A. Wilson. 2001. Analyses of gonadal cycling by oyster broodstock, *Crassostrea virginica* (Gmelin), in Louisiana. *Journal of Shellfish Research* 20:215–20.

Tsangaris, C., K. Kormas, E. Strogyloudi, I. Hatzianestis, C. Neofitou, B. Andral et al. 2010. Multiple biomarkers of pollution effects in caged mussels on the Greek coastline. *Comparative Biochemistry and Physiology C-Toxicology and Pharmacology* 151:369–78.

Viarengo, A., D. Lowe, C. Bolognesi, E. Fabbr, and A. Koehler. 2007. The use of biomarkers in biomonitoring: A 2-tier approach assessing the level of pollutant-induced stress syndrome in sentinel organisms. *Comparative Biochemistry and Physiology C-Toxicology and Pharmacology* 146:281–300.

Volety, A. K., L. M. Oliver, F. J. Genthner, and W. S. Fisher. 1999. A rapid tetrazolium dye reduction assay to assess the bactericidal activity of oyster *Crassostrea virginica* hemocytes against *Vibrio parahaemolyticus*. *Aquaculture* 172:205–22.

Volety, A. K., S. Savarese, G. Tolley, P. Sime, P. Goodman, P. Doering et al. 2009. Eastern oysters (*Crassostrea virginica*) as an indicator for restoration of Everglades Ecosystems. *Ecological Indicators* 9: S120-S136. doi:10.1016/j.ecolind.2008.06.005.

Winstead, J. T. and J. A. Couch. 1988. Enhancement of protozoan pathogen *Perkinsus marinus* infections in American oysters *Crassostrea virginica* exposed to the chemical carcinogen n-nitrosodiethylamine (DENA). *Diseases of Aquatic Organisms* 5:205–13.

Xia, K., G. Hagood, C. Childers, J. Atkins, B. Rogers, L. Ware et al. 2012. Polycyclic aromatic hydrocarbons (PAHs) in Mississippi seafood from areas affected by the *Deepwater Horizon* oil spill. *Environmental Science and Technology* 46:5310–8.

Yeats, P., F. Gagne, and J. Hellou. 2008. Body burden of contaminants and biological effects in mussels: an integrated approach. *Environment International* 34:254–64.

Zaroogian, G. and P. Yevich. 1993. Cytology and biochemistry of brown cells in *Crassostrea virginica* collected at clean and contamination stations. *Environmental Pollution* 79:191–7.

15

Occurrence of Lemon Sharks (Negaprion brevirostris) *at the Chandeleur Islands, Louisiana, before and after the 2010* Deepwater Horizon *Disaster*

Jonathan F. McKenzie, Christopher Schieble, Patrick W. Smith, and Martin T. O'Connell

CONTENTS

Introduction

In the United States, the Gulf Coast is known for its extensive chains of barrier islands. Louisiana's barrier islands have attracted the most attention in recent years because of impacts to them by an increase in tropical storm activity, primarily the occurrence of Hurricanes Katrina (2005), Rita (2005), and Gustav (2008). These storms caused significant land loss on Louisiana barrier islands (Kulp 2005; Fearnley et al. 2009). Tropical storm activity is not the only cause of land loss observed at these barrier islands. During the past 215 years, the Southwest Pass (the main Mississippi River shipping channel) has advanced into the Gulf of Mexico by 15 km (Dean 2006). Without restoration efforts, the future of the Louisiana barrier islands is uncertain. The barrier island and wetland systems of Louisiana will never fully return to their previous state (Kulp 2005; Blum and Roberts 2009). Due to the degradation of these systems, they have become increasingly susceptible to inundation from relative sea-level rise. The Louisiana region is experiencing the highest relative sea-level rise of any area within the United States, with local rates in excess of or equal to 1.00 cm/y (Kolker et al. 2011).

Barrier islands serve as important nursery grounds for multiple species (Norcoss et al. 1995; Layman 2000; Dorenbosch et al. 2004). When barrier islands are remote and mostly isolated from mainland impacts, they confer increased benefits to juvenile organisms. Benefits include greater habitat quality and reduced predation pressure. In Louisiana, the Chandeleur Islands are the oldest and most remote barrier island chain in the state. Unfortunately, these islands are threatened by erosion, global sea-level rise, hurricane

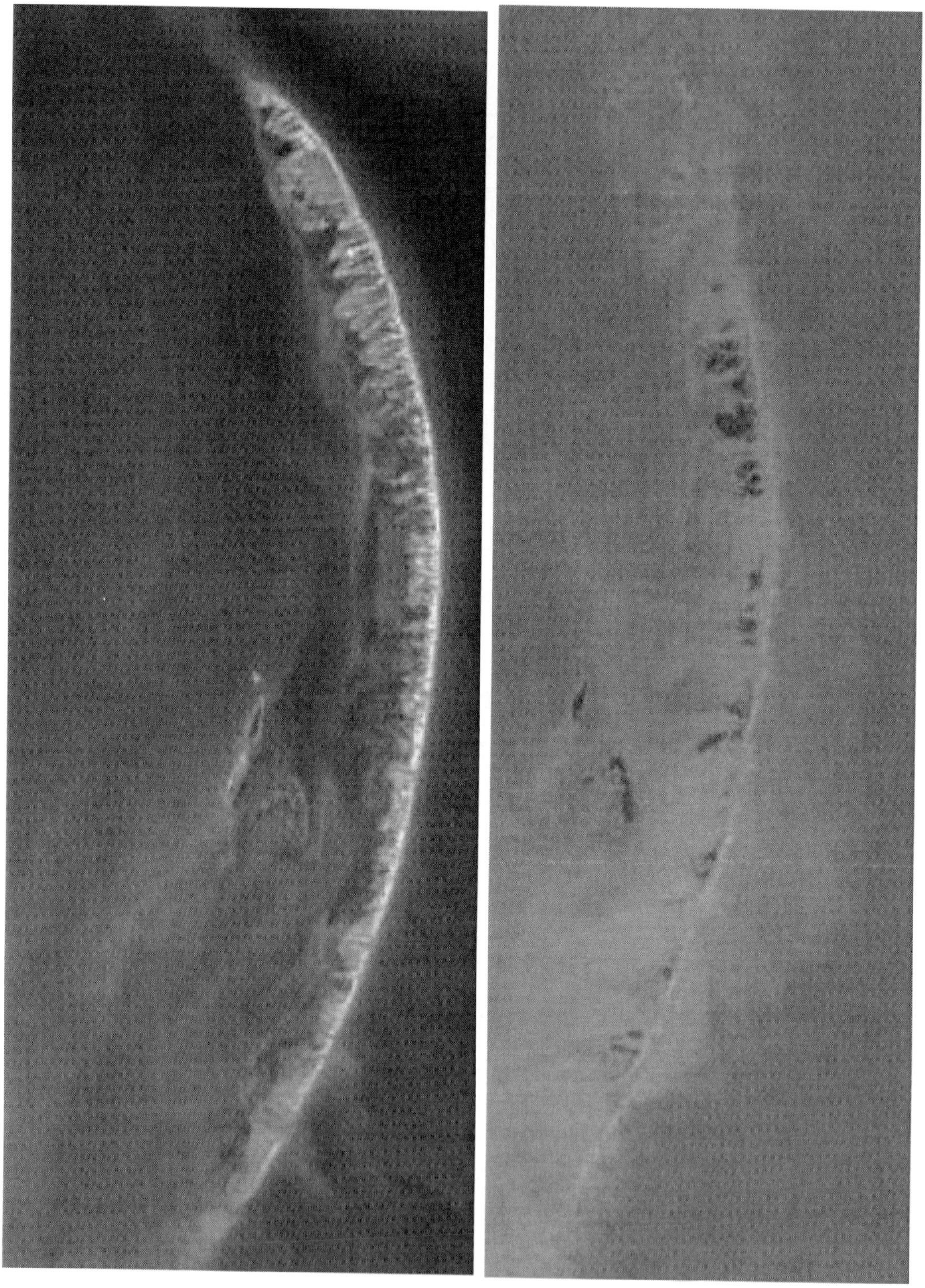

FIGURE 15.1
Location of the Chandeleur Islands, Louisiana, in the northern Gulf of Mexico.

activity, and now oil and dispersant contamination (Figure 15.1), which may jeopardize efforts to restore them (Kulp 2005; Blum and Roberts 2009). Preliminary research at the Chandeleur Islands suggested that juvenile lemon sharks (*Negaprion brevirostris*) occurred in shallow habitats (Laska 1973), though this was based on the capture of only two individuals. More than 30 years later, in 2008, sampling produced 34 juvenile *N. brevirostris* in these same habitats. These sharks were collected from several sites around the islands and all were measured, tagged, and released (C. Schieble, University of New Orleans, unpublished data). Both studies suggest that not only has *N. brevirostris* been using Chandeleur Island habitats for many decades, but that their numbers have increased locally in recent years. Because the majority of *N. brevirostris* collected in 2008 were densely clustered in shallow habitats, it has been suggested that these sharks are using these areas to avoid predation from larger sharks. These shallow habitats were created as a result of recent hurricanes that created cuts and passes across the Chandeleur Islands. Much like *N. brevirostris* nursery habitats in other parts of the world (e.g., Bimini), these shallow areas exclude large predators (Ruple 1984; Luettich et al. 1999; Trnski 2001; Brown et al. 2004). It is possible that these recently formed habitats have the potential to serve as *N. brevirostris* nursery grounds in Louisiana.

Young sharks will benefit when shallow nursery habitats provide not only protection from predators but also an ample food supply (Beck et al. 2001). The northern sections of the Chandeleur Islands remain the most undisturbed of the island chain, retaining a large back barrier area characterized by extensive sea grass beds. Globally, sea grass beds are recognized as being highly productive habitats supporting multiple tropic levels (Duffy 2006; Orth et al. 2006). Sea grass beds should support an abundance of prey items that are easily accessible from nearby sand flats, habitats that have been shown to be preferred by young *N. brevirostris* in other locations (Springer 1950; Morrissey and Gruber 1993a,b). Connectivity between these sea grass beds and the surf zone increased after recent storms, improving access for any *N. brevirostris* that may have been pupped in the surf zone, possibly augmenting their growth and survival. Unfortunately, in summer 2010, both habitat types were inundated with oil from the *Deepwater Horizon* oil spill disaster (*DWH*). Since that time, little research has been conducted on the effect of the oil and dispersants on the health of sharks and shark populations in the region, making it especially important to track the numbers of sharks caught during 2010 and 2011. Following the disaster, a large sand berm was constructed on the Gulf side of the island using material dredged from the north end of the islands (Figure 15.2). This artificial feature filled in numerous aquatic habitats used by *N. brevirostris* and reduced connectivity between sea grass and surf zone habitats, along with producing other unwanted ecological impacts (Martinez et al. 2012).

The purpose of our research was to use *N. brevirostris* occurrence and catch per unit effort (CPUE) data taken at the Chandeleur Islands before, during, and after *DWH* to determine if the oil or construction of the artificial berm had an impact on this population. Differences in habitat use between sexes and different size classes of *N. brevirostris* were also investigated.

Methods

Lemon Shark Sampling

Between April 2009 and August 2011, *N. brevirostris* were sampled at the Chandeleur Islands in Southeast Louisiana (Figure 15.1) on nine separate occasions. These targeted sampling efforts were conducted during pupping periods (May through October).

FIGURE 15.2
Photograph of the oil protection sand berm built at the Chandeleur Islands following the *Deepwater Horizon* disaster. The berm closed off many of the naturally formed cuts and passes, two of which can be seen in the lower right.

Pupping periods were targeted so that both offspring and mothers had the potential to be collected. Sampling consisted of two researchers using a small skiff (4.88 m) outfitted with a ladder (2.44 m) for spotting the sharks. One researcher searched for sharks from the top of the ladder, whereas the other poled the skiff through shallow water (0.5–2 m). Each researcher was equipped with a 2.13 m spinning rod outfitted with 13.6 kg braided fishing line. A steel leader was attached to the fishing line and a 3/0 J-hook was connected to the leader. Lines were primarily baited with either juvenile spot (*Leiostomus xanthurus*) or pinfish (*Lagodon rhomboides*). The sampling effort took place from 0800 hours and generally continued until 1800 hours, unless interrupted by storm activity.

When a shark was sighted, bait was cast and placed in the vicinity of the individual until the subject took the hook. The shark was allowed to eat the bait and remain on the line until the hook could be set. Coordinates were taken with a handheld GPS-72h (Garmin International Inc., Chicago, Illinois) when sharks were sighted. Once each shark was captured, it was brought aboard the boat. It was then sexed, measured (total length, fork length, and precaudal length), and weighed. The shark was then scanned with a Pocket Reader (Biomark Inc., Boise, Idaho) passive integrated transponder (PIT) tag reader. If the individual was untagged, it was then injected with a HPT12 PIT tag (Biomark Inc.) at the bottom left side of their front dorsal fin. When biological data were taken from the shark and the tag was injected, it was released. Water quality characteristics were also recorded at the shark's initial location. Salinity, temperature, conductivity, specific conductivity, dissolved oxygen, and pH were measured using an YSI 556 Multiprobe (Yellow Springs Instruments Inc., Yellow Springs, Ohio). All sharks were collected and handled under UNO-IACUC protocol #09–009.

Data Analysis

To test for possible differences in habitat use between *N. brevirostris* males and females and different size classes, sharks were divided into three different size classes: (1) size class 1, (2) size class 2, and (3) size class 3. Size class 1 was defined as all sharks retaining open umbilical scars. The largest neonate collected was 74 cm tail length (TL). This became the largest size for sharks to be considered size class 1. Using the 90 cm TL size denoted by Chapman et al. (2009) as a larger immature (generally, >3 years old in Bimini), all sharks ranging from 74 cm to 90 cm were designated to be size class 2 individuals. Sharks larger than 90 cm TL were defined as size class 3. Data were recorded and input into Microsoft Excel (Microsoft Corporation, Redmond, Washington) and analyzed using R statistical software (version 2.13.1). Data were tested for homogeneity of variances using Shapiro–Wilk tests of normality ($\alpha < 0.05$). A generalized linear model (GLM) was used to analyze relationships among sizes, sexes, and environmental characteristics (water parameters, distance from shore, depth).

To determine if *N. brevirostris* at the Chandeleur Islands were showing a similar habitat preference to those at other populations, researchers throughout the Gulf of Mexico were contacted to request occurrence data they may have collected on *N. brevirostris*. Water depth and size of shark were plotted to ascertain if any commonalities existed between the Chandeleur Islands population and populations from other known nursery grounds.

Due to a lack of continuous monitoring of environmental characteristics (data only taken when a shark was captured), it was necessary to perform several different analyses to determine if there was a change in shark numbers before and after *DWH*. If such a change occurred, it was important to determine if the change was a result of this event or natural variation in environmental characteristics. The first method of analysis calculated the probability of catching a lemon shark before or after the event. This test was completed using a logistic model and a binomial assumption of catch or no catch during sampling.

The second method analyzed the change in shark abundance following the disaster. The first step was to remove all noncatch data due to the lack of environmental values when no shark was captured. Multiple regressions were then run to determine if correlations existed between any of the predictor variables (GLM, $\alpha = 0.05$). Hour, day, and dissolved oxygen were all removed from further analysis due to significant correlation among these response variables, which violates the assumption of independence among independent variables in multiple regression models. A step-wise GLM reduction was then performed using Akaike information criterion (AIC) scores on log-transformed CPUE data. Factors used in this analysis were (1) month of capture (to protect against seasonal influence), (2) year (a surrogate for oil presence/absence), (3) salinity, (4) depth, (5) temperature, and (6) distance from shore.

A modified form of CPUE for this project was developed based on dividing the number of sharks caught by the total number of hours spent fishing. This CPUE was calculated on a daily basis and then averaged per trip to determine a monthly CPUE value (each trip lasted 4 days with the exception of May 2010, which lasted 3 days). CPUE numbers were compared across years and pre- and postdisaster spill/berm construction. Daily and monthly CPUE were tested for homogeneity of variances using a Shapiro–Wilk test of normality ($\alpha = 0.05$). CPUE was then compared using a two-way analysis of variance (ANOVA, $\alpha = 0.05$). This made it possible to determine if there was any change in the number of sharks being sampled over the course of the study. Although there have been a number of studies that suggest that CPUE is not proportional to abundance, these studies tend to investigate commercial techniques (Punt et al. 2000; Harley et al. 2001; Bigelow et al. 2002; Walters 2003; Kleiber and Maunder 2008). There have been several studies that have looked at the relationship between CPUE with rod-reel fishing and abundance and have found a positive correlation (Tsubois and Endoua 2008; Alford and Jackson 2010). Sight fishing is comparable to encounter data in that only visible individuals can be captured. Encounter data are currently used in studies looking at the abundance of both terrestrial and aquatic vertebrates (Hochachka et al. 2000; Holmberg et al. 2009; Wiley and Simpfendorfer 2010).

Results

During 2009, four-day trips were made in May, June, July, and August. In 2010, sampling efforts were impeded by the *DWH*, which began in April. One sampling trip was conducted immediately after the oil spill in May and again in August and October, after most of the area had been inundated with oil (Figure 15.3). In 2011, trips were conducted in May and August. From May 2009 until August 2011, 115 sharks were collected, measured, and tagged (52 males and 63 females, Table 15.1). Sharks ranged in size from 56.8 to 177.0 cm TL. There were no significant differences between the sexes and any water characteristics or substrate composition (ANOVA, $p > .05$, Table 15.2). The only differences found among size classes and water characteristics were with size class 1 (TL < 74 cm) and size class 2 (74 > TL < 90 cm, Table 15.3). These size classes showed a significantly higher occurrence in shallower water ($F = 11.719$, $p < .001$) than the largest size class (Figure 15.4). No differences were found between size classes or sex and water quality.

When shark collection data from throughout the Gulf of Mexico were analyzed, patterns were similar to those we observed at the Chandeleur Islands. These data from various researchers yielded catch and size information on 912 lemon sharks. Size class 2 sharks were typically the most abundant age class in these collections (Figure 15.5). It is doubtful that gear selectivity biased the numbers or sizes of sharks, because various forms of sampling gear (longline, gill net, and hook and line) were used at different sites (Table 15.4). The depth and size relationship observed at the Chandeleur Islands was investigated further and, similar to our results, a high correlation exists between shark size and depth throughout the Gulf of Mexico ($R^2 = 0.64$, Figure 15.6).

The probability of catching a shark before and after the disaster showed a significant decrease between 2009 and 2011 (logistic regression model, $Z = -2.34$, $p = .02$). The GLM comparing hourly CPUE among years and interaction effects of environmental variables yielded complex results that may suffer from a lack of a long-term data set. The model showed correlation between and among several environmental variables including year and salinity, temperature, distance, and depth (Table 15.5). To determine the extent of the interaction between CPUE and these variables, graphs were made for each predictor variable and then the slopes and R^2 values were analyzed. From these values, it does not appear that there is a strong relationship between hourly CPUE and any of the tested environmental variables (Table 15.6).

Due to the complexity and uncertain results provided by the hourly CPUE model, it was necessary to investigate a monthly CPUE. This analysis revealed that CPUE decreased significantly (ANOVA, $F = 9.04$, $p = .02$) following the oil spill (Figure 15.7). On the basis of collection locality data, the number of sharks collected in back barrier marshes appeared to have decreased since the construction of the oil prevention sand berm, which began construction in June 2010 (Figure 15.8).

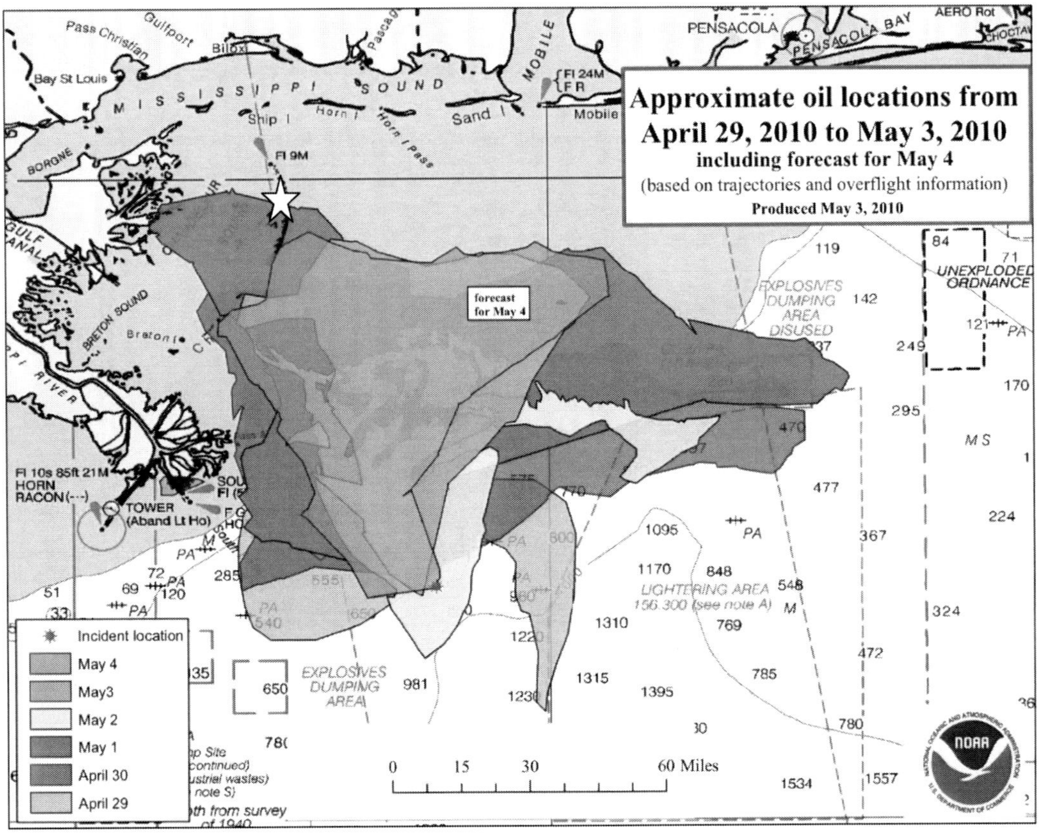

FIGURE 15.3
(See color insert.) Map of oil locations resulting from the *Deepwater Horizon* disaster from April 29, 2010 to May 3, 2010. The Chandeleur Islands (upper left corner, starred) were inundated with oil on May 1, 2010. (Courtesy of NOAA.)

TABLE 15.1
Results of Monthly *Negaprion brevirostris* Sampling Efforts at the Chandeleur Islands, Louisiana from 2009 to 2011

Dates	*N*	Males	Females	Size Class 1	Size Class 2	Size Class 3
May-09	14	3	11	9	4	1
June-09	14	10	4	3	4	7
July-09	10	4	6	8	0	1
August-09	18	11	7	8	4	5
May-10	25	11	14	13	5	6
August-10	15	6	9	10	2	3
October-10	3	0	3	2	1	0
May-11	5	2	3	1	4	0
August-11	11	5	6	10	0	1

Note: Size class 1 < 74 cm, size class 2 < 90 cm, > 74 cm, and size class 3 > 90 cm (N = 115 sharks).

TABLE 15.2

Results from Generalized Linear Model Comparing Sex of *Negaprion brevirostris* Caught at the Chandeleur Islands, Louisiana to Water Quality and Habitat Characteristics

	Estimate	Std. Error	*z* Value	Pr (≥ |z|)
(Intercept)	−2.12	3.11	−0.68	0.50
Temperature	0.02	0.10	0.25	0.81
Salinity	−0.01	0.05	−0.14	0.89
Dissolved oxygen	0.07	0.13	0.54	0.59
Depth	2.14	1.09	1.95	0.05
Distance	0.002	0.01	0.36	0.72

Note: Characteristics include temperature, salinity, dissolved oxygen, water depth where shark was sighted, and distance from shore that sharks were sighted. None of these characteristics significantly affected the sex distribution.

TABLE 15.3

Results from Generalized Linear Model Comparing Size Classes of *Negaprion brevirostris* Caught at the Chandeleur Islands, Louisiana to Water Quality and Habitat Characteristics

	Estimate	Std. Error	*t* Value	Pr (≥ |t|)
(Intercept)	2.79	1.16	2.41	0.02*
Temperature	−0.03	0.04	−0.94	0.35
Salinity	−0.03	0.02	−1.40	0.17
Dissolved oxygen	−0.01	0.05	−0.24	0.81
Depth	1.91	0.40	4.86	<0.001*
Distance	−0.0002	0.002	−0.08	0.93

Note: Characteristics include temperature, salinity, dissolved oxygen, water depth where shark was sighted, and distance from shore that sharks were sighted. Depth was the only parameter that affected size distribution.

Cells labeled with * indicate a significant ($p < .05$) effect.

Discussion

The Chandeleur Islands are a unique habitat for neonatal and juvenile *N. brevirostris*, as unlike other known habitats they lack major structures such as mangroves. Sea grass beds have also been in decline due to increased tropical activity; for example, following Hurricane Katrina, sea grass cover decreased by 20% (Bethel et al. 2007). Mangroves and sea grass beds act as feeding grounds and refuge from predators for juveniles of many species including sharks (Nagelkerken et al. 2000a; Newman et al. 2007; DeAngelis et al. 2008). Previous studies of fish assemblages have shown a shift in species dominance following disturbance of feeding and refuge habitat (Minello and Rozas 2002; Greenwood et al. 2006). A lack of normal habitat results in the use of shallow water areas (cuts and passes) created by tropical activity important to the survival of neonatal and juvenile *N. brevirostris*. These shallow areas are the only protection offered to these animals from larger sharks (bull sharks, black tips, and larger lemon sharks) often seen in deeper waters at the Chandeleur Islands. The creation of the oil prevention berm following the *DWH* disaster filled in many of these shallow water habitats, which reduces available habitat and may also inhibit sharks pupped on the windward side of the island from reaching the protected sea grass beds on the leeward side.

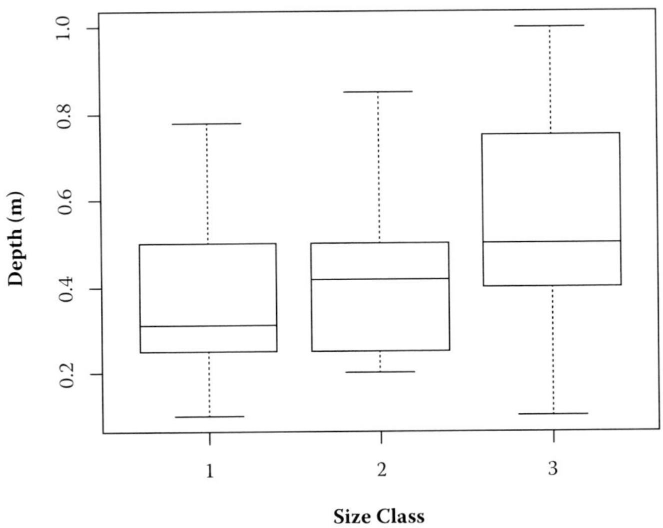

FIGURE 15.4
Association between depth at which lemon sharks at the Chandeleur Islands were sighted and their respective sizes during sampling efforts from 2009 to 2011. Size class 1 < 74 cm, size class 2 < 90 cm, > 74 cm, and size class 3 > 90 cm. Size classes 1 and 2 exhibited a significant preference for shallow water ($p < .001$).

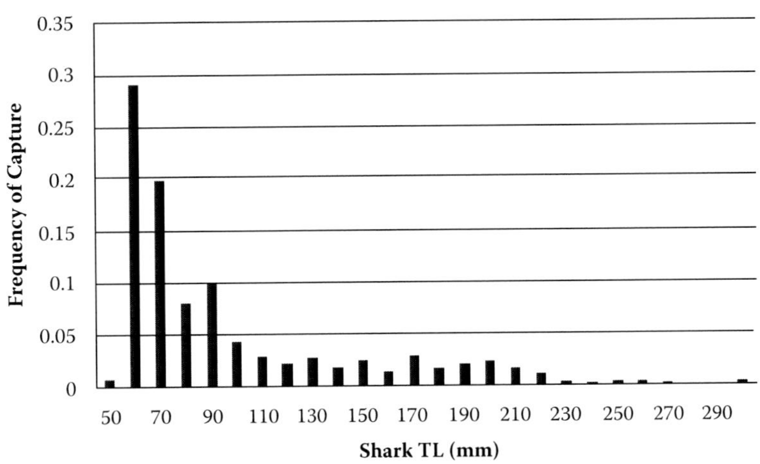

FIGURE 15.5
Frequency distributions of *Negaprion brevirostris* sizes collected around the Gulf of Mexico. Graph includes data collected from the Chandeleur Islands, Louisiana; Texas; and Florida.

The smallest size classes of *N. brevirostris* preferred shallower habitats than the larger size classes. This behavior was also observed in this species at the Virgin Islands (DeAngelis et al. 2008). In Fish Bay, U.S. Virgin Islands, researchers found neonatal and YOY *N. brevirostris* using shallow areas as a means of avoiding predators. Reducing predation pressure by occupying shallow habitats is a likely behavior for young *N. brevirostris* at the Chandeleur Islands due to the number of large sharks seen in the area (bull sharks, black tips, and larger lemon sharks) and the unique conditions (lack of mangroves) found at this location. In other known nursery grounds, *N. brevirostris* have access to mangroves

TABLE 15.4

Sources of *Negaprion brevirostris* Data Used in Analysis for This Study, Associated Locations, Sample Size, Total Length Range, Years of Study, and Sampling Techniques

Source	Location	Sharks	TL Range	Years	Gear
Texas Parks and Wildlife	Texas Coast	192	57–114	1982–2010	Gill net
Nekton Research Lab	Chandeleur Islands	143	57–177	2008–2011	Rod and reel
GULFSPAN	Northeastern Gulf	4	79–170	1998–2000	Longline
NOAA	Florida Coast	5	NA	1995–2011	Longline
Neil Hammerschlag	Florida Bay	95	130–300	2009–2011	Longline
Bimini Filed Station	Marquesas, FL	383	58–272	1998–2010	Gill net/longline
Rookery Bay	Florida Gulf Coast	97	67–190	2000–2012	Gill net/longline

NA, no information available.
TL, tail length; NOAA, National Oceanic and Atmospheric Administration.

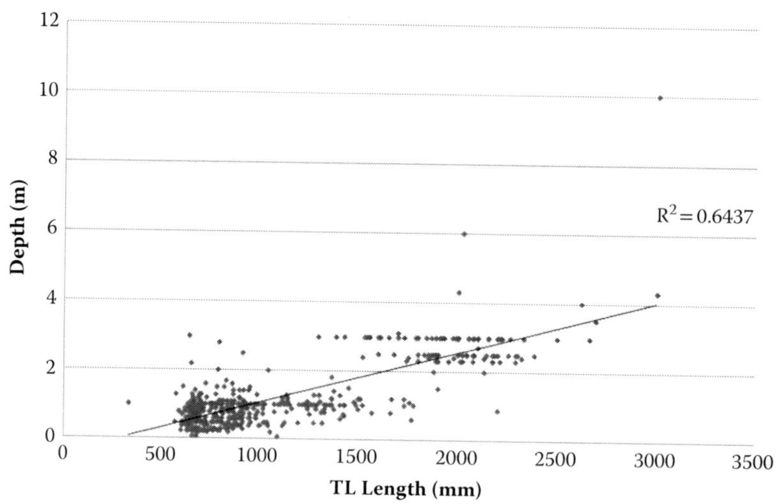

FIGURE 15.6
Relationship between the size and depth of capture of *Negaprion brevirostris* throughout the Gulf of Mexico ($N = 484$).

TABLE 15.5

Step-Wise Model Reduction Results from the Generalized Linear Model

Source of Variation	Df	Deviance	AIC
	1	1.57	−13.35
Year:depth	1	1.64	−12.51
Temp:depth	1	1.68	−10.80
Distance	1	1.68	−10.50
Month:salinity	3	1.85	−7.97
Month:depth	3	1.90	−6.36
Month:year	2	1.86	−5.77
Month:temperature	3	1.98	−3.49
Salinity:depth	1	1.89	−2.53

Note: The table shows the results for *Negaprion brevirostris* hourly catch per unit effort at the Chandeleur Islands, Louisiana, from 2009 to 2011 and associated date and habitat characteristics.

TABLE 15.6

Slope and R^2-Values for Time and Environmental Characteristics When Plotted against Hourly Catch per Unit Effort of *Negaprionbrevirostris* Collected at the Chandeleur Islands, Louisiana from 2009 to 2011

Variable	Slope	R^2
Month	−0.301	0.0022
Year	−0.153	0.0021
Temperature	−0.997	0.0004
Depth	−0.426	0.0123
Dissolved oxygen	−1.226	0.0064
Salinity	1.2606	0.0009
Distance	−24.5	0.022

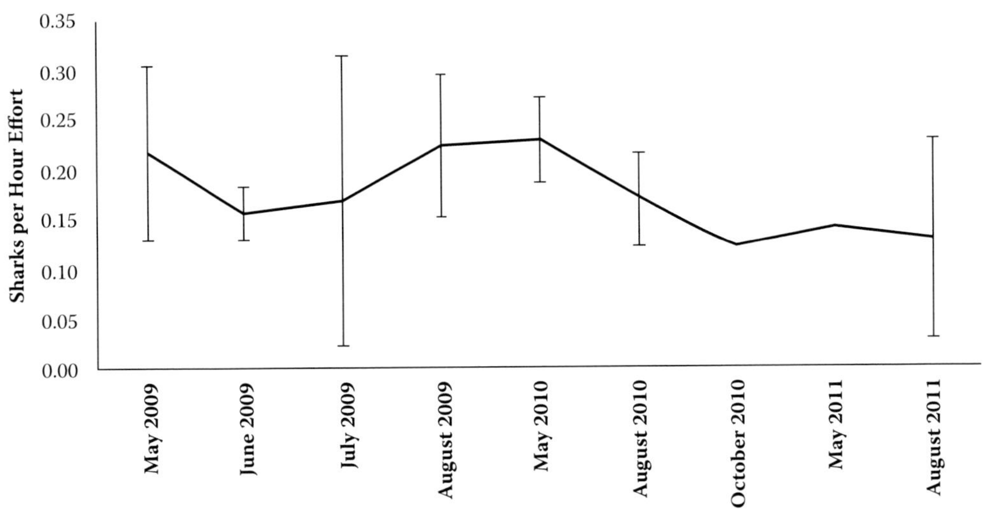

FIGURE 15.7
Modified catch per unit effort of lemon sharks at the Chandeleur Islands from 2009 to 2011 with standard deviation error bars. Catch per unit was calculated by dividing the number of sharks caught per month by total hours spent fishing.

as a form of structure to use for protection from larger predators (Morrissey and Gruber 1993a,b). The Chandeleur Islands lack these large mangroves and shallow habitats without protective structures are the most likely predation refugia at the Chandeleur Islands. From the data collected from other researchers, it is clear that young *N. brevirostris* around the Gulf of Mexico are showing a preference for shallow water and remain in these habitats until they reach a larger size before moving to deeper water.

Oil and dispersants from *DWH* inundated the Chandeleur Islands during May 2010. Following the oil spill, prevention measures were introduced to keep more oil from reaching farther beyond the islands into Chandeleur Sound. The most significant was a 35 km sand berm created from dredge material from the northern end of the islands (Martinez et al. 2012). Having studied lemon sharks in this area during the year before the oil spill, this study was in a unique position to compare shark numbers at the Chandeleur Islands before and after the disaster. On the basis of the monthly CPUE and bionomial logistic model, the oil spill appeared to have had a short-term impact on the relative abundance

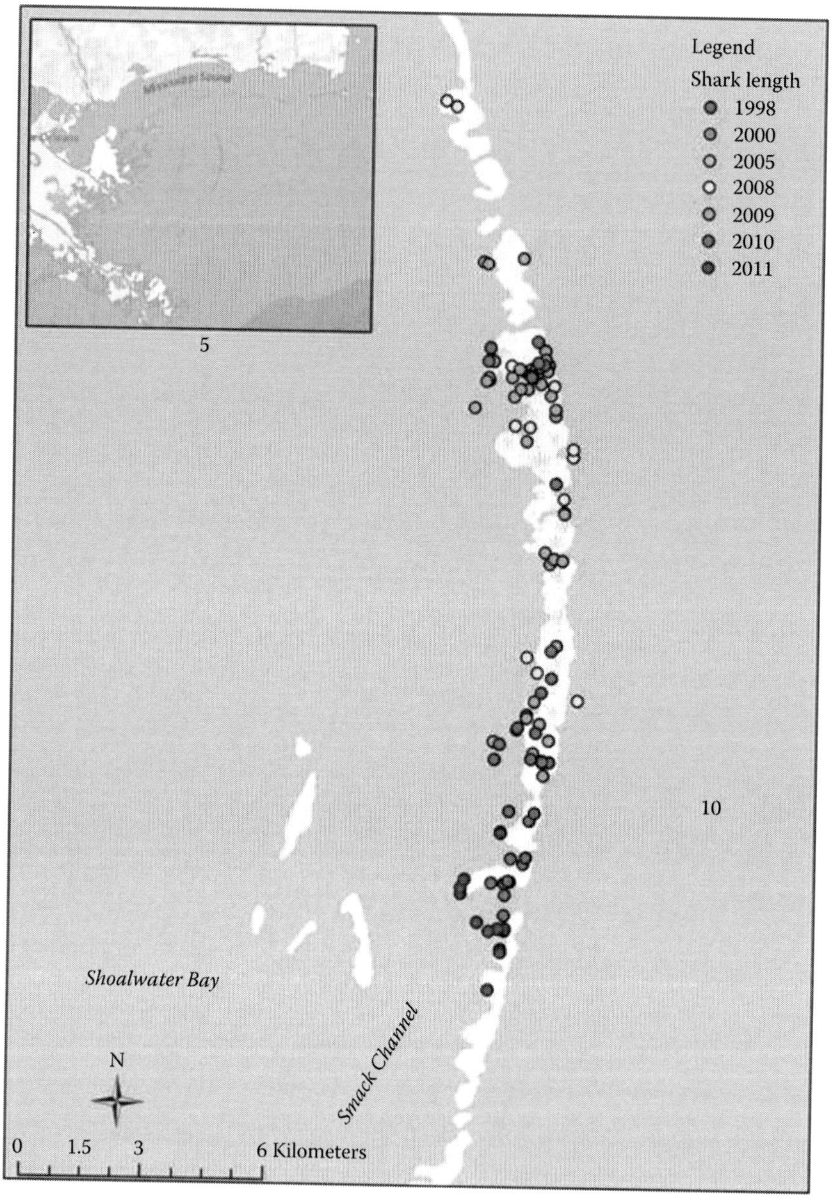

FIGURE 15.8
(See color insert.) Map of the Chandeleur Islands showing locations of lemon shark collections from 1998 to 2011. Oil spill occurred in 2010 and the number of lemon sharks collected in back barrier marshes appears to have decreased since the creation of the oil prevention sand berm.

of *N. brevirostris* occurring at the Chandeleur Islands. The sand berm may have reduced the number of cuts and passes around the islands, areas that provide essential shallow water habitat for juvenile *N. brevirostris*. A lack of suitable protected nursery habitat may result in a further reduction of *N. brevirostris* numbers around the Chandeleur Islands. It is important to continue monitoring the population numbers and the occurance patterns of these animals to determine the long-term effects this disaster and the subsequent oil prevention measures may have on the lemon sharks of the Chandeleur Islands.

In 2011, the Chandeleur Islands were again influenced by anthropogenic impacts. The Bonnet Carré Spillway was opened on May 23 in reponse to heavy flooding in the Lower Mississippi River. Spillway openings result in freshwater flow through Lakes Pontchartrain and Borgne into Biloxi Marsh and eventually the Chandeleur Islands. During the May 2011 sampling trip, which took place a week after the opening of the spillway, salinity levels associated with the captured sharks were the lowest recorded over the 3 years of the project (21.08 average). A YOY (76 cm) lemon shark was collected on the surf side of the islands (opposite of the freshwater influx) in waters with a salinity of 14.9, well below the normal preference levels of the species (Morrisey and Gruber 1993). The fewest number of sharks (5) were also collected during this trip. It is possible that the influx of freshwater and resulting salinity decrease during the early summer months of 2011 resulted in the decreased CPUE mentioned earlier. Salinity was found to be a significant predictor in the hourly CPUE analysis and further analysis indicated a positive correlation. However, a low R^2 value indicates that this positive correlation may not be very strong ($p < .001$). When salinities returned to normal levels in August 2011, monthly CPUE reduced to lower levels than those recorded the previous May. If normal CPUE values were found for this trip, then it would be more likely that decreased salinity affected habitat occurrence, but this was not the case. Lower CPUE values indicates that the oil spill may have resulted in fewer sharks returning to the Chandeleur Islands despite the possible confounding factors resulting from the opening of the spillway. Results from the logistic model appear to support this statement as the entire year of 2011 showed a signicantly lower probability of catching a shark compared to the other sampling years.

There are other possibilities that may have affected the CPUE calculations. This statistic was based on a targeted sampling effort, which involved researchers seeking out the areas with the highest shark numbers. This results in researchers spending a large portion of their time searching for areas where the sharks are most prevalent. Over the course of this experiment, there were a number of stressors and routine changes that may have affected the search effort. The *DWH* disaster, subsquent oil prevention measures, and the opening of the Bonnet Carré Spillway may have caused the sharks to move from the areas where they were previously found in high numbers (most heavily searched areas) to more suitable habitat around the islands. The area used as a research staging site was also changed during the sampling period. During all of 2009 and in May 2010, researchers staged at a barge near the northern end of the islands; following the *DWH* spill staging took place at the southern porition of the islands. However, these issues and changes would have a minimal impact on CPUE numbers as collectors surveryed the entire length of the islands on each sampling trip in an effort to locate *N. brevirostris*.

Another problem that can occur when sampling with rod and reel is the effect of angler ability. This sampling technique requires the researchers to be adequely skilled with the sampling gear and this ability varies from person to person. It is plausible that angler ability could have caused the reduction in monthly CPUE as several members of the crew were changed throughout the sampling process. This, however, is highly unlikely as a majority of the sampling team remained the same and their angling ability increased over the duration of the research, counteracting any angling deficiencies of newer research team members (Chris Schieble, pers. comm., May 20, 2013).

The long-term impacts the disaster will have on *N. brevirostris* are difficult to determine because of limited research on the effect of oil on sharks. Research was conducted on salmon sharks (*Lamna ditropis*) in Prince William Sound following the *Exxon Valdez* oil spill (Hulbert and Rice 2002). The research did not look at the direct effects of oil but rather examined the change in population numbers from before, during, and after the oil spill.

This research investigated population numbers from the 1980s to the 1990s. These data show that during the early 1990s, there was a marked increase in *L. ditropis* numbers in Prince William Sound. The authors suggest that a combination of factors played a role in this increase: (1) a moratorium of fishing practices that previously included *L. ditropis* as bycatch, (2) a trophic shift due to increased prey fish populations, (3) increased salmon hatchery production, and (4) ontological shifts in habitat range. The only impact of oil to this shark population was that the commercial and recreational fisheries were closed, an indirect consequence of a human response and prevention measure. No further studies evaluated the direct impacts of oil contamination on the physiology or population impacts to this species of shark.

Two other studies have looked at levels of polycyclic aromatic hydrocarbons (PAHs) and polychlorinated biphenyls (PCBs) in sharks, but neither of the studies were able to quantify any physiological damage associated with these pollutants (Al-Hassan et al. 2000; Storelli and Marctrigiano 2000). Al-Hassan et al. (2000) found that sharks collected from several locations within the Arabian Gulf accumulated noticeable levels of PAHs and aliphatic hydrocrabons regardless of their mobility, feeding methods, or location of capture. PAH levels up to 72.96 μg/g wetweight were documented in saw-toothed reef sharks. They also found that high levels (up to 11.87 μg/g wetweight) of PAH were passed down from mother to offspring (Al-Hassan et al. 2000). Researchers in Italy determined that several PCBs were measured at high levels ranging from 0.364 to 2.57 μg/g wetweight in the muscles, liver, and eggs in both the gulper shark (*Centrophorus granulosus*) and the longnose spurdog (*Squalus blainvillei*).

Perhaps more worrisome than the oil from *DWH* could be the application of the dispersants Corexit 9500 and 9527. Although the dispersants on their own have not been found to be overly lethal to bony fishes such as juvenile rainbow trout (Ramachandran et al. 2003), studies have shown that their use, when combined with oil and water, result in an increase in PAH uptake by fish exposed to this mixture (Ramachandran et al. 2003). This information, combined with the studies showing sharks already are storing background levels of PAH in their tissues, provides evidence that *N. brevirostris* exposed to the oiled water from *DWH* could potentially be storing PAHs that are higher compared to background levels had *DWH* not occurred.

With knowledge that these animals could be subjected to high levels of PAHs, but noting the fact that there are few studies on the physiological impacts of oil on any cartilaginous species, it was necessary to look at petroleum effects on bony fishes. Experiments on Atlantic cod (*Gadus morhua*) showed that prolonged exposure to crude oil resulted in decreased food consumption, increased gallbladder size, and a decreased rate of gametogensis in males (Kiceniuk and Khan 1987.) A study on cunners (*Tautogolabrus adspersus*) exposed to crude oil for a period of 6 months found that chronic exposure resulted in changes in testis somatic index, lens diameter, and plasma chloride (Payne et al. 1978). Mortality in both of these studies was low (<5%).

Although limited research could be located on larger apex predators, there may be some evidence to predict what may happen to a large-bodied migratory species. Two pods of killer whales (*Orcinus orca*) (one resident and one transient) were observed swimming in oil following the *Exxon Valdez* spill in 1989. These pods were monitored 5 years prior and 16 years after the oil spill, and both saw reductions in populations of 33% and 41% (Matkin et al. 2008). With both pods having distinct genetics and life histories, the only link between the heavy losses and the two pods is their interaction with the oil spill. Unlike whales, sharks are equipped with well-developed olfactory senses that may allow them to avoid or escape oiled waters, however even the migratory, transient pod of whales appeared to be affected by the oil (Matkin et al. 2008; Meredith and Kajiura 2010). This suggests that even short-term exposure to oil or oiled prey may have a long-term impact on an apex predator such as *N. brevirostris*.

All of the known lemon shark nursery grounds are associated with island habitats: Glover's Reef, Belize, Bimini, Bahamas, Marquesas Keys, Florida, and now the Chandeleur Islands, Louisiana. This makes them susceptable to habitat loss as sea level rises. Although most of these habitats are not in immediate danger of innundation, it has been estimated that these areas are vulnerable to complete submergence and abandonment by the end of the twenty-first century (Nicholls and Cazenave 2010). With the Chandeleur Islands already experiencing high rates of relative sea-level rise (≥1.00 cm/y), it is important to monitor its population of *N. brevirostris* and its response to habitat partitioning and eventually habitat loss (Nicholls and Cazenave 2010). The information that can be obtained over the next decade at the Chandeleur Islands has the potential to provide insight to the future that other known nursery grounds may be facing over the next century.

Restortation of barrier islands off the coast of Louisiana would have a number of benefits both to the local ecosystems and for human interests. Restored barrier islands and associated habitat (sea grass beds) are beneficial to human interests in that future offshore oil spills would be slowed by the natural defense provided by these islands. These same natural defenses buffer inshore cities against tropical activity and assoicated storm surge. From an ecological standpoint a healthy barrier island habitat provides refuge and food sources for a number of species from migratory birds to commercially and recreationally important fish species.

References

Alford, J. B., and D. C. Jackson. 2010. Associations between watershed characteristics and angling success for sport fishes in Mississippi wadeable streams. *North American Journal of Fisheries Management* 30:12–120.

Al-Hassan, J. M., M. Afzal, C. V. N. Rao, and S. Fayad. 2000. Petroleum hydrocarbon pollution in sharks in the Arabian Gulf. *Bulletin of Environmental Contamination and Toxicology* 65:391–398.

Beck, M. W., K. L. Heck, K. W. Able, and D. L. Childers. 2001. The identification, conservation, and management of estuarine and marine nurseries for fish and invertebrates. *BioScience* 51:633–641.

Bethel, M. B., L. A. Martinez, and S. P. O'Brien. 2007. Chandeleur Islands, Louisiana: Seagrass bed map. Change analysis: January 2005 to October 2005. Available at http://www.ladigitalcoast.uno.edu/chadE/CI-seagrass-change-2005.pdf.

Bigelow, K. A., J. Hampton, N. Miyabe. 2002. Application of habitat-based model to estimate effective longline fishing effort and relative abundance of Pacific bigeye tuna (*Thunnus obesus*). *Fisheries Oceanography* 11:143–155.

Blum, M. D. and H. H Roberts. 2009. Drowning of the Mississippi delta due to insufficient sediment supply and global sea-level rise. *Nature Geoscience* 2:488–491.

Brown, C. A., S. A. Holt, G. A. Jackson, D. A. Brooks, and G. J. Holt. 2004. Simulating larval supply to estuarine nursery: How important are physical processes to the supply of larvae to the Aransas Pass inlet. *Fisheries Oceanography* 13:181–196.

Chapman, D. D., E. A. Babcock, S. H. Gruber, J. D. DiBattista, B. R. Franks, S. A. Kessel, T. Guttridge, E. K. Pikitch, and K. A. Feldheim. 2009. Long-term natal site-fidelity by immature lemon sharks (*Negaprion brevirostris)* at a subtropical island. *Molecular Ecology* 18:3500–3507.

Dean, R. G. 2006. New Orleans and the wetlands of southern Louisiana. *The Bridge* 36:35–43.

DeAngelis, B. M., C. T., McCandless, N. E. Kohler, C. W. Recksiek, and G. B. Skomal. 2008. First characterization of shark nursery habitat in the United States Virgin Islands, evidence of habitat partitioning by two shark species. *Marine Ecology Progress Series* 358:257–271.

Dorenbosch, M., M. C. Verweij, I. Nagelkerken, N. Jiddawi, and G. Velde. 2004. Homing and daytime tidal movements of juvenile snappers (Lutjanidae) between shallow-water nursery habitats in Zanzibar, Western Indian Ocean. *Envrionmental Biology of Fishes* 70:203–209.

Duffy, J. E. 2006. Biodiversity and the functioning of seagrass ecosystems. *Marine Ecology Progress Series*. 311:233–250.

Fearnley, S. M., M. D. Miner, M. Kulp, C. Bohling, and S. Penland. 2009. Hurricane impact and recovery shoreline change analysis of the Chandeleur Islands, Louisiana, USA: 1855 to 2005. *Geo-Marine Letters* 29:455–466.

Greenwood, M. F. D., P. W. Stevens, and R. E. Matheson, Jr. 2006. Effects of the 2004 hurricanes on the fish assemblages in two proximate southwest Florida estuaries: Change in the context of interannual variability. *Estuaries and Coasts* 29(6A):985–996.

Harley, S. J., R. A. Myers, and A. Dunn. 2001. Is catch-per-unit-effort proportional to abundance? *Canadian Journal of Fisheries and Aquatic Sciences* 58:1760–1772.

Hochachka, W. M., K. Martin, F. Doyle, and C. J. Krebs. 2000. Monitoring vertebrate populations using observation data. *Canadian journal of Zoology* 78:521–529.

Holmberg, J., B. Norman, and Z. Arzoumanian. 2009. Estimating population size, structure, and residency time for whale sharks *Rhinocodon typus* through collaborative photo-identification. *Endangered Species Research* 7:39–53.

Hulbert, L. B. and S. D. Rice. 2002. Salmon shark, *Lamna ditropis*, movements, diet, and abundance in the Eastern North Pacific Ocean and Prince William Sound, Alaska. *Exxon Valdez* Oil Spill Restoration Project 02396 Final Report. NOAA Fisheries, Auke Bay Laboratory, Juneau, Alaska.

Kiceniuk, J. W. and R. A. Khan. 1987. Effect of petroleum hydrocarbons on Atlantic cod, *Gadus morhua*, following chronic exposure. *Canadian Journal of Zoology* 65:490–494.

Kleiber, P, and M. N. Maunder. 2008. Inherent bias in using aggregate CPUE to characterize abundance of fish species assemblages. *Fisheries Research* 93:140–145.

Kolker, A. S., M. A. Allison, and S. Hameed. 2011. An evaluation of subsidence rates and sea-level variability in the northern Gulf of Mexico. *Geophysical Research Letters* 38:L21404.

Kulp, M. A. 2005. Punctuated coastal reorganization: A lesson learned from Hurricane Katrina? Paper presented at the Annual Meeting of the Geological Society of America, Salt Lake City, Utah.

Laska, A. L. 1973. Fishes of the Chandeleur Islands, Louisiana. PhD Dissertation. Tulane University, New Orleans, Louisiana.

Layman, C. A. 2000. Fish assemblage structure of the shallow ocean surf-zone on the eastern shore of Virginia barrier islands. *Estuarine, Coastal and Shelf Science* 51:210–213.

Luettich, R. A. Jr., J. L. Hench, C. W. Fulcher, F. E. Werner, B. O. Blanton, and J. H. Churchill. 1999. Barotropical tidal and wind-driven larval transport in the vicinity of a barrier island inlet. *Fisheries Oceanography* 8:190–209.

Martínez, M. L., R. A. Feagin, K. M. Yeager et al. 2012. Artificial modifications of the coast in response to the *Deepwater Horizon* oil spill: Quick solutions or long-term liabilities? *Frontiers in Ecology and the Environment* 10:44–49.

Matkin, C. O., E. L. Saulitis, G. M. Ellis, P. Olesluk, and S. D. Rice. 2008. Ongoing population-level impacts on killer whales (*Orcinus orca*) following the *Exxon Valdez* oil spill in Prince William Sound, Alaska. *Marine Ecology Progress Series* 356:269–281.

Meredith T. L. and S. M. Kajiura. 2010. Olfactory morphology and physiology of elasmobranchs. *The Journal of Experimental Biology* 213:3449–3456.

Minello, T. J. and L. P. Rozas. 2002. Nekton in Gulf Coast wetlands: Fine-scaled distributions, landscape patterns, and restoration implications. *Ecological Applications* 12:441–455.

Morrissey, J. F. and S. H. Gruber. 1993a. Habitat selection by juvenile lemon sharks, *Negaprion brevirostris. Environmental Biology of Fishes* 38:311–319.

Morrissey, J. F. and S. H. Gruber. 1993b. Home range of juvenile sharks, *Negaprion brevirostris. Copeia* 1993:425–435.

Nagelkerken, I., M. Dorenbosch, W. C. E. P. Verberk, E. Cocheret de la Morinière, and G. van der Velde. 2000a. Day-night shifts of fishes between shallow-water biotopes of a Caribbean bay, with emphasis on the nocturnal feeding of Haemulidae and Lutjanidae. *Marine Ecology Progress Series* 194:55–64.

Newman, S. P., R. D. Handy, and S. H. Gruber. 2007. Spatial and temporal variations in mangrove and seagrass faunal communities at Bimini, Bahamas. *Bulletin of Marine Science* 800:529–553.

Nicholls, R. J. and A. Cazenave. 2010. Sea-level rise and its impact on coastal zones. *Science* 328:1517.

Norcross, B. L., B. A. Holladay, and F. J. Muter. 1995. Nursery area characteristics of pleuronectids in coastal Alaska, USA. *Netherlands Journal of Sea Research* 34:161–175.

Orth, R. J., T. J. B. Carruthers, W. C. Dennison et al. 2006. A global crisis for seagrass ecosystems. *BioScience* 56:987–996.

Payne, J. F., J. W. Kiceniuk, W. R. Squires, and G. L. Fletcher. 1978. Pathological changes in a marine fish after a 6-month exposure to petroleum. *Journal of Fisheries Research Board of Canada* 35:665–667.

Punt, A. E., T. I. Walker, B. L. Taylor, and F. Pribac. 2000. Standardization of catch and effort data in a spatially-structured shark fishery. *Fisheries Research* 45:129–145.

Ramachandran, S. D., P. V. Hodson, C. W. Khan, and K. Lee. 2003. Oil dispersant increases PAH uptake by fish exposed to crude oil. *Ecotoxicology and Environmental Safety* 59:300–308.

Ruple, D. L. 1984. Occurrence of larval fishes in the surf zone of a northern Gulf of Mexico barrier island. *Estuarine, Coastal, and Shelf Science* 18:191–208.

Springer, S. 1950. Natural history notes on the lemon shark *Negaprio brevirostris. Texas Journal of Science* 2:349–359.

Storelli, M. M. and G. O. Marcotrigiano 2001. Persistent organochlorine residues and toxic evaluation of polychlorinated biphenyls in sharks from the Mediterranean Sea (Italy). *Marine Pollution Bulletin* 42:1323–1329.

Trnski, T. 2001. Diel and tidal abundance of fish larvae in a barrier-estuary channel in New South Wales. *Marine Freshwater Research* 52:995–1006.

Tsuboi, J. and S. Endou. 2008. Relationships between catch per unit effort, catchability and abundance based on actual measurements of salmonids in a mountain stream. *Transactions of the American Fisheries Society* 137:496–502.

Walters, C. 2003. Folly and fantasy in the analysis of spatial catch rate data. *Canadian Journal of Fisheries and Aquatic Sciences* 60:1433–1436.

Wiley, T. R. and C. A. Simpfendorfer. 2010. Using public encounter data to direct recovery efforts for the endangered smalltooth sawfish *Pristis pectinata. Endangered Species Research* 12:179–191.

Index

D